Interdisciplinary Biotechnological Advances

Series Editors

Jayanta Kumar Patra, Research Institute of Integrative Life Sciences, Dongguk University, Ilsandong, Goyang, Korea (Republic of)

Gitishree Das, Research Institute of Integrative Life Sciences, Dongguk University, Goyang, Korea (Republic of)

This series is an authoritative resource that explains recent advances and emerging directions in biotechnology, reflecting the forefront of research clearly and reliably, without excessive hype. Each volume is written by authors with excellent reputations and acknowledged expertise in the topic under discussion. The volumes span the entire field from an interdisciplinary perspective, covering everything from biotechnology principles and methods to applications in areas including genetic engineering, transgenic plants and animals, environmental problems, genomics, proteomics, diagnosis of disease, gene therapy, and biomedicine. The significance of these applications for the achievement of UN Sustainable Development Goals is highlighted. The series will be highly relevant for Master's and PhD students in Biotechnology, Nanochemistry, Biochemical Engineering, and Microbiology, medical students, academic and industrial researchers, agricultural scientists, farmers, clinicians, industry personnel, and entrepreneurs.

Vasudeo Zambare • Mohd Fadhil Md Din
P. Saranraj • Latika Shendre
Editors

Biotechnological Advances and Applications in Microbial Proteases

Springer

Editors
Vasudeo Zambare
R&D Department
Balaji Enzyme and Chemical Pvt Ltd
Mumbai, Maharashtra, India

P. Saranraj
Department of Microbiology
Sacred Heart College (Autonomous)
Tirupattur, Tamil Nadu, India

Mohd Fadhil Md Din
Center for Environmental Sustainability and Water Security (IPASA), Department of Water and Environmental Engineering, School of Civil Engineering
University of Technology Malaysia
Johor Bahru, Johor, Malaysia

Latika Shendre
Dr. D. Y. Patil Biotechnology and Bioinformatics Institute
Dr. D. Y. Patil Vidyapeeth
Pune, Maharashtra, India

ISSN 2730-7069 ISSN 2730-7077 (electronic)
Interdisciplinary Biotechnological Advances
ISBN 978-981-95-9010-0 ISBN 978-981-95-9011-7 (eBook)
https://doi.org/10.1007/978-981-95-9011-7

This Springer imprint is published by the registered company Springer Nature Singapore Pte Ltd.
The registered company address is: 152 Beach Road, #21-01/04 Gateway East, Singapore 189721, Singapore

Foreword

The advancement of microbial proteases as industrial biocatalysts has reached a pivotal juncture, marked by significant progress in enzyme engineering, sustainable bioprocessing, and integrative omics technologies. This edited volume, *Biotechnological Advances and Applications in Microbial Proteases*, systematically presents state-of-the-art developments in microbial protease research, contextualized within emerging global demands for eco-efficient and industrially scalable biotechnologies.

The book encompasses a broad spectrum of contemporary topics, including but not limited to the molecular evolution of protease-producing strains, genetic modifications through CRISPR-Cas9, and intelligent process control strategies supported by artificial intelligence. Chapters such as those on halophilic protease producers and dermatological applications reflect the expansion of protease relevance into niche and interdisciplinary domains, including cosmeceuticals and clinical biotechnology.

A particular emphasis is placed on the concluding chapter, "Future Research Directions: Sustainable Production, Scalability, and Safety Considerations," which synthesizes critical technological trajectories and process innovations necessary for next-generation enzyme manufacturing. This chapter delineates a roadmap for integrating systems biology and metabolic engineering to construct robust microbial cell factories. It highlights the transition from traditional submerged fermentation to solid-state systems and continuous fermentation platforms, thereby optimizing protease yield, process economics, and environmental sustainability.

Advanced computational tools such as CFD for bioreactor design and machine learning-driven high-throughput screening are recognized as essential instruments for bridging the lab-to-industry translational gap. Moreover, the work delves into innovations in downstream processing—membrane chromatography, aqueous two-phase systems, and low-energy purification—providing viable solutions to the longstanding challenges of enzyme recovery and formulation.

Of equal importance are the safety frameworks and regulatory landscapes addressed within the text. The inclusion of comprehensive strategies for toxicological assessments, occupational exposure controls, and enzyme formulation for stability and biosecurity underscores the industrial and societal readiness of microbial protease applications.

Collectively, this volume stands as a critical resource for microbiologists, biochemical engineers, enzymologists, and industrial stakeholders. It not only encapsulates the current scientific consensus but also projects the future of protease biotechnologies as central components of sustainable industrial ecosystems, particularly within the context of circular bioeconomy frameworks.

As microbial proteases continue to evolve from ancillary enzymatic tools to cornerstone biocatalysts in sectors ranging from pharmaceuticals to environmental engineering, this work provides a timely and rigorous foundation upon which future innovations will be built.

In an era of sustainable industrial goals and circular bioeconomy frameworks, this book bridges enzyme innovation and real-world impact. It encapsulates current scientific consensus and positions proteases as cornerstone biocatalysts across sectors from pharmaceuticals to industrial manufacturing, serving as a rigorous foundation for future advances. May it inspire your work and fuel next-generation breakthroughs.

Professor of Environmental Biotechnology, Changsha University of Science and Technology
Changsha, China

Hong Chen

Preface

In the ever-evolving landscape of industrial biotechnology, enzymes stand as indispensable biocatalysts, driving efficiency, sustainability, and innovation across diverse sectors. Among these, microbial proteases occupy a position of paramount importance, representing over 60% of the global enzyme market. Their remarkable versatility, specificity, and compatibility with green chemistry principles have cemented their role as transformative agents in industries ranging from food processing and detergents to pharmaceuticals, waste management, and beyond. This book, *Biotechnological Advances and Applications in Microbial Proteases*, is conceived as a comprehensive exploration of the cutting-edge science, technology, and applications that define this dynamic field.

The accelerating global shift toward sustainable and circular bioeconomies has placed unprecedented emphasis on the development of efficient, eco-friendly biocatalysts. Microbial proteases, with their ability to function under mild conditions and utilize renewable feedstocks, are at the forefront of this transition. However, harnessing their full potential requires a deep integration of multidisciplinary knowledge spanning Microbiology, Molecular Biology, Fermentation Technology, Protein Engineering, and Process Optimization. This volume aims to bridge these domains, offering readers a holistic view of the protease lifecycle: from novel discovery and strain improvement to advanced production, purification, and tailored application.

The book is structured into two thematic sections. The first, Recent Advances in Microbial Protease-Based Technologies for Production and Properties, delves into the foundational and emerging methodologies that underpin modern protease science. Chapters systematically cover sources and classification, innovative fermentation strategies (submerged and solid-state), the promise of extremophiles, and the revolutionary impact of genetic engineering tools like CRISPR-Cas9 and machine learning. Critical downstream processes, including novel partitioning and purification techniques, are examined in detail, highlighting pathways to more sustainable and economical production.

The second section, Emerging Trends in Microbial Protease Applications and Future Prospects, translates these technological advances into real-world impact. It showcases the expanding horizon of protease utility, from traditional strongholds in dairy, meat, baking, and detergent industries to pioneering roles in dermatology, cosmetics, biofilm mitigation, wound healing, and biorefineries. A dedicated focus on food processing and preservation illustrates their dual role in enhancing quality and ensuring safety. The volume concludes by critically addressing the essential pillars of sustainable production, scalability, and safety—key considerations for the responsible and commercially viable deployment of next-generation enzymes.

This work is the fruit of a truly international and collaborative effort, bringing together the expertise of distinguished scientists, researchers, and industry specialists from across the globe. Each chapter has been meticulously crafted to provide not only a state-of-the-art review but also to identify future research trajectories and unresolved challenges. It is our sincere hope that this compilation will serve as an invaluable resource for academic researchers, graduate students, industrial biotechnologists, and professionals in allied fields. Our ultimate goal is to inspire further innovation, fostering the development of microbial proteases that are not only high-performing but also aligned with the principles of environmental stewardship and economic sustainability, thereby contributing to a greener and more resilient future.

Mumbai, Maharashtra, India — Vasudeo Zambare
Johor Bahru, Johor, Malaysia — Mohd Fadhil Md Din
Tirupattur, Tamil Nadu, India — P. Saranraj
Pune, Maharashtra, India — Latika Shendre

Acknowledgments

We, the editors, would like to express our sincere gratitude to all the contributors and editors from around the world, including India, Brazil, Malaysia, Algeria, West Africa, Azerbaijan, Indonesia, Jordan, Iraq, and Syria, for their valuable chapters in *Biotechnological Advances and Applications in Microbial Proteases*. This publication reflects the collective effort of editors and contributors who advanced a timely discourse on microbial innovations for sustainable environmental remediation and industrial applications, aligned with the growing global demand for resilient, eco-efficient solutions.

Special thanks to Mr. Abhay Kainya, Director of Balaji Enzyme & Chemical Pvt Ltd, Mumbai, India, for his all-time encouragement. We are also thankful to management of Sacred Heart College (Autonomous), Tirupattur, India, University Teknologi Malaysia's (UTM) Campus Sustainability Centre (CCS), the Vice Chancellor Office (SAGE), and the Centre for Environmental Sustainability and Water Security (IPASA), for their steadfast institutional support, enabling environment, and strategic stewardship that strengthened the successful publication of this book. My special thanks to my mentors Dr. Smita Nilegaonkar, Dr. Pradnya Kanekar and Dr. Prashant Dhakephalkar from Agharkar Research Institute, Pune, India who helped me for the development of protease enzyme related tools and techniques. We appreciate the efforts of the Springer team, reviewers, Dr. Lee Mei, and Dr. Jayanta Kumar Patra, Book Series Editor of Interdisciplinary Biotechnological Advances, whose expertise and commitment enhanced the quality, relevance, and integrity of this volume.

Thank you all for making this book a reality!

Contents

Contributors

J. Albino Wins Department of Botany, Holy Cross College (Autonomous), Nagercoil, Tamil Nadu, India

Ola Alnaddaf Department of Horticultural Sciences, Damascus University, Swaida, Syria

Samer Saleem Alshkarchy Department of Pathology, Al-Qasim Green University, Al-Qasim City, Babylon Province, Iraq

Muthanna N. K. Altaee Department of Biology, College of Science, University of Wasit, Wasit, Iraq

Abdel Razzaq Mohammad Al-Tawaha National Agricultural Research Center (NARC), Baq'a, Jordan

Jyothi Judith Alvares Microbiology Programme, School of Biological Sciences and Biotechnology, Goa University, Taleigao Plateau, Goa, India

B. U. Ananya Marine Biotechnology and Bioproducts Laboratory, School of Biosciences and Technology, Vellore Institute of Technology, Vellore, Tamil Nadu, India

Andhere Animish Marine Biotechnology and Bioproducts Laboratory, Department of Bio Medical Sciences, School of Biosciences and Technology, Vellore Institute of Technology, Vellore, Tamil Nadu, India
Department of Bio-Medical Sciences, School of Biosciences and Technology, Vellore Institute of Technology, Vellore, Tamil Nadu, India

Sai Sankeerthana Bedadakota Marine Biotechnology and Bioproducts Laboratory, School of Biosciences and Technology, Vellore Institute of Technology, Vellore, Tamil Nadu, India

Alexander Machado Cardoso Department of Biology, Rio de Janeiro State University (UERJ), Rio de Janeiro, RJ, Brazil

M. Charumathy PG and Research Department of Biochemistry, Marudhar Kesari Jain College for Women (Autonomous), Vaniyambadi, Tamil Nadu, India

S. Dayana PG and Research Department of Microbiology, Sacred Heart College (Autonomous), Tirupattur, Tamil Nadu, India

M. Dharshinn Agricultural College and Research Institute, Tamil Nadu Agricultural University (TNAU), Coimbatore, Tamilnadu, India

Sumitha Elayaperumal Department of Biotechnology and Bioinformatics, JSS Academy of Higher Education and Research, Mysore, Karnataka, India

Israa M. Essa Department of Public Health, College of Veterinary Medicine, University of Basrah, Basra, Iraq

Glaucio Dire Feliciano Laboratory of Chemical and Biological Analysis (LAQB), State University of Rio de Janeiro (UERJ), Rio de Janeiro, Brazil

Irene Furtado Microbiology Programme, School of Biological Sciences and Biotechnology, Goa University, Taleigao Plateau, Goa, India

Sanket K. Gaonkar Department of Microbiology, P.E.S's R.S.N College of Arts and Science, Ponda, Goa, India
Microbiology Programme, School of Biological Sciences and Biotechnology, Goa University, Taleigao Plateau, Goa, India

K. Gayathri PG and Research Department of Microbiology, Sacred Heart College (Autonomous), Tirupattur, Tamil Nadu, India

Hasanain A. J. Gharban Department of Internal and Preventive Veterinary Medicine, College of Veterinary Medicine, University of Wasit, Wasit, Iraq

Rajat Kani Ghosal Marine Biotechnology and Bioproducts Laboratory, Department of Bio Medical Sciences, School of Biosciences and Technology, Vellore Institute of Technology, Vellore, Tamil Nadu, India

Goverdhini Marine Biotechnology and Bioproducts Laboratory, Department of Bio Medical Sciences, School of Biosciences and Technology, Vellore Institute of Technology, Vellore, Tamil Nadu, India

Mustafa Salah Hasan College of Veterinary Medicine, University of Fallujah, Fallujah, Iraq

Endang Ismail Department of Fishery Product Technology, Faculty of Fisheries and Marine Sciences, Brawijaya University, Malang, East Java, Indonesia
Secondary Fisheries Business School, Kotaagung, Lampung, Indonesia

Anne Jana Marine Biotechnology and Bioproducts Laboratory, Department of Bio Medical Sciences, School of Biosciences and Technology, Vellore Institute of Technology, Vellore, Tamil Nadu, India

P. R. Janani Department of Microbiology, Padmavani Arts and Science College for Women (Autonomous), Salem, Tamil Nadu, India

M. A. Jayasri Marine Biotechnology and Bioproducts Laboratory, School of Biosciences and Technology, Vellore Institute of Technology, Vellore, Tamil Nadu, India

L. Charlie Jelura PG and Research Department of Microbiology, Sacred Heart College (Autonomous), Tirupattur, Tamil Nadu, India

C. Mabel Joshaline Department of Rural Development Science, Arul Anandar College, Madurai, Tamil Nadu, India

Ravi-Kumar Kadeppagari Centre for Incubation Innovation Research and Consultancy, Food Technology Department, Jyothy Institute of Technology, Bengaluru, Karnataka, India

K. Kesavardhini PG and Research Department of Microbiology, Sacred Heart College (Autonomous), Tirupattur, Tamil Nadu, India

Anshu Kumari Dr. D. Y. Patil Biotechnology and Bioinformatics Institute, Dr. D. Y. Patil Vidyapeeth, Pune, Maharashtra, India

B. Lokeshwari PG and Research Department of Microbiology, Sacred Heart College (Autonomous), Tirupattur, Tamil Nadu, India

Sejal Mali Department of Bio-Medical Sciences, School of Biosciences and Technology, Vellore Institute of Technology, Vellore, Tamil Nadu, India

Turana Mammadova Department of Plant Protection, Azerbaijan State Agricultural University, Ganja, Azerbaijan

Sonali Mandal Marine Biotechnology and Bioproducts Laboratory, Department of Bio Medical Sciences, School of Biosciences and Technology, Vellore Institute of Technology, Vellore, Tamil Nadu, India

M. Manigandan PG and Research Department of Microbiology, Sacred Heart College (Autonomous), Tirupattur, Tamil Nadu, India

Sofia Mohanty Department of Bio-Medical Sciences, School of Biosciences and Technology, Vellore Institute of Technology, Vellore, Tamil Nadu, India

Slimane Mokrani Department of Agronomy, Laboratory of Research on Biological Systems and Geomantic (L.R.S.B.G.), University of Mustapha Stambouli, Mascara, Algeria
Laboratoire de Maitrise des Energies Renouvelables, Faculté des Sciences de la Nature et de la Vie, Université de Bejaia, Bejaia, Algeria

A. Mubarak Ali PG and Research Department of Microbiology, Sacred Heart College (Autonomous), Tirupattur, Tamil Nadu, India

El-hafid Nabti Laboratoire de Maitrise des Energies Renouvelables, Faculté des Sciences de la Nature et de la Vie, Université de Bejaia, Bejaia, Algeria

Ashish Kumar Nayak Department of Civil Engineering, Prasad V. Potluri Siddhartha Institute of Technology, Vijayawada, Andhra Pradesh, India

Asharfi Pathela Department of Bio-Medical Sciences, School of Biosciences and Technology, Vellore Institute of Technology, Vellore, Tamil Nadu, India

Asep Awaludin Prihanto Department of Fishery Product Technology, Faculty of Fisheries and Marine Sciences, Brawijaya University, Malang, East Java, Indonesia

Shruti Rangari Department of Bio-Medical Sciences, School of Biosciences and Technology, Vellore Institute of Technology, Vellore, Tamil Nadu, India

Pranav Rao Marine Biotechnology and Bioproducts Laboratory, School of Biosciences and Technology, Vellore Institute of Technology, Vellore, Tamil Nadu, India

Chandan Kumar Sahu Centre for Incubation Innovation Research and Consultancy, Food Technology Department, Jyothy Institute of Technology, Bengaluru, Karnataka, India

P. Saranraj PG and Research Department of Microbiology, Sacred Heart College (Autonomous), Tirupattur, Tamil Nadu, India
Department of Microbiology, Sacred Heart College (Autonomous), Tirupattur, Tamil Nadu, India

Latika Shendre Dr. D. Y. Patil Biotechnology and Bioinformatics Institute, Dr. D. Y. Patil Vidyapeeth, Pune, Maharashtra, India

Richa Sikligar Marine Biotechnology and Bioproducts Laboratory, School of Biosciences and Technology, Vellore Institute of Technology, Vellore, Tamil Nadu, India

Amit Kumar Singh Center of Excellence in Tissue Engineering, Department of Biotechnology and Medical Engineering, National Institute of Technology, Rourkela, Odisha, India

Pallavi Singh Department of Biotechnology, Graphic Era Deemed to be University, Dehradun, Uttarakhand, India

Neha Sinha Department of Bio-Medical Sciences, School of Biosciences and Technology, Vellore Institute of Technology, Vellore, Tamil Nadu, India

P. Sivasakthivelan Tamil Nadu Agricultural University, Agricultural Research Station, Bhavanisagar, Tamil Nadu, India

Marius K. Somda Laboratory of Biochemistry, Biotechnology, Food Technology and Nutrition (LABIOTAN), Department of Biochemistry and Microbiology, University Joseph KI-ZERBO, Ouagadougou, Burkina Faso

U. Subhalakshmi PG and Research Department of Microbiology, Sacred Heart College (Autonomous), Tirupattur, Tamil Nadu, India

Bandapalli Sujatha Department of Zoology, KVR Govt College for Women (Autonomous), Cluster University, Kurnool, Andhra Pradesh, India

Sukoso Sukoso Department of Fishery Product Technology, Faculty of Fisheries and Marine Sciences, Brawijaya University, Malang, East Java, Indonesia

K. Suthindhiran Marine Biotechnology and Bioproducts Laboratory, Department of Bio Medical Sciences, School of Biosciences and Technology, Vellore Institute of Technology, Vellore, Tamil Nadu, India
Department of Bio-Medical Sciences, School of Biosciences and Technology, Vellore Institute of Technology, Vellore, Tamil Nadu, India

M. Swathi Marine Biotechnology and Bioproducts Laboratory, School of Biosciences and Technology, Vellore Institute of Technology, Vellore, Tamil Nadu, India

Shashank Vasanthkumar Centre for Incubation Innovation Research and Consultancy, Food Technology Department, Jyothy Institute of Technology, Bengaluru, Karnataka, India

Vasundhara Marine Biotechnology and Bioproducts Laboratory, Department of Bio Medical Sciences, School of Biosciences and Technology, Vellore Institute of Technology, Vellore, Tamil Nadu, India

G. Vithiya PG and Research Department of Microbiology, Sacred Heart College (Autonomous), Tirupattur, Tamil Nadu, India

Archita Vyas Marine Biotechnology and Bioproducts Laboratory, Department of Bio Medical Sciences, School of Biosciences and Technology, Vellore Institute of Technology, Vellore, Tamil Nadu, India

Vasudeo Zambare R&D Department, Balaji Enzyme & Chemical Pvt Ltd, Mumbai, Maharashtra, India

Abbreviations

$(NH_4)_2SO_4$	Ammonium Sulfate
3D	Three-Dimensional
A.I.S.E.	International Association for Soaps, Detergents and Maintenance Products
ACE	Angiotensin-Converting Enzyme
AI	Artificial Intelligence
AIDS	Acquired Immunodeficiency Syndrome
AMC	7-Amino-4-Methylcoumarin
AMFEP	Association of Manufacturers and Formulators of Enzyme Products
AMPs	Antimicrobial Peptides
ANFs	Anti-Nutritional Factors
APC	Advanced Process Control
APP	Amyloid Precursor Protein
AprE	Alkaline protease E (*Bacillus subtilis*)
ASTM	American Society for Testing and Materials
ATP	Adenosine Triphosphate
ATPS	Aqueous Two-Phase Partitioning Systems
BACE1	β-Secretase
BACs	Bacterial Artificial Chromosomes
bFGF	Basic Fibroblast Growth Factor
BGCs	Biosynthetic Gene Clusters
BKC	Benzalkonium Chloride
BLAST	Basic Local Alignment Search Tool
BOD	Biological Oxygen Demand
BODIPY	Boron-Dipyrromethene
BPN'	Subtilisin BPN'
BSA	Bovine Serum Albumin
$CaCl_2$	Calcium Chloride
CAGR	Compound Annual Growth Rate

CAPDE	Computer-Aided Protein-Directed Evolution
CAPEX	Capital Expenditure
Cas9	CRISPR-associated protein 9
CBE	Cytidine Base Editor
CdrA	Cyclic di-GMP Regulated Adhesin A
cDSP	Continuous Downstream Processing
CFD	Computational Fluid Dynamics
CHP	Combined Heat and Power
CIBRC	Central Insecticides Board and Registration Committee (India)
CIP	Clean-in-Place
CNNs	Convolutional Neural Networks
CoA	Certificate of Analysis
COD	Chemical Oxygen Demand
COPD	Chronic Obstructive Pulmonary Disease
CPPs-	Casein Phosphopeptides
CRISPR	Clustered Regularly Interspaced Short Palindromic Repeats
CRISPR/Cas9	Clustered Regularly Interspaced Short Palindromic Repeats/CRISPR-associated protein 9
CRISPRa	CRISPR activation
CRISPRi	CRISPR interference
CRL	*Candida rugosa* Lipase
CTAB	Cetyltrimethylammonium Bromide
DBTL	Design-Build-Test-Learn
dCas9	Catalytically dead Cas9
DES	Deep Eutectic Solvents
DFP	Di-isopropylfluorophosphate
DH	Degree of Hydrolysis
DHN	Dihydroxynaphthalene
DIAMOND	Double Index Alignment of Next-generation sequencing Data
DMF	Dimethylformamide
DMS	Deep Mutational Scanning
DMSO	Dimethyl Sulfoxide
DNA	Deoxyribonucleic Acid
DOE	Design of Experiments
DOPA	3,4-Dihydroxyphenylalanine (also L-DOPA)
DSB	Double-Strand Break
DSP	Downstream Processing
DTT	Dithiothreitol
EC	Enzyme Commission
ECHA	European Chemicals Agency
ECM	Extracellular Matrix
eDNA	environmental DNA
EDTA	Ethylenediaminetetraacetic Acid

EFSA	European Food Safety Authority
EGF	Epidermal Growth Factor
EGTA	Ethylene Glycol Tetraacetic Acid
ELISA	Enzyme-Linked Immunosorbent Assay
EMA	European Medicines Agency
EPA	Environmental Protection Agency (USA)
EPS	Extracellular Polymeric Substances
ERA	Environmental Risk Assessment
ETA	Enzyme Technical Association
EU	European Union
FACS	Fluorescence-Activated Cell Sorting
FAO	Food and Agriculture Organization
FDA	Food and Drug Administration (USA)
FGF	Fibroblast Growth Factor
FITC	Fluorescein Isothiocyanate
FRET	Fluorescence Resonance Energy Transfer
GANs	Generative Adversarial Networks
GHG	Greenhouse Gas
GMMs	Genetically Modified Microorganisms
GMOs	Genetically Modified Organisms
GMP	Good Manufacturing Practices
GNNs	Graph Neural Networks
GOS	Galacto-oligosaccharides
GRAS	Generally Recognised as Safe
GST	Glutathione S-Transferase
HDR	Homology-Directed Repair
HEPA	High-Efficiency Particulate Air
HFCS	High Fructose Corn Syrup
HIV	Human Immunodeficiency Virus
HIV-1	Human Immunodeficiency Virus Type 1
hK5, hK7	Human Kallikrein 5, Human Kallikrein 7
HMM	Hidden Markov Model
HMW	High-Molecular-Weight
HPLC	High-Performance Liquid Chromatography
HPP	High-Pressure Processing
I&I	Industrial and Institutional
IBCs	Institutional Biosafety Committees
ICLR	International Conference on Learning Representations
ICML	International Conference on Machine Learning
IDRs	Intrinsically Disordered Regions
IgE	Immunoglobulin E
IGF-1-	Insulin-like Growth Factor-1
IL-1β	Interleukin-1 Beta
IL-6	Interleukin-6

ILs	Ionic Liquids
IMG/M	Integrated Microbial Genomes & Microbiomes
IMP	Inosine Monophosphate
Indels	Insertions and Deletions
IoT	Internet of Things
ISO	International Organization for Standardization
IUBMB	International Union of Biochemistry and Molecular Biology
JECFA	Joint FAO/WHO Expert Committee on Food Additives
K_2CO_3	Potassium Carbonate
K_2HPO_4	Dipotassium Hydrogen Phosphate
$K_3C_6H_5O_7 \cdot H_2O$	Potassium Citrate Tribasic Monohydrate
KCl	Potassium Chloride
KGF	Keratinocyte Growth Factor
Km	Michaelis constant (enzyme affinity)
KPs	Kallikrein Proteases
LAB	Lactic Acid Bacteria
Las B	*Pseudomonas aeruginosa* Elastase B
LCA	Life Cycle Assessment
LCDs	Low Complexity Domains
L-DOPA	L-Dihydroxyphenylalanine
LSTM	Long Short-Term Memory
M1/M2	Macrophage Phenotypes (Pro-inflammatory/Anti-inflammatory)
MAGs	Metagenome-Assembled Genomes
MBM	Meat-and-Bone Meal
MBP	Maltose-Binding Protein
MC1R	Melanocortin 1 Receptor
MCA	7-Methoxycoumarin-4-Acetic Acid
MD	Molecular Dynamics
MDR	Multidrug Resistance
MF	Microfiltration
$MgSO_4$	Magnesium Sulfate
MIC	Minimum Inhibitory Concentration
ML	Machine Learning
MMPs	Matrix Metalloproteinase
MPA	Metalloprotease A
MPC	Milk Protein Concentrate
MPS	Mononuclear Phagocyte System
MSW	Municipal Solid Waste
Na_2CO_3	Sodium Carbonate
$Na_3C_6H_5O_7$	Trisodium Citrate
NaCl	Sodium Chloride
nCas9	Nickase Cas9
NCBI NR	National Center for Biotechnology Information Non-Redundant

NGS	Next-Generation Sequencing
NHEJ	Non-Homologous End Joining
NIH	National Institutes of Health (USA)
NIR	Near-Infrared
NMR	Nuclear Magnetic Resonance
NprE	Neutral protease E (*Bacillus subtilis*)
NPs	Neutral Proteases
NS	Not Specified
NSPs	Non-Starch Polysaccharides
OECD	Organisation for Economic Co-operation and Development
OEL	Occupational Exposure Limit
OFAT	One Factor at a Time
OP	Organophosphorus
OPA	Ortho-Phthaldialdehyde
OPEX	Operational Expenditure
ORFs	Open Reading Frames
OSHA	Occupational Safety and Health Administration (USA)
P.P.E.	Personal Protective Equipment
PACE	Phage-Assisted Continuous Evolution
PAIPCR	Preamplification Inverse-PCR
PAM	Protospacer Adjacent Motif
PAP	Prolyl Aminopeptidase
PAR-2	Protease-Activated Receptor 2
PAT	Process Analytical Technology
PCMB	p-Chloromercuribenzoate
PCR	Polymerase Chain Reaction
PDI	Protein Disulfide Isomerase
PEF	Pulsed Electric Fields
PEG	Polyethylene Glycol
pegRNA	Prime Editing Guide RNA
pI	Isoelectric Point
PICS	Proteomic Identification of Cleavage Specificity
PID	Proportional-Integral-Derivative
PKU	Phenylketonuria
pLMs	Protein Language Models
PMSF	Phenylmethylsulfonyl Fluoride
pNA	p-Nitroaniline
PNAG	Poly-N-Acetylglucosamine
PNPs	Polymeric Nanoparticles
POPs	Persistent Organic Pollutants
PVP-I	Povidone-Iodine
PWMs	Position-Weight Matrices
QC	Quality Control
R&D	Research and Development

rDNA	Recombinant DNA
REACH	Registration, Evaluation, Authorisation and Restriction of Chemicals
RNA	Ribonucleic Acid
RNP	Ribonucleoprotein
ROS	Reactive Oxygen Species
RSM	Response Surface Methodology
RT	Room Temperature
SAR	Structure-Activity Relationship
SARS	Severe Acute Respiratory Syndrome
SDM	Site-Directed Mutagenesis
SDS	Sodium Dodecyl Sulfate
SDS-PAGE	Sodium Dodecyl Sulfate–Polyacrylamide Gel Electrophoresis
sgRNA	Single-Guide RNA
SIGEX	Substrate-Induced Gene Expression Screening
SMART	Simple Modular Architecture Research Tool
SmB	Submerged Fermentation
SmF	Submerged Fermentation
SOD	Superoxide Dismutase
SSD	Silver Sulfadiazine
SSF	Solid-State Fermentation
ssODN	Single-Stranded Oligodeoxynucleotide
SUMO	Small Ubiquitin-Related Modifier
SVMs	Support Vector Machines
SwissProt	Swiss Protein Database
Tat pathway	Twin-arginine Translocation pathway
t-butanol	Tert-Butanol
TCA	Trichloroacetic Acid
TDDS	Transdermal Drug Delivery Systems
TEA	Techno-Economic Analysis
TEV	Tobacco Etch Virus
TGase-	Transglutaminase
TGF-β	Transforming Growth Factor Beta
THN	Tetrahydroxynaphthalene
TIGRFAMs	The Institute for Genomic Research Functional Families
TIMPs	- Tissue Inhibitors of Metalloproteinases
TLCK	- Tosyl-L-Lysine Chloromethyl Ketone
TNBS	- 2,4,6-Trinitrobenzenesulfonic Acid
TNF-α	- Tumor Necrosis Factor-Alpha
t-PA	Tissue Plasminogen Activator
TPP	Three-Phase Partitioning
Triton X-100	Octyl Phenol Ethoxylate (surfactant)
TSCA	Toxic Substances Control Act (USA)

Tween 20/40/60/80	Polyoxyethylene Sorbitan Esters (surfactants)
UF	Ultrafiltration
UHT	Ultra-High Temperature
USD	United States Dollar
USDA	United States Department of Agriculture
UTR	Untranslated Region
UV	Ultraviolet
VAEs	Variational Autoencoders
VEGF	Vascular Endothelial Growth Factor
Vmax	Maximum reaction velocity
WHO	World Health Organization
WPH	Whey Protein Hydrolysates
XAI	Explainable AI
Zn^{2+}	Zinc ion

Part I
Recent Advances in Microbial Protease Based Technologies for Production and Properties

Chapter 1
Microbial Protease: Sources, Classification, Detection, and Applications

P. Saranraj, B. Lokeshwari, Vasudeo Zambare, M. Manigandan, P. Sivasakthivelan, and Glaucio Dire Feliciano

Abstract Microbial proteases, which hydrolyze protein peptide bonds, predominate the global enzyme market. Biocatalysts are primarily derived from bacteria, such as *Bacillus* sp. (especially the Alkaline protease-producing *B. licheniformis* and *B. amyloliquefaciens*), and fungi, like *Aspergillus* sp., *Penicillium* sp., and *Mucor* sp. Microbial proteases are classified by their catalytic mechanisms, their site of action on the protein substrates, and their optimal pH of activity. These enzymes must be detected and quantified using qualitative methods, such as the protein agar plate assay and radial diffusion assay, as well as highly sensitive quantitative methods that utilize chromogenic or fluorogenic synthetic substrates for characterization and application. Microbial proteases are widely utilized in various industries, enhancing their commercial value. They are essential detergents in bio-formulation additives for protein-based stain degradation. They aid in meat tenderization, dairy protein hydrolysis (for cheese and debittering), soy product processing, and beverage clarification. They also assist the leather industry in dehairing and bating hides, manage waste by treating protein-rich waste, facilitate bioremediation, and enhance the digestibility of animal feed and nutrient availability. The future of microbial proteases is promising due to advances in Biotechnology, including protein engineering for improved stability and specificity, metagenomic screening for novel enzymes from previously unculturable microbes, and more efficient fermentation processes. Molecular proteases are vital biocatalysts whose broad specificity, high

P. Saranraj (✉) · B. Lokeshwari · M. Manigandan
PG and Research Department of Microbiology, Sacred Heart College (Autonomous), Tirupattur, Tamil Nadu, India

V. Zambare
R&D Department, Balaji Enzyme & Chemical Pvt Ltd, Mumbai, Maharashtra, India

P. Sivasakthivelan
Tamil Nadu Agricultural University, Agricultural Research Station, Bhavanisagar, Tamil Nadu, India

G. D. Feliciano
Laboratory of Chemical and Biological Analysis (LAQB), State University of Rio de Janeiro (UERJ), Rio de Janeiro, Brazil

V. Zambare et al. (eds.), *Biotechnological Advances and Applications in Microbial Proteases*, Interdisciplinary Biotechnological Advances,
https://doi.org/10.1007/978-981-95-9011-7_1

catalytic efficiency, and sustainable production methods make them essential in many industrial sectors. This chapter examines the microbial proteases, their diverse sources, systematic classification, advanced detection methods, and purification techniques, as well as their wide-ranging industrial applications.

Keywords Microbial proteases · Enzyme classification · Detection · Enzyme purification · Industrial applications

1.1 Introduction

Protease enzymes, such as microbial proteases, represent an essential group of biotechnological and industrial enzymes with diverse applications in waste management, food processing, pharmaceuticals, leather, detergents, and bioremediation (Zambare 2008; Zambare and Nilegaonkar 2016). They are necessary to microbial physiology, nutrition, pathogenesis, and industrial biocatalysts (Choudhary and Jain 2011). Proteases are classified based on their molecular structure, evolutionary relationship, optimum pH, site of action, and mechanism of catalysis. The leading producers of microbial proteases are bacteria, fungi, and actinomycetes, due to their fast growth rates, genetic manipulation, ability to extracellularly release enzymes into the medium, and potentially scalable fermentation processes. Bacteria such as *Bacillus* sp., *Pseudomonas* sp., *Streptomyces* sp., and *Clostridium* sp.; fungi, including *Aspergillus* sp., *Penicillium* sp., *Rhizopus* sp., and *Mucor* sp.; and yeasts are known to produce them. Microbial protease screening and detection are fundamental steps in generating novel, productive enzymes. Various methods, including quantitative spectrophotometric, chromatographic, and zymographic techniques, as well as qualitative assays, are employed (Anbu et al. 2015).

Metagenomics, protein engineering, site-directed mutagenesis, and recombinant DNA technology are examples of contemporary molecular techniques that are being increasingly used to identify, alter, and optimize proteases with enhanced thermostability, altered substrate affinity, solvent tolerance, and inhibitor resistance (Patil and Chaudhari 2010). Microbial proteases are utilized in an ever-increasing range of applications. Proteases hydrolyze proteinaceous stains, such as blood, food residues, and water droplets, in detergent formulations, which is the largest market for these products. They additionally play an essential part in leather processing, assisting with softening, bathing, and removing hair coats as environmentally friendly alternatives to harsh chemicals. Proteases are critical to the food industry for brewing, baking, cheese production, tenderizing meat, protein hydrolysate production, and flavor enhancement. Microbial proteases play a crucial role in waste management and environmental sustainability, as they facilitate the bioremediation of protein-rich wastewater from the food and leather industries, converting waste into valuable products such as fertilizers, bioactive peptides, and amino acids (da Silva 2017).

Microbial proteases are used in agriculture as biocontrol agents, animal feed additives, and hay processing to improve nutrient availability and digestibility. Although they have a vast array of applications, production costs, stability in industrial processing conditions, and regulatory issues remain. These challenges may be addressed, nevertheless, by the growing interest in genetic engineering, metabolic pathway optimization, and bioprocess improvement. In addition to the advancement of industrial biotechnology, microbial protease studies are crucial for addressing issues related to food, health, the environment, and economic growth (Adrio and Demain 2014). Microbial proteases are an essential class of enzymes that facilitate a wide range of biological processes and technological advancements. Due to their exceptional catalytic efficiency, specificity, and wide range of peptide bond-hydrolyzing activity in protein substrates, they account for over 60% of the industrial enzyme market. A diverse range of microorganisms, including bacteria, fungi, yeasts, and actinomycetes, found in numerous locations such as deep-sea hydrothermal vents, alkaline soda lakes, acidic hot springs, and hypersaline areas, are the primary source of biocatalysts (Bhunia et al. 2012).

The enzyme-specific properties, such as optimal pH and temperature range, stability, substrate specificity, and insensitivity to inhibitors, are solely dependent on the microbial source used. Microbial proteases are classified into a complex system that is regulated by their catalytic mechanism, active site amino acid residue, optimum pH, location of action on a protein substrate, evolutionary relationship, and similarities in amino acid sequence. This is required to be classified to match the enzyme with the optimal pH for use. The discovery, characterization, and quality assurance of these enzymes necessitate several sophisticated detection and assay methods. Protease-producing microbial colonies are visually identified in the primary screening method through the development of clear zones of hydrolysis or precipitation in protein-rich media, such as skim milk, casein, or gelatin. Quantitative assays of enzyme activity, specificity, and inhibition characteristics can be performed accurately through quantitative analysis in liquid culture using a defined substrate, such as casein or artificial chromogenic/fluorogenic peptides. Sophisticated techniques, such as mass spectrometry to determine cleavage sites and perform more detailed structural analysis, as well as enzyme-linked immunosorbent assay (ELISA), zymography, and molecular biology methods like metagenomic screening of microbial communities in environmental samples and polymerase chain reaction (PCR) with degenerate primers targeting preserved Protease sequences (Pokhrel et al. 2013).

The widespread use of Protease across various industrial sectors is the result of its discovery and characterization. Biological washing powders and liquids utilize *Bacillus* species Alkaline serine proteases to hydrolyze protein-based stains, such as blood, egg, grass, and sweat, at low temperatures. They also play essential roles in the production of cheese, protein hydrolysates, meat tenderization, flavor enhancers, baking, and brewing. The pharmaceutical and therapeutic industries utilize microbial proteases for waste management, wound debridement, the production of anti-inflammatory drugs, digestive aids, biotherapeutics, the silk industry, molecular biology labs, and the emerging field of peptide synthesis (Chantawannakul et al. 2002). Microbial proteases demonstrate the microbial world's incredible catalytic power, from their diverse ecological origins and complex taxonomic organization to their sophisticated identification methods and significant impact on human endeavors.

1.2 Sources of Microbial Proteases

The most abundant and diverse source of proteases is the extensive microbial world, which scientists have explored in the discovery of efficient and versatile proteolytic enzymes. Microorganisms are superior to plant or animal sources for protease production due to their rapid growth on inexpensive substrates, high yield potential resulting from their metabolic diversity, ease of genetic manipulation for strain improvement, and ability to be produced in controlled fermentation vessels, despite seasonal or geographical constraints (Contesini et al. 2018). Extremophiles that live in alkaline lakes, acidic hot springs, or cold marine depths produce enzymes, including proteases, that are active in function under those extreme conditions, and are of immense industrial value. This section explains the significant bacterial and fungal agents in the protease industry and their effects. They are the primary agents in the industrial production of proteins due to their high secretion capacity and resilience properties, particularly those of the genus *Bacillus*. The *B. licheniformis* is a Gram-positive, thermotolerant bacterium that is highly recognized to produce a lot of highly stable alkaline proteins, to mention a few, subtilisin (serine proteins), which are most active in high pH levels (8–11) and high temperatures (50–60 °C) (Devi et al. 2008).

The microbial protease has also been extensively applied in the food and leather industries, particularly in protein hydrolysis procedures and in cleaning and hair removal processes, respectively, as alternatives to highly hazardous chemicals. *Bacillus* species, specifically *B. amyloliquefaciens*, often produce a complex cocktail of proteins, including neutral metalloproteins and alkaline serine proteins (Johnvesly and Naik 2001). Its enzymes are particularly sought after in the food and beverage sector for the production of soy sauce and sake, where their ability to produce flavor under controlled protein breakdown is utilized. Additionally, they are used in the manufacture of protein hydrolysates as nutritional supplements due to their specific cleavage patterns, which produce peptides with defined functional properties and are not bitter-tasting. *Lactobacillus* species produce proteases that are essential for dairy fermentation, while *Pseudomonas* species are known to produce cold-adapted proteases for low-temperature applications. Fungal proteases, derived from eukaryotic sources, hold a significant market share. They have significance for their generally recognized as safe (GRAS) status, their capacity to secrete enzymes directly into the culture medium, and their ability to produce a wide variety of protease types, including a high concentration of acidic proteases (Chakraborty et al. 2011).

The most crucial fungal source is *Aspergillus* sp., which produces a potent combination of acidic, neutral, and alkaline proteases that hydrolyze soy and wheat proteins into amino acids and peptides, giving miso, soy sauce, and sake their umami flavor. *Aspergillus niger* is a workhorse for citric acid production, and

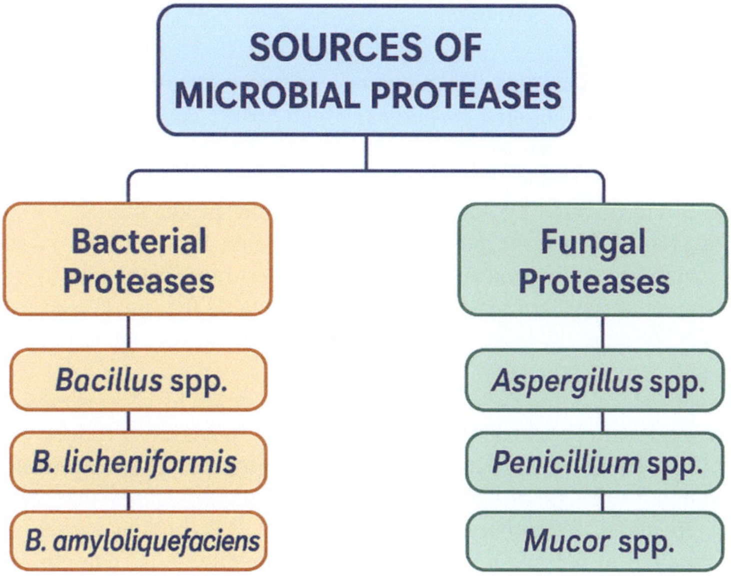

Fig. 1.1 Microbial sources of protease

other species, like *Penicillium chrysogenum*, are being investigated for producing proteases with applications in waste treatment and the production of bioactive peptides. *Penicillium* species, such as *P. roqueforti* and *P. camemberti*, are well-known for their role in cheese production, as well as the extensive proteolysis that occurs during ripening, resulting in the development of intense, characteristic flavors and a soft texture. *Mucor* spp., particularly *M. miehei* and *M. pusillus*, are known for producing effective milk-clotting agents (rennets) for cheese production. These microbial rennets hydrolyze κ-casein protein in milk, causing it to coagulate. They include a viable alternative to animal rennet (from calf stomachs), meeting vegetarian demands and providing a consistent, cost-effective supply. The strategic selection of these microbial sources, often achieved through sophisticated screening of natural environments or genetic engineering, enables the improvement of protease yield and properties for a wide range of applications, from industry to medicine (Khandelwal et al. 2014). Proteases are sourced from a wide array of microorganisms, each contributing enzymes with unique properties. Figure 1.1 provides a systematic overview of these microbial sources, categorizing them into bacterial, fungal, and yeasts known for their high enzyme yields.

1.2.1 Bacterial Proteases

Bacterial proteases dominate the industrial enzyme industry due to their superior physiological and biochemical properties, which enable cost-effective and scalable production. Bacteria, such as prokaryotes, have relatively simple genetic systems that are easily manipulable, allowing scientists to use techniques such as random mutagenesis and directed evolution to generate hyper-producing strains capable of producing grams of enzyme per liter of fermentation broth, which is critical for economic viability in large-scale production. In addition, compared to eukaryotic systems, the rapid development times of bacteria enable the quick establishment of high-density varieties in large-scale fermenters, significantly reducing production duration and costs (Banerjee and Ray 2017).

Bacteria produce a diverse array of proteases to adapt to various environmental conditions on Earth. Psychrophilic bacteria from polar regions produce cold-active proteases that remain functional at refrigeration temperatures. In contrast, thermophiles from hot springs produce extremely versatile proteases that can tolerate industrial processes at temperatures above 70 °C. Alkaliphiles secrete proteases that are effective at pH levels of 10 and higher. Additionally, halophiles produce enzymes that are stable in high-salt environments. This biological variety is rich in biocatalysts. Most commercial bacterial proteases are extracellular, so they are secreted into the growing environment. This eliminates the need to extract the enzyme from within the cell, significantly simplifying downstream processing and purification. *Bacillus* is a spore-forming, nonpathogenic bacterium that is widely used for its ability to release large amounts of strong enzymes, making it ideal for fermentation technology (Gopinath et al. 2013).

1.2.1.1 *Bacillus* spp.

The manufacture of industrial proteases is closely related to the genus *Bacillus*, which comprises a wide assortment of rod-shaped, Gram-positive, and endospore-forming bacteria. This is facilitated by the formation of persistent endospores, which enable bacterial strains to be stored for longer and endure the severe environment of large-scale fermentation, thereby guaranteeing a continuous and consistent supply of inoculum. The *Bacillus* species are hyper-secretive of enzymes into the extracellular medium, which is not new; this has been built over decades of strain development into industrial applications, and can often lead to concentrations capable of contributing to up to 1% of the total protein of the culture broth (Kalpana and Rajeswari 2019). Proteins that liberate internal peptide bonds are often found as the dominant ones in the complex protease mixture that Bacilli secretes. These endopeptidases are neutral metalloproteases and alkaline serine proteases of the subtilisin family. Subtilisin are especially important due to their extensive substrate specificity and superior performance in detergents and organic solvents. The production of proteases in Bacilli is controlled by elaborate mechanisms that involve carbon, nitrogen, and phosphate starvation. This means that the production of proteases is optimized in nutrient-restricting environments, which can readily be established in a fed-batch fermentation procedure. The *Bacillus* genes can be engineered into expression vectors, native protease genes can be removed to prevent degradation of other enzyme applications in the background, and powerful and functional promoters can be used to induce overexpression. *B. licheniformis* and *B. amyloliquefaciens* have become important vital organisms, and they each have their own niche and range of specialized genes (Liu et al. 2003).

1.2.1.1.1 *Bacillus Licheniformis*

Bacillus licheniformis is a Gram-positive, thermotolerant bacterium with a GRAS designation, making it a safe and recommended host for food and feed enzyme synthesis. It naturally degrades keratinaceous and other complex proteinaceous materials, as it is present in soil and bird feathers. The ability of *B. licheniformis* to thrive at high temperatures (up to 50–55 °C) and secrete extremely high levels of enzymes, particularly a subtilisin-type alkaline serine protease that is referred to as "subtilisin Carlsberg" or "subtilisin Lentus" depending on the strain, is responsible for its industrial dominance in protease production. Serine, histidine, and aspartate residues form a catalytic group in this 27 kDa primary protease. Its resilience, high specific activity, and stability across a wide pH range (7–11), with an optimum around 9–10, and its tolerance to temperatures of up to 60 °C, make it relevant in industry. It is most significantly resistant to denaturation by chelating agents, such as EDTA, oxidizing agents like hydrogen peroxide and sodium perborate, and surfactants. Its stability renders it suitable for use in modern washing and dishwashing detergents, where it dissolves protein-based stains such as blood, egg, grass, and gravy at low wash temperatures, thereby reducing energy consumption. Along with their use in detergents, *B. licheniformis* proteases are of great use in the leather industry to debond the hair follicles during the dehairing process and soften skins during drying, thereby serving as a sustainable alternative to damaging agents such as lime and sodium sulfide. Its proteases are applied in the food industry to produce protein hydrolysates in sports nutrition and baby formula, to modify the solubility and emulsification of proteins, and to enhance the flavor of fermented foods. *B. licheniformis* has already established its own status as a phenomenon in industrial biotechnology, through the application of conventional methods of mutagenesis and selection in the production of mutants with a many-fold increase in protease production compared to wild-type organisms (Patel et al. 2005).

1.2.1.1.2 *Bacillus amyloliquefaciens*

Bacillus amyloliquefaciens is closely related to *B. subtilis* and *B. licheniformis*, but it possesses a distinct repertoire of enzymes and diverse industrial applications. It is a sporulating bacterium with the capacity to secrete high-level extracellular enzymes, similar to its related organisms, and has GRAS status. Alkaline proteases are secreted, but their enzyme repertoire is usually more expansive, including an essential neutral metalloprotease (Haddar et al. 2009). Neutral protease with a pH optimum of 7.0 and a temperature optimum near 60 °C is more heat labile than serine proteases and readily inactivated by chelating chemicals that remove its necessary metal ion. Due to its ability to be inactivated by heating or chelators, its specificity is not a drawback, making it valuable in applications that require regulated proteolysis. This is crucial for the food and beverage industries. In the processing of soy sauce and other Asian fermented foods, *B. amyloliquefaciens* proteases hydrolyze soy proteins to peptides and amino acids.

Selective hydrolysis of a combination of proteins maximizes flavor and texture. The baking industry uses low dosages of its protease to hydrolyze gluten proteins mildly, making dough less elastic, highly able to distribute cookies, and reducing shrinking, but not fully dispersing the protein networks. Furthermore, *B. amyloliquefaciens* is a major producer of subtilisin, which is a model enzyme for protein engineering studies. This protease has been enhanced through site-directed mutagenesis to improve its oxidative stability, substrate specificity, and efficiency in organic solvents, making it a valuable tool in peptide synthesis and chemical processes. *B. amyloliquefaciens* supplies precise proteolysis tools and an environment for specific enzymes, enhancing the industrial protease environment (Kaur et al. 2018).

1.2.2 Fungal Proteases

Fungal proteases represent a novel and useful submarket of the microbial protease segment because they have eukaryotic biology and are GRAS, required in food and pharmaceutical use (Chandra et al. 2020). Fungi are multicellular or filamentous and secrete enzymes, some of which are diverse proteases, to decompose complex organic substances and obtain nutrients. These inherent properties are utilized in both submerged and solid-state fermentations (SSF). Fungi grow well in wet solid media, such as wheat bran or rice husks, and SSF is an economical process that generates less wastewater compared to liquid fermentation. It usually has high enzyme yields. Fungal proteins contain all four catalytic classes of proteases, including serine, cysteine, aspartic, and metalloproteases. However, they also comprise a large proportion of acidic aspartic proteases, which are common in bacteria, fruit processing, and the reaction of fermented beverages. Fungi synthesize acidic proteases in large quantities. Fungal proteases tend to have high specificity for a particular peptide bond, allowing protein hydrolysates with specific functional or nutritional properties to be produced without generating acidic peptides. *Trichoderma reesie* was also explored for protease production and its strain improvement using physicochemical mutation (Zambare 2010a). The most critical fungal proteases are those produced by the *Aspergillus* sp., *Penicillium* sp., and *Mucor* sp., which have been utilized for decades in the food industry and are currently employed in biotechnology for the production of biomedical drugs (Bari et al. 2010).

1.2.2.1 *Aspergillus* spp.

Aspergillus sp. is the most economically important genus of fungus in biotechnology, and it has a giant influence on protease technology. *Aspergillus oryzae* (the koji mold) have been used for millennia in East Asian food fermentations and are being used in industrial enzyme production. *Aspergillus oryzae* is a Japanese microorganism used in the manufacture of sake, soy sauce, and miso (Ghorbel-Bellaaj et al. 2011). Its acid aspartic, neutral metalloproteases, and alkaline serine proteases secretions combine into a complex process. This multipronged proteolytic attack is necessary to completely break down soy and rice proteins into a rich blend of natural amino acids, the source of umami, and small peptides, which are responsible for the savory flavor of these products. These proteases are closely regulated by nitrogen supply and pH levels to achieve optimal productivity during fermentation. *Aspergillus niger* is well-known for its production of citric acid. Nonetheless, it also yields food-grade enzymes, including pepsin, an enzyme produced by microorganisms used in cheese production, and fungal acid proteases used in baking. The proteins are gluten-hydrolyzing proteins used in dough baking to make it more flexible, easier to manipulate, and form a finer, more uniform dough structure. *Aspergillus niger* proteases are used to produce protein hydrolysates in nutritional supplements, such as those derived from pricier plant and milk sources. Besides playing a vital role in the field, *Aspergillus* species have evolved into cell factories owing to their safety and genetic plasticity. The strains of the fungus have been engineered to express proteases or heterologous proteins of other organisms (Prakasham et al. 2006).

1.2.2.2 *Penicillium* spp.

The genus *Penicillium* is renowned for producing the antibiotic penicillin. Still, several of its species are also indispensable in the food industry, particularly in cheese production, where their proteases are responsible for defining the texture and flavor of some of the world's most beloved cheeses. *Penicillium roqueforti* is the blue mold used in Roquefort, Gorgonzola, and Stilton cheeses. During the ripening process, it grows internally within the cheese, secreting a potent combination of proteases (mainly metallo- and aspartic proteases) and lipases (Charbonneau and Beauregard 2013). The proteolytic activity is responsible for the extensive breakdown of casein proteins, leading to the characteristic soft, crumbly, and moist texture of blue cheese, as well as the generation of a vast array of peptides and free amino acids that are precursors for flavor compounds. Similarly, *Penicillium camemberti*, the white, fluffy mold on the surface of Brie and Camembert, secretes proteases and lipases that act from the rind inward. This proteolysis softens the cheese interior, creating a creamy consistency and a milder, buttery flavor compared to the more pungent blue cheeses.

Penicillium sp. are explored for their proteolytic potential. *Penicillium chrysogenum* and *Penicillium citrinum* are studied for their ability to produce alkaline proteases suitable for detergent formulations and for the bioremediation of protein-rich waste, such as feathers and leather industry by-products. The proteases from these strains often exhibit good stability and activity under alkaline conditions. Furthermore, fungal proteases from *Penicillium* are being investigated for their ability to hydrolyze various food proteins, releasing bioactive peptides with antihypertensive, antioxidant, or antimicrobial properties for their nutritional applications (Khan and Singh 2011).

1.2.2.3 *Mucor* spp.

The genus *Mucor*, comprising saprophytic fungi, has found its most critical industrial niche in the dairy industry, specifically through species like *Mucor miehei* and *Mucor pusillus*, which are prolific producers of aspartic proteases that serve as highly effective microbial rennets. The widespread adoption of microbial rennets, which rely on chymosin extracted from unweaned calves, a supply that was variable, costly, and raised ethical concerns. The aspartic proteases from *Mucor* spp., often referred to as mucorpepsin, provided a perfect solution. The *Mucor*-derived rennets, which are microbial proteases, can not only replace traditional animal-based products but can also be refined and optimized through biotechnology to meet precise industrial specifications (Arulmani et al. 2007).

1.3 Classification of Microbial Proteases

The variety of microbial proteases and their different mechanisms requires a systematic approach to classifying them. This is the protease classification of the enzyme, specifically identifying its application, activity under process conditions, and plans to regulate or inhibit activity as needed. The categorization of the sources is according to the phylogeny and functional and structural according to the catalytic machinery of the enzyme, the way it attacks the protein substrate, and also its activity profile in terms of pH. The most common system is known as the catalytic system, on the basis of which proteins are classified into four different groups, namely: serine, cysteine, aspartic, and metalloproteases, depending on the chemical groups present in the active site of the enzyme that are responsible for nucleophilic attack on the peptide. Structural biology and biochemistry support this classification, and it has dramatic consequences on an enzyme and its sensitivity to inhibitors and stability. Diisopropyl fluorophosphate (DFP) and metal chelators such as EDTA inhibit serine and metalloproteins, respectively. A second major system of classification depends on the location of action on the protein substrate that differentiates between

Table 1.1 Classification and properties of microbial proteases

Classification	Categories	Key characteristics	Common sources and examples	Primary inhibitors
Catalytic mechanisms	Serine proteases	Catalytic triad (Ser, His, Asp); alkaline pH optimum; covalent intermediate.	*Bacillus* spp. (subtilisin)	PMSF, DFP
	Cysteine proteases	Catalytic dyad (Cys, His); neutral/acidic pH; requires reducing environment	Papain (plant), some fungi, and bacteria	Iodoacetate, oxidizing agents
	Aspartic proteases	Two Asp residues; acidic pH optimum; no covalent intermediate	Fungi (*Aspergillus*, *Mucor*), pepsin	Pepstatin
	Metalloproteases	Require metal ion (e.g., Zn_2^+); neutral pH optimum; no covalent intermediate	*Bacillus* spp. (thermolysin)	EDTA, 1,10-phenanthroline
Site of action	Endopeptidases	Cleave internal peptide bonds; produce a mixture of peptides	Subtilisin, thermolysin, pepsin	Varies by mechanism
	Exopeptidases	Cleave terminal amino acids; release free amino acids	Aminopeptidases, carboxypeptidases	Varies by mechanism
Optimal pH	Acidic proteases	pH 2.0–5.0; mainly aspartic proteases	Fungi (*Aspergillus*, *Mucor*)	Pepstatin
	Neutral proteases	pH 5.0–8.0; mainly metalloproteases	*Bacillus* spp. (thermolysin)	EDTA
	Alkaline proteases	pH 8.0–11.0; mainly serine proteases (subtilisins)	Alkaliphilic *Bacillus* sp.	PMSF, DFP

endopeptidases, which cut internal peptide bonds of the polypeptide chain, and exopeptidases, which cut off terminal amino acids. This is assorted on the basis of the optimum pH of activity that determines the environmental conditions under which an enzyme works, such as acidic stomach environments to neutral food systems and alkaline detergent solutions (Mehta et al. 2006). Their classification, which is primarily based on catalytic mechanism, pH optimum, and evolutionary relationships, is detailed in Table 1.1.

1.3.1 *Based on Catalytic Mechanism*

The scientifically most rigorous and informative system of classification for proteases is one that focuses on enzymes that catalyze the hydrolysis of thermodynamically stable peptide bonds. The proteins are classified into four clans of proteins, namely, serine, cysteine, aspartic, and metalloproteins, with the fifth group (threonine proteins, e.g., the proteasome) not being as abundant in microbial industrial settings. The catalytic triad in the active site of serine, histidine, and aspartate identifies serine proteases. This occurs by the serine hydroxyl group acting as a potent nucleophile that attacks the carbonyl carbon atom of the peptide bond (Oskouie et al. 2008). One of the histidine residues serves as a general base to abstract a proton from serine, increasing its nucleophilicity, and the other aspartate stabilizes the charged histidine. This results in the development of a covalent acyl-enzyme intermediate that is then hydrolyzed by a water molecule to create the cleaved peptide and restore the free enzyme. This class includes the hugely important subtilisins from *Bacillus* (used in detergents) and trypsin-like enzymes. They are typically active at neutral to alkaline pH levels and are inhibited by specific agents, such as phenylmethylsulfonyl fluoride (PMSF) and DFP. Cysteine proteases (or thiol proteases) employ a catalytic dyad of cysteine and histidine. Here, the thiol group (-SH) of the cysteine residue acts as the nucleophile. The histidine residue activates the cysteine thiol by abstracting a proton, allowing it to attack the peptide bond and form a thioester acyl-enzyme intermediate (Niyonzima and More 2015). This intermediate is then hydrolyzed by water. These enzymes are generally active in neutral to mildly acidic conditions and are notoriously sensitive to oxidation. Their activity is absolutely dependent on reducing the environment to keep the cysteine thiol group in its reduced state; alkylating agents, such as iodoacetate, and oxidizing agents inhibit them.

Papain is a plant enzyme, while microbial equivalents are found in certain bacteria and fungi. Metalloproteases are a diverse class that requires a divalent metal ion, typically zinc, for catalytic activity. The metal ion is held in place by amino acid ligands and activates a water molecule as a nucleophile, directly attacking the peptide bond. This process polarizes the substrate's carbonyl group, making it more susceptible to attack. Unlike serine and cysteine proteases, no covalent intermediate is formed, and hydrolysis occurs via a single-step mechanism. This class includes neutral proteases from *Bacillus* and many human enzymes, which are active at neutral pH ranges. Aspartic proteases, also known as acid proteases, utilize a pair of aspartic acid residues to catalyze peptide bond hydrolysis (Gaur et al. 2015). The enzymes, including pepsins, rennet, and fungal proteases, are primarily active at low pH (2–5), where the aspartic acid residues are in the correct protonation state. These enzymes, which include pepsins, rennet, and fungal proteases from *Aspergillus* sp. and *Mucor* sp., are used in cheese making and food processing (Kumar et al. 2008a). Pepstatin, a potent transition-state analogue, is inhibited explicitly by this enzyme.

1.3.2 Based on the Site of Action

In this type of classification, proteins are classified by their point of attack on the polypeptide chain, which immediately dictates the result of the proteolytic reaction:

either a pool of smaller peptides or the release of free amino acids. This is a significant difference in the industrial applications where the desired end-product is specific. Endopeptidases are enzymes that break down bonded peptides (internal) anywhere along the polypeptide chain, other than at the ends (Deb et al. 2013). Industrial proteases, subtilisin, thermolysin, and pepsin, initiate protein digestion by breaking down large protein to smaller peptides. These enzymes find use in areas such as detergent cleaning, in leather processing as well as in the production of protein hydrolysates. They decrease the size of proteins, viscosity, and molecular weight, which are beneficial in the different industries. They may be very broad or narrow in their substrate specificity, but in any case, they do give peptide fragments. The exopeptidases, in turn, only work at the terminal ends of the polypeptide chains (Hmidet et al. 2009).

Two classes of enzymes involved in the digestion of proteins are the aminopeptidases and the carboxypeptidases. Aminopeptidases peel off single amino acids at the N-terminus of the peptides, whereas carboxypeptidases peel off single amino acids at the C-terminus. They are required in treating protein hydrolysates to remove bitter flavors and in the ripening of cheese to develop flavors. Carboxypeptidases excrete individual amino acids via C-terminus cleavage, and certain ones are selective to a type of amino acid. An endo-exopeptidase synergistic combination is used in various natural processes and industries (Al-Dhabi et al. 2020). Endopeptidases cleave larger peptides into smaller sizes, exposing more termini, and then exopeptidases cleave free amino acids, which is vital in flavor production in fermented foods, savory extracts, and in the digestive supplement market.

1.3.3 Based on Optimal pH of Activity

This is the most operational and application-driven classification system, grouping proteases based on the pH range in which they exhibit their maximum activity. This property is a direct reflection of the enzyme's evolutionary origin and the environment in which the producing microorganism resides, and it is the first parameter considered when matching an enzyme to a process (Al-Dhabi et al. 2020). Acidic proteases, primarily aspartic and cysteine proteases, are active in the low pH range of 2.0 to 5.0. They are suited for applications in highly acidic environments and are used in cheese manufacturing, digestive aid, brewing, and silver recovery. Their activity depends on the protonation state of key acidic residues in their active site. Neutral proteases, on the other hand, exhibit optimal activity in a pH range of 5.0 to 8.0. Their primary uses include coagulating milk at pH ~6.6, supplementing gastric digestion, preventing chill haze, and digesting the gelatin layer of used photographic films (Madan and Dhillon 2018).

Metalloproteases and serine proteases are a class of enzymes that require a neutral environment and are sensitive to metal chelators or inhibitors. They are used in the food industry for baking, protein hydrolysis, brewing, and functional modification of proteins. Alkaline proteases are most active at high pH, typically between 8.0 and

11.0, and are dominated by serine proteases of the subtilisin family (Haddar et al. 2010a). These enzymes are essential in the leather industry for dehairing and bating, the textile industry for silk degumming, and waste treatment processes for degrading proteinaceous waste. The pH profile of a protease is crucial for its applications, as it can lead to drastic reductions in activity or irreversible inactivation.

1.3.3.1 Acidic Proteases

Acidic proteases, also known as aspartic proteases or acid proteinases, represent a critically important class of enzymes characterized by their optimal catalytic activity within a highly acidic pH range, typically between 2.0 and 5.0, a trait directly stemming from their unique catalytic mechanism, which relies on the precise protonation state of two essential aspartic acid residues within the active site. The mechanism of pepsin, unlike serine, cysteine, or metalloproteases, involves activating a water molecule between two aspartate residues. The nucleophilic attack of the water molecule on the carbonyl carbon of the scissile peptide bond forms a gem-diol intermediate, which collapses to release two peptide products without the formation of covalent acyl-enzyme intermediates (Dodia et al. 2008). Acidic proteases are essential enzymes produced by microorganisms, such as fungi, yeasts, and bacteria, that enable them to thrive in acidic, protein-rich environments. The most renowned and historically significant acidic proteases are pepsins and chymosin. These proteases revolutionize the global cheese industry by providing a consistent, cost-effective, and vegetarian-friendly coagulant that hydrolyzes the κ-casein protein in milk, destabilizing casein micelles and causing them to coagulate into curds. Acidic proteases are also indispensable in the food and beverage sector, such as baking, where they modify gluten strength, produce soy sauce, miso, and other fermented foods, and prevent chill haze in the brewing industry. These enzymes hydrolyze haze-forming proteins, ensuring long-term clarity and stability of beer. Overall, acidic proteases play a crucial role in various sectors, including dairy, food, and beverages. (Paranthaman et al. 2009).

Furthermore, their application extends into the realm of health and nutrition, where they are used as digestive aids to supplement natural gastric pepsin, particularly for individuals with pancreatic insufficiency, and in the production of protein hydrolysates for hypoallergenic infant formulas and specialized nutritional supplements. In the industrial domain, acidic proteases are utilized in the recovery of silver from photographic and X-ray films by digesting the gelatin layer, which contains the silver particles. They are also being explored for the treatment of protein-rich industrial waste. The production of these enzymes is predominantly achieved via submerged fermentation (SmF) or SSF of fungal cultures, with SSF often yielding higher titers as it mimics the natural habitat of filamentous fungi. Downstream processing involves filtration, precipitation, and chromatographic techniques to achieve the desired purity, with the final product often standardized and sold as a liquid concentrate or a lyophilized powder (Basso and Serban 2019). Acidic proteases, despite their limited pH optimum, can be modified through techniques like enzyme immobilization and chemical modification. This is used to explore novel acidophilic microbial sources, protein engineering to enhance stability and specificity, and to develop new applications in peptide synthesis and therapeutic areas. This solidifies the role of acidic proteases as versatile biocatalysts in various industrial processes, thereby enhancing their versatility.

1.3.3.2 Neutral Proteases

Neutral proteases, as the name implies, are a class of proteolytic enzymes that exhibit their peak catalytic activity within a narrow, neutral pH range, typically between 5.0 and 8.0, a characteristic that makes them uniquely suited for a vast array of applications in the food, brewing, and pharmaceutical industries, where processes are often conducted at or near physiological pH, thus avoiding the drastic pH adjustments required for acidic or alkaline proteases and preserving the native functionality and sensory properties of the final products (de Souza et al. 2015). The pH optimum of neutral proteases is due to their catalytic mechanism, which is primarily that of metalloproteases. These enzymes, which are primarily metalloproteases, rely on a divalent metal ion, typically zinc, coordinated by histidine, glutamate, or aspartate residues. This activates a water molecule as a nucleophile, directly attacking the carbonyl carbon of the peptide bond. Metal-chelating agents, such as EDTA and 1,10-phenanthroline, inhibit this process. While some neutral serine and cysteine proteases exist, their industrial prevalence is less. The primary industrial producers of neutral metalloproteases are bacteria, particularly those from the genus *Bacillus*, with species like *Bacillus subtilis, Bacillus amyloliquefaciens*, and *Bacillus thermoproteolyticus* being the most significant (Gohel and Contractor 2015).

Fungi, like *Aspergillus* sp. and *Penicillium* sp., produce neutral proteases as part of a broader protease cocktail. A notable example is thermolysin, a thermostable neutral metalloprotease from *B. thermoproteolyticus*. It is known for its ability to cleave peptide bonds on the amino side of hydrophobic amino acids, making it useful in peptide synthesis and the production of aspartame. In the food industry, neutral proteases are utilized for controlled and limited proteolysis, such as baking to hydrolyze gluten proteins in dough, cheese ripening to accelerate flavor development, and the production of protein hydrolysates for use as flavor enhancers and functional food ingredients. They are also crucial in the dairy industry for ripening cheese without excessive bitterness and in the brewing industry for adjunct liquefaction. In the leather industry, hides are softened by removing non-collagenous proteins after dehairing. They are also used in the pharmaceutical and cosmetic industries for the production of bioactive peptides and wound cleaning ointments. Their production via bacterial fermentation is highly optimized, and their relative instability is managed through careful process control (Deng et al. 2010). The future of protein engineering lies in improving stability and specificity, as well as its increasing use in green chemistry for peptide and speciality chemical synthesis, due to its crucial role as precise and gentle biocatalytic tools in neutral environments (Ellaiah et al. 2002).

1.3.3.3 Alkaline Proteases

Alkaline proteases are a class of proteolytic enzymes that demonstrate their maximum catalytic activity under alkaline pH conditions, typically within a range of 8.0 to 11.0, and are often characterized by significant thermostability and a remarkable resistance to denaturation by surfactants, oxidizing agents, and chelating compounds, a combination of properties that has cemented their status as the most critical class of industrial enzymes, dominating the global market primarily due to their ubiquitous use as key additives in household and industrial detergents. Industrial alkaline proteases, primarily serine proteases, are highly functional in harsh environments due to their unique catalytic mechanism and structural biology. These proteases, part of the subtilisin family, utilize a triad of serine, histidine, and aspartate residues to catalyze peptide bond hydrolysis through a two-step mechanism, which is optimized for high-pH operation due to its surface charge and amino acid composition (Rai and Mukherjee 2011).

Alkaline metalloproteases are enzymes sourced from alkaliphilic microorganisms, primarily bacteria from the genus *Bacillus*. These enzymes are used in the detergent industry, accounting for over 30% of the global enzyme market. They efficiently hydrolyze stubborn protein-based stains, such as blood, egg, grass, gravy, and human sweat, at lower wash temperatures, thereby enhancing cleaning efficiency and allowing for eco-friendly formulations with reduced phosphate content and lower washing temperatures (Fakhfakh-Zouari et al. 2010). In addition to detergents, alkaline proteases are essential in the leather industry for dehairing and bating animal hides, facilitating easy removal and softening of hides. In the textile industry, they are used for degumming silk and biopolishing cotton to improve softness, color brightness, and resistance to pilling. They also play a role in the production of protein hydrolysates and peptones for microbial culture media, silver recovery from X-ray films, waste management, and nucleic acid isolation.

1.4 Detection and Assay of Microbial Proteases

Enzymology relies on the accurate detection and quantification of protease activity at every stage of protease research and application, from screening microbial isolates in environmental samples to monitoring enzyme production during fermentation, to rigorous quality control of the final commercial enzyme preparations, and comprehensive kinetic characterization of purified enzymes for specific industrial or therapeutic uses. Qualitative and quantitative methods, with their respective advantages and disadvantages, are employed to study proteases. Qualitative methods, such as the protein agar plate assay and the radial diffusion assay, which form clear zones of

Table 1.2 Methods for the detection and quantification of proteases

Methods	Specific method	Principle	Advantages	Applications
Qualitative	Protein agar plate assay	Microbial growth on a protein-rich (e.g., skim milk) agar plate Protease secretion creates a clear zone of hydrolysis	Simple, cheap, high-throughput, visual result	Initial screening of microbial isolates for proteolytic activity
	Radial diffusion assay	Enzyme samples are placed in wells in a protein-agar gel Protease diffusion creates clear zones after acid fixation	Separates growth from activity; semi-quantitative; can use buffers/ inhibitors	Comparing activity in preformed samples (e.g., culture broths, purification fractions)
Quantitative	Chromogenic assays	Uses synthetic peptides conjugated to a chromophore (e.g., p-nitroaniline) Cleavage releases the dye, increasing absorbance	Real-time kinetics are specific, cost-effective, and easy to use	Routine kinetics, inhibitor screening, and quality control of enzyme preparations
	Fluorogenic assays	Uses peptides with a fluorophore and a quencher Cleavage separates them, resulting in a significant increase in fluorescence	Extremely sensitive, ideal for low-abundance enzymes and high-throughput applications	Drug discovery, detecting low-concentration proteases, and characterizing potent inhibitors

growth, are crucial for high-throughput screening processes that require rapid evaluation of numerous isolates. Quantitative assays measure enzymatic activity in micromoles of substrate hydrolyzed per minute under specified conditions. They are critical for determining enzyme kinetics (K_m, V_{max}, k_{cat}), enzyme concentration in a crude extract or formulation, specific activity during purification steps, and the effects of pH, temperature, inhibitors, and other components (Pant et al. 2015). The assay for protease analysis is determined by sample type, equipment, and the need for precise measurement. The substrate must match the protease's specificity for developing a sensitive, robust, and high-throughput assay is crucial for identifying novel enzymes and optimizing their use. These methods, summarized in Table 1.2, range from classic colorimetric assays to advanced fluorescent and bioluminescent sensors, each with unique advantages in sensitivity, throughput, and applicability to complex biological samples.

1.4.1 Qualitative Detection Methods

The initial stages of research and development utilize powerful, low-cost, and visually simple qualitative detection methods for microbial proteases to rapidly screen large numbers of microbial colonies or environmental samples for proteolytic potential, offering a preliminary, comparative assessment of enzyme activity levels without the need for expensive equipment or tedious analysis. The hydrolysis of a substrate protein in a solid growth medium, such as an agar plate, is visualized as a clear zone or halo around the enzyme source against an opaque or dark background (Melo et al. 2021). The size and clarity of the halo are proportional to the amount and potency of the protease produced. Microbiologists and enzymologists seeking novel protease-producing microorganisms from extreme environments, such as hot springs, alkaline lakes, or deep-sea sediments, can quickly identify potential matches from thousands of isolates using these assays due to their simplicity and scalability. The protein agar plate assay and the radial diffusion assay are the two most common methods in this category. The former is used to detect protease production by growing microbial colonies directly. In contrast, the latter is used to test preformed enzyme samples, such as culture supernatants or column fractions from a purification process. These qualitative methods are essential to protease discovery and characterization due to their low cost, high throughput, and ability to provide quick visual observations on proteolytic activity (Agrawal et al. 2005).

1.4.1.1 Protein Agar Plate Assay

The protein agar plate assay, also known as the skim milk agar assay or casein hydrolysis test, is one of the oldest, simplest, and most widely used qualitative methods for detecting extracellular protease production by microorganisms. Its simplicity, cost-effectiveness, and dramatic visual result allow for the simultaneous screening of thousands of microbial isolates on a single petri dish. The procedure involves preparing a sterile growth medium, usually nutrient agar, into which a substrate protein, such as skim milk powder or purified casein, is added. Gelatin, albumin, or collagen can be incorporated at a concentration of 1–2% (w/v) before the agar solidifies, creating opaque, milky-white plates (Ahmad and Sardar 2015). Test microorganisms secrete extracellular proteases into the agar matrix, hydrolyzing insoluble casein molecules into soluble peptides that release amino acids. This specific digestion eliminates the medium, forming a transparent zone around the proficient protease-producing colony. This area is a semiquantitative indicator of the protease produced, with a large zone indicating highly active enzymes and a small or no zone indicating weak or absent proteolytic activity (Zambare 2010b, c).

The skim milk plate assay is a flexible method used to analyze the activity of endopeptidase enzymes in casein. A clear halo indicates broad-specificity endopeptidase activity, while a turbid or hazy zone may indicate specific exopeptidases that may not fully solubilize the substrate. The assay can be adjusted to obtain further clarification. Protease inhibitors can be added to the agar to determine the enzyme's catalytic class. However, the method cannot provide absolute quantitative data due to factors, like protease diffusion rate, microbial growth rate, and incubation time (Giongo et al. 2007). The results are qualitative and often require determination. The assay is essential for the rapid, visual, and inexpensive initial analysis of novel proteolytic microbes and for evaluating strain improvement programs aimed at increasing protease yields (Nilegaonkar et al. 2004; Zambare et al. 2011a).

1.4.1.2 Radial Diffusion Assay

The radial diffusion assay is a qualitative and semiquantitative method used to evaluate prepared protease samples, such as cell-free culture supernatants, partially purified enzyme fractions, or industrial products. It separates microbial growth from enzyme activity measurement, allowing for a more controlled assessment of proteolytic capacity. A substrate protein is suspended in an opaque matrix, and enzyme-containing samples are added in equal amounts. The samples are then incubated under controlled temperature and duration, often in a humidified room to prevent gel drying (Nascimento and Martins 2004). The radial diffusion assay is a method used in enzymology labs to compare the relative potency of enzyme samples, track protease activity during purification, and estimate enzyme titer in fermentation broth. Protease molecules diffuse outward from the well into the surrounding agarose matrix, where they hydrolyze the substrate protein. The reaction is stopped by submerging the plate in a fixing and precipitating solution, such as trichloroacetic acid (TCA) or acetic acid, which denatures and precipitates any residual protein, making the plate background opaque and white. The clear zones around the original wells are directly proportional to the logarithm of the enzyme concentration or activity in the sample well. However, the primary disadvantages are the diffusion time and the potential of non-radial diffusion when the well is inadequate (Daroit and Brandelli 2014).

1.4.2 Quantitative Assay Methods

Quantitative assay methods for microbial proteases are crucial in enzymology, providing precise, numerical, and reproducible measurements of enzymatic activity for various scientific and industrial applications. These methods enable the calculation of kinetic parameters, determination of inhibitor constants, standardization of commercial enzyme preparations, and rigorous monitoring of biotechnological production processes. Quantitative assays track the time-dependent formation of products or substrate disappearance under monitored conditions, accurately measuring the initial rate of the reaction and expressing activity in precise international units (Mothe and Sultanpuram 2016). Modern quantitative assays utilize synthetic peptide substrates conjugated to chromogenic or fluorogenic dyes, with natural proteins, such as casein, still being employed. These substrates have high specificity for specific protease types, superior solubility, and the ability to conduct continuous, real-time kinetic measurements without disrupting the reaction. They offer high-quality data with minimal manual intervention and high sensitivity. Fluorogenic assays are more sensitive than chromogenic assays, but chromogenic assays are durable, simple, and readily available. These quantitative approaches are crucial for characterizing, validating, and using microbial proteases in various fields of science and industry (Beg and Gupta 2003).

1.4.2.1 Chromogenic Substrate Assays

Quantitative protease analysis uses chromogenic substrate assays, which hydrolyze synthetic peptides conjugated to a chromophore to generate a change in absorbance directly proportional to enzymatic activity. These assays are reliable and widely available, allowing for accurate real-time monitoring of the reaction. The peptide is covalently linked to a synthetic chromogenic leaving group, usually p-nitroaniline (pNA). The absorbance spectrum is masked or shifted in the unchanged substrate; however, after enzymatic cleavage, the pNA group is released into solution, displaying an intense yellow color with a maximum absorbance at 405–410 nm. The rate of absorbance increase is converted to a product production rate using the coefficient-free pNA, allowing for international unit calculations of enzyme activity. This method can modify the peptide sequence to match the primary specificity of almost any protease, such as chymotrypsin-like enzymes and trypsin-like serine proteases. The ability to quantify one protease type in a crude extract while detecting others is a significant advantage over tests using natural proteins (Naveed et al. 2021). The pNA-based continuous assay is a widely used and reliable method for routine enzyme kinetics studies, high-throughput screening of enzyme inhibitors, and quality control, despite its lower sensitivity compared to fluorogenic assays and potential interference from colored or turbid sample components. Its simplicity, cost-effectiveness, and widespread availability make it the preferred method in quantitative protease analysis.

1.4.2.2 Fluorogenic Substrate Assays

Fluorogenic substrate assays are sensitive and versatile quantitative protease assays that use synthetic peptides labelled with fluorophores. These peptides enhance fluorescence through enzyme cleavage, detecting tiny and fetal protease activity. They are used in drug discovery screens and to detect low-abundance proteins in biological test samples. The peptide is synthesized with a fluorescent group and a quenching group bound, resulting in a high fluorescence intensity that can be monitored in real-time using a fluorescence spectrophotometer or microplate reader (Benkiar et al. 2013). Fluorescence rate ($\Delta F/\text{min}$) is a sensitive method that correlates with substrate hydrolysis and enzyme activity, allowing researchers to use very low substrate and enzyme concentrations. It is 100 to 1000 times more sensitive than chromogenic assays, allowing for the characterization of tight-binding inhibitors at enzyme K_m conditions. Fluorogenic substrates are specific for protease classes, making them essential for biomedical studies. Alternative designs are used in complex applications, such as fluorescence polarization or time-resolved fluorescence. Although more expensive per sample due to substrate costs and specialized instrumentation, fluorogenic assays offer unparalleled sensitivity, a broad dynamic range, and adaptability for ultrahigh-throughput screening in 384- or 1536-well plates (Haddar et al. 2010b).

1.5 Applications for Microbial Proteases

Microbial proteases are indispensable across numerous industries. In detergents, they efficiently break down protein-based stains, enabling effective low-temperature washing. The food industry utilizes them for cheese production, meat tenderization, beverage clarification, and the creation of protein hydrolysates and flavor enhancers. They have revolutionized the leather industry by replacing hazardous chemicals in dehairing and bating processes, significantly reducing environmental pollution. Furthermore, proteases play a vital role in waste management by degrading protein rich organic waste, enhancing digestibility in animal feed and mitigating environmental impact, and in bioremediation to break down pollutants such as organophosphorus compounds, highlighting their versatility and sustainability. The commercial value of microbial proteases is demonstrated by their widespread use across multiple sectors. As illustrated in Fig. 1.2, these enzymes are critical catalysts in industries ranging from detergents and food processing to leather treatment and pharmaceuticals, highlighting their versatility and economic importance.

1.5.1 Detergent Industry

The detergent industry is a major consumer of microbial proteases, which are used to replace harsh chemicals with environmentally friendly, biodegradable, and efficient protein-degrading agents. The development of these enzymes began in the 1930s with pancreatic trypsin in preparation solutions. In the 1960s, Novo Industries from the A/S commercialized "Alcalase" a subtilisin-type alkaline serine protease from *Bacillus licheniformis*. These enzymes hydrolyze stubborn, insoluble proteinaceous stains, such as blood, egg, grass, gravy, sweat, and food residues, by breaking down peptide bonds into smaller, soluble peptides and free amino acids. The suitability of a microbial protease for detergent formulations depends on its compatibility with hazardous washing environment chelating substances (EDTA), oxidizing agents, anionic and nonionic surfactants, and a broad temperature range. Modern detergent

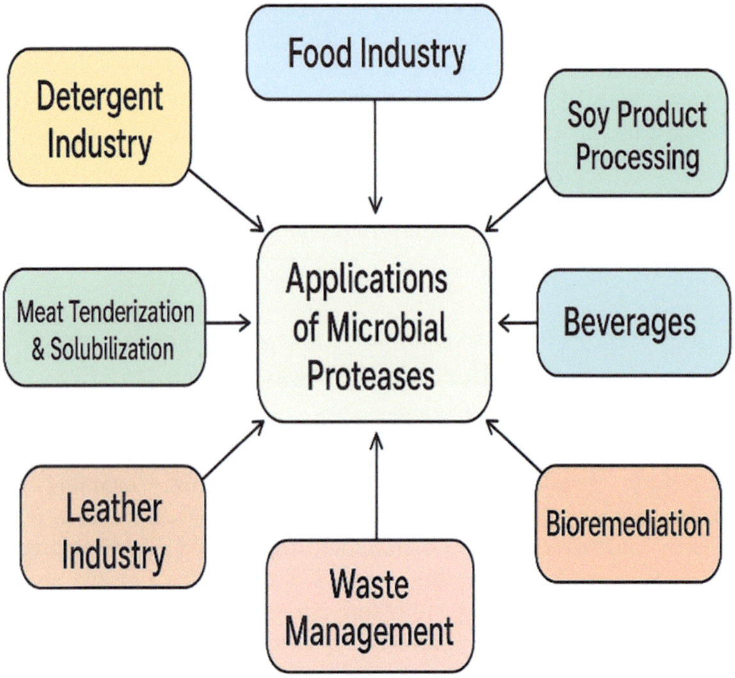

Fig. 1.2 Industrial applications of microbial proteases

proteases are mainly produced from alkaliphilic *Bacillus* species and are found in powders, liquids, and single-unit doses. Polyols, like enzyme-based detergents, promote effective cleaning at lower temperatures, reducing household energy, greenhouse gas emissions, and fabric durability. Microbial proteases are essential components of modern detergent science, and future improvements aim to enhance their cold-water activity, compatibility with fresh bleach activators and surfactants, and substrate specificity to tackle more complex stains (Najafi et al. 2005; Nilegaonkar et al. 2006; Zambare et al. 2014).

1.5.2 Food Industry

The food industry relies on microbial proteases to modify the functional, nutritional, and sensory properties of proteins in a particular and controlled manner, enabling processes that would be impossible or impractical with chemical or physical methods. This enhances texture, flavor compounds, digestibility, and yield across various food sectors. They are ideal for food processing applications that require safety, consistency, and efficiency due to their consistent and scalable fermentation production, as well as a wide range of catalytic properties, including pH, temperature, and GRAS status (Hossain et al. 2016).

1.5.2.1 Meat Tenderization and Solubilization

Microbial proteases tenderize tough meat pieces and solubilize meat proteins to improve the taste and reduce waste in meat processing. Endogenous proteases (calpains and cathepsins) slow down postmortem tenderization during ageing. Still, the extensive structure of connective tissues, particularly collagen, is a structural protein, and myofibrillar proteins in muscle fibers make meat tough. Neutral and alkaline proteases from fungi like *Aspergillus oryzae* (e.g., fungal rennet) and bacteria like *Bacillus subtilis* are applied exogenously to accelerate this process by selectively hydrolyzing connective tissue collagen and elastin and myofibrillar proteins, disrupting the rigid protein structure and improving tenderness, juiciness, and flavor (Gupta et al. 2002). Some proteases have been modified to be inactivated at specific temperatures, ensuring they remain active during the initial stages of cooking before being denatured. To avoid overtreatment, spraying or dipping enzyme solutions onto the meat surface is necessary. After tenderization, proteases solubilize proteins and produce meat extracts, savory flavor enhancers, and protein-rich ingredients for soups, sauces, and gravies from meat residues, delivering value to low-value materials while assisting in the sustainable utilization of animal carcasses.

1.5.2.2 Dairy Protein Hydrolysis

Microbial proteases benefit cheese, whey, and other nutritious products in the dairy sector. In cheese coagulation, chymosins that are derived from calf stomachs and microbial rennet substitutes from *Rhizomucor miehei, Rhizopus niveus*, and *Cryphonectria parasitica* hydrolyze κ-casein protein in milk, causing micelles of milk protein to coagulate into a gel (curd) that is processed into cheese. Microbial rennets are an ethical, economical, and reliable alternative to rennets. In addition to coagulation, proteases (e.g., neutral proteases from *Aspergillus oryzae*) break down caseins into smaller peptides and free amino acids, which are precursors to flavor and aroma compounds, reducing cheese ripening time from months to weeks and storage costs (Annamalai et al. 2014). In whey processing, proteases break down β-lactoglobulin and α-lactalbumin to reduce allergenicity, improve digestibility, and prevent gelation during heat concentration. Hydrolyzed milkshakes are essential for infants with cow's milk protein allergies. Bioactive peptides derived from milk proteins, produced by proteases, exhibit antihypertensive, antibacterial, antioxidant, and immunomodulatory properties, making dairy components suitable for high-value, functional foods.

1.5.2.3 Soy Product Processing

Microbial proteases hydrolyze soy proteins to alter their functional properties and eliminate undesirable traits, transforming soybeans into a variety of traditional and modern food products. A koji mold (*Aspergillus oryzae* or *Aspergillus sojae*) secretes a complex cocktail of enzymes, including proteases, amylases, and lipases, which hydrolyze soy proteins into oligopeptides and eliminate amino acids, which are fermented with lactic acid and yeast to produce the complex, savory, salty flavor of soy sauce. The fermentation of miso and tempeh utilizes fungal proteases that modify the flavor and texture. Proteinases improve the solubility, emulsification, foaming, and water-binding properties of soy protein isolates and concentrated forms in modern food technology, making them more versatile ingredients for meat substitutes, baked goods, and beverages. Soy protein hydrolysates can be made less bitter by removing anti-nutritional components through enzymatic hydrolysis, thereby improving the quality and consumption of soy-based foods (Jellouli et al. 2011).

1.5.2.4 Beverages

The beverage industry utilizes microbial proteases for clarification, stabilizing fruit juices and beverages, and producing protein hydrolysates for functional drinks. Apple and grape juices contain colloidal suspensions of proteins, pectins, and tannins that can cause haze after packaging. Proteases are used in conjunction with pectinases to hydrolyze the haze-forming proteins into soluble fragments, thereby clarifying the juice and preventing it from producing turbidity afterward. Proteases degrade gluten proteins into amino acids and smaller peptides, promoting healthy yeast that contributes to strength, foam stability, and beer taste during the brewing stage (Liu et al. 2017). Proteases are also used to produce translucent protein hydrolysates and bioactive peptides from whey, soy, pea, and rice for use in sports drinks, nutritional shakes, and wellness beverages, thereby eliminating sedimentation and viscosity issues. This process yields easily digestible protein supplements with improved nutritional profiles and enhanced health benefits.

1.5.3 Leather Industry

Microbial proteases have eliminated toxic and hazardous chemicals, such as sodium sulfide and lime, in roof construction processes, making the leather industry more environmentally friendly and safer for workers (Nilegaonkar et al. 2007a, b). After slaughtering, the animal skin or hides were preserved in slats, where it removes the water and preserves the skin or hide. Soaking is one of the important processes that removes the excess salt and soaks the skin in water. Proteases are used in the soaking operation to remove albuminous protein, which minimizes the further leather processing time (Zambare et al. 2013). Proteases are primarily used in a process called "bating," which removes specific non-collagenous proteins, mucins, sugars, and elastin from the fabric after dehairing and liming (Zambare 2010b; Bansal et al. 2012). This process cleans the skin, leaving it smooth, soft, and pliable, and imparts the desired grain texture and quality to the finished leather product. Microbial proteases from *Bacillus* species precisely hydrolyze unwanted proteins without damaging the leather's collagen structure, a balance that requires precise time, temperature, and enzyme concentration. Pancreatic enzymes from animals (dog or pigeon dung) were previously used. Proteases also weaken hair follicles by breaking down keratinous proteins and hydrolyzing the base of the hair shaft and surrounding tissue, allowing for the mechanical removal of hair with drastically reduced sulfide, thereby reducing the toxic load and biological oxygen demand (BOD) of effluent wastewater (Nilegaonkar et al. 2007c). Enzymatic dehairing and bating produce higher-quality leather with a cleaner grain, increased dye uptake, and improved softness, while reducing processing time and water consumption. Most significantly, this process lowers environmental pollution by integrating tanning with innovative sustainable practices (Mahajan et al. 2010).

1.5.4 Waste Management

Waste management relies on microbial proteases to degrade and valorize protein-rich organic waste streams from diverse sectors, decreasing pollution and transforming trash into valuable resources. Tannery effluents (hair, fleshings, and proteinaceous sludge), poultry processing waste (feathers, which are over 90% keratin), animal hair, fish processing plant waste (scales, fins, guts), and hazardous cheese whey are resistant to degradation due to their complex protein structures. They can cause significant disposal issues, high BOD/COD in water bodies, and foul odors (Negi and Banerjee 2010). Keratinases from *Bacillus licheniformis* and *Streptomyces* species can hydrolyze millions of tons of discarded chicken feathers into a valuable, digestible protein hydrolysate that can be used as a high-nitrogen organic fertilizer or nutrient-rich animal feed. Protease treatment of fish waste recovers protein hydrolysates and oils, whereas sewage sludge treatment reduces volume and viscosity, enhancing handling and dewatering. Proteases enable the upcycling of waste into economically valuable products, thereby solving waste disposal problems while generating value from waste materials (Cang et al. 2010). X-ray photographic films from hospitals generate another kind of waste, and some proteases hydrolyze the gelatin layer and valorize it in terms of silver recovery (Zambare et al. 2011b).

1.5.5 Bioremediation

Microbial proteases are used in bioremediation as effective biological agents to clean up environments contaminated by proteinaceous waste, along with specific hazardous substances. In food processing plants, slaughterhouses, leather tanneries, and gelatin factories, they promote protein degradation to reduce organic matter and prevent eutrophication of water sources. In addition, specific potent microbial serine proteases (e.g., prolidases and related enzymes) can hydrolyze the peptide-like bonds in toxic organophosphorus (OP) nerve stimulants, such as sarin, soman, and VX, rendering them less harmful (Oberoi et al. 2001). Phosphodiesterases are generally liable for the degradation of these compounds. These enzymes can be utilized in bioremediation techniques to clean up water, soil, and equipment contaminated with these substances. Protease-producing microorganisms can also be utilized to clean municipal wastewater and treat textile and dairy effluents, which often contain protein-based reducing agents. This allows them to break down complex proteins into simpler, biodegradable units that can be processed by conventional wastewater treatment systems, thereby improving environmental mitigation strategies (Dandavate and Madamwar 2010).

1.5.6 Animal Feed Industry

Microbial proteases in animal feed are an effective means of enhancing nutrition, promoting growth, and minimizing environmental impact. Anti-nutritional factors (ANFs), such as trypsin inhibitors and lectins, are found in many conventional feed ingredients, particularly plant-based proteins like soybean meal, rapeseed meal, and sunflower meal, which are not fully digestible by monogastric animals, including poultry and swine. Exogenous microbial proteases in the feed supplement break down complex proteins and protein-ANF complexes in the gastrointestinal tract, thereby increasing amino acid availability, protein digestibility, and intestinal protein absorption. Microbial enzymes can enhance animal agriculture by improving feed conversion ratio, reducing nitrogenous waste excretion in manure, and allowing formulators to use cheaper, lower-quality protein resources. This leads to enhanced farmer economic efficiency, reduced ammonia emissions, and reduced water pollution and eutrophication (Huang et al. 2003). This application exemplifies that microbial enzymes can directly support more efficient and sustainable animal agriculture.

1.6 Future Prospects

Microbial proteases have the potential for revolutionary growth due to advances in biotechnology, environmental awareness, and novel applications in high-value sectors. Metagenomics enables scientists to tap into the vast reservoir of unculturable microorganisms from extreme environments, potentially uncovering new categories of proteases with remarkable stability and novel characteristics. New proteases with tailored properties will be the foundation of future development (Kumar et al. 2008b). Additionally, the development of "designer" proteases with improved catalytic efficiency, modified substrate specificity, and superior stability under complex industrial conditions, such as in the presence of high concentrations of organic solvents for peptide synthesis or novel detergent formulations, will be rendered feasible through protein engineering techniques, particularly directed evolution, which uses iterative rounds of mutagenesis and screening to evolve desired traits. Proteases are being investigated for targeted drug delivery, wound cleaning, and cystic fibrosis treatment. Proteases are used in gentle exfoliants, antiaging creams, and anti-dandruff shampoos. Multi-enzymes enhance biorefining efficiency by converting biomass into biofuels and biochemicals. As global demand for sustainable industrial processes grows, microbial proteases will evolve from industrial essentials into sophisticated, precision tools that support innovations in manufacturing, medicine, and sustainable development (Kandasamy et al. 2012).

1.7 Conclusion

Microbial proteases are essential biocatalysts that are crucial in various biotechnological and industrial processes. They are primarily sourced from bacterial genera like *Bacillus* and GRAS-status fungal species like *Aspergillus*, *Penicillium*, and *Mucor*. These enzymes are known for their high yield, ease of genetic manipulation, and scalable production through advanced fermentation technologies. Their functional diversity is captured through a classification system based on catalytic mechanism, site of action on the protein substrate, and optimal pH activity. Detection and characterization of these enzymes rely on sophisticated methods, from simple qualitative assays to highly sensitive quantitative chromogenic and fluorogenic substrate assays. Microbial proteases have a significant commercial value, with applications across critical sectors, such as detergents, food, leather, waste management, bioremediation, and animal feed. Their advancements in metagenomics, protein engineering, and directed evolution. They are also being developed in pharmaceuticals, cosmetics, and peptide synthesis. Microbial proteases demonstrate the integration of microbial catalysis into sustainable industrial practices, for innovation and economic growth in the world, increasingly toward green and efficient technological solutions.

References

Adrio JL, Demain AL (2014) Microbial enzymes: tools for biotechnological processes. Biomol 4(1):117–139. https://doi.org/10.3390/biom4010117

Agrawal D, Patidar P, Banerjee T, Patil S (2005) Alkaline protease production by a soil isolate of *Beauveria* sp. Lett Appl Microbiol 41(1):90–96. https://doi.org/10.1111/j.1472-765X.2005.01685.x

Ahmad R, Sardar M (2015) Enzyme immobilisation: an overview on nanoparticles as immobilisation matrix. Biochem Anal Biochem 4(1):1. https://doi.org/10.4172/2161-1009.1000178

Al-Dhabi NA, Esmail GA, Ghilan AKM, Arasu MV, Duraipandiyan V, Ponmurugan K (2020) Isolation and purification of alkaline protease from *Bacillus cereus* strain S8 and its application as a debriding agent. Saudi J Biol Sci 27(1):474–479. https://doi.org/10.1016/j.sjbs.2019.11.008

Anbu P, Gopinath SCB, Chaulagain BP, Tang TH, Citartan M (2015) Microbial enzymes and their applications in industry and medicine. Biomed Res Int 2015:816419. https://doi.org/10.1155/2015/816419

Annamalai N, Rajeswari MV, Balasubramanian T (2014) Extraction, purification and application of thermostable alkaline protease from *Bacillus alveayuensis* CAS 5 using marine wastes. Food Bioprod Process 92(4):335–342. https://doi.org/10.1016/j.fbp.2013.10.006

Arulmani M, Aparanjini K, Vasanthi K, Arumugam P, Arivuchelvi M, Kalaichelvan PT (2007) Purification and partial characterisation of serine protease from thermophilic *Bacillus* sp. Afr J Biotechnol 6(5):447–450. https://doi.org/10.5897/AJB2007.000-2028

Banerjee G, Ray AK (2017) The impact of microbial proteases on biotechnological industries. Biotechnol Genet Eng Rev 33(2):119–143. https://doi.org/10.1080/02648725.2017.1408256

Bansal N, Kanwar SS, Azmi W (2012) Purification and characterisation of a microbial protease produced by *Bacillus* sp. APCMST-RS7. J Basic Microbiol 52(6):824–832. https://doi.org/10.1002/jobm.201100529

Bari MR, Alizadeh M, Farbeh F (2010) Optimising protease production from an extremely halophilic bacterium, *Salinivibrio* sp. AF-2004. Rom Biotechnol Lett 15(5):5597–5606

Basso A, Serban S (2019) Industrial applications of immobilised enzymes—A review. Mol Catal 479:110607. https://doi.org/10.1016/j.mcat.2019.110607

Beg QK, Gupta R (2003) Purification and characterisation of an oxidation-stable, thiol-dependent serine alkaline protease from *Bacillus mojavensis*. Enzym Microb Technol 32(2):294–304. https://doi.org/10.1016/S0141-0229(02)00293-4

Benkiar A, Nadia ZJ, Badis A, Rebzani F, Soraya BT, Rekik H, Naili B, Ferradji FZ, Bejar S (2013) Biochemical and molecular characterisation of a thermo-and detergent-stable alkaline serine keratinolytic protease from *Bacillus circulans* strain DZ100 for detergent formulations. Int J Biol Macromol 58:103–113. https://doi.org/10.1016/j.ijbiomac.2013.03.054

Bhunia B, Basak B, Dey A (2012) A review on the production of serine alkaline protease by *Bacillus* spp. J Biochem Technol 3(4):448–457

Cang C, Liu Y, Wang Y, Yu W (2010) A novel extracellular protease from *Pseudomonas* sp. XJ-2: purification and characterisation. Appl Biochem Biotechnol 160(7):2107–2118. https://doi.org/10.1007/s12010-009-8703-4

Chakraborty S, Khopade A, Biao R, Jian W, Liu XY, Mahadik K, Zhang L, Kokare C (2011) Characterisation and stability studies on surfactant, detergent and oxidant stable α-amylase from marine haloalkaliphilic *Saccharopolyspora* sp. A9. J Mol Catal B Enzym 68(1):52–58. https://doi.org/10.1016/j.molcatb.2010.09.008

Chandra P, Enespa, Singh R, Arora PK (2020) Microbial lipases and their industrial applications: a comprehensive review. Microb Cell Fact 19(1):169. https://doi.org/10.1186/s12934-020-01428-8

Chantawannakul P, Oncharoen A, Klanbut K, Chukeatirote E, Lumyong S (2002) Characterisation of proteases of *Bacillus subtilis* strain 38 isolated from traditionally fermented soybean in northern Thailand. Sci Asia 28(3):241–245. https://doi.org/10.2306/scienceasia1513-1874.2002.28.241

Charbonneau DM, Beauregard M (2013) Role of key salt bridges in thermostability of G. thermodenitrificans EstGtA2: distinctive patterns within the new bacterial lipolytic enzyme family XV. PLoS One 8(10):e76675. https://doi.org/10.1371/journal.pone.0076675

Choudhary V, Jain PC (2011) Production and characterization of alkaline protease from *Aspergillus niger*. J Microbiol Biotechnol Res 1(4):75–86

Contesini FJ, Melo RR, Sato HH (2018) An overview of *Bacillus* proteases: from production to application. Crit Rev Biotechnol 38(3):321–334. https://doi.org/10.1080/07388551.2017.1354354

da Silva RR (2017) Bacterial and fungal proteases: production, purification, and therapeutic applications. Front Bioeng Biotechnol 5:81. https://doi.org/10.3389/fbioe.2017.00081

Dandavate V, Madamwar D (2010) Production of alkaline protease by a newly isolated *Bacillus cereus* strain MCM B-326 and its application as a dehairing agent. World J Microbiol Biotechnol 26(2):351–360. https://doi.org/10.1007/s11274-009-0178-0

Daroit DJ, Brandelli A (2014) A current assessment on the production of bacterial keratinases. Crit Rev Biotechnol 34(4):372–384. https://doi.org/10.3109/07388551.2013.794768

de Souza PM, Bittencourt ML d A, Caprara CC, de Freitas M, de Almeida RPC, Silveira D, Fonseca YM, Filho EXF, Pessoa Junior A, Magalhães PO (2015) A biotechnology perspective of fungal proteases. Braz J Microbiol 46(2):337–346. https://doi.org/10.1590/S1517-838246220140359

Deb P, Talukdar SA, Mohsina K, Sarker PK, Sayem SA (2013) Production and partial characterization of extracellular amylase enzyme from *Bacillus amyloliquefaciens* P-001. Springerplus 2:154. https://doi.org/10.1186/2193-1801-2-154

Deng A, Wu J, Zhang Y, Zhang G, Wen T (2010) Purification and characterization of a surfactant-stable high-alkaline protease from *Bacillus* sp. B001. Bioresour Technol 101(18):7100–7106. https://doi.org/10.1016/j.biortech.2010.04.013

Devi MK, Banu AR, Gnanaprabhal GR, Pradeep BV, Palaniswamy M (2008) Purification, characterization of alkaline protease enzyme from native isolate *Aspergillus Niger* and its compatibility with commercial detergents. Indian J Sci Technol 1(7):1–6

Dodia MS, Rawal CM, Bhimani HG, Joshi RH, Khare SK, Singh SP (2008) Purification and stability characteristics of an alkaline serine protease from a newly isolated Haloalkaliphilic bacterium sp. AH-6. J Ind Microbiol Biotechnol 35(2):121–131. https://doi.org/10.1007/s10295-007-0273-x

Ellaiah P, Adinarayana K, Bhavani Y, Padmaja P, Srinivasulu B (2002) Optimization of process parameters for glucoamylase production under solid state fermentation by a newly isolated *Aspergillus* species. Process Biochem 38(4):615–620. https://doi.org/10.1016/S0032-9592(02)00175-3

Fakhfakh-Zouari N, Haddar A, Hmidet N, Frikha F, Nasri M (2010) Application of statistical experimental design for optimization of keratinases production by *Bacillus pumilus* A1 grown on chicken feather and some biochemical properties. Process Biochem 45(5):617–626. https://doi.org/10.1016/j.procbio.2009.12.007

Gaur R, Tiwari S, Tiwari KL (2015) Extracellular production of alkaline protease by a newly isolated *Bacillus cereus* SIU1. J Appl Biol Biotechnol 3(6):021–025. https://doi.org/10.7324/JABB.2015.3604

Ghorbel-Bellaaj O, Hmidet N, Jellouli K, Younes I, Maâlej H, Hachicha R, Nasri M (2011) Shrimp waste fermentation with *Pseudomonas aeruginosa* A2: optimization of chitin extraction and protease production. Process Biochem 46(11):2133–2140. https://doi.org/10.1016/j.procbio.2011.08.014

Giongo JL, Lucas FS, Casarin F, Heeb P, Brandelli A (2007) Keratinolytic proteases of *Bacillus* species isolated from the Amazon basin showing remarkable de-hairing activity. World J Microbiol Biotechnol 23(3):375–382. https://doi.org/10.1007/s11274-006-9233-2

Gohel HR, Contractor CN (2015) A review on protease enzyme and their applications. Int J Sci Res 4(12):1755–1760

Gopinath SCB, Anbu P, Lakshmipriya T, Hilda A (2013) Strategies to characterize fungal lipases for applications in medicine and dairy industry. Biomed Res Int 2013:154549. https://doi.org/10.1155/2013/154549

Gupta R, Beg QK, Lorenz P (2002) Bacterial alkaline proteases: molecular approaches and industrial applications. Appl Microbiol Biotechnol 59(1):15–32. https://doi.org/10.1007/s00253-002-0975-y

Haddar A, Bougatef A, Agrebi R, Hmidet N, Nasri M (2009) A novel surfactant-stable alkaline serine-protease from a newly isolated *Bacillus mojavensis* A21. Purification and characterization. Process Biochem 44(1):29–35. https://doi.org/10.1016/j.procbio.2008.09.011

Haddar A, Hmidet N, Ghorbel-Bellaaj O, Fakhfakh-Zouari N, Sellami-Kamoun A, Nasri M (2010a) Alkaline proteases produced by *Bacillus licheniformis* RP1 grown on shrimp wastes: application in chitin extraction, chicken feather degradation and as a dehairing agent. Biotechnol Bioprocess Eng 15(4):669–678. https://doi.org/10.1007/s12257-009-3107-z

Haddar A, Sellami-Kamoun A, Fakhfakh-Zouari N, Hmidet N, Nasri M (2010b) Characterization of a detergent-stable protease from a newly isolated *Bacillus pumilus*. J Ind Microbiol Biotechnol 37(8):813–820. https://doi.org/10.1007/s10295-010-0729-2

Hmidet N, Ali NEH, Haddar A, Kanoun S, Alya SK, Nasri M (2009) Alkaline proteases and thermostable α-amylase co-produced by *Bacillus licheniformis* NH1: characterization and potential application as detergent additive. Biochem Eng J 47(1–3):71–79. https://doi.org/10.1016/j.bej.2009.07.005

Hossain MT, Khan A, Anwar M (2016) Purification and characterization of an extracellular alkaline protease from *Bacillus subtilis* MKK. J Genet Eng Biotechnol 14(2):227–232. https://doi.org/10.1016/j.jgeb.2016.06.004

Huang Q, Peng Y, Li X, Wang H, Zhang Y (2003) Purification and characterization of an extracellular alkaline serine protease with dehairing function from *Bacillus pumilus*. Curr Microbiol 46(3):169–173. https://doi.org/10.1007/s00284-002-3835-1

Jellouli K, Ghorbel-Bellaaj O, Ayed HB, Manni L, Agrebi R, Nasri M (2011) Alkaline-protease from *Bacillus licheniformis* MP1: purification, characterization and potential application as a detergent additive and for shrimp waste deproteinization. Process Biochem 46(6):1248–1256. https://doi.org/10.1016/j.procbio.2011.02.014

Johnvesly B, Naik GR (2001) Studies on production of thermostable alkaline protease from thermophilic and alkaliphilic *Bacillus* sp. JB-99 in a chemically defined medium. Process Biochem 37(2):139–144. https://doi.org/10.1016/S0032-9592(01)00191-1

Kalpana DM, Rajeswari V (2019) Production, optimization and characterization of alkaline protease by *Aspergillus versicolor* PF/F/107. J Genet Eng Biotechnol 17:11. https://doi.org/10.1186/s43141-019-0009-0

Kandasamy N, Velmurugan R, Thangasamy S (2012) Screening and characterization of protease enzyme produced by *Bacillus* sp. isolated from soil sample. J Microbiol Biotechnol Res 2(1):138–144

Kaur I, Kocher GS, Gupta VK (2018) Production of alkaline protease by *Bacillus circulans* using agricultural residues: a statistical approach. Indian J Exp Biol 56:370–378

Khan JA, Singh SK (2011) Microbial proteases: production and application. Int J Curr Res 3(4):23–28

Khandelwal AK, Nigam V, Choudhary M (2014) Alkaline protease production by *Bacillus subtilis* MTCC 10321 using wheat bran as a substrate and its application in tannery wastewater treatment. Braz Arch Biol Technol 57(5):702–709. https://doi.org/10.1590/S1516-8913201402170

Kumar AG, Nagesh N, Prabhakar TG, Sekaran G (2008a) Purification of extracellular acid protease and analysis of fermentation metabolites by *Synergistes* sp. utilizing proteinaceous solid waste from tanneries. Bioresour Technol 99(7):2364–2372. https://doi.org/10.1016/j.biortech.2007.05.006

Kumar D, Savitri, Thakur N, Verma R, Bhalla TC (2008b) Microbial proteases and application as laundry detergent additive. Res J Microbiol 3(9):661–672. https://doi.org/10.3923/jm.2008.661.672

Liu BL, Kao PM, Tzeng YM, Feng KC (2003) Production of chitinase from *Verticillium lecanii* F091 using submerged fermentation. Enzym Microb Technol 33(4):410–415. https://doi.org/10.1016/S0141-0229(03)00134-0

Liu X, Kokare C, Zhou X (2017) Microbial enzymes of use in industry. In: Brahmachari G (ed) Biotechnology of microbial enzymes. Academic, pp 267–298. https://doi.org/10.1016/B978-0-12-803725-6.00010-X

Madan M, Dhillon S (2018) Microbial proteases in commercial applications. J Pharma Chem Biol Sci 6(1):29–39

Mahajan RT, Badgujar SB, Ghag SM (2010) Investigation of potential protease producing moderate halophiles from saline habitats. Int J Biol Chem 4(6):312–321. https://doi.org/10.3923/ijbc.2010.312.321

Mehta VJ, Thumar JT, Singh SP (2006) Production of alkaline protease from an alkaliphilic actinomycete. Bioresour Technol 97(14):1650–1654. https://doi.org/10.1016/j.biortech.2005.07.023

Melo RR, de Oliveira JR, da Silva RR, Cabral H (2021) Proteases from actinobacteria: a new approach to biocatalysis for industrial applications. Biocatal Agric Biotechnol 35:102095. https://doi.org/10.1016/j.bcab.2021.102095

Mothe T, Sultanpuram VR (2016) Production, purification and characterization of a thermotolerant alkaline serine protease from a novel species *Bacillus caseinilyticus*. Biotech 6:53. https://doi.org/10.1007/s13205-016-0373-2

Najafi MF, Deobagkar D, Deobagkar D (2005) Potential application of protease isolated from *Pseudomonas aeruginosa* PD100. Electron J Biotechnol 8(2):197–203. https://doi.org/10.2225/vol8-issue2-fulltext-7

Nascimento WCA, Martins MLL (2004) Production and properties of an extracellular protease from thermophilic *Bacillus* sp. Braz J Microbiol 35(1–2):91–96. https://doi.org/10.1590/S1517-83822004000100015

Naveed M, Nadeem F, Mehmood T, Bilal M, Anwar Z, Amjad F (2021) Protease—a versatile and ecofriendly biocatalyst with multi-industrial applications: an updated review. Catal Lett 151(2):307–323. https://doi.org/10.1007/s10562-020-03316-7

Negi S, Banerjee R (2010) Optimization of culture parameters to enhance production of amylase and protease from *Aspergillus awamori* in a single fermentation. Afr J Biochem Res 4(3):73–80

Nilegaonkar SS, Zambare VP, Kanekar PP (2004) Extracellular protease from *Bacillus* sp. BSA-26: application in dehairing of buffalo hide. In: Reddy MS, Khanna S (eds) Biotechnological approaches for sustainable development. Allied Publishing Pvt Ltd, New Delhi. (ISBN-81-7764-669-9), pp 186–191

Nilegaonkar SS, Zambare VP, Kanekar PP, Dhakephalkar PK, Sarnaik SS, Chandrababu NK, Rajaram R, Ramanaiah B, Ramasami T, Saikumari YK, Balaram P (2006) A novel protease for industrial application. Indian Patent, 2471DEL2006

Nilegaonkar SS, Zambare VP, Kanekar PP, Dhakephalkar PK, Sarnaik SS, Chandrababu NK, Rajaram R, Ramanaiah B, Ramasami T, Saikumari YK, Balaram P (2007a) A novel protease for industrial application. German Patent, 102007013950.2

Nilegaonkar SS, Zambare VP, Kanekar PP, Dhakephalkar PK, Sarnaik SS, Chandrababu NK, Rajaram R, Ramanaiah B, Ramasami T, Saikumari YK, Balaram P (2007b) A novel protease for industrial application *US Patent*, 11/931537

Nilegaonkar SS, Zambare VP, Kanekar PP, Dhakephalkar PK, Sarnaik SS (2007c) Production of dehairing protease by *Bacillus cereus* MCM B-326. Bioresour Technol 98(6):1238–1245. https://doi.org/10.1016/j.biortech.2006.05.003

Niyonzima FN, More SS (2015) Detergent-compatible proteases: microbial production, properties, and stain removal analysis. Prep Biochem Biotechnol 45(3):233–258. https://doi.org/10.1080/10826068.2014.907183

Oberoi R, Beg QK, Puri S, Saxena RK, Gupta R (2001) Characterization and wash performance analysis of an SDS-stable alkaline protease from a *Bacillus* sp. World J Microbiol Biotechnol 17(5):493–497. https://doi.org/10.1023/A:1011913802920

Oskouie SFG, Tabandeh F, Yakhchali B, Eftekhar F (2008) Response surface optimization of medium composition for alkaline protease production by *Bacillus clausii*. Biochem Eng J 39(1): 37–42. https://doi.org/10.1016/j.bej.2007.08.013

Pant G, Prakash A, Pavani JVP, Bera S, Deviram GVNS, Kumar A, Panchpuri M, Prasuna RG (2015) Production, optimization and partial purification of protease from *Bacillus subtilis*. J Taibah Univ Sci 9(1):50–55. https://doi.org/10.1016/j.jtusci.2014.07.001

Paranthaman R, Alagusundaram K, Indhumathi J (2009) Production of protease from rice mill wastes by *Aspergillus niger* in solid state fermentation. World J Agric Sci 5(3):308–312

Patel R, Dodia M, Singh SP (2005) Extracellular alkaline protease from a newly isolated haloalkaliphilic *Bacillus* sp.: production and optimization. Process Biochem 40(11): 3569–3575. https://doi.org/10.1016/j.procbio.2005.03.049

Patil U, Chaudhari A (2010) Production of alkaline protease by solvent-tolerant alkaliphilic *Bacillus circulans* MTCC 7942 isolated from hydrocarbon contaminated habitat: process parameters optimization. Indian J Exp Biol 48(1):58–65

Pokhrel S, Joo JC, Yoo YJ, Kim YH (2013) Shifting the optimum pH of *Bacillus circulans* xylanase towards acidic side by introducing arginine. Biotechnol Bioprocess Eng 18(1):35–42. https://doi.org/10.1007/s12257-012-0195-y

Prakasham RS, Subba Rao C, Sarma PN (2006) Green gram husk—an inexpensive substrate for alkaline protease production by *Bacillus* sp. in solid-state fermentation. Bioresour Technol 97(13):1449–1454. https://doi.org/10.1016/j.biortech.2005.07.015

Rai SK, Mukherjee AK (2011) Optimization of production of an oxidant and detergent-stable alkaline β-keratinase from *Brevibacillus* sp. strain AS-S10-II: application of enzyme in laundry detergent formulations and in leather processing. Biodegradation 22(4):787–802. https://doi.org/10.1007/s10532-010-9409-1

Zambare V (2008) Biotechnological applications of proteases in leather processing: a green technology. In: Proceedings of national conference on green technology (NCGT-2008), 10—12 November 2008. Association with AICTE. Organized By Government Polytechnic, Nashik

Zambare VP (2010a) Optimization of nutritional factors for extracellular protease production from *Bacillus* sp. using response surface resistance. Int J BioEng Technol 1(1):43–47

Zambare V (2010b) Purification and characterization of neutral serine protease from *Bacillus* sp. Asiatic J Biotechnol Res 3:183–192

Zambare V (2010c) Strain improvement of alkaline protease from *Trichoderma reesei* MTCC-3929 by physical and chemical mutagen. IIOAB J 1(1):25–28

Zambare VP, Nilegaonkar SS (2016) Proteases in leather processing. In: Thangadurai D, Sangeetha J (eds) Industrial biotechnology, sustainable production and bioresource utilization. Apple Academic Press/CRC Press., (e-Book ISBN-13: 978-1-77188-262-0), pp 209–238

Zambare VP, Nilegaonkar SS, Kanekar PP (2011a) Production optimization and purification of a novel extracellular protease from *Pseudomonas aeruginosa* MCM B-327. New Biotechnol 28(2):173–181. https://doi.org/10.1016/j.nbt.2010.10.002

Zambare VP, Nilegaonkar SS, Kanekar PP (2011b) Use of agroresidues for protease production and application in degelatinazation. Res J BioTechnol 6(2):62–65

Zambare VP, Nilegaonkar SS, Kanekar PP (2013) Protease production and enzymatic soaking of salt-preserved buffalo hides for leather processing. IIOAB Lett 3(1):1–4. https://doi.org/10.5195/iioablett.2013.19

Zambare VP, Nilegaonkar SS, Kanekar PP (2014) Scale up production of protease using *Pseudomonas aeruginosa* MCM B-327 and its detergent compatibility. J Biochem Technol 5(2): 698–707

Chapter 2
Emerging Trends in Microbial Protease Application

J. Albino Wins, M. Dharshinn, K. Gayathri, Mustafa Salah Hasan, A. Mubarak Ali, and P. R. Janani

Abstract Fermentation techniques are appealing sources of proteases as large amounts of microorganisms can be produced in a short period of time. Microbial proteases constitute approximately two-thirds of the detergent market in the world's enzyme industry. They are also used as soil/environmental stress indicators and chemical substitutes. Little information is known about the reasons for differences in the selectivity of acidic, neutral, and alkaline proteases, even though their phylogeny is studied based on their protein sequences. Microbial enzymes have received a great deal of attention with respect to their broad applications in industry and medicine, thanks to their excellent catalytic efficiency, stability, and convenient production and engineering when compared with plant- or animal-derived ones. Enzymes are low-cost, natural, nontoxic products, energy-efficient, and very fast to process; hence, they have a wide application in foodstuff manufacturing, agricultural organisms' protection, chemical production, and medicine. Depending on their composition, toxic substances may be solubilized or transformed by microbial enzymes (amines, nitriles, or phenolic compounds are some of those present in waste, domestic, and suspect waste). Microbial proteases can be exploited in many industries, such as cleaning, leather, silver recovery, milk processing, dairy industry, baking, beverages, and pharmaceuticals. These enzymes are already widely used

---

J. Albino Wins
Department of Botany, Holy Cross College (Autonomous), Nagercoil, Tamil Nadu, India

M. Dharshinn
Agricultural College and Research Institute, Tamil Nadu Agricultural University (TNAU), Coimbatore, Tamilnadu, India

K. Gayathri (✉) · A. Mubarak Ali
PG and Research Department of Microbiology, Sacred Heart College (Autonomous), Tirupattur, Tamil Nadu, India

M. S. Hasan
College of Veterinary Medicine, University of Fallujah, Fallujah, Iraq

P. R. Janani
Department of Microbiology, Padmavani Arts and Science College for Women (Autonomous), Salem, Tamil Nadu, India

V. Zambare et al. (eds.), *Biotechnological Advances and Applications in Microbial Proteases*, Interdisciplinary Biotechnological Advances,
https://doi.org/10.1007/978-981-95-9011-7_2

in the food industry for waste minimization, both at home and on industrial sites, as well as for improving nutritive value, digestibility, palatability, flavor, and reduction of allergens. They also function in protein synthesis and in reading the genetic code.

Keywords Microorganisms · Microbial protease · Protein sequences · Industries and medical · Nutritional values

2.1 Introduction

Microbial enzymes constitute a critical class of biocatalysts utilized across the food, agricultural, chemical, medical, and energy sectors. Their widespread adoption arises from their operational efficacy, rapid catalytic turnover, low energy demand, and overall eco-compatibility. Proteases, which are ubiquitously distributed in microorganisms, plants, and animals, catalyze the hydrolytic cleavage of peptide bonds within polypeptide substrates (Tkaczewska 2020). These enzymes participate in protein synthesis, degradation, posttranslational regulation, and physiological modulation, thereby influencing developmental processes ranging from cellular growth to organismal senescence. Single amino acid alterations in protease-encoding genes have been linked to heritable disorders in more than fifty human proteases, underscoring their medical significance.

Through the effect of microbial fermentation, highly active and substrate-selective proteases can be efficiently manufactured (Assad et al. 2020). In addition to their physiological functions, proteases are now the major industrial enzymes in use—ranging from such diverse areas as detergents and food processing to bioremediation, pharmaceuticals, and leather production. Microbial proteases are more cost-effective and techno-efficient than those obtained from plant/animal sources as they are produced in higher yields and can be easily cultured while also being genetically amenable (Apweiler et al. 1998; Yachnin et al. 2022). Current industrial estimations show that microbial proteases represent the main share of enzyme sales and are indeed key components in a variety of biotechnological processes (Pessela et al. 2023; El-Fatah et al. 2023).

2.2 Microorganisms Enzymes

The industrial exploitation of microbial enzymes dates back to early fermentation-driven processes, such as cheese production. Microbial systems are now recognized as strategic platforms for generating enzymes, antibiotics, and bioactive compounds due to their lower production cost and scalable fermentation capabilities (Mahnashi et al. 2022). Soil-derived isolates serve as reservoirs for identifying novel microbial enzymes optimized for specific industrial reactions. Screening and subsequent strain characterization enable the selection of robust microbial candidates for targeted enzyme biosynthesis.

2.2.1 Microbial Enzymes

Soil samples can be examined from across the earth to find a new enzyme or biological process for a specific industrial process. "It can involve a lot of lab work to find out which is the best for producing a particular type of enzyme because in one bacterium you may have more than 1,000 enzymes—that's the complexity." And once the right microbe is found, work continues. The microorganism will need to be characterized and potentially engineered, in order to generate useful amounts of the enzyme or process desired. In the industry, the bacteria are grown in large vats and produce the desired product. One other process, fermentation (not to be confused with metabolic fermentation), made it possible to produce enzymes, vitamins, and other products in quantities that are virtually limitless at a price that is even reasonable. One extractant that can be used to isolate the desired ingredients is a culture broth, which is the end product of fermentation. The solids in the fermentation broth are filtered off or separated by centrifugation. The resultant biomass, usually referred to as sludge, could also be widely applied as fertilizer; it is made up of residual microorganisms and raw materials (Omoniyi et al. 2024).

The application of these enzymes has gained importance due to their industrial potential and operational role in metabolic functions. The proteins produced through the market are microbially sourced as they are usable in biotechnological work, have higher product formation in less time/space, have high genetic manipulation, and have low cost. They are microbial proteases, such as having an advantage over those of plants or animals, for they possess all the qualities required in the industries. Small, compact, and globular proteolytic enzymes are found in the mammalian system as well as in microbes. One of the most important producers of alkaline protease is the *Bacillus* species. Most of the dilutions for these products are also taken from these microorganisms and used in industry. Carbon alkaline proteases are typically sourced from the manufacturer in a liquid broth; these culture broths are then subjected to downstream purification and concentration in order to produce an end product (Matter 2023).

Proteases of microorganisms do not require as much effort as it does for those of plants and humans do. Microorganism-derived proteins are categorized in a number of different ways depending upon whether the protein concerned is basic or acidic. They are also of different types on the basis of the position of peptide bonds and the presence of functional groups. World trade-class enzyme is the microbial protease. A variety of intracellular proteases are produced, which play a significant role in differentiation, protein turnover, hormone regulation, and the cellular protein pool. However, protein denaturation is dictated by the circumstances of extracellular proteins (Solanki et al. 2021).

2.3 Detection of Microbial Protease

2.3.1 Endopeptidase Detection

2.3.1.1 Observation of Halos

Protease production on solid media is typically indicated by the appearance of clear zones (halos) around microbial colonies inoculated on protein-containing agar. These halos form when microorganisms secrete extracellular endopeptidases capable of degrading the protein substrate incorporated into the medium. Protein substrates, such as skim milk, casein, BSA, gelatin, keratin, fibrin, or elastin, are mixed with a suitable growth medium, poured into Petri dishes, and subsequently inoculated. For liquid-culture assays, either cell-free supernatants containing extracellular proteases or cell lysates containing intracellular proteases are applied to wells cut into the protein-supplemented agar, allowing halo formation to be assessed in the same manner (Razzaq et al. 2019).

A simple and widely used qualitative approach for protease detection is halo visualization; however, this method is limited to endopeptidases or enzymes with sufficient activity to produce a visible clearance zone. More commonly, protease activity is quantified by measuring either the breakdown products (hydrolysates) or the reduction of intact substrate resulting from enzymatic hydrolysis. Because proteases differ in their catalytic properties, substrate specificity, and hydrolysis patterns, a variety of detection strategies are required. Substrates currently employed for protease assays are generally classified into native substrates and modified substrates, with the latter, including chromogenic and fluorogenic forms, that allow instrument-based detection. Different substrates are selected, depending on the assay principle and the requirements of the analytical method (Song et al. 2023).

2.3.1.2 Detection by Natural Protein Substrates

Natural protein substrates are those that exist in nature (plant, animal, microbial proteins). Casein is the most commonly used substrate for measuring protease activity. Under suitable temperature and pH conditions, protease hydrolyzes casein, producing acid-soluble peptides or amino acids. When acid deposition occurs, the resulting product goes into solution under the upper acid condition, and a precipitatable mass of unhydrolyzed protein is formed. The supernatant is recovered, following centrifugation, and protease activity can be measured from the peptides or amino acids generated using one of several folic reagents, ninhydrin, TNBS, or OPA (each with pros and cons) (Gaur et al. 2022).

2.3.1.3 Detection by Modified Protein Substrates

2.3.1.3.1 Detection by Chromogenic Substrates

In some of the protease activity detection methods, a modified protein substrate is employed to increase its solubility and sensitivity of detection, which must be a compound to give a colored end product upon proteolysis, or an initially formed product should have the potential for reaction into a colored complex. An example is azo-casein colored by staining with p-amino benzenesulfonic acid suspension, which, digested with proteases, absorbs at 440 nm, and is soluble in trichloroacetic acid. There are chemically succinate amino groups in casein, except caseins (Klebe 2013), while mineral optimization allows better dissolve effect at pH values above 4.

Depending on the substrate specificity, the protease type screened can be deduced from the nature of the synthetic substrate used as Tosyl-Gly-Pro-Arg-pNA for trypsin, Suc-Ala-Ala-Pro-Phe-pNA for chymotrypsin, and Suc-Ala-Ala-Pro-Val-pNA for elastase. N-Cbz-Ala-Ala-Leu-pNA and N-Cbz-Gly-Gly-Leu-pNA are, however, good substrates of subtilisin. The assay is based on the amide bond hydrolysis properties of proteases, where the linkage between p-nitroaniline (pNA) and the adjacent amino acid residue is broken by proteases, resulting in free pNA, which shows a specific absorption at 405 nm wavelength (Du et al. 2025).

2.3.1.3.2 Detection by Fluorogenic Substrates

Protease assays often require detection of enzymes at extremely low concentrations, typically in the picomolar to nanomolar range, which is below the detection limits of conventional colorimetric methods. In this context, highly sensitive fluorescent peptide-based substrates are employed because they can respond to very low levels of proteolytic activity.

Fluorescent labeling strategies for protease substrates can generally be categorized into three approaches:

(1) Single-fluorescence labeling, where a fluorescent dye is attached to a specific site of the protein or peptide substrate. Upon proteolysis, the labeled fragment generates a measurable fluorescence signal.
(2) Dual-fluorescence labeling, where two different fluorophores are incorporated into the peptide. One of the two serves as donor while the second functions as acceptor; proteolytic cleavage abolishes this interaction, leading to changes in emission properties characteristic for Förster type energy transfer.
(3) Mechanism (FRET-based) of phototransfer labeling using a FRET pair system consisting of a fluorophore and quencher. The fluorescence in the intact substrate is quenched, but when enzymatically hydrolyzed, the pair separates, and all quenching disappears, creating a strong fluorescent signal (Aehle et al. 2009).

2.3.2 Exopeptidase Detection

Endopeptidases cleave peptide bonds within proteins and mainly produce peptide pieces, while exopeptidases digest the terminal residues from protein chains into free amino acids. Due to these mechanistic differences, different strategies need to be employed for the detection and discrimination of endo- and exopeptidase activities, and assays optimized for endopeptidases are typically not sensitive toward exopeptidases (Bansal et al. 2023).

Synthetic peptide substrates are the most frequently employed if exopeptidase activity has been measured. For aminopeptidase tests, short oligopeptides (e.g., di- and tripeptides) are used as the test substrates. Sensitivity is increased by attachment of a chromogenic or fluorogenic group, e.g., p-nitroaniline (pNA) or 7-methoxycoumarin-4-acetic acid (MCA), to the C-terminal end of the peptide. Enzymatic hydrolysis liberates free pNA or MCA in the assay buffer, and absorption or fluorescence of such products is a quantitative index for aminopeptidase activity (López-Otín and Bond 2008).

2.4 Microbial Protease in Food Industries

2.4.1 Meat Tenderization

Alkaline proteases can break down structural muscle fiber and connective tissue proteins. This property is used in the softening of meat, which can be carried out by soaking meat in protease solutions or by applying powders of enzymes. Furthermore, alkaline elastase or thermophilic alkaline proteases have been introduced into the vascular system of animals a few minutes before slaughtering (10–30 min) for tenderization purposes (Czapinska and Otlewski 1999).

2.4.2 Meat Solubilization

Value-added products that originate from the meat industry can include soluble meat protein hydrolysates or meat-flavored hydrolysates. Common protein ingredients include bone, organ (e.g., raw lung), and MBM derived from residual bone after the mechanical removal of all usable meat tissue. Alkaline proteases have been found particularly attractive to achieve this effect. However, enzymatic hydrolysis of fish proteins by alkaline proteases is extremely efficient, cost-effective, and overall practical when utilized for the production of fish protein hydrolysates (Bizzozero and Dutler 1981).

2.4.3 *Soy Products*

There has been a long tradition in the use of fungal proteases in the preparation of soy sauce and other fermented soy products in Asian cultures. Alkaline and neutral proteases from *Aspergillus* species are critical for hydrolyzing soybean proteins during fermentation and exert important effects on the special flavor of traditional soy sauce. These enzymes also improve the overall quality of fermented soy products. Members of the genus *Bacillus* have likewise been evaluated for industrial applications and are recognized as predominant alkalophilic microorganisms, noted for their high alkaline protease productivity. Species such as *Bacillus amyloliquefaciens* and *Bacillus licheniformis* and related strains are particularly important in this regard. Many fungi also secrete extracellular alkaline proteases, with *Aspergillus* species being among the most widely utilized (Elsässer and Goettig 2021).

2.5 Classification of Protease

The amino acid sequences and evolutionary relationships of proteins are used to classify proteins into several families and clans. Evolutionarily, clans, or groups of families, have no apparent sequence similarity, but there is a connection between them. Due to the same order of their catalytic-site residues, as well as tertiary folds, families of various catalytic classes can also be classified into a clan. One of the similarities is that the amino acid sequence of proteolytic enzymes in one family is significantly similar, which means that there are similar enzymes. They can be recognized by a particular kind of enzyme that belongs to the family or by a protein that is similar to the kind of enzyme and so belongs to the family. The MEROPS database is an extensive database that carries information about different proteases according to such classification. Glutamic and asparagine, cysteine, threonine, serine, metalloproteases, aspartic, and unknown proteins are the categories into which all proteins in groups and families can be categorized based on their phylogenetic links and modes of action. These natural enzymes are selective and diverse, as indicated by their characterization. Two criteria allow categorizing proteases (Blay and Pei 2019).

2.5.1 *Based on Active Sites*

2.5.1.1 Exopeptidases

Exopeptidases hydrolyze peptide bonds located near the terminal ends of protein substrates, acting either at the amino terminus or the carboxyl terminus. Based on their site of cleavage, they are broadly classified as aminopeptidases or carboxypeptidases, which target the N-terminal and C-terminal residues, respectively (Troncoso et al. 2022).

2.5.1.2 Carboxypeptidases

Carboxypeptidases function as catalytic enzymes that release dipeptides or free amino acids from the C-terminal end of peptide chains. Unlike endopeptidases, they do not cleave internal peptide bonds but instead act exclusively at the protein's terminal region. This property also enables their use in removing affinity tags that are engineered at the C-terminus of recombinant proteins. Type A carboxypeptidases, which are metalloproteases, preferentially hydrolyze aromatic or branched side-chain residues, while type B carboxypeptidases show specificity toward basic amino acids (Díaz-Cervantes et al. 2023).

2.5.1.3 Serine Proteases

Serine proteases are widely distributed across cellular organisms and also occur in numerous viral genomes. They constitute nearly one-third of all known proteolytic enzymes. Most of these enzymes serve as endopeptidases, which break internal peptide bonds; however, some go through a stage in which they may become exopeptidases, removing the terminal amino acids from polypeptide chains. They are so-called because of the nucleophilic serine in their active site. This serine residue attacks the peptide bond of the substrate during catalysis to form a transient acyl-enzyme intermediate (Mótyán et al. 2013).

2.5.2 *Depending on the Active Optimal pH*

2.5.2.1 Alkaline Proteases

Alkaline proteases are secreted by microorganisms living in soil, water, and many other very alkaline environments. They can be found in various environmental sources, such as detergent-contaminated wastewater, dried seafood processing places, sandy or saline soils, and slaughterhouse waste. Of serine proteases, those from alkaliphiles are of particular industrial interest, with detergents being their main application. Due to stability and catalytic activity in the presence of alkaline conditions, various food, pharmaceutical, and other formulations also include these enzymes. Continued work is being carried out to discover or design new strains that can yield higher active alkaline proteases and other distinct characteristics with respect to certain applications. *Bacillus* species have also been reported to exhibit hydrolytic, elastolytic, and keratinolytic activities that are relevant in the dehairing of animal hides and skins (Gimenes et al. 2019).

2.5.2.2 Acidic Proteases

Acidic proteases are important because of their stability and high catalytic efficiency in the pH range of 3.8–5.6, which makes them suitable for processing soy sauce, protein hydrolysates, digestive aids, and flavor enhancers. Such enzymes present an isoelectric point of 3.0–4.5, and a molecular mass in the range from 30 to 45 kDa, with optimal activity at pH values between 3.0 and 4.0. Acid proteases are used to tenderize meats, clear fruit juices and beer, and enhance the rheology of doughs. Unlike alkaline proteases, most of the acid proteases are produced by fungi, such as *Aspergillus niger*, *A. oryzae*, *A. awamori*, *A. fumigatus*, and *A. saitoi.* Aspartic proteases have an active site typically formed by long polypeptide chains of 380–420 amino acids; the enzymes are often called aspergillopepsins because of their acidic nature. Acid endopeptidase there are three major families of acid endopeptidases, which can be found in the CA clan, those being pepsins, retropepsins, and pararetroviral protease. One characteristic structural property is the active-site cleft, which lies between the lobes of the enzyme. They are highly specific to aromatic residues on either side of the peptide bond, although they are generally less selective than pepsin for those amino acid residues. Acidic proteases, as a whole, the acidic proteases can be divided into two categories: (i) those structurally similar to pepsin and (ii) rennin-type proteases, including species of *Rhizopus, Endothia, Aspergillus, Penicillium*, and *Mucor* (Dyer and Weiss 2021).

2.5.2.3 Neutral Proteases

Neutral proteases are defined by their action at or near neutral pH. The neutral proteases from *Bacillus* species usually have low thermostability and work best at pH 5–8. Their low hydrolysis rate also imparts them with some usefulness in the food industry, where their presence serves to limit bitterness in protein hydrolysates. They are also used widely in brewing, primarily because they are not inhibited by plant proteinase inhibitors (Naveed et al. 2020). CONTROLS The general affinity of neutral proteases for hydrophobic amino acid residues and their relatively low thermostability may be useful in guiding enzymatic hydrolysis during the modification of food proteins. Indeed, proteases in general are of great interest in the food industry, since they can be used to catalyze protein hydrolysis and thus develop protein-rich foods, tenderized meat, or simply processed dairy products. In addition, bioactive peptides, which have health benefits, are liberated by proteolytic enzymes. The growing interest in proteases is also an expression of the new future trends in consumers' preferences toward products that are hardly processed, sustainable, and health-related (Vojcic et al. 2015).

2.6 Protease Engineering

Protease engineering has gained attention in a wide range of fields, such as biosynthesis, the energy industry, the food industry, human healthcare, animal husbandry, and environmental protection. The development of genetic modifications will facilitate the production of therapeutic, industrial, and biomedical proteins and enzymes on a global scale, also benefiting from combating strategies for diseases. Recombinant *E. coli* strains were pivotal in promoting the development of the recombinant protein expression system, and site-directed mutagenesis was applied for the isolation and improvement of proteases (Vojcic et al. 2016).

Through engineering, detergent-quality proteases have been active at low temperatures while achieving an increase in thermal and operational stabilities. Together, these findings offer a comprehensive understanding of protease structure-function relationships and serve to provide concepts for the design of improved enzyme variants for several applications. Several subtilisin-like serine proteases depend on an N-terminal pro-sequence for folding and the processing into a functionally active enzyme. Pro-sequence engineering, using site-directed or random mutagenesis within the serine protease class of autoprocessing enzymes, has been popular for optimizing the performance of autoprocessing proteases. Historically, mutations in the protease domains were utilized to change catalytic properties, while pro-sequence engineering offers a new route for developing novel active site proteases. Apart from its use for investigating protein folding, this approach has practical implications in the development of novel enzymes with defined biochemical properties. It is also an essential tool in studies aimed at investigating the linkage between gene structure and the organization of biological functions, as well as aiding industrial-scale production of enzymes from microbial sources. Microbial proteases have been purified and genetically modified to (1) enhance enzyme production, (2) define primary structures of amino acids, and (3) produce protein engineering evidenced materials. The protease genes have been cloned and sequenced recently, making these applications possible (Yang et al. 2023).

2.7 Emerging Protease Technology

Biochemical and molecular aspects of proteolytic systems (proteases) have become particularly well studied because of their enormous industrial interest and high economic importance. As a consequence, a search for new and highly efficient bacterial enzymes has been conducted. It is anticipated that protein engineering will be a key technology for the design of enzymes with superior or novel properties. There are many commercial applications for alkaline bacterial proteases, and the potential is great, with it expected to increase even more in the future. In order to fast-forward the improvement of the enzymes, modern techniques like protein and genetic engineering and molecular and computational biology are being utilized to improve protease-producing strains. In vitro evolution has been used to alter the primary structure of proteins, as a tool for in vivo selection of bacterial strains with desirable properties. Specifically, increased enzyme yield, improved substrate specificity and thermal stability, changes in optimal pH, and decreased auto-proteolytic degradation are important aims for development (Zhao and Liu 2024).

2.8 Protease and Yield Improvement

The wide range of substrates for microbial proteases and the demand for good-quality/feed value protein sources have prompted genetic manipulation as an approach to improving the yield of protease. Contemporary biotechnological approaches offer clusters of regulatable promoters that enable site-directed genome editing in bacteria, thus allowing improved production and expression of proteases with defined functional characteristics. Due to the malleability of this microbial population and recent developments in biotechnology, novel routes for the establishment of attractive industrial production strains have become available. In industry, optimization strategies often combine several technologies, including fed-batch and chemostat fermentation and screening for strains, cloning, etc. Statistical techniques, such as response surface methodology, have also been utilized to optimize growth conditions and composition of the media. Conventional mutagenesis (e.g., UV or chemical treatment) and recombinant DNA (rDNA) strategies are used to create protease hyper-producing strains (Tavano et al. 2018).

The recombinant DNA technique is useful in placing a gene of interest into the genomes of microbes, plants, or animals to produce transgenic (or genetically modified) organisms that can produce the desired product. For example, microorganisms, including *Escherichia coli*, can be transformed by recombinant protease genes that have been altered to encode alternate proteases. Although proteases have wide applications, their industrial production is inhibited by the high cost of production and low stability during recovery and processing. To overcome these limitations, several immobilization technologies have been examined to enhance pH stability, solvent resistance, and thermal stability. Alkaline protease immobilization in solid supports (most frequently alginate-chitosan beads) has been reported to increase the enzyme yield and stability up to ~47 °C, so genetic modifications are being sought for the development of stable and economical alkaline proteases in the supply volume needed to support worldwide biotechnological markets.

Bioengineered cultured proteases generated by rDNA technology have been used in the detergent industry to enhance the stability of enzymes in the process. Genes for proteases in *Bacillus stearothermophilus*, *B. subtilis*, and *E. coli* can be cloned and kept in various vector systems, which permit expression under extreme conditions (Abril et al. 2023).

2.9 Uses of Microbial Enzymes

Application of new enzymes may lead to considerable savings in different areas, such as environmental impact, process economy (better yield), production costs, energy consumption, and so forth. In particular, microbial enzymes are increasingly more important in the food, pharmaceuticals, paper, leather, and textile industries, among others. Of the microbial enzymes, hydrolases are the only selectively commercially important group, with a few exceptions. The wide variety of microbial enzymatic activity has compelled thorough bioprospecting investigation in the enzyme sector, where soil and water samples from different global habitats (including sites subject to extreme physical and chemical environments) are tested for microorganisms with the ability to yield certain enzymes (Kumar et al. 2024).

2.9.1 Use of Protease in the Treatment of Garbage

Microbial enzymes have been utilized in waste management increasingly because of their potential to degrade toxic pollutants. Effluents from industry and households are frequently vehicles of dangerous components, which are considered extremely dangerous for human health as well as the natural environment. Microbially derived enzymes can be used individually or in combination to transform the toxic compounds, phenols and substituted phenols, aromatic amines, nitriles, and other pollutants into less toxic or environmentally safe molecules. Popular enzymes used in waste treatment are amylases, glucoamylases, cellulases, pectinases, lipases, nitrile hydratases, and proteases (Souza et al. 2015).

Oxidoreductases play a major role in detoxification by catalyzing oxidative coupling reactions that neutralize toxic compounds. Industrial wastewater frequently contains chlorinated phenolic compounds, which can be degraded by enzymes, such as lignin peroxidase, manganese peroxidase, tyrosinase, and laccase. Microbial enzymes also contribute to resource recovery by enabling the conversion or reuse of waste materials—for example, enhancing oil extraction from oilseeds, converting starch to sugar, or transforming whey into value-added products. Microbial monooxygenases and dioxygenases (MIOs) display broad substrate specificity and can metabolize chlorinated aliphatic compounds and other halogenated organic pollutants frequently encountered in industry. Such chemicals include plasticizers, hydraulic, and heat transfer fluids; agrochemicals, such as herbicides, insecticides, and fungicides; and other intermediates used in chemical synthesis (Zhang et al. 2018).

2.10 Use of Protease in Industries

Proteases find applications in industry in baking, brewing, detergents, fermented products, medications, textiles, and leather processing. The use of enzymes is depicted by the list of things that follow in Fig. 2.1.

2.10.1 Pharmaceutical and Analytical Business

Enzymes are important in the pharmaceutical and diagnostic industries. They are often employed as a diagnostic test, e.g., by ELISA and diabetes testing kits, and therapeutic drugs to correct enzyme deficiencies and digestive issues. Enzyme use in medicine is common as in industry, and the use is escalating rapidly. The most generalized uses of microbial enzymes in the field of medicine nowadays have been the exfoliation of dead skin and the healing of burns, either with the help of proteolytic enzymes or the disintegration of the clots by fibrinolytic enzymes. A potent fibrinolytic enzyme may be applied as a thrombosis therapy. Rhodamine, dextranase, and acid protease can be used to treat cyanide poisoning, tooth decay,

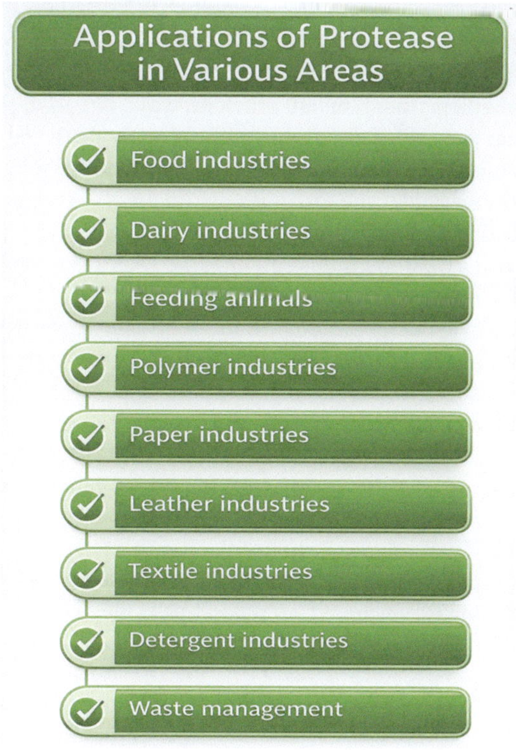

Fig. 2.1 Uses of protease in industries

and alimentary dyspepsia. Lipases are the enzymes that are the most commonly used in organic synthesis and are applied to obtain lactones, acids, alcohols, and optically active esters (Gouseti et al. 2023).

Diltiazem intermediate and polyphenol oxidase produce 3, 4-dihydroxylphenylalanine (DOPA) for Parkinson's disease. Tyrosinase is a necessary oxidase enzyme for melanogenesis and the synthesis of L-dihydroxyphenylalanine (L-DOPA). Dopamine's precursor, L-DOPA, is a powerful drug that also treats Lesch-Nyhan syndrome and blood levels of prolactin; it inhibits the release of norepinephrine. Such enzymes include urate oxidase of uric acid, creatinine and sarcosine oxidases of creatinine, lipase, carboxyl esterase, glycerol ionize of triglycerides, urease, glutamate dehydrogenase of urea, and also glucose oxidase to quantitatively obtain your diabetes mellitus in clinical diagnosis. Also, cholesterol oxidase applications have been described to be applied in biotechnological use, with many advantages of identification and transformation of cholesterol. Biogenic amines are identified using putrescine oxidase in putrescine, which is one of the signs of food spoilage (Ghoraba 2023).

2.10.2 Food Industries

Enzymes can be used to address the problem of a high-quality food supply in the food business. These biomolecules improve food production and its characteristics, like taste, smell, color, touch, look, and nutritional value, which are required to augment them in an efficient way. An in-depth understanding of the action of enzymes in the activity of food manufacturing and food ingredients industry has enhanced the key processes of delivering safer and quality products to the markets. Furthermore, the enzymes also drew the attention of the new areas like fat modification and technology of sweeteners (Shettar et al. 2025). It is forecasted that by the year 2020, the industrial enzyme sector that now dominates the food and beverage business will have a value of 2.3 billion dollars. Enzymes are also added to animal and vegetable proteins in the food and beverage industry in order to augment their nutritional value and functionality in the food through enzyme hydrolysis, which produces better color and aroma in the juice. Enzymes have numerous uses in the food industry, among them brewing and dairy, baking, and juice production. They are also applied to control the brewing process and also deliver high-quality beer on a permanent basis. Microbial enzymes have been widely applied in bakeries, where the main market in the food industry is the shelf life, crumb softness and structure of the dough, and stability of the dough. The dairy industry is the next major consumer of microbial enzymes after the beverage industry, with large quantities of microbial enzymes consumed by the industry because of the rising application in the production of cheeses (Bharti et al. 2020).

2.10.3 Dairy Industries

Dairy enzymes constitute a significant part of the food enzyme industry since they aid in the flavor, color, and aroma of milk products, besides increasing the productivity of milk products. Enzymes (proteases, lipases, esterase, lactase, aminopeptidase, lysozyme, lactose peroxidase, transglutaminase, catalase, etc.) are regularly used in the dairy industry to make the dairy products safer and to prolong their shelf life. These enzymes include bio-protective enzymes and coagulant enzymes. Cheese dairy enzymes are used in the processing of cheese, yoghurt, and other products of milk. The mixture of chymosin and pepsin solidifies milk into curds that are used in cheese and liquid whey products. In the present day, microbial rennet is utilized to produce over 33 percent of the cheese that is consumed worldwide. Other proteases are also used to speed up the cheese-making process and also minimize the allergenicity of the milk products. Lipases are used today in the lipolyzing of milk fat, fat enhancement of taste, accelerating the process of cheese production, and also in the making of custom milk items. Transglutaminase is a protein polymerizing enzyme that improves the performance of the dairy products (Calva-Estrada et al. 2019).

2.10.4 Beverages Industry

In addition to two major groups in the beverage industry, there are eight smaller ones. The nonalcoholic product line includes tea, coffee, fruit juices, syrups, soft drinks, and bottled water. The individuals included spirits, wine, and beer that were distilled. Enzymes in the industrial circle are used in the brewing industry to streamline the processing and to produce quality and dependable products. Microbial enzymes have been used in the brewing sector in the extraction of the plant materials to break their cell walls and increase their output, color, and smell andz to come up with clearer products. Enzymes are used in the current fruit and vegetable juice market as an integrating factor (Ramkumar et al. 2016). Enzymes used in the processing of the fruit and vegetable juice industry are used in enhancing the efficacy of the process, such as clarification, extraction, peeling, and juicing, in addition to improving the quality of the products. The application of cellulases, amylases, and pectinases to help in maceration, liquefaction, and clarification of fruit juices enhances productivity and lowers the costs. Enzymes are enriched to enhance the stability and the quality of the juices produced. The enzymes are applied in fruits and vegetables in order to degrade pectin, carbohydrates, proteins, and cellulose to increase revenues, faster processing, and sensorial improvements. Amylases are added during the clarification to improve the volume of clear or hazy juice that is generated (Simplicio et al. 2023).

Pectinases and cellulases are used to increase texture, cloud stability, yield, and juice extraction. Hydrolysis of the bitter constituents with these debittering enzymes, limonin, and naringin is used to improve citrus juices. Pectin heteropoly saccharide is a structural component, a component of polygalacturonate, pectin esterase, pectin lyase, and so on, to regulate the cloudy fluids. Pectin can be found in nearly all fruits. To reduce the starch-induced turgidities, the distilled alcoholic drinks may be fermented using a microbial amylase to convert starch to sugars. Enzymes have also been used to hydrolyze the unsalted barley and other starch adjuncts to reduce the cost of production of the beer. Chill haze in beer can be controlled by incorporating proteins when brewing the beer, and proteases are the technologies that are used (Strat 2022).

2.10.5 Feed Industry

When animal food is prepared, enzymes that are available in the feeds can be used. One such instance is that they are used to decompose specific feed materials, which would otherwise be dangerous or unbalanced. In addition, the increase in the quantity of protein in chicken feeds could be due to the treatment of feeds with enzymes (Craik et al. 2011). The major feed enzymes that are used in poultry are phytases, proteases, α-galactosidases, gluconates, xylanases, amylases, and polygalacturonate. The largest feed enzyme market is the phytase, which makes use of the natural phosphorus that is entrapped in the phytic acid present in the feed. High cellulose and hemicellulose diets that are of plant-based foods cannot be digested by monogastric animals. They are added to their meals because the xylanase and β-glucans are total digesters and degraders of massive starch. Animals are also fed on proteins that do not contain anti-nutrient factors by means of using proteases, which digest the protein into separate amino acids. The value of such feed enzymes is rising on the basis of the possibility of decreasing the feed cost, improving the quality of meat, and enhancing the nutritional quality of the feed to help animals convert them to greater efficiency (Mokrani and Nabti 2024).

2.10.6 Polymer Industry

These growing issues over ecological soundness and human health have seen the application of microbial enzymes in making biodegradable polymers. In vitro enzyme-catalyzed polymer synthesis is an environmentally safe process, as opposed to the traditional chemical processes, and has numerous benefits. Green biodegradable materials are biopolymers that increase the biological degradation of the compound after its usage and its disposal to the environment as biomass and CO2, since they are prepared using renewable carbon sources through biological processes. Biomedical application of biopolymers is in the creation of tissue engineering, adhesion barriers, regulation of drug delivery systems, orthopedic devices, and occurrences that are infinite in nature, as they employ the use of polyesters, polycarbonates, and polyphosphates. Biodegradable polymers can be used to satisfy the increasing demand for both the packaging materials and environmental safety. Lipases are applied in the catalysis of polymerization of lactones, cyclic diesters, and cyclic carbonates to produce polyester and polycarbonates. The polycondensation and poly-addition of the polymerization processes done by catalyzing the process using lipase is a good method that can be used to make acceptable polyesters as a way of environmental technique. Other enzymes utilized in the biopolymer industry include transglutaminase, peroxidase, and laccase. They create cross-links on biopolymers to produce materials on-site through polymerization processes (Palacios-Mateo et al. 2025).

2.10.7 *Paper and Pulp Industry*

With the increased awareness of sustainability issues, microbial enzymes have been actively used in pulp and paper industries to reduce the adverse environmental impact. Enzymes save on the amount used, processing time, and energy. Enzymes also favor the waste treatment and deinking and bleaching of pulp and paper industry, increasing the chemical and biological oxygen demand (COD and BOD, respectively). In pulp and paper industries, the xylanases and lignin are used in order to add value to the pulp by removing lignin and hemicelluloses. A lot of industries use amylases to boost the coating of starch, drainage, deinking, and cleaning of paper. Whereas lipases are used in deinking and improved control on pitch, cellulases are used in deinking, softness, and drainage. In the use of cellulase, a bioprocess to recycle used printed sheets that had been recycled was also developed. Laccase is utilized rather than chlorine that is used in huge quantities in the chemical pulping process. This will lower the level of garbage that wastes the ozone layer and causes acidification. The paper sector takes advantage of mannoses in the breakdown of glucomannan and enhances the brightness of the same (Cuq et al. 1995).

2.10.8 *Leather Industries*

The presence of waste and discharges in any of the steps of leather production is increasingly posing a challenge to the environment and human health due to the increasing traditionalism of the leather industry. There is an opportunity to use profitability bio-enzymes in order to increase the quality of leather in terms of reducing the wastes. The initial application of enzymes in the leather industry was the dehairing operation, which is the largest process of the leather production procedure that employs a significant amount of enzymes, such as lipases, amylases, and proteases (Verhamme et al. 2018). Application of enzymatic dehairing is attractive because it can have the potential of retaining hair and decreasing the number of organic debris released into the effluent. By using enzymatic dehairing, the use of some harmful chemicals like amines, lime, and sulfur is decreased or eliminated. In order to enhance the quality of the leather, enzymes should be applied in order to hasten the various processes of processing leather, and this includes curing, soaking, liming, dehairing, batting, plucking, degreasing, and tanning. The enzymes that have been used in the leather industry are lipases, alkaline proteins, and neutral proteins. The alkaline proteins are used to dry soaked leather and dissolve non-fibrillar proteins to give the leather the soft, elastic, and pliable properties. During dehairing, the amount of water spent is reduced with the help of neutral and alkaline protease (Zhang et al. 2021).

2.10.9 Textile Industries

The textile industry, which generates much waste in terms of dye, bleaching agents, and desired materials, is among the key pollutants of the environment. The enzymes in these industries are employed to enable them to devise environmentally friendly fiber processing steps and to enhance the final product quality. Hydrolase and oxidoreductase are the two key types of enzymes that are used in the pretreatment and finishing of cotton. Hydrolases are used in diverse processes, such as biopolishing and bioscouring of textiles, wool antifelting, cotton softening, denim finishing, desiring, wool finishing, and the modification of synthetic fibers; horologists are involved in producing amylase, cellulase, cutinase, protease, pectinase, and lipase/esterase. The oxidoreductase group of enzymes, which involves catalase, laccase, peroxidase, and lignans, is involved in many processes, such as bio-bleaching, bleach termination, dye decolorization, and fabric and wool finishing (Biškauskaitė et al. 2021).

2.10.10 Organic Synthesis Industry

The possibility of enzyme-based procedures to produce high-purity products at a reduced cost in a manner that is well-regarded regarding the environmental aspect is rapidly widening its practical value in the generation of fine chemicals. The excellence of selectivity of enzymes has made them the method of choice in the synthetic process of industrial chemistry as compared to the conventional methods. A high level of selectivity, such as this one, is especially useful in the field of chemical synthesis, as it may lead to fewer environmental issues, reduced separation, and minimal to no generation of byproducts. Other advantages of enzyme-based commercial uses are high catalytic activity and low operating temperatures. The catalytic ability of microbes has been applied for hundreds of years to the production of cheese and alcohol in the context of industrial synthetic chemistry (Wanyonyi and Mulaa 2019). One of the most common enzymes that are used in organic synthesis is lipases, which are used to form a diverse array of optically active alcohols, acids, esters, and lactones. Diltiazem involves the production of (S, R)-2, 3- methoxyphenyl alicyclic acid by the action of lipases. 3, 4-Dihydroxylphenylalanine (DOPA) is a precursor to treating Parkinson's disease and is produced by oxidoreductases, including polyphenol oxidase. Highly region- and stereoselective glycosyltransferases are also used in the process of industrial production of polysaccharides and oligosaccharides, required in cellular communication and recognition. Lyases are involved in the organic synthesis of cyanohydrins in ketones, acrylamide in acrylonitrile, and malic acid in fumaric acid. A Japanese-based firm, Nitto Chemical firm operates through a process known as the nitrile hydratase-based process to produce approximately 40,000 tons of acrylamide annually. Glucose isomerase is commercially produced every year to produce a multimillion-ton batch of high fructose corn syrup (HFCS), a sweetener that is used in the food and beverage industries in place of sucrose (Bizerra et al. 2023).

2.10.11 Protease in Laundry Businesses

The most frequent use of enzymes is in detergents (approximately 25–30% of total enzyme sales). This market would expand by an average growth rate of approximately 11.5% between 2015 and 2020, according to projections. One of the most important applications of enzymes in the current times is the invention and growth of industrial detergents. Detergents are applied in homes, offices, and institutions for cleaning, laundry, and dishwashing. One can use enzymes in detergent products to remove protein, carbohydrate, and lipid-based stains (Couture 2022). The enzymes contained in laundry detergents are effective in the destruction of smashed cotton fibers, increasing the whiteness, color, and care of fabrics. The most popular enzymes are currently protease and amylase, which belong to the hydrolase group and are widely found in detergent products. Sometimes a mixture of enzymes, such as lipases, cellulases, amylases, pectinases, and proteases, is employed to increase the efficacy of stain removal and fabric care. Lipases can remove the stains of fatty foods successfully, whereas amylases are able to remove the deposits of starchy foods. Protease dissolves biological stains, like blood, grass, eggs, and sweat, but cellulases dissolve fabrics, give color, and collect minute fibers without destroying the major threads in the fabric. Dishwasher detergents contain protease and amylase to break protein and carbohydrate-containing food particles. Since enzymes lower the concentration of phosphates and bleaching products in detergent products, they are healthy for human and environmental health (Lasoń-Rydel et al. 2024).

2.11 Protease Special Application

2.11.1 Contact Lens Cleaning

Plant (as papain) and animal (as pancreatin, trypsin, and chymotrypsin) proteins are widely used as contact lens cleaning agents. Most of these cures either emit a nasty smell after a few hours of use or one develops. These cleaning solutions are odorless and safe; however, certain microbial proteases have been demonstrated to be capable of removing the material on the contact lenses and tear film as well. Let us consider the example of the proteases of *Aspergillus* species, *Streptomyces* species, and *Bacillus* species, which do not produce irritation of the eye and allergic reactions (Hasan et al. 2022).

2.11.2 Biological Process

In biotechnology processes, some proteases are utilized to cleave a number of fusion tags following the formation of protein fusions. In order to enhance the soluble expression and solubility of the target protein, it can be fused with a special form of ubiquitin-related protein called a small ubiquitin-related modifier (SUMO). Following the expression, SUMO protease can be able to identify and cleave the target protein sequence with high precision (Welbourn and Pause 2006). This protease is utilized in the extraction of genomic DNA in most common molecular biology and cell biology studies, as well as in the digestion and elimination of enzymes. This endopeptidase is an enzyme that breaks large proteins down into smaller peptides by hydrolyzing the lysine and arginine C-terminal peptide bonds. Trypsin is widely used in many biotechnological processes, including stem cell therapy, cancer cell therapy, enzymatic protein hydrolysis and sequencing, cell separation of different tissues in cell culture studies, and degrading denatured proteins. There is a wide source of microbial proteases of any kind of microorganism. The usage of fungal proteases in food has been applied because of their safe and enzymatic nature. Among other things, acid proteases can replace the functions of renin, papain, and pepsin activities (Gao et al. 2023).

2.11.3 Protease in Cosmetics Products

Toothpaste, mouthwashes, sunscreen cream, hair coloring, and waving products find use as enzymes as free radical scavengers. Superoxide dismutase is applied in the prevention of free radicals and in curing skin damage due to infection, air water, and pollutants as well as other infectious agents. Being free radical scavengers, SOD and peroxidases are incorporated together in the sunscreen cream to minimize erythema.

To clean and flatten the skin, skin creams use proteins to exfoliate the damaged or dead skin. Two additional enzymes commonly used in toothpaste and mouthwash that are applied to whiten teeth, remove plaque, and eliminate odor-causing deposits on teeth and gum tissue are enzymes, known as endoglycosidase and papain. Hair is colored with laccase, oxidases, peroxidases, and polyphenol oxidases, whereas protein disulfide isomerase, glutathione sulfhydryl oxidase, transglutaminase, lipase, catalase, papain, bromelain, and subtilisin are applied to the skin care and hair care, respectively. Protein deposits are also eliminated using enzymes in contact lens cleaners (Khambhaty 2020).

2.11.4 Photographic Industries

Alkaline proteins secreted by *B. subtilis*, *Streptomyces avermectin,* and *Conidiobolus coronatus* have been used to recover silver using X-ray film; the procedure is eco-friendly compared to using chemicals. It has also been reported that silver can be recovered with the help of mutant alkaline protease of *Bacillus* sp. that is thermally stable (Singh and Bajaj 2017).

2.11.5 Silk Degumming

In the process of alkaline treatment of raw silk in a solution of alkaline soap, ancient methods require the extraction of a type of proteinaceous substance known as sericin or silk gum. Alkaline protease is the best option for disposing sericin without leaving any damage to the fiber. It has been proven that fiber breaking is unavoidable and silk threads are found to be much stronger than those prepared with the help of the old methods of traditional processes (Hao et al. 2023).

2.12 Conclusion

Microbial proteases form a vital component of industrial biotechnology, and their market presence as well as the scope of application can only grow. The varied biochemical properties, genetic manipulability, low cost, and manufacturing efficiency of these proteins make them advantageous compared with plant and animal proteins. These enzymes today play an important role in a wide variety of other industries, including detergents, food processing (dairy, baking, and meat tenderization), leather processing, waste management, and pharmaceuticals. The future of microbial proteases is directly connected with the development of biotechnology. Protein and genetic engineering play a significant role in the production of next-generation proteases with a higher specific activity, changed pH optima, and increased thermal stability to satisfy the high commercial demands. Proteases with special and strong properties could also be obtained by investigating new microbial strains in harsh environments. Microbial proteases is an innovative and sustainable device that promotes the process of creating greener processes in industries. Their growing use in the processes of bioremediation, biofuel production, and the production of small molecules highlight its role in meeting the needs of the world in environmental safety and resource management. Engineering and strain enhancement responsibility research and development on enzyme engineering and strain enhancement will surely open the door to new opportunities and establish their key role in the global shift to a more sustainable bio-based economy.

References

Abril B, Bou R, García-Pérez JV, Benedito J (2023) Role of enzymatic reactions in meat processing and use of emerging Technologies for Process Intensification. Foods 12:1940

Aehle W, Bott R, Graycar TP (2009) Proteolytic cleavage, reaction mechanisms. In: Encyclopedia of industrial biotechnology, vol 1

Assad I, Bhat SU, Gani A, Shah A (2020) Protein based packaging of plant origin: fabrication, properties, recent advances and future perspectives. Int J Biol Macromol 164:707

Bansal P, Sharma AK, Kumar R, Chanalia P, Gupta B (2023) Chapter 1 serine proteases: classification, catalytic mechanism, types, and industrial applications, p 1

Bharti SK, Pathak V, Alam T, Arya A, Basak G, Awasthi MG (2020) Materiality of edible film packaging in muscle foods: a worthwhile conception. J Packag Technol Res 4:117. https://doi.org/10.1007/s41783-020-00087-9

Biškauskaitė R, Valeikiene V, Valeika V (2021) Enzymes for leather processing: effect on pickling and chroming. Materials 14:1480. https://doi.org/10.3390/ma14061480

Bizerra V d C, Sales MB, Melo RLF, do Nascimento JGA, Brandão JA, Silva MPF, dos Santos KM, Sousa P d S, da Fonseca AM, de Souza MCM, dos Santos JCS (2023) Opportunities for cleaner leather processing based on protease enzyme: current evidence from an advanced bibliometric analysis. Renew Sust Energ Rev 191:114162. https://doi.org/10.1016/j.rser.2023.114162

Bizzozero SA, Dutler H (1981) Stereochemical aspects of peptide hydrolysis catalyzed by serine proteases of the chymotrypsin type. Bioorg Chem 10:46. https://doi.org/10.1016/0045-2068(81)90042-0

Blay V, Pei D (2019) Serine proteases: how did chemists tease out their catalytic mechanism? ChemTexts 5. https://doi.org/10.1007/s40828-019-0093-4

Calva-Estrada S d J, Jiménez-Fernández M, Lugo-Cervantes E (2019) Protein-based films: advances in the development of biomaterials applicable to food packaging. Food Eng Rev 11:78. https://doi.org/10.1007/s12393-019-09189-w

Couture F (2022) Therapeutic targeting of the proteolytic enzymes. Int J Mol Sci 24:521. https://doi.org/10.3390/ijms24010521

Craik CS, Page MJ, Madison EL (2011) Proteases as therapeutics. Biochem J 435:1

Cuq B, Aymard C, Cuq J-L, Guilbert S (1995) Edible packaging films based on fish myofibrillar proteins: formulation and functional properties. J Food Sci 60:1369. https://doi.org/10.1111/j.1365-2621.1995.tb04593.x

Czapinska H, Otlewski J (1999) Structural and energetic determinants of the S1-site specificity in serine proteases. Eur J Biochem 260:571

Díaz-Cervantes E, Robles J, Solà M, Swart M (2023) The peptide bond rupture mechanism in the serine proteases: an in silico study based on sequential scale models. Phys Chem Chem Phys 25: 8043. https://doi.org/10.1039/d2cp04872h

Du S, Kretsch RC, Parres-Gold J, Pieri E, Cruzeiro VWD, Zhu MC, Pinney MM, Yabukarski F, Schwans JP, Martínez TJ, Herschlag D (2025) Conformational ensembles reveal the origins of serine protease catalysis. Science 387. https://doi.org/10.1126/science.ado5068

Dyer RP, Weiss GA (2021) Making the cut with protease engineering. Cell Chem Biol 29:177

El-Fatah RAA, Rozan MA, Ziena HM, Imre K, Morar A, Herman V, Abdel-Naeem HHS (2023) Improvement of microbial quality, physicochemical properties, fatty acids profile, and shelf life of Basa (Pangasius bocourti) fillets during chilling storage using pepsin, rosemary oil, and citric acid. Foods 12:4170. https://doi.org/10.3390/foods12224170

Elsässer B, Goettig P (2021) Mechanisms of proteolytic enzymes and their inhibition in QM/MM studies. Int J Mol Sci 22:3232

Gao M, Song J, Zhang X, Zhang C, Peng B, Chattha SA (2023) Key mechanism of enzymatic dehairing technology for leather-making: permeation behaviors of protease into animal hide and the mechanism of charge regulation. Collagen and Leather 5. https://doi.org/10.1186/s42825-023-00117-7

Gaur VK, Gupta S, Sharma P, Gupta P, Varjani S, Srivastava JK, Chang J, Bui X (2022) Metabolic cascade for remediation of plastic waste: a case study on microplastic degradation. Curr Pollut Rep 8:30. https://doi.org/10.1007/s40726-021-00210-7

Ghoraba W (2023) Understanding the molecular mechanisms of proteases in bioprocessing: a review on the future of the food industry. Zenodo (CERN European Organization for Nuclear Research). https://doi.org/10.4172/2320

Gimenes NC, Silveira E, Tambourgi EB (2019) An overview of proteases: production, downstream processes and industrial applications. Sep Purif Rev 50:223. https://doi.org/10.1080/15422119.2019.1677249

Gouseti O, Larsen ME, Amin A, Bakalis S, Petersen IL, Lametsch R, Jensen PE (2023) Applications of enzyme technology to enhance transition to plant proteins: a review. Foods 12:2518

Hao M, Shi C, Gong W, Liu J, Meng X, Liu F, Lu F, Zhang H (2023) Heterologous expression and characterization of an M4 family extracellular metalloprotease for detergent application. J Gen Appl Microbiol 69:309. https://doi.org/10.2323/jgam.2023.09.002

Hasan M d J, Haque P, Rahman MM (2022) Protease enzyme based cleaner leather processing: a review. J Clean Prod 365:132826

Khambhaty Y (2020) Applications of enzymes in leather processing. Environ Chem Lett 18:747. https://doi.org/10.1007/s10311-020-00971-5

Klebe G (2013) Inhibitors of hydrolases with an acyl–enzyme intermediate, p 493

Kumar A, Dhiman S, Krishan B, Samtiya M, Kumari A, Pathak N, Kumari A, Aluko RE, Dhewa T (2024) Microbial enzymes and major applications in the food industry: a concise review. Food Prod Process Nutr 6

Lasoń-Rydel M, Sieczyńska K, Gendaszewska D, Ławińska K, Olejnik TP (2024) Use of enzymatic processes in the tanning of leather materials. AUTEX Res J 24. https://doi.org/10.1515/aut-2023-0012

López-Otín C, Bond J (2008) Proteases: multifunctional enzymes in life and disease. J Biol Chem 283:30433

Mahnashi MH, Muddapur UM, Turakani B, Shaikh IA, Awadh AAA, Alshahrani MM, Almazni IA, Malpani J, Goudar SV, Sindagimath V, Alsaikhan F, Khan AA, Mannasaheb BA, Ahmed MZ, Iqubal SMS (2022) A review on versatile eco-friendly applications of microbial proteases in biomedical and industrial applications. Sci Adv Mater 14:622

Matter IR (2023) Industrial applications of microbial protease: a review. Acad Sci J 1:141

Mokrani S, Nabti E (2024) Recent status in production, biotechnological applications, commercial aspects, and future prospects of microbial enzymes: a comprehensive review. Intl J Agricult Sci Food Technol 10:6

Mótyán JA, Tóth F, Tőzsér J (2013) Research applications of proteolytic enzymes in molecular biology. Biomolecules 3:923. https://doi.org/10.3390/biom3040923
Naveed M, Nadeem F, Mehmood T, Bilal M, Anwar Z, Amjad F (2020) Protease—A versatile and ecofriendly biocatalyst with multi-industrial applications: an updated review. Catal Lett 151: 307. https://doi.org/10.1007/s10562-020-03316-7
Omoniyi OAO, Moro DD, Afolabi OB (2024) Microbial proteases: sources, significance and industrial applications. Int J Curr Microbiol App Sci 13:1. https://doi.org/10.20546/ijcmas.2024.1306.001
Palacios-Mateo C, Lwanga EH, Harings J, Blank LM (2025) Enzymatic remediation of polyester microfibers in sewage sludge and green compost samples. Microplast Nanoplast 5. https://doi.org/10.1186/s43591-025-00132-x
Pessela BC, Okamoto DN, Cabral H (2023) Editorial: microbial proteases: biochemical studies, immobilization and biotechnological application. Front Microbiol 14:1126989. https://doi.org/10.3389/fmicb.2023.1126989
Ramkumar A, Sivakumar N, Victor R (2016) Fish waste-potential low cost substrate for bacterial protease production: a brief review. Open Biotechnol J 10:335
Razzaq A, Shamsi S, Ali A, Ali Q, Sajjad M, Malik A, Ashraf M (2019) Microbial proteases applications. Front Bioeng Biotechnol 7:110
Shettar SS, Bagewadi ZK, Alasmary MY, Mannasaheb BA, Shaikh IA, Khan AA (2025) Comprehensive biochemical, molecular and structural characterization of subtilisin with fibrinolytic potential in bioprocessing. Bioresour Bioprocess 12:21. https://doi.org/10.1186/s40643-025-00860-1
Simplicio JLF, de Sousa BM, Pereira HO d S, do Nascimento YM, de Freitas AC (2023) Produção, Pré-Purificação e Encapsulamento de Enzima Proteolítica Secretada por Fungo Filamentoso. J Eng Exact Sci 9:15785. https://doi.org/10.18540/jcecvl9iss3pp15785-01e
Singh S, Bajaj BK (2017) Potential application spectrum of microbial proteases for clean and green industrial production. Energy Ecol Environ 2:370. https://doi.org/10.1007/s40974-017-0076-5
Solanki P, Putatunda C, Kumar A, Bhatia RK, Walia A (2021) Microbial proteases: ubiquitous enzymes with innumerable uses. *3 Biotech* 11:428
Song P, Zhang X, Wang S, Xu W, Wang F, Fu R, Feng W (2023) Microbial proteases and their applications. Front Microbiol 14
Souza PM, Bittencourt ML d A, Caprara CC, de Freitas MM, de Almeida RPC, Silveira D, Fonseca YM, Filho EXF, Pessoa A, Magalhães PO (2015) A biotechnology perspective of fungal proteases. Braz J Microbiol 46:337
Strat YL (2022) Study of the fungal community associated to the red macroalga Palmaria palmata for the production of cell wall degrading enzymes of biotechnological interest. HAL (Le Centre pour la Communication Scientifique Directe)
Tavano OL, Berenguer-Murcia Á, Secundo F, Fernández-Lafuente R (2018) Biotechnological applications of proteases in food technology. Compr Rev Food Sci Food Saf 17:412. https://doi.org/10.1111/1541-4337.12326
Tkaczewska J (2020) Peptides and protein hydrolysates as food preservatives and bioactive components of edible films and coatings a review. Trends Food Sci Technol 106:298
Troncoso FD, Sánchez DA, Ferreira ML (2022) Production of plant proteases and new biotechnological applications: an updated review. ChemistryOpen 11
Verhamme IM, Leonard SE, Perkins RC (2018) Proteases: pivot points in functional proteomics. Methods Mol Biol 313
Vojcic L, Pitzler C, Körfer G, Jakob F, Martínez R, Maurer K, Schwaneberg U (2015) Advances in protease engineering for laundry detergents. New Biotechnol 32:629
Vojcic L, Jakob F, Martínez R, Hellmuth H, O'Connell TD, Mühl H, Lorenz MG, Schwaneberg U (2016) Engineering proteases for industrial applications, p 101
Wanyonyi WC, Mulaa F (2019) Alkaliphilic enzymes and their application in novel leather processing technology for next-generation tanneries. Adv Biochem Eng Biotechnol 172:195. https://doi.org/10.1007/10_2019_95

Welbourn S, Pause A (2006) HCV NS2/3 protease. Europe PMC (PubMed central)
Yachnin BJ, Azouz LR, White RE, Minetti CASA, Remeta DP, Tan VM, Drake JM, Khare SD (2022) Massively parallel, computationally guided design of a proenzyme. Proc Natl Acad Sci 119. https://doi.org/10.1073/pnas.2116097119
Yang W, Lu F, Liu Y (2023) Recent advances of enzymes in the food industry. Foods 12:4506. https://doi.org/10.3390/foods12244506
Zhang Y, He S, Simpson BK (2018) Enzymes in food bioprocessing – novel food enzymes, applications, and related techniques. Curr Opin Food Sci 19:30. https://doi.org/10.1016/j.cofs.2017.12.007
Zhang X, Gao M, Chattha SA, Zhu Y, Peng B, Ye Y (2021) Application of acidic protease in the pickling to simplify the pelt bating process. J Leather Sci Eng 3. https://doi.org/10.1186/s42825-021-00068-x
Zhao J, Liu F (2024) Enzymes in food industry: fermentation process, properties, rational design, and applications. Foods 13:3196. https://doi.org/10.3390/foods13193196

Chapter 3
Fermentation Technologies for Protease Production: Submerged and Solid-State Fermentation

Samer Saleem Alshkarchy, Ashish Kumar Nayak, Turana Mammadova, B. Lokeshwari, Sumitha Elayaperumal, and Bandapalli Sujatha

Abstract Proteases are among the most commercially important industrial enzymes, with wide-ranging applications in detergents, food processing, pharmaceuticals, textiles, leather processing, and environmental management. The large-scale microbial production of proteases is primarily based on two fermentation processes which are submerged fermentation (SmF) and solid-state fermentation (SSF). SmF entails the growth of microorganisms in a liquid nutrient medium with microorganisms, such as the *Bacillus subtilis*, *Bacillus licheniformis*, and *Aspergillus niger*. Process optimization focuses on the supply of carbon, nitrogen, and minerals, while efficient bioreactor design aims to control parameters such as aeration, agitation, temperature, and foam management in order to develop a high enzyme titer. SSF utilizes low-moisture conditions and agro-industrial residues as solid substrates, promoting the growth of microorganisms such as *Aspergillus oryzae*, *Rhizopus* spp., and *Bacillus* spp. Optimization of solid-state fermentation (SSF) process parameters, such as moisture content, temperature, and aeration,

S. S. Alshkarchy
Department of Pathology, Al-Qasim Green University, Al-Qasim City, Babylon Province, Iraq

A. K. Nayak
Department of Civil Engineering, Prasad V. Potluri Siddhartha Institute of Technology, Vijayawada, Andhra Pradesh, India

T. Mammadova
Department of Plant Protection, Azerbaijan State Agricultural University, Ganja, Azerbaijan

B. Lokeshwari (✉)
PG and Research Department of Microbiology, Sacred Heart College (Autonomous), Tirupattur, Tamil Nadu, India

S. Elayaperumal
Department of Biotechnology and Bioinformatics, JSS Academy of Higher Education and Research, Mysore, Karnataka, India

B. Sujatha
Department of Zoology, KVR Govt College for Women (Autonomous), Cluster University, Kurnool, Andhra Pradesh, India

V. Zambare et al. (eds.), *Biotechnological Advances and Applications in Microbial Proteases*, Interdisciplinary Biotechnological Advances,
https://doi.org/10.1007/978-981-95-9011-7_3

significantly improves productivity and cost-effectiveness. Comparative analysis of submerged fermentation (SmF) and SSF reveals varying advantages and disadvantages impacting yield and scalability. Enhanced protease production involves optimizing media formulation, pH, and moisture content, along with filtration and chromatography for recovery. Further strategies are needed to increase yields and manage issues like contamination in SmF and process variability in SSF. Future prospects include utilizing new microbial strains, designing innovative bioreactors, and integrating enzyme production into sustainable bioeconomy systems. This chapter examines the production of microbial proteases using SmF and SSF, optimization strategies, downstream processing, industrial application of proteases, and challenges and future perspectives.

Keywords Microbial proteases · Submerged fermentation (SmF) · Solid-state fermentation (SSF) · Bioprocess optimization · Downstream processing

3.1 Introduction

Proteins are hydrolytic enzymes, which accelerate the breaking of peptide bonds in proteins, therefore making them necessary in many biological processes and industrial practices. Their roles are important in the physiological processes, including protein metabolism, cellular signaling, apoptosis, digestion, and immune defense of microorganisms, plants, and animals (Razzaq et al. 2019). *Bacillus subtilis* and *Bacillus licheniformis* are also characterized by a high rate of growth and extracellular release of enzymes, and they can survive under harsh environmental conditions. *Aspergillus niger* and *Aspergillus oryzae* are also applicable in the food and drinks sector because of their acid preservation (Jisha et al. 2013). The actinomycetes have proteases that have specific substrates, and the archaea proteins have extremophilic proteins, which are active under extreme salinity, alkalinity, and temperature.

Proteases have wide applications in diverse industries, such as detergents, food and beverage, leather and textiles, as well as the pharmaceutical industry. The increasing need for environmentally friendly processes has aggravated the emphasis on the efficient production of microbial proteases, mainly via fermentation (SmF and SSF) processes. SmF also enables a strict control of the environment in which the growth takes place, which results in high productivity and uniform quality, and SSF utilizes solid materials and reduces costs of production through the use of agro-industrial residues (Røyseth et al. 2023). The best temperatures, aeration, and nutrient sources are important in maximizing the activity and yield of the enzymes. The third step after fermentation is downstream processing, which is essential in making sure that the enzymes are pure, active, and stable (Kittler et al. 2025).

Recent developments have advanced protease production beyond conventional processes. Innovations in genetic engineering, including recombinant DNA technology, promoter engineering, and CRISPR-Cas genome editing, are now employed to create strains that are more yielding, more efficient, and more stable (Marín et al. 2018). Although they have strengths, issues like the control of contamination and enzyme stability are some of the challenges presented by these methods. These barriers can only be overcome through a multidisciplinary approach in which strain improvement, novel bioreactor designs, and efficient downstream processing are considered. New microbial systems, as well as metabolic engineering, can be used to advance the future (Yachnin et al. 2022). Agro-industrial residues can be used to generate sustainable production through a circular bioeconomy. This will increase efficiency in production and reduce adverse environmental effects to provide high-yield stable proteases to a wide range of industries.

3.2 Types of Fermentation for Protease Production

3.2.1 Submerged Fermentation (SmF)

Submerged fermentation (SmF) is a highly eminent biotechnological approach in which microorganisms are grown in controlled conditions in a liquid nutrient medium in order to generate metabolites, encompassing enzymes, which include proteases. Technology has developed through the transfer of traditional batch-type culture to the modern stirred reactors in bioreactors, where the key variables like the pH, dissolved oxygen, and temperature could be measured and controlled, which directly affected the enzyme harvests (Ahmed et al. 2011). SmF facilitates the high growth of microbial strains like *Bacillus subtilis*, *Bacillus licheniformis*, and *Aspergillus niger* that are popular due to the elevated extracellular protease release. The homogeneity of the liquid media enhances the uniform distribution of nutrients and oxygen, hence eliminating the stratification of nutrients. SmF can be scaled, which allows transferring laboratory shake flasks to pilot and industrial-scale bioreactors and, thus, predicting and reproducing the level of production.

Moreover, SmF may be performed in batch, fed-batch, or continuous as well, providing an industry-specific production of proteases. In spite of its benefits, SmF has such disadvantages as high operation in the form of aeration, agitation, and sterilization of large volumes of liquids and possible foaming, which may disrupt oxygen delivery and require antifoaming agents. These agents can interfere with the metabolism of microbes or make the recovery of the enzymes difficult. However, SmF is the most preferable of all large-scale protease production due to the high reproducibility and ease of control of the process, which can also be extended to compatibility with the downstream purification process. Continuous optimization programs and bioreactor technologies have also facilitated the boosting of the enzyme yield and stabilization, and this has made SmF a critical concern in enzyme biotechnology (Mukherjee et al. 2008). Figure 3.1 illustrates the submerged fermentation process, whereby the microbial cells proliferate in a liquid nutrient medium that is continuously stirred to ensure even distribution of substrates and oxygen. They are fed with air at the bottom to form bubbles, which improve aeration, and a temperature control system is there to keep the growth at optimal conditions. The mixed and aerated nature of the bioreactor, along with the input of the substrates, enables the effective metabolism and formation of the products by the microbes in the bioreactor.

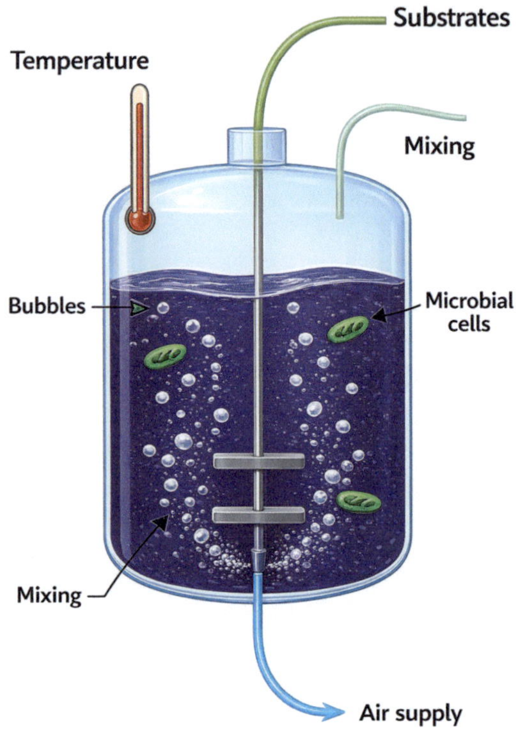

Fig. 3.1 Submerged fermentation process

3.2.1.1 Microorganisms Used in SmF

3.2.1.1.1 *Bacillus subtilis*

Bacillus subtilis is a Gram-positive, rod-shaped bacterium that has great industrial applications because it produces high yields of extracellular protease in SmF systems. It produces various proteases, including neutral, alkaline, and serine proteases, which are used in the manufacture of detergents, food, and leather products. Its strains have the potential to be genetically engineered to increase yield and alter enzyme properties, and this is made easy by the fact that it is nonpathogenic and genetically tractable. Enzyme yield can be optimized by taking advantage of the bacterium's robust growth characteristics through adjustments to carbon and nitrogen sources, pH, and dissolved oxygen levels (Shettar et al. 2025). The use of enzymes can also be done using cheap raw materials due to the metabolic plasticity of this bacterium. In addition, *B. subtilis* is employed in coculture systems and recombinant technologies to increase the utility of protease with the assistance of molecular biology as a preferred option for scalable and varied protein production.

3.2.1.1.2 *Bacillus licheniformis*

Bacillus licheniformis is a Gram-positive bacterium, which is employed in submerged fermentation to produce alkaline proteases, well-known for their high secretion of proteases at various pH and temperature conditions, which makes it quite useful in the detergent, textile, and leather sectors. Carbon and nitrogen sources, aeration, and mineral salts (calcium and magnesium) affect the synthesis of its thermostable enzymes (Bensmail et al. 2023). The recent advances in the selection of strains through mutagenesis and recombinant technology have enhanced the isolation of proteases that may be used in bioremediation and food processing, where enzyme stability is of utmost importance. It can also be used to produce bio-proteases applicable in the industries due to its potential to utilize cost-effective feedstock and can also be cultured in large-scale bioreactors.

3.2.1.1.3 *Aspergillus niger*

Aspergillus niger is a filamentous fungus that produces acidic proteases, with a high yield and generally recognized as safe (GRAS) status. It has a high capacity to absorb nutrients and to release enzymes, and thus, it is applicable in the manufacture of enzymes in the food, beverage, and pharmaceutical industries. *A. niger* has the ability to ferment different types of carbon, such as cellulose and starch, and thus, it can be considered cost-effective. Key fermentation control parameters include pH, temperature, dissolved oxygen, and agitation, the latter being crucial to avoid shear stress on the fungal mycelium. The balance of nutrients is important, such as the sources of nitrogen and minerals, to get maximum production of proteases. Classical mutagenesis and modern genetic engineering have enhanced the stability and production of such enzymes (Vijayaraghavan et al. 2012). Also, the stability of *A. niger* in acidic environment is crucial to the usage of *A. niger* in food processing and fermentation, as well as its highly developed purification methods that make enzyme recovery and concentration feasible, which supports *A. niger* in industrial fermentation.

3.2.1.2 Substrate and Nutrients

3.2.1.2.1 Carbon Sources

Carbon sources play a crucial role in submerged fermentation (SmF), where they supply the energy and carbon required by microbial growth and enzyme synthesis. The common sources are glucose, sucrose, starch, and molasses, whose concentrations influence microbial growth rates and protease synthesis. Simple sugars promote rapid growth but inhibit protease gene expression, while complex carbohydrates tend to induce it (Betchem et al. 2024). The choice of carbon source significantly affects microbial metabolism. Simple sugars promote rapid growth but inhibit protease gene expression, while complex carbohydrates tend to induce it. Optimizing carbon sources is essential for maximizing protease production in large-scale fermenters, with strategies like co-substrates and sequential feeding being crucial for enhancing production efficiency.

3.2.1.2.2 Nitrogen Sources

Nitrogen is vital in amino acid and protein synthesis, particularly in the production of proteases. The organic sources, such as peptone, yeast extract, and soybean meal, are better than inorganic sources such as ammonium sulfate and ammonium nitrate. The nature and amount of nitrogen have an influence on the production of proteases, with the more intricate organic sources tending to produce better enzymes (Swarna and Gnanadoss 2021). Enzyme production by the utilization of low-cost nitrogen-based agro-industrial by-products, such as soybean meal or corn steep liquor, is more economically viable.

3.2.1.2.3 Mineral Salts and Trace Elements

The presence of trace elements and mineral salts is essential to the process of submerged fermentation, serving as cofactors of the enzyme and providing stability of membranes in microorganisms and maintenance of osmotic pressure. Major minerals such as magnesium, calcium, potassium, and phosphate help in growth, sporulation, and enzyme secretion of microbes, and minor elements such as zinc, iron, manganese, and cobalt play crucial roles in enzyme metabolism and stability (Barros and Gouvinhas 2020). Lack of these minerals may cause poor growth of microbes, low enzyme synthesis, and low-quality proteases. The best interaction of the medium components and mineral salts should be observed in order to stimulate protease production, where magnesium aids in the synthesis of proteases and the release of enzymes. Optimization of the types and amounts of minerals to maximize protease production is done using assessments such as design of experiments (DOE) and artificial intelligence modeling. The use of agro-industrial residues, which are mineral-enriched, can be useful in increasing the economic efficiency and the regulation of minerals and trace elements that are required to increase submerged fermentation.

3.2.1.3 Bioreactor Design and Parameters

In submerged fermentation (SmF), bioreactor design plays a vital role because it determines oxygen transfer and distribution of nutrients, temperature, and foam that are important in the growth of microorganisms and proteases generation. Stirred-tank bioreactors are favored due to their flexibility and scalability as well as homogenization of the culture environment. Some of the important design considerations are the type and location of impellers, sparging, design of the baffles, and geometry of the vessel, which affect mixing efficacy, aeration, and shear stress. Recent bioreactors use real-time sensors of dissolved oxygen, pH, and temperature and make automated modifications to optimize the yield of proteases.

Stable fermentation and high enzyme yields require foam control, which is managed using mechanical foam breakers and antifoaming agents. However, these agents can complicate downstream processing (Pérez-Rodríguez et al. 2014). Other essential parameters in SmF are agitation, heat, and pH. Agitation enhances nutrient diffusion and oxygen dissolution. Temperature regulation is essential to prevent enzyme denaturation and maintain productivity. Moreover, stabilization of pH is important in the metabolism of the microbes and stability of enzymes, especially alkaline proteases, which function best in moderately alkaline conditions. The environmental conditions of the bioreactor and the medium composition are very crucial in that the synergistic control of the environment, and the medium is required to achieve large yields of proteases in the industrial process.

3.2.1.4 Advantages and Limitations of SmF

Submerged Fermentation (SmF) provides a number of benefits in protease production in the industry that include a uniformity of the environment such that there are equal distribution of nutrients, oxygen, and temperature. This assists with consistent enzyme production as well as simplified parameter monitoring, including pH and temperature through automation. Several downstream processing methods can be applied with the use of SmF, which leads to the productivity of filtration and purification. It is capable of operating in batch, fed-batch, or continuous processes, and as such, it provides flexibility in production. Other microorganisms include *Bacillus subtilis, Bacillus licheniformis,* and *Aspergillus niger,* which can grow in large bioreactors and produce a high number of proteases. The method can tap into alternative sources of carbon and nitrogen, such as agro-industrial wastes, which further enhances the cost-efficiency and sustainability in the business of detergents, food, pharmaceuticals, and leather treatment. These are the high cost of operation, foaming, shear stress of fungal cells, and risk of contamination. These shortcomings are overcome by the efficient design of bioreactors and control of the process (Pawar et al. 2023).

3.2.2 Solid-State Fermentation (SSF)

Solid-state fermentation (SSF) is a bioprocess involving the cultivation of microorganisms on a solid substrate with low moisture content. This environment naturally replicates the habitat of many fungi, particularly filamentous fungi, and some bacteria. SSF is particularly suitable for producing enzymes and metabolites, where high cell densities or specific substrate diffusion properties are advantageous.

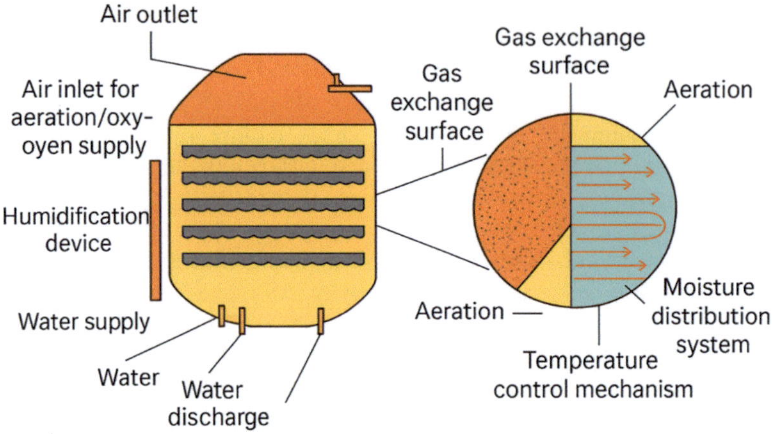

Fig. 3.2 Solid-state fermentation process

Among the salient features of SSF, the concentration of the enzyme is achieved in the solid matrix as a result of the low water activity that minimizes the process of enzyme dilution and allows the secretion to continue. Microorganisms grow in biofilms and mycelium networks on the solid surface and enhance the uptake of nutrients. SSF is applicable in industries such as detergents, food, and leather processing, offering economic and ecological advantages, including reduced water consumption in downstream processes (Matei et al. 2021).

Despite the above advantages, SSF has scaling as well as process control issues and particularly the uniformity of the nutrients, the oxygen content, and the temperature of the large-scale solid matrices, leading to the disparate production of enzymes. In addition, SSF is harder to follow and automatize compared to submerged fermentation due to the heterogeneity of the solid material. Nevertheless, recent advancements in the creation of reactor design, substrate synthesis, and optimization of processes have made SSF increasingly economical as an alternative to liquid fermentation systems, and it is a sustainable technology (Sharma and Bagshawe 2017). Figure 3.2 illustrates a solid-state fermentation system, where microorganisms grow on moist solid substrates with controlled aeration, humidity, and temperature. Air is supplied through inlets and distributed across gas-exchange surfaces to maintain oxygen availability; a humidification device regulates moisture levels. The integrated temperature control and moisture distribution mechanisms ensure optimal microbial activity throughout the solid matrix.

3.2.2.1 Microorganisms Used in SSF

3.2.2.1.1 *Aspergillus oryzae*

Aspergillus oryzae is a filamentous fungus that is often utilized in solid-state fermentation (SSF) to generate proteases since it is fast in enzyme secretion, and it can proliferate in solid media. GRAS (generally recognized as safe), it can be used in a very broad range of industries, including food and pharmaceuticals. *A. oryzae* is a low water activity acid and neutral proteases that are active in SSFs and feeds on agro-industrial wastes like wheat and rice bran (Danilova and Sharipova 2020). The moisture of the substrate and the size of the particles are the crucial factors that affect the fungal growth and the enzyme production. Hence, it is necessary to optimize the production process to ensure that it yields maximum and reduces contamination. Mutagenesis and recombinant DNA technology can be used to further improve the rate of protease production. The resultant proteases are also economically applied in food, such as soy sauce and cheese production, and in the production of bioactive peptides, which is also an economic advantage of the use of low-cost solid media in SSF an advantage of the application of inexpensive solid media in SSF. *A. oryzae* is used as the best microbe in protease production in SSF due to its high secretion capacity of enzymes and also due to its ability to utilize a range of substrates.

3.2.2.1.2 *Rhizopus* spp.

Rhizopus oryzae and *Rhizopus stolonifer* are filamentous fungi, which are vital in solid-state fermentation (SSF) to generate proteases through the use of their fast mycelial proliferation and the ability to use various agro-industrial wastes. They synthesize thermostable extracellular proteases that can be used in the food and fermentation industries with acidic or neutral pH. The composition of substrate, particle size, and moisture content are the key factors that influence the protein productivity and affect oxygen diffusion and nutrient access. Stabilization and enhancement of the nutrient production are achieved by the inclusion of wheat bran and sugarcane bagasse as the substrates for the production of enzymes. Additionally, the temperature and pH are needed to regulate fungi growth and protease production (Nawab et al. 2018). The transfer of oxygen and heat removal are improved by fermentation engineering techniques, which include the utilization of tray and packed-bed reactors, thus improving the efficiency of SSF. It is applied in the biosynthesis of bioactive peptides and extracted proteins. Extracted proteins are viable, provided they can be compatible with inexpensive substrates and have a future in enzyme biotechnology.

3.2.2.1.3 *Bacillus* spp.

The *Bacillus* species (particularly *Bacillus subtilis* and *Bacillus licheniformis*) are now being employed in solid-state fermentation (SSF) because of their ability to grow in solid media, and the expression of thermostable alkaline proteases. These proteases are useful in many industries, including the detergents and leather processing industries. The *Bacillus* spp. use agro-industrial wastes and optimize the output of enzymes resting on the composition of the substrate, moisture, and aeration. The essential organic sources, including soybean meal, wheat bran, and rice bran, are the most prevalent sources of nutrients, and the environmental conditions, including the temperature and pH, contribute to a larger extent to growing the proteases (Abraham et al. 2014). The use of low-cost solid materials through SSF is a means not only of increasing the enzyme concentration but also of economic viability to allow *Bacillus* spp. to complement fungal systems in the production of proteases in a multitude of industrial applications.

3.2.2.2 Substrate Selection in SSF

Substrate is another major factor in solid-state fermentation (SSF) as it offers nutrients and support structures that the microorganisms need to survive. The application of waste materials, such as wheat bran, rice bran, and corn cob, used in the agricultural sector, is desirable because it contains a high carbohydrate concentration, level of protein, and level of minerals. These are the kind of substrates that determine the yielding of enzymes, stability, and cost-effectiveness. The significant characteristics of physical properties, like the size of particles and porosity, affect the colonization by microbes and enzymes (Haider et al. 2021). These processes can be adopted to improve the characteristics of the substrate and include milling or nutrient supplementation to improve the output of proteases. The circular bioeconomy trend and the substrate selection based on the considerations of sustainability and cost.

3.2.2.3 Process Parameters in SSF

The moisture content, temperature, and level of aeration are critical points of the solid-state fermentation (SSF) process that indirectly influence the growth of the organism, enzyme formation, and product stability. Water is essential in the metabolism of microbes and the prevention of contamination. Aeration and temperature maintain the optimum activity of the enzyme and microbial viability, respectively, and maintain the conditions of the substrate to be favorable. The interplay of these processes influences such processes as oxygen diffusion and temperature distribution. The frequent production and quality of proteins, particularly on large-scale bioreactors, necessitates an optimization approach that can contain statistical designs (Cerda et al. 2017). Some of the methods providing sufficient aeration and heat dissipation are forced air circulation and intermittent mixing. The characteristics of the substrates and the environment are significant in colonization of microbes and enzymes release regarding the establishment of economically viable SSF systems in industrial practices.

3.2.2.3.1 Moisture Content

The moisture content is an essential factor in solid-state fermentation (SSF) because it influences the metabolism of microorganisms, the swelling of the substrate, nutrient dissolution, and the release of enzymes. However, excessively moist conditions may reduce substrate porosity and oxygen diffusion, favoring the growth of anaerobic bacteria. The best moisture content is 40–70% (w/w) and depends on microorganisms and substrates (Naik et al. 2023a). Humidity is critical to the maximum production of protease, and thus, methods such as periodic spraying or humidification in large bioreactors are necessary. Moisture measurement devices of high resolution are used to stabilize the moisture content to achieve high yields of protease in SSF systems.

3.2.2.3.2 Temperature Control

Solid-state fermentation (SSF) requires temperature regulation because it determines the development of microorganisms, their metabolic processes, and the liberation of proteases. The optimal temperature for enzyme production by protease-producing fungi and *Bacillus* species is between 25 °C and 40 °C, for any deviation may prevent or enhance contaminants. Forced air exchange practices and insulated reactors are some of the methods that are applied by SSF in dealing with heat produced by the microbial metabolism processes in order to avoid overheating. The temperature distribution is also influenced by substrate composition and moisture level and enhances heat tolerance. The sophisticated SSF systems have enriched reproducibility, enzyme yield, and scalability through real-time temperature control and automatic feedback control (Göğüş and Tarı 2019).

3.2.2.3.3 Aeration

In solid-state fermentation (SSF), aeration plays a vital role since it dictates the growth of the microbe, metabolism, and activity of the enzymes and, in particular, proteases. The characteristics of the substrate, including porosity and moisture, can inhibit the diffusion of oxygen, and aeration should be controlled. This is due to the sufficient oxygen pressure that facilitates aerobic metabolism, elimination of carbon dioxide, and dissipation of heat, which are important in enzyme stability (El-Bakry et al. 2015). The aeration types include forced air circulation, periodic substrate mixing, and rotating drum reactors in an attempt to enhance the availability of oxygen and in order to maintain the homogeneous conditions of growth. The rate of airflow needs to be optimized to meet the oxygen needs of microbes, and excess or deficiency of airflow can negatively impact growth and proteolytic activity. Diffusion of oxygen is also very important in industrial applications in which the depth of substrate and compaction prevent aeration. It must be aerated with the necessary moisture and temperature so that it can achieve a homogenous growth of the microbe and maximize the production of enzymes. Modern bioreactor systems are provided with controlled airflow, trays, and multilayered substrates to stimulate the rates of oxygen transfer and efficient enzyme production in SSF.

3.2.2.4 Advantages and Limitations of SSF

Solid-state fermentation (SSF) has some benefits to protease production, including high concentration of the enzymes, owing to low degree of moisture content; cost-effective, owing to low cost of agro-industrial by-products utilized; and energy saving, owing to low use of water as compared to submerged fermentation (SmF). SSF offers a natural way of developing a favorable environment for filamentous fungi, which enhances stability and the amount of enzymes. It also has certain issues, such as the scale limitation, control of the processes, unequal distribution of moisture and nutrients, as well as potential contamination issues (Niego et al. 2023). Nevertheless, the recent progress in reactor design and process optimization is enhancing the feasibility of SSF to produce proteins, in large quantities, using a protease.

3.3 Comparison of Submerged and Solid-State Fermentation

Submerged fermentation (SmF) and solid-state fermentation (SSF) are the two common forms of fermentation used to produce proteases, and the two types of processes operate differently, with their advantages and disadvantages. SmF occurs in an aqueous medium, where nutrients are distributed evenly, and sufficient oxygen dissolution is achievable, as well as the pH and temperature are maintained. It is suitable in bacterial strains like *Bacillus subtilis* and *Bacillus licheniformis*, which secrete proteases out of cell surface. SSF involves growing microbes on dry solid substances, which are either naturally occurring, and the filamentous fungi, such as *Aspergillus oryzae* and *Rhizopus* spp., which grow in massive networks of mycelia and produce concentrated fungal enzyme products. The properties of the enzymes produced differ between the two fermentation types. SmF tends to yield enzymes that are readily extractable from the liquid medium but are less concentrated due to dilution. SSF forms concentrated products of enzymes on the surface of the substrate but may complicate extraction due to the presence of a solid matrix (Ahmed and Abood 2017).

Table 3.1 Comparison of submerged and solid-state fermentation for protease production

S. no	Parameters	Submerged fermentation (SmF)	Solid-state fermentation (SSF)
1	Microorganisms	Bacteria (e.g., *Bacillus subtilis*, *Bacillus licheniformis*)	Filamentous fungi (e.g., *Aspergillus oryzae*, *Rhizopus* spp.)
2	Process control	Control parameters (pH, temperature, aeration, nutrient distribution)	Control parameters uniformly (aeration, moisture, heat)
3	Enzyme concentration & yield	Lower enzyme concentration due to dilution in the liquid medium. High volumetric productivity is possible	Higher enzyme concentration and product yield per gram of substrate
4	Downstream processing	Easier; enzymes are readily purified from the liquid medium	More difficult; requires extraction from the solid matrix
5	Enzyme stability	Standard stability	Enhanced stability; reduced water activity prevents denaturation and microbial contamination
6	Cost-effectiveness	High capital and operational costs (energy, water, sterilization)	Lower costs (low-cost agro-industrial substrates, less water, fewer effluents)
7	Scalability	Highly scalable with large, controlled bioreactors	Scalability is a challenge due to heat buildup and uneven aeration/moisture in large solid beds
8	Major advantages	Excellent process control & reproducibility, easy scale-up and downstream processing	High product concentration, use of low-cost substrates, lower energy and water consumption and enhanced enzyme stability
9	Major limitations	High water and energy consumption, lower enzyme concentration, and higher effluent generation	Difficult process control and monitoring, complex downstream extraction
10	Applications	High-volume, continuous production where process control and reproducibility are critical	Production of high-concentration proteases using renewable solid substrates, especially with fungi

This can stabilize the proteases as the activity of water is low and averts the denaturation and degradation. The choice between SmF and SSF is determined by the microorganism, the desired enzyme properties, and the requirements of the industrial process (Vittaladevaram 2017). The two methods are optimizable to increase enzyme concentration and downstream processing efficiency, which can be achieved by hybrid methods. Factors that determine the choice of the two fermentation techniques in an endeavor to ensure that a considerable quantity of proteases is produced to be applied in the industry also include scalability, the availability of the substrates, and the economical nature. Table 3.1 illustrates the comparative analysis of the SmF and SSF in the process conditions and efficiency. It demonstrates that SF is based on liquid media of high moisture and provides easier control of the parameters, whereas SSF is based on low-moisture solid substrates and frequently provides greater enzyme productivity.

3.4 Optimization Strategies

These are methods used to maximize protease production in terms of yield, stability of the enzyme, and cost-effectiveness. Some of the major strategies are media optimization, where nutrient compositions are optimized so that microbial growth and enzyme production are promoted, and process optimization, where environmental conditions are adjusted to suit microbial growth. These methods are related because the composition of the medium has a direct effect on the metabolism of the microorganism. Both classical and modern optimization methods ensure that the production of protease that will be utilized in industries is done in large quantities. Response surface methodology, factorial designs, and artificial intelligence modeling are powerful approaches that enable the simultaneous evaluation of multiple variables, thereby improving experimental efficiency while reducing resource consumption and labor (Sheikh et al. 2024). This is necessary because strain-specific optimization is important, and different microorganisms will respond differently to environmental and nutrient conditions. Therefore, the methods must be tailored to substrate, microorganism, and industrial application requirements. Strain, media optimization, as well as process control: This is used to enhance the productivity of the proteases on a sustainable and cost-effective basis.

3.4.1 Media Optimization

In order to obtain the maximum production of proteases, the media should be optimized and achieve a balance between carbon, nitrogen, mineral salts, and trace elements that have a profound influence on the growth of microorganisms and their production of enzymes. Carbon and nitrogen sources are crucial for energy production, cellular structure, and protein synthesis, whereas mineral salts and trace elements are enzyme cofactors, structure stabilizers, and better metabolic rates (Yachnin et al. 2022). Optimization entails the utilization of low-cost agricultural industrial residues and experimental design, such as one-factor-at-a-time (OFAT) and response surface methodology (RSM), to determine the optimal ratios of the nutrients. It is therefore very important to focus on complex nitrogen and sufficient carbon with proper intake of trace elements in order to have enzyme activity. Also, using low-cost feeds such as wheat and rice bran can be used to increase sustainability. Therefore, not only does optimum media preparation increase the generation of proteases, but it also enhances economic and ecological benefits in the industry.

3.4.2 *Process Optimization*

Process optimization plays a significant role in enhancing microbial growth, enzyme release, and protease concentration through the control of environmental conditions, such as pH, temperature, aeration, and moisture. The SmF method can be automated, controlling agitation, temperature, and dissolved oxygen, whereas SSF needs close attention to moisture, aeration, and substrate characteristics to be colonized by microorganisms (Al-mansoori et al. 2020). Contemporary statistical methods like factorial design and response surface methodology assist in the evaluation of numerous factors at once, so as to optimize protein production to improve productivity by optimizing media and improving the strain. It results in better microbial metabolism, release of proteases, reproducibility, lower variability, and generation of economically viable industrial enzymes that can be scaled.

3.4.2.1 pH

pH plays an important role in the growth of microorganisms, the functioning of enzymes, and the stability of proteases, where each microorganism possesses an optimal pH that allows growth. As an example, *Bacillus* species grow at pH 8–10, whereas *Aspergillus oryzae* grows on slightly acidic to neutral pH. Their variations above and below these levels may affect the enzymatic activity, denature proteases, and prevent the growth of microbes. Buffering agents and automated pH control systems are used in SmF to stabilize the conditions in the fermentation process, with substrate selection being used to address pH changes in SSF. Enzyme stabilization and productivity of proteases in industrial applications are improved with proper control of pH.

3.4.2.2 Temperature

Temperature has a great influence on microbial metabolism, growth, and production of enzymes, especially on the protease-producing microorganisms that grow between 25 C and 40 C. These enzymes can be denatured and create greater chances of contamination. SmF bioreactors have automated temperature control, which is beneficial over Solid-State Fermentation because there are minimal temperature variations. The enzyme stability depends on the optimum secretion temperature, especially to prevent the slow folding and inactivity problems (Moradbeygi et al. 2025). The use of thermostable proteases, temperature optimization, and strain screening are crucial during industrial applications so as to achieve robust proteases. The key to the successful process design is a deep insight into the growth needs of microorganisms and thermal characteristics of fermentation medium to achieve a steady production of protease and at the same time make it scalable.

3.4.2.3 Aeration

Aeration also plays a vital role in providing oxygen to the aerobic organisms and also in the elimination of the heat and carbon dioxide produced by the organisms. In SmF, aeration is achieved through sparging and agitation, while in SSF, it is achieved through forced-air circulation or periodic mixing, for example, in specialized bioreactors such as packed beds (Melo et al. 2023). The most appropriate aeration rates will ensure that the oxygen will not be deficient, and consequently, the growth will not be discouraged and the aeration of the substrate will not dry it up.

3.4.2.4 Moisture

The moisture content is also vital in solid-state fermentation (SSF), as it affects the production of enzymes, as well as the growth of microbes, by maintaining sufficient water activity. The best moisture content (40–70 w/w) is the one that facilitates substrate swelling, solubility of nutrients, and the colonization of microorganisms without causing anaerobic microbial growth. Too much or too little moisture may reduce the porosity of substrates and the oxygen supply and limit microbial growth and enzyme release. Good moisture management will improve the transfer of heat and mass, homogenous growth of microbes, and eliminate dry and wet spots (Sriniva et al. 2023). Methods such as presoaking and humidification assist in the preservation of moisture levels in large-scale systems, which is important in high protease production and the reproducibility of the process. Sophisticated SSF systems that sense moisture increase the yield and cost-effectiveness.

3.5 Downstream Processing

Downstream processing is a crucial stage in protease production, involving the recovery, purification, and formulation of enzymes for optimal industrial applications. After fermentation, the proteases are found within a complex medium containing microbial cells and residual substrate. Downstream processing is required to remove these impurities and purify the enzymes. This multistage process involves solid-liquid separation, concentration, purification, and stabilization, all aimed at maximizing enzyme recovery without denaturation (Benluvankar et al. 2016). The extraction method depends on the type of fermentation, e.g., submerged fermentation (SmF) or solid-state fermentation (SSF). Modern processes, such as ultrafiltration, membrane separation, and chromatography, are complementary to traditional processes. This enzyme stability, type, and intended use are essential to scalable, reproducible, and cost-effective recovery schemes, which are also important in commercial protease production.

3.5.1 Recovery and Purification of Proteases

The recovery and purification of proteases from fermentation broths begin with the removal of cells and solid residues. Physical separation (filtration, centrifugation), chemical precipitation (salting out, organic solvents), and chromatography are some of the commonly used ways of purifying a substance (Colla et al. 2023). In the pharmaceutical industry, high-resolution chromatography is a must-have, since it is necessary to attain the desired purity of enzymes. Mild conditions are required so that there is no denaturation, and the catalytic activity is maintained. Such a strategy of a mixture of filtration, centrifugation, precipitation, and chromatography is effective and can be scaled to commercial-level production of proteases.

3.5.2 Filtration

The most common technique for isolating cellular microorganisms, solid debris, and particulates in enzyme-containing solutions in submerged fermentation (SmF) and solid-state fermentation (SSF) is the filtration technique. Fermentation broth is clarified using different filtration methods, such as cloth, membrane, and vacuum-assisted filtration. Filtration can be used together with extraction techniques in SSF, and mesh sieves or press filters may be used to separate solid and liquid enzyme extracts (Zaier et al. 2021). The type and size of the filter play a vital role in the recovery of the enzyme and the ability to remove impurities, as coarse filtration removes larger particles, while microfiltration and ultrafiltration remove finer cellular debris. Viscosity, substrate composition, and colloidal particles are some of the factors that determine filtration efficiency. Its benefits are its simplicity, scalability, and low effect on the activity of enzymes, and thus, it is a common downstream processing step, followed by concentration and purification, which leads to high enzyme recovery and quality while reducing membrane fouling.

3.5.3 Centrifugation

Isolation of microbial biomass and particulates is done through the centrifugation process by taking advantage of density variations in the enzyme extracts in submerged fermentation. The high-speed centrifugation effectively partially purifies enzyme solutions, but it requires careful control of force and time to prevent enzyme denaturation and protein damage. Protease purification is commonly done using batch centrifuges in industry, whereas continuous centrifuges have become popular due to their mass production efficiency. This step enhances filtration by removing deposited particulates, thereby improving downstream purity (Martău et al. 2021). The medium may be precipitated, unfiltered, for chromatography to obtain high enzyme recovery and quality after post-centrifugation.

3.5.4 Precipitation

Protein extracts should be concentrated by precipitation, which may be by salting out with ammonium or potassium sulfate, or organic solvents like ethanol or acetone. It is an effective way of reducing the number of enzyme solutions, removing contaminants, and enhancing the storage or further purification of enzymes (Yimer 2018). Factors that influence the effectiveness of precipitation include salt concentration, the type of solvent, temperature, and pH. Precipitation using ammonium sulfate has been preferred as it has minimum effects on the activity of the enzyme, and it is easily desalinated. The enzymes are centrifuged or filtered after the precipitation and then redissolved in the corresponding buffers. It is a crucial step in the isolation and partial purification of enzymes and is an intermediate step between crude extracts and other purification techniques, such as chromatography (Dar et al. 2017). Parameters of precipitation are very critical to maximize the yield and enzyme stability.

3.5.5 Chromatography

Chromatography is a high-resolution purification method that is employed to achieve the purity of proteins employing ion-exchange, gel filtration, and affinity chromatography methods. These separation techniques are also employed to make both pharmaceutical-grade biochemicals and sensitive biochemical experiments, separating enzymes that are dependent upon their charge, size, and binding properties. The principle of chromatography is the property of the enzyme, such as the isoelectric point and molecular weight, and sequential chromatography is applied to enhance the purity of the product with the help of orthogonal separation principles. Some of the most critical parameters, whose values are strictly controlled in order to maintain the activity of the enzyme and to prevent its denaturation, include its composition, flow rate, and temperature (Webb 2017). Chromatography is resource-intensive, despite being efficient, and is not used with low-value proteins; instead, filtration, centrifugation, and precipitation are used as alternative large-scale processes in the extraction of enzymes in the case of detergents and leather processing, among other industries. Chromatography plays an essential role in the purification and stability of enzymes.

3.5.6 Enzyme Stability

Enzyme stability plays an important role in downstream processing, affecting its storage, transportation, and application in industries. Proteases are sensitive to various factors, such as temperature and pH, and therefore, good means of stabilizing them is required, such as protective agents and lyophilization. Optimal stability will retain catalytic activity and shelf-life. The minimal denaturation may be avoided by using gentle methods of handling, such as filtration and centrifugation, and avoiding physical and chemical stresses by providing buffer systems and cryoprotectants (Kabir et al. 2025). Industries favor pH and thermostable enzymes to enhance protease performance in sectors such as detergents, food processing, leather treatment, and pharmaceuticals. During the formulation and packaging development, stability testing is required to avoid inactivation during storage.

3.5.7 *Industrial Formulations*

Proteases are formulated to be stable in different forms, like liquids, powders, granules, or immobilized enzymes. They use stabilizers and surfactants to increase shelf-life and effectiveness. Liquid formulations are favored in the detergent and food processing industries for their ease of mixing and in fermentation due to the quick enzyme reactions (Yao et al. 2023). The solution to this is through enzyme immobilization on solid supports so as to be reused in bioreactors, which lowers the cost of operation and processing. They also improve enzyme performance by enhancing their transportation, storage, and profitability across various industrial applications.

3.6 Industrial Applications for Proteases

Protein enzymes, such as proteases, play an important role in the industries by catalyzing the breakdown of proteins into constituent amino acids and peptides. They are employed in several sectors, including detergents, food and beverages, leather, textile, pharmaceuticals, and environmental management. The choice of a particular protease depends on the catalytic ability, stability, and suitability for the process conditions (Kalaiselvan et al. 2025). The high demand of proteases is influenced by their efficiency, cost-effectiveness, and friendliness to the environment in comparison with the traditional chemical methods. Different proteins have been engineered to fit specific purposes such as alkaline proteins as detergents and thermostable proteins as bioreactors. The stability of proteins and resistance to inhibitors of operation and proteases have led to the advancement in the industries. It has been improved with the help of microbial strain development, fermentation, and downstream processing that enhance their productivity and stability (Negi et al. 2020). They also make the process more efficient, and the effect on the environment is also reduced because harsh chemicals are reduced to a minimum (Yazid et al. 2016). The standard formulations and immobilization techniques have helped to introduce proteases in industrial processes, hence expanding their applications.

3.6.1 Detergent Industry

The detergent industry utilizes proteases to remove proteinaceous stains, such as blood, sweat, and food residues (Pessela et al. 2023). Alkaline proteases, especially those active at high pH, are ideal for laundry and dishwashing detergents as they improve stain removal efficiency, reduce cleaning time, and enable low-temperature washing, leading to energy savings. These enzymes must be compatible with other detergent components, like surfactants, bleaches, and stabilizers, to maintain their activity during use. Additionally, enzyme formulations are designed to remain stable across varying temperatures and pH levels to ensure effectiveness in diverse washing conditions (Lu et al. 2020). Proteases involved in detergents are screened and evolved selectively to be both resistant to chemical oxidants and adverse conditions to maintain the cleaning activity. Mass production results in the production of the enzymes by a process termed high-level fermentation, and then the enzymes are processed downstream to be used in powdered or liquid detergents. The use of effective and cost-efficient protease enzymes has transformed the detergent industry tremendously, such that enzyme-based detergents are gaining prominence among the traditional chemical-based washers since they are sustainable.

3.6.2 Food and Beverage Industry

Applications of proteases in the food and beverage industry rely on their ability to hydrolyze proteins, which improves the food texture and alters the flavor profile. Proteases are used to coagulate casein in cheese production. They are also important in the manufacture of protein hydrolysates that contribute to the augmented nutritional worth and digestive properties. Proteases are utilized in meat production and brewing and are useful in clarifying beverages, tenderness of meat, and overall increase in yield (Amin et al. 2025). Acid and neutral proteases are particularly suitable for the food industry because they are incorporated with food safety requirements and can operate under low pH conditions. The importance of controlling enzyme activity to prevent excessive hydrolysis, which could negatively impact product quality. It emphasizes that the stabilization process and the correct dosage of proteases are critical factors for achieving consistent results in food processing (Nema et al. 2019). Proteinases play a crucial role in the food and biotechnology industries, where they are widely utilized for protein hydrolysis, process optimization, and improvement of product quality. In addition, this is because they are environmentally friendly, and hence, enzymatic processes would be a good alternative to chemical processing that promotes safer and more sustainable food production practices.

3.6.3 Leather and Textile Processing

The use of proteases in the leather and textile industry significantly enhances processes such as dehairing, bating, and finishing by replacing hazardous chemicals, like lime and sodium sulfide. In textiles, they effectively remove protein-based sizing agents and improve the quality of protein fibers, like wool and silk, enhancing attributes like softness and dyes (Ooi et al. 2021). Alkaline proteases are typically used in the finishing of leather, whereas neutral or acid proteases are used in the finishing of textiles. To optimize enzyme activity and minimize damage to fibers or hides, critical parameters, such as enzyme concentration, treatment duration, and temperature, must be carefully regulated. The use of proteases offers significant environmental and economic benefits, including lower chemical usage, decreased wastewater toxicity, and shortened processing times (Khatoon et al. 2023). Also, enzymatic treatments have been gaining momentum as they are biodegradable, with green industrial developments.

3.6.4 Pharmaceutical Applications

The pharmaceutical preparations, including diagnostic and therapeutic, have involved proteases and, in particular, proteases. Their anti-inflammatory effects and wound healing effects, as well as the treatment of digestive disorders, have their uses as such proteolytic enzymes (trypsin, chymotrypsin, and bromelain). Moreover, proteases are important in antithrombotic therapy, fibrinolysis, and enzyme replacement therapy and are active compounds (Pereira et al. 2019). Proteases are also applied in the manufacture of biopharmaceuticals and drug formulation to digest protein substrates in order to enhance the bioavailability of drugs and ease the downstream processing. Pharmaceutical-grade proteases are essential for wide applications, requiring purification, stability, and reproducibility. They must demonstrate precisely determinable activity under physiological conditions, during storage, and upon administration, complying with regulatory standards to ensure therapeutic efficacy through appropriate downstream processing and formulations (da Silva et al. 2016). The emergence of new technology and design of proteases in recent times in recombinant DNA technology has been in a position to produce selective proteases that are characterized by high stability, specificity, and low immunogenicity of the proteases. These innovations have rendered proteases as biocatalysts in medical and biomedical frameworks, which has rendered in the drug industry.

3.6.5 Waste Management and Environmental Applications

The significance of proteases in environmental purification is also that they facilitate the degradation of proteinaceous waste in industrial effluent, wastewater, and organic waste streams. They work particularly well with tannery effluents, food production by-products, and slaughterhouse waste, in which they are employed to break down the complex proteases into smaller peptides and amino acids and are hence more susceptible to biodegradation (Moradbeygi et al. 2024). The application of protease in the processes can help to reduce major environmental concerns of the chemical oxygen demand, total suspended solids, and foul odor, to even more sustainable waste management systems (Mircella et al. 2017).

Bioremediation by the use of proteases is preferable due to the fact that it is compatible with the environment, specific, and generates other pollutants like those produced during the conventional technique of chemical remediation. Proteases play a crucial role in composting, organic fertilizer production, and the bioconversion of agro-industrial residues, contributing to nutrient recovery in a sustainable bioeconomy. Their use in waste treatment requires careful regulation regarding enzyme type, formulation, and dosage in industrial settings (Chandrasekaran et al. 2015). Enzymes function effectively under varying environmental conditions, such as temperature and pH changes, and the presence of inhibitors. Incorporating protease processes can promote environmentally friendly industrial practices by offering a sustainable, cost-effective alternative to traditional chemical methods.

3.7 Strategies for Enhanced Protease Production

3.7.1 Strain Improvement

Strain improvement is essential for enhancing microbial protease production, aiming to achieve high enzyme titers, stability, and efficient substrate utilization while maintaining cell viability and metabolic performance. Bacteria and fungi are subject to some improvement methods, such as classical mutagenesis and recombinant DNA technology. Classical mutagenesis is the approach based on exposing microbcs to mutagens in order to obtain random genetic mutations that increase the production of proteases, whereas recombinant DNA technology is the method of obtaining a specific gene modification to produce, secrete, and regulate proteases (Ramkumar et al. 2016). It is usually performed by mutation, selection, and fermentation tests to determine high-yield variants in high-throughput screening assays. Special strains promote production of protease on an industrial scale, for the fermentation time and enhancing economic viability (Lim et al. 2024).

3.7.1.1 Mutagenesis

Mutagenesis refers to a technique employed to develop microbial strains with the aim of introducing random genetic mutations with the help of physical or chemical agents. The process modifies the sequence of nucleotides with the help of mutagens such as UV radiation, gamma rays, or chemicals such as ethyl methanesulfonate. These effects may include the production of more protease and adaptation to the different environmental conditions. The exposure time, the concentration of mutagen, and the environment are important issues that need to be properly controlled to achieve maximum mutation rates with cell viability. In most cases, several mutagenesis and screening steps are required for the improvement in the yield of proteases (Pilli and Siddalingeshwara 2016). Mutagenesis is popular due to their cost-effectiveness and application in different microorganisms. It may also be combined with the current molecular technology for the improved production of proteases to be used in industries.

3.7.1.2 Genetic Engineering

Genetic engineering enhances protease production by using methods such as recombinant DNA technology, gene overexpression, promoter engineering, and CRISPR-Cas genome editing. Engineered strains produce higher levels of enzymes with better stability and versatility as opposed to the conventional procedures. Metabolic engineering enhances protease production by generating proteases in different hosts, boosting signal peptide release, and knocking out protease inhibitor genes. It optimizes resource distribution in protease production while reducing by-products, particularly benefiting the industrial applications requiring stability in harsh conditions (Sheikhs et al. 2020). However, the use of nonpathogenic and genetically stable strains is a regulatory and safety requirement in food, pharmaceutical, or environmental applications.

3.7.2 *Media and Process Optimization*

Protease production entails optimization of media and processes that will result in the optimal yield of enzymes (Singh and Bajaj 2017). Important aspects are carbon, nitrogen, minerals, and temperature, which influence microbial metabolism and enzyme secretion. Factorial designs and response surface methodology (RSM) are statistical designs that can be used to determine these critical variables in productivity maximization (Shah et al. 2024). Protease can be produced significantly by the nutrient content, aeration, and temperature. Furthermore, it is crucial to employ strains with high yield in large-scale fermentation to scale-up production to meet an industrial scale and at a cost-effective level (Yepuru 2018). The combination of strain improvement and media and process optimization is the best plan toward sustainable and commercially viable production of proteases.

3.8 Challenges and Limitations

Although fermentation and enzyme engineering have advanced in several ways, the main challenges of producing proteases include the control of the growth of microbes, avoiding contamination, and ensuring the stability of the enzyme during production, recovery, and storage (Indhur et al. 2023). SmF and SSF have issues in these areas, requiring innovative ideas on how to achieve cost-effective production. Other factors like pH and temperature and chemical inhibitors should also be well taken care of, especially in large scale, where it requires stability and repetition. Enzyme stabilization can be done through strain engineering, optimization of the media and process, better bioreactor design, and optimization of the processing, to help stabilize the enzymes and make productions of proteases sustainable. Table 3.2 identifies the key drawbacks of submerged and solid-state fermentation in the production of proteases, including excess energy and water consumption in the submerged system and the inability to manipulate moisture and aeration in solid-state systems.

Table 3.2 Key limitations of submerged and solid-state fermentation in protease production

S. no	Parameters	Submerged fermentation (SmF)	Solid-state fermentation (SSF)
1	Microbial growth control	Easier to monitor; risk of foaming, shear stress, and contamination	Uneven growth on solid substrates; monitoring is challenging
2	Nutrient & oxygen distribution	Homogeneous nutrients and efficient aeration; suitable for aerobic bacteria	Limited diffusion; nutrient and oxygen gradients may reduce yield
3	Temperature & pH control	Precise control enhances enzyme stability	Difficult to maintain uniform conditions; hotspots may denature enzymes
4	Moisture content	High (~80–90%); contamination risk higher	Moderate (~40–60%); lower contamination risk, but stress on microbes is possible
5	Contamination risk	High; requires strict sterility	Lower due to low water activity; spores may persist
6	Substrate & cost	Requires refined, often expensive media	Uses agro-industrial wastes; cost-effective but variable
7	Product recovery	Easy enzyme extraction from broth	Labor-intensive extraction from solids; lower efficiency
8	Scalability	Large-scale reactors; reproducible	Due to heat and mass transfer limitations
9	Enzyme stability	Susceptible to shear, pH shifts, and degradation	Often more stable within solid matrix; less microbial degradation
	Process duration	Faster	Slower, suited for fungi and specific proteases

3.8.1 Contamination Risks in SmF and SSF

Both SmF and SSF face significant contamination challenges, which can influence the choice of microorganism. The SmF environment is favorable to the quick growth of bacteria and fungi, requiring high aseptic conditions. On the contrary, although SSF has a lower moisture content due to which the risk of microbial development is eliminated, contamination may still occur at different stages (Leite et al. 2020). Preventive measures, such as strict aseptic techniques, equipment sterilization, and environmental management, are essential in fermentation processes. In industrial submerged fermentation (SmF), risks are mitigated through air filtration and automated cleaning systems, while solid-state fermentation (SSF) can avoid contamination by managing moisture and utilizing its microbial strains. Monitoring and timely interventions are crucial, which can lead to decreased enzyme yield and purity, increasing processing costs (Abu-Tahon et al. 2023).

3.8.2 Process-Control Difficulties in SSF

The heterogeneity of substrates, low water activity, distribution of nutrients, and oxygen are the challenges of solid state fermentation (SSF) that make the process hard to control. Large reactors are challenging to maintain at the best temperature and moisture levels, resulting in possible local overheating and drying out, thus preventing the growth of microbes and enzyme activity. Despite the improved design of the SSF bioreactor in the form of a tray and rotating drum type, the process is not easily controlled relative to submerged fermentation (SmF). The main parameters, such as the size of substrate particles, the degree of packing, and the rate of air flow, require optimization of microbial colonization and enzyme production to remain stable (Mokrani and Nabti 2024). There are challenges associated with the scalability and reproducibility of SSF to produce protease. The implementation of improved reactor designs, predictive modeling, and automated controls has resulted in enhanced productivity, reduced variability in batches, and more cost-effective production of proteases in industries.

3.8.3 Enzyme Stability and Purification Process

Enzyme stability is critical in the production of proteases due to the risk of denaturation and degradation during the fermentation and storage process. Enzyme activity must be maintained by controlling ionic strength, pH, temperature, and protease inhibitors need to be controlled to maintain the activity of enzymes affecting productivity and shelf-life factor. The challenges with the downstream purification process also require an application of the techniques that do not only preserve the enzyme activity but also achieve the desired purity (Vojnović et al. 2024). Modifications that can be used to enhance the stability include addition of the salts and lyophilization. These issues are essential in the manufacturing of the industrial proteases on the mild recovery conditions and stabilized processes to increase the enzyme activity, consistency, and viability.

3.9 Future Perspectives

The future of protease production hinges on the development of high-yield, low-cost, and environmentally sustainable processes. Recent innovations in biotechnology, bioprocess engineering, and systems biology are designed to overcome the shortcomings of the existing technologies of microbial enzyme production (Mayer et al. 2004). The combination of novel microbial strains with recombinant technologies and advanced bioprocessing strategies will lead to improved enzyme productivity and stability in industrial settings. The use of cheap raw materials and the principles of the circular bioeconomy that revolves around the reuse of resources for their sustainable production (Naik et al. 2023b). Novel downstream processing and fermentation techniques are required to enhance scalability and operational efficiency. The high-tech bioreactor designs and automated control systems will enable better control of the growth and enzyme production by the microbes and will eliminate variation and enhance yields. The synergistic sustainability model integrates multiple ecological, technological, and economic factors to promote efficient resource utilization and long-term environmental sustainability (Nkemnaso 2019).

3.9.1 Novel Microbial Strains and Recombinant Technology

Novel microbial strains, including extremophiles, marine organisms, and soil bacteria, are promising sources for industrial protease production as they are capable of surviving under extreme conditions. The screening of these strains improves the protein production in specialized industries. The recombinant DNA technology plays a central role, as it is possible to specifically manipulate microbial genomes to overexpress protease genes, to enhance secretion pathways, and to remove feedback inhibitions. More sophisticated methods, such as CRISPR-Cas genome editing and promoter engineering, enhance enzyme production, which provides greater yield, specificity, and stability (Moharram et al. 2021). These processes allow changes in the processing parameters, and they are used in different industries, such as detergents, pharmaceuticals, and food processing.

3.9.2 Emerging Bioreactor Designs and Bioprocess Intensification

The developments in bioreactor technology aim at improving mass transfer, heat transfer, aeration, and environmental regulation with the aim of maximizing production. Submerged and solid-state fermentation are carried out in innovative designs, such as perfusion bioreactors, airlift reactors, packed-bed, and rotating drum fermenters, to provide uniform growth of microorganisms, effective use of substrates, and high rates of enzyme production (Palacios-Mateo et al. 2025). The main strategies of bioprocess intensification involve augmenting the productivity per volume and cutting down the operational expenses. Techniques used to ensure the production of proteases with a minimum of downtime include fed-batch, continuous fermentation, and immobilized cell systems. Sensors are used to ensure that the fermentation conditions are at the optimum. The industrial use of proteases depends on scalability and reproducibility, which are important factors, and these technological innovations ensure the production of large quantities of enzymes, energy efficiency, and low levels of waste, which sustainability contribute to financial and environmental sustainability (Song et al. 2023).

3.9.3 Sustainable Low-Cost Production Using Agro-industrial Residues

Proteases can be prepared using agro-industrial wastes as a remedy to economic and environmental issues. Fermentation material of high nutritional value, such as lignocellulosic wastes, bran, husks, and food processing by-products, reduces the cost of production and pollution of the environment. This is done with the concept of circular bioeconomy, where waste is used to generate useful enzymes. These wastes are rich in carbon, nitrogen, and minerals; these are the elements that guarantee growth of microbes and increase their enzyme production. Pretreatment processes increase the nutrient availability. It is a sustainable solution to worldwide production of enzymes and the process of achieving ecologically friendly bioprocesses (Guillaume et al. 2019).

3.9.4 Integration with Biorefineries and the Circular Bioeconomy

Protease production in the biorefineries is a circular bioeconomy that transforms biomass into value-added products, including biofuels and enzymes, which are more economically viable and less ecological in their impacts because they optimize resources and reduce waste. By-products are used as food by microbe, which results in a closed-loop system that facilitates efficiency of operations and source of revenue (Assad et al. 2020). The future biorefineries are likely to make use of inventive microbial engineering and effective waste transformation to create sustainable proteases and bio products, as per the principles of long-term environmental stewardship and green chemistry.

3.10 Conclusion

Both submerged (SmF) and solid-state fermentation (SSF) are important industrial systems for protease production. SmF offers a high degree of physicochemical control, excellent reproducibility, ease of automation, and compatibility with bacterial systems such as *Bacillus subtilis* and *Bacillus licheniformis* that grow in liquid media and secrete large amounts of extracellular enzymes. Conversely, SSF is an efficient and environmentally friendly technique that employs low-moisture substrates and agro-industrial waste, offering benefits such as low cost, greater enzyme stability, and high product concentration. However, SSF faces challenges related to heat and mass distribution, monitoring complications, and scale-up issues. The composition of the media, aeration, and temperature management are some of the key factors that are important in maximizing the production of protease in the two systems. The subsequent downstream methods, such as filtration and chromatography, are considered important in purifying enzymes, although SSF extracts demand rigorous extraction procedures because of the solid matrix. Proteases are widely used in detergents, food processing, pharmaceutical, and bioremediation sectors as a universal biocatalyst capable of replacing chemical processes as a sustainable solution. Challenges such as risks of contamination, nonhomogeneous SSF operations, and SmF operational requirements highlight the necessity of novel bioreactor designs, automation of the control systems, and improvements of microbial strains and genetic engineering. Protease biotechnology in the future is in the implementation of molecular technologies, green practice, and bioprocessing efficiency to satisfy the global industrial requirements of scalable, low-waste systems.

References

Abraham J, Gea T, Sánchez A (2014) Substitution of chemical dehairing by proteases from solid-state fermentation of hair wastes. J Clean Prod 74:191–198. https://doi.org/10.1016/j.jclepro.2014.03.035

Abu-Tahon MA, Abdel-Mageed AM, Ghareib M, Housseiny MM, Abdallah WE (2023) Examination of thrombolytic and anticoagulant activities of purified fibrinolytic enzyme extracted from *Cochliobolus hawaiiensis* under solid state fermentation. Res Sq. https://doi.org/10.21203/rs.3.rs-2503121/v1

Ahmed SA, Abood NH (2017) Effect of some growth factors on protease production by *Rhizopus oryzae*. J Al-Nahrain Univ-Sci 20(2):90–99. https://doi.org/10.22401/jnus.20.2.12

Ahmed I, Zia MA, Iftikhar T, Iqbal HMN (2011) Characterization and detergent compatibility of purified protease produced from *Aspergillus Niger* by utilizing agro wastes. Bioresources 6(4): 4505–4522. https://doi.org/10.15376/biores.6.4.4505-4522

Al-mansoori L, Bashraheel SS, Qahtani ADA, O'Connor CD, Elsinga P, Goda SK (2020) In vitro studies on CNGRC-CPG2 fusion proteases for ligand-directed enzyme prodrug therapy for targeted cancer therapy. Oncotarget 11(6):619–632. https://doi.org/10.18632/oncotarget.27478

Amin OE, Aboul-Enein AM, Abd-Elsalam IS, Wahba MI, Helmy YS (2025) Optimization of protease production by newly isolated *Bacillus* sp. from the red sea using defatted soybean cake. Sci Rep 15(1):3855. https://doi.org/10.1038/s41598-025-14643-3

Assad I, Bhat SU, Gani A, Shah A (2020) Protein based packaging of plant origin: fabrication, properties, recent advances and future perspectives. Int J Biol Macromol 164:707–716. https://doi.org/10.1016/j.ijbiomac.2020.07.140

Barros A, Gouvinhas I (2020) Innovation in the food sector through the valorization of food and agro-food by-products. In: IntechOpen eBooks. IntechOpen. https://doi.org/10.5772/intechopen.91078

Benluvankar V, Priya SE, Gnanadoss JJ (2016) Medium formulation and its optimization for increased protease production by *Penicillium* sp. LCJ228 and its potential in blood stain removal. J Appl Biol Biotechnol 4(1):1–8. https://doi.org/10.7324/jabb.2016.40104

Bensmail S, Boudjema K, Naimi-Fazouane F, Bensmail S, Djouahra-Fahem D, Ferhoum F, Bourfis N (2023) Factors affecting acid protease production by *Mucor circinelloides* MG603064.1 through SmF process: characterization and fromage frais making. Biotechnologia 104(4): 333–348. https://doi.org/10.5114/bta.2023.132770

Betchem G, Monto AR, Lü F, Billong LF, Ma H (2024) Prospects and application of solid-state fermentation in animal feed production—A review. Ann Anim Sci 24(4):1123–1141. https://doi.org/10.2478/aoas-2024-0029

Cerda A, Gea T, Vargas-García M d C, Sánchez A (2017) Towards a competitive solid state fermentation: cellulases production from coffee husk by sequential batch operation and role of microbial diversity. Sci Total Environ 589:56–65. https://doi.org/10.1016/j.scitotenv.2017.02.184

Chandrasekaran S, Sundaram S, Kumaresan P, Murugan M (2015) Production and optimization of protease by filamentous fungus isolated from paddy soil in Thiruvarur district Tamilnadu. J Appl Biol Biotechnol 3(6):58–63. https://doi.org/10.7324/jabb.2015.3610

Colla LM, Reinehr CO, Manfredini PG, Cavanhi VAF, Costa JAV (2023) Simultaneous production of proteases and antioxidant biopeptides by solid-state fermentation. Sustain Food Technol 1(6): 874–884. https://doi.org/10.1039/d3fb00077j

da Silva OS, de Oliveira RL, de Souza-Motta CM, Porto ALF, Porto TS (2016) Novel protease from *Aspergillus tamarii* URM4634: production and characterization using inexpensive agroindustrial substrates by solid-state fermentation. Adv Enzyme Res 4(4):125–143. https://doi.org/10.4236/aer.2016.44012

Danilova IV, Sharipova MP (2020) The potential of bacilli and their enzymes for industrial production. Front Microbiol 11:1782. https://doi.org/10.3389/fmicb.2020.01782

Dar MA, Pawar KD, Pandit RS (2017) Gut microbiome analysis of snails: a biotechnological approach. In: InTech eBooks. https://doi.org/10.5772/68133

El-Bakry M, Abraham J, Cerda A, Barrena R, Ponsá S, Gea T, Sánchez A (2015) From wastes to high value added products: novel aspects of SSF in the production of enzymes. Crit Rev Environ Sci Technol 45(18):1999–2042. https://doi.org/10.1080/10643389.2015.1010423

Göğüş N, Tarı C (2019) Effect of substrate concentration and scale up on the polygalacturonase production. J Food 44(5):802–812. https://doi.org/10.15237/gida.gd19054

Guillaume A, Thorigné A, Carré Y, Vinh J, Levavasseur L (2019) Contribution of proteases and cellulases produced by solid-state fermentation to the improvement of corn ethanol production. Bioresour Bioprocess 6(1):13. https://doi.org/10.1186/s40643-019-0241-0

Haider S, Haider A, Catalá Á (2021) Hydrolases. In: Biochemistry. IntechOpen. https://doi.org/10.5772/intechopen.95149

Indhur R, Amoah ID, Bux F, Kumari S (2023) Nanomaterials for microplastic removal from wastewater: current state of the art nanomaterials and future prospects. ACS EST Water 3(12):3741–3754. https://doi.org/10.1021/acsestwater.3c00203

Jisha VN, Smitha RB, Pradeep S, Sreedevi S, Unni KN, Sajith S, Priji P, Josh MKS, Benjamin S (2013) Versatility of microbial proteases. Adv Enzyme Res 1(3):39–51. https://doi.org/10.4236/aer.2013.13005

Kabir MF, Ovi AQ, Ju L (2025) Real-time pH and temperature monitoring in solid-state fermentation reveals culture physiology and optimizes enzyme harvesting for tailored applications. Microb Cell Factories 24(1):32. https://doi.org/10.1186/s12934-025-02820-y

Kalaiselvan P, Devi NC, Deepti M, Devi AA, Akamad K, Dheeran P, Debbarma S, Vadivel D, Vekariya RD (2025) Solid-state fermentation—a sustainable future technology in aquafeeds? Front Mar Sci 12:1669719. https://doi.org/10.3389/fmars.2025.1669719

Khatoon N, Ullah N, Sarwar A, Rahman SU, Khan AAH, Aziz T, Alharbi M, Alshammari A (2023) Isolation and identification of protease producing *Bacillus* strain from cold climate soil and optimization of its production by applying different fermentation conditions. Appl Ecol Environ Res 21(4):3391–3401. https://doi.org/10.15666/aeer/2104_33913401

Kittler S, Müller F, Elshazly M, Wandrey G, Klein T, Daub A, Spadiut O, Kopp J (2025) Transferability of bioprocessing modes for recombinant protease production: from fed-batch to continuous cultivation with *Bacillus licheniformis*. BMC Biotechnol 25(1):13. https://doi.org/10.1186/s12896-025-00947-9

Leite P, Sousa D, Fernandes H, Ferreira M, Costa AR, Filipe D, Gonçalves M, Perés H, Belo I, Salgado JM (2020) Recent advances in production of lignocellulolytic enzymes by solid-state fermentation of agro-industrial wastes. Curr Opin Green Sustain Chem 27:100407. https://doi.org/10.1016/j.cogsc.2020.100407

Lim IKJ, Chong CS, Chong EZH, Lam MQ, Ee KY (2024) Advances in employing omics and bioinformatics to reveal the potential of halophilic bacteria for biodegrading proteinaceous high-salt food waste. Asia Pac J Mol Biol Biotechnol:88–101. https://doi.org/10.35118/apjmbb.2024.032.4.10

Lu P, Hu B, Wang Y, Zhu M (2020) Combined water extraction and sodium chlorite pretreated spent mushroom compost for protease production by separate hydrolysis and fermentation and simultaneous saccharification and co-fermentation. Bioresources 15(2):3201–3218. https://doi.org/10.15376/biores.15.2.3201-3218

Marín M, Artola A, Sánchez A (2018) Production of proteases from organic wastes by solid-state fermentation: downstream and zero waste strategies. 3 Biotech 8(4):205. https://doi.org/10.1007/s13205-018-1226-y

Martău GA, Unger PD, Schneider R, Venus J, Vodnar DC, López-Gómez JP (2021) Integration of solid state and submerged fermentations for the valorization of organic municipal solid waste. J Fungi 7(9):766. https://doi.org/10.3390/jof7090766

Matei JC, Oliveira JA d S, Pamphile JA, Polonio JC (2021) Agro-industrial wastes for biotechnological production as potential substrates to obtain fungal enzymes. Cienc Nat 43:e30. https://doi.org/10.5902/2179460x63133

Mayer A, Sharma SK, Tolner B, Minton NP, Purdy D, Amlot P, Tharakan G, Begent RHJ, Chester KA (2004) Modifying an immunogenic epitope on a therapeutic protein: a step towards an improved system for antibody-directed enzyme prodrug therapy (ADEPT). Br J Cancer 90(12):2402–2410. https://doi.org/10.1038/sj.bjc.6601888

Melo AM d O, Silva TP, Tavares IM d C, Silva FN, Bilal M, Salay LC, Franco M, de Oliveira JR (2023) New protease produced by solid state fermentation of *Penicillium camemberti* 0798100075 using coffee hull applied in milk coagulation. Res Sq. https://doi.org/10.21203/rs.3.rs-2834065/v1

Mircella MA, Salomao RM, Bianca C d S, Maria FST (2017) *Aspergillus flavo furcatis*: aflatoxin test and milk-clotting protease production in submerged and solid state fermentation. Afr J Microbiol Res 11(7):312–321. https://doi.org/10.5897/ajmr2016.8400

Moharram AM, Zohri AA, Hesham AE, Maher MA-A, Al-Bedak OAM (2021) Production of cocktail enzymes by three *Cladosporium* isolates and bioconversion of orange peel wastes into valuable enzymes. J Pure Appl Microbiol 15(4):2336–2351. 10.22207/jpam.15.4.58

Mokrani S, Nabti E (2024) Recent status in production, biotechnological applications, commercial aspects, and future prospects of microbial enzymes: a comprehensive review. Int J Agric Sci Food Technol 10(1):006–015. https://doi.org/10.17352/2455-815x.000202

Moradbeygi F, Ghasemi Y, Farmani AR, Hemmati S (2024) Glucarpidase (carboxypeptidase G2): biotechnological production, clinical application as a methotrexate antidote, and placement in targeted cancer therapy. Biomed Pharmacother 166:115292. https://doi.org/10.1016/j.biopha.2023.115292

Moradbeygi F, Ghoshoon MB, Sadeghian I, Moradi A, Ghasemi Y, Morowvat MH, Hemmati S (2025) Toward the development of a biosimilar variant of Glucarpidase (carboxypeptidase G2): secretory production, optimization, and immobilization. Mol Biotechnol. https://doi.org/10.1007/s12033-025-01450-4

Mukherjee AK, Adhikari H, Kumar S (2008) Production of alkaline protease by a thermophilic *Bacillus subtilis* under solid-state fermentation (SSF) condition using *Imperata cylindrica* grass and potato peel as low-cost medium: characterization and application of enzyme in detergent formulation. Biochem Eng J 39(2):353–361. https://doi.org/10.1016/j.bej.2007.09.017

Naik B, Kumar V, Goyal S, Tripathi AD, Mishra S, Saris PEJ, Kumar A, Rizwanuddin S, Kumar V, Rustagi S (2023a) Pullulanase: unleashing the power of enzyme with a promising future in the food industry. Front Bioeng Biotechnol 11:1139611. https://doi.org/10.3389/fbioe.2023.1139611

Naik B, Kumar V, Rizwanuddin S, Chauhan M, Gupta AK, Rustagi S, Kumar V, Gupta S (2023b) Agro-industrial waste: a cost-effective and eco-friendly substrate to produce amylase. Food Prod Process Nutr 5(1):8. https://doi.org/10.1186/s43014-023-00143-2

Nawab S, Aasim M, Saddique H, Rabi F, Ali W, Khan NZ (2018) Utilization of integrative technique for partial recovery of proteases from soil microbes. Periodica Polytech Chem Eng 63(3):469–476. https://doi.org/10.3311/ppch.12637

Negi S, Jain S, Raj A (2020) Combined ANN/EVOP factorial design approach for media screening for cost-effective production of alkaline proteases from *Rhizopus oryzae* (SN5)/NCIM-1447 under SSF. AMB Express 10(1):30. https://doi.org/10.1186/s13568-020-00996-7

Nema A, Patnala SH, Mandari V, Sobha K, Kumar DS (2019) Production and optimization of lipase using *Aspergillus niger* MTCC 872 by solid-state fermentation. Bull Natl Res Cent 43(1):82. https://doi.org/10.1186/s42269-019-0125-7

Niego AGT, Lambert C, Mortimer PE, Thongklang N, Rapior S, Große M, Schrey H, Charria-Girón E, Walker A, Hyde KD, Stadler M (2023) The contribution of fungi to the global economy. Fungal Divers 121(1):95–137. https://doi.org/10.1007/s13225-023-00520-9

Nkemnaso C (2019) Solid state fermentation: substrates uses and applications in biomass and metabolites production – a review. South Asian Res J Biol Appl Biosci 1(1):20–29. https://doi.org/10.36346/sarjbab.2019.v01i01.004

Ooi CK, Rasit N, Rafizah W, Abdullah WNW (2021) Optimization of protease from *Aspergillus niger* under solid-state fermentation utilizing shrimp shell substrate. Biointerf Res Appl Chem 11(6):14809–14824. https://doi.org/10.33263/briac116.1480914824

Palacios-Mateo C, Lwanga EH, Harings J, Blank LM (2025) Enzymatic remediation of polyester microfibers in sewage sludge and green compost samples. Microplast Nanoplast 5(1):8. https://doi.org/10.1186/s43591-025-00132-x

Pawar KS, Singh PN, Kumar S (2023) Fungal alkaline proteases and their potential applications in different industries. Front Microbiol 14:1138401. https://doi.org/10.3389/fmicb.2023.1138401

Pereira GV d S, Pereira GV d S, de Araujo EF, Neves EMPX, Joele MRSP, Lourenço L d FH (2019) Optimized process to produce biodegradable films with myofibrillar proteases from fish byproducts. Food Packag Shelf Life 21:100364. https://doi.org/10.1016/j.fpsl.2019.100364

Pérez-Rodríguez N, Oliveira FMA, Pérez-Bibbins B, Belo I, Torrado A, Domínguez JM (2014) Optimization of Xylanase production by filamentous fungi in solid-state fermentation and scale-

up to horizontal tube bioreactor. Appl Biochem Biotechnol 173(3):803–825. https://doi.org/10.1007/s12010-014-0895-1
Pessela BC, Okamoto DN, Cabral H (2023) Editorial: microbial proteases: biochemical studies, immobilization and biotechnological application. Front Microbiol 14:1126989. https://doi.org/10.3389/fmicb.2023.1126989
Pilli R, Siddalingeshwara KG (2016) Effect of fermentation kinetics for the biosynthesis of protease from *Aspergillus* sp. J Drug Deliv Ther 6(3):1–5. https://doi.org/10.22270/jddt.v6i3.1281
Ramkumar A, Sivakumar N, Victor R (2016) Fish waste-potential low cost substrate for bacterial protease production: a brief review. Open Biotechnol J 10(1):335–341. https://doi.org/10.2174/1874070701610010335
Razzaq A, Shamsi S, Ali A, Ali Q, Sajjad M, Malik A, Ashraf M (2019) Microbial proteases applications. Front Bioeng Biotechnol 7:110. https://doi.org/10.3389/fbioe.2019.00110
Røyseth V, Hurysz BM, Kaczorowska A, Dorawa S, Fedöy A, Arsın H, Serafim MSM, Myers SA, Werbowy O, Kaczorowski T, Stokke R, O'Donoghue AJ, Steen IH (2023) Activation mechanism and activity of globupain, a thermostable C11 protease from the arctic mid-ocean ridge hydrothermal system. Front Microbiol 14:1199085. https://doi.org/10.3389/fmicb.2023.1199085
Shah TA, Li Z, Zhang A, Akhtar HMS, Salamatullah AM, Bourhia M (2024) Production of alkaline protease from *Bacillus amyloliquefaciens* TAS-2 and optimizing fermentation conditions. Res Sq. https://doi.org/10.21203/rs.3.rs-4296273/v1
Sharma SK, Bagshawe KD (2017) Antibody directed enzyme prodrug therapy (ADEPT): trials and tribulations. Adv Drug Deliv Rev 118:2–7. https://doi.org/10.1016/j.addr.2017.09.009
Sheikh H, Anand GG, Shivanna GB (2024) Fermentation: a potential strategy for microbial metabolite production. In: IntechOpen eBooks. IntechOpen. https://doi.org/10.5772/intechopen.114814
Sheikhs NN, Qurat-ul-Ain Q-A, Altaf S (2020) Production of extracellular protease from bacterial co-cultures using solid state fermentation. BioSci Rev 2(4):13–23. https://doi.org/10.32350/bsr.0204.02
Shettar SS, Bagewadi ZK, Alasmary MY, Mannasaheb BA, Shaikh IA, Khan AA (2025) Comprehensive biochemical, molecular and structural characterization of subtilis in with fibrinolytic potential in bioprocessing. Bioresour Bioprocess 12(1):7. https://doi.org/10.1186/s40643-025-00860-1
Singh S, Bajaj BK (2017) Potential application spectrum of microbial proteases for clean and green industrial production. Energy Ecol Environ 2(6):370–386. https://doi.org/10.1007/s40974-017-0076-5
Song P, Zhang X, Wang S, Xu W, Wang F, Fu R, Feng W (2023) Microbial proteases and their applications. Front Microbiol 14:1236368. https://doi.org/10.3389/fmicb.2023.1236368
Sriniva S, Paulraj M, Velladurai S, Govindan K, Puvanakrishnan R, Esackimuthu P, Dhandapani P, Gopalan V (2023) Bacterial fermentation for protease production in fermenters: impact of oxygen transfer characteristics and rheological features on scale up process. Res Sq. https://doi.org/10.21203/rs.3.rs-3118616/v2
Swarna D, Gnanadoss JJ (2021) Optimization of protease production by *Streptomyces* SP. LCJ1A using response surface methodology. Int J Life Sci Pharma Res 11(1):P218–P232. https://doi.org/10.22376/ijpbs/lpr.2021.11.1.l218-232
Vijayaraghavan P, Vijayan A, Arumugaperumal A, Jenisha JK, Vincent SGP (2012) Cow dung: a potential biomass substrate for the production of detergent-stable dehairing protease by alkaliphilic *Bacillus subtilis* strain VV. Springerplus 1(1):76. https://doi.org/10.1186/2193-1801-1-76
Vittaladevaram V (2017) Fermentative production of microbial enzymes and their applications: present status and future prospects. J Appl Biol Biotechnol 5(4):90–98. https://doi.org/10.7324/jabb.2017.50414

Vojnović S, Aleksić I, Ilic-Tomic T, Stevanović M, Nikodinović-Runić J (2024) *Bacillus* and *Streptomyces* spp. as hosts for production of industrially relevant enzymes. Appl Microbiol Biotechnol 108(1):185. https://doi.org/10.1007/s00253-023-12900-x

Webb C (2017) Design aspects of solid state fermentation as applied to microbial bioprocessing. J Appl Biotechnol Bioeng 4(1):00094. https://doi.org/10.15406/jabb.2017.04.00094

Yachnin BJ, Azouz LR, White RE, Minetti CASA, Remeta DP, Tan VM, Drake JM, Khare SD (2022) Massively parallel, computationally guided design of a proenzyme. Proc Natl Acad Sci 119(15):e2116097119. https://doi.org/10.1073/pnas.2116097119

Yao H, Liu S, Liu T, Ren D, Zhou Z, Yang Q, Mao J (2023) Microbial-derived salt-tolerant proteases and their applications in high-salt traditional soybean fermented foods: a review. Bioresour Bioprocess 10(1):87. https://doi.org/10.1186/s40643-023-00704-w

Yazid NA, Barrena R, Sánchez A (2016) The immobilisation of proteases produced by SSF onto functionalized magnetic nanoparticles: application in the hydrolysis of different protein sources. J Mole Catal B Enzym 133:S567–S574. https://doi.org/10.1016/j.molcatb.2017.01.009

Yepuru SK (2018) Production of alkaline protease from *Aspergillus oryzae* isolated from seashore of bay of Bengal. J Appl Nat Sci 10(4):1210–1215. https://doi.org/10.31018/jans.v10i4.1905

Yimer D (2018) Microbial biotechnology review in microbial enzymeproduction methods, assay techniques and protein separation and purifications. J Nutr Health Food Eng 8(1):00249. https://doi.org/10.15406/jnhfe.2018.08.00249

Zaier H, Maktouf S, Roussos S, Rhouma A (2021) Filamentous fungi isolated from Tunisian olive mill wastes: use of solid-state fermentation for enzyme production. Notulae Botanicae Horti Agrobotanici Cluj-Napoca 49(1):12125. https://doi.org/10.15835/nbha49112125

Chapter 4
Recent Advances in Proteases from Halophilic Bacteria

Sanket K. Gaonkar, Jyothi Judith Alvares, and Irene Furtado

Abstract Protease are important groups of enzymes in biotechnology. In the past, most industries were dependent on proteases, which were mesophilic in nature which remained nonfunctional in the presence of stressful environmental conditions. The invention of extreme microbes, specifically those are under halophilic bacteria, is source of commercially important halophilic enzymes. This chapter deals with the introduction and significance of the subject of enzymology with/in extremely halophilic bacteria. It also unveils how globally, studies on halophilic bacteria are taken up as front-line research to study extracellular halophilic proteases, keeping in mind their biotechnological application. A brief account of the production methodology used in harnessing proteases and their characterization are described, Furthermore, this chapter highlights the contribution of genetic engineering and synthetic biology for advancements in overproduction and challenges on purification and research opportunities. This comprehensive study will represent the choice of salt-stable proteases for various industrial biotransformation as well as biocatalytic processes in biotechnological applications.

Keywords Halophilic · Extremozymes · Proteases · Biotechnology

S. K. Gaonkar (✉)
Department of Microbiology, P.E.S's R.S.N College of Arts and Science, Ponda, Goa, India

Microbiology Programme, School of Biological Sciences and Biotechnology, Goa University, Taleigao Plateau, Goa, India
e-mail: micro.sanket@unigoa.ac.in

J. J. Alvares · I. Furtado
Microbiology Programme, School of Biological Sciences and Biotechnology, Goa University, Taleigao Plateau, Goa, India

V. Zambare et al. (eds.), *Biotechnological Advances and Applications in Microbial Proteases*, Interdisciplinary Biotechnological Advances,
https://doi.org/10.1007/978-981-95-9011-7_4

4.1 Introduction

Microbial enzymes are secreted by microbes as an adaptation to thrive in their habitats, utilizing different sources of carbon and nitrogen. For humans, enzymes offer applications in various sectors of biotechnology. As per analysis, by the end of the year 2033, the global enzyme market is expected to show a steep rise at the rate of 6.6% CAGR, projecting to reach USD 24,672.1 million(https://www.grandviewresearch.com/industry-analysis/enzymes-industry). Extremophiles represent diverse groups of microorganisms that are known to produce extremozymes, which show added advantages over enzymes from mesophilic microbes (Kour et al. 2025; Nwankwo et al. 2023; Yavari-Bafghi and Amoozegar 2025). Extremophilic proteases represent the core share of the enzyme market and are found in food processing, pharmaceuticals, the leather industry, as additives in laundry detergents, and in reagents of diagnostic kits (Zambare and Nilegaonkar 2016; Amoozegar et al. 2017). From an environmental perspective, plant and animal waste of proteinaceous origin is hydrolyzed by proteases into easily metabolized molecules, thus aiding the release of assimilable nitrogen, contributing toward biogeochemical cycles and waste bioremediation (da Silva 2017).

Halophilic microbes are mostly native to hypersaline environments like salt lakes, salt pans, salt marshes, deep-sea brines, and even salted food. Based on optimal salt requirement for growth, halophiles are classified into slight halophiles (1–3% NaCl), moderate halophiles (3–15% NaCl), and extreme halophiles (>15% NaCl). Salt-in strategy, which includes KCl accumulation to balance the high NaCl concentration, and salt-out strategy, which is related to the exclusion of salts and compatible solute accumulation, are the unique adaptation possess by halophilic microorganisms. These two strategies help the intracellular proteins and other biomolecules to efficiently function in the presence of NaCl concentrations inside the cells.

Globally, proteases are sought from halophilic microbes due to their excellent performance at high salt concentrations. In addition, halophilic protease is tolerant to low water activity, elevated temperature, pH, and in the presence of solvents (Nwankwo et al. 2023). The majority of these salt-stable enzymes showed optimal activity between pH 5 and 10 and temperatures from 40 to 75 °C (Kour et al. 2025; Mokashe et al. 2018; Nwankwo et al. 2023; Yavari-Bafghi and Amoozegar 2025). Halophilic enzymes are catalytically active in the range of 4–5 M NaCl while denaturing at lower concentrations (Lanyi 1974). Recent reviews have highlighted the major members of halophilic bacteria producing halophilic proteases and other hydrolytic enzymes functional at different extreme conditions. For instance, major halophilic bacterial members producing proteases belong to *Proteobacteria*, *Bacteroidetes*, *Firmicutes*, and *Actinobacteria* with dominant genera of *Bacillus*, *Pseudomonas*, and *Pseudoalteromonas* (Nwankwo et al. 2023). In 1938, subtilisin Carlsberg from *Bacillus alcalophilus was* pioneered as a halophilic bacteria producing extracellular protease secretion (Kour et al. 2025; Nwankwo et al. 2023; Yavari-Bafghi and Amoozegar 2025). Later, the expansion is observed in other genera of the phyla, such as *Alicyclobacillus*, *Virgibacillus*, *Pseudoalteromonas*, *Vibrio*,

Alteromonas, Planococcus, Stenotrophomonas, Streptomyces, Marinirhabdus, Shewanella, etc. (Nwankwo et al. 2023). Recently, with the aid of high-throughput methods and genetic engineering interventions, obstacles for maximum production, purification, and characterization of halophilic enzymes have been overcome that has led to their maximum discovery (Kour et al. 2025; Nwankwo et al. 2023; Yavari-Bafghi and Amoozegar 2025).

Considering the extraordinary features and high diversity of proteases from halophilic bacteria, they are currently gaining actual application in industries, bypassing the major hurdles and challenges faced during their production and purification.

4.2 Proteases from Halophilic Eubacteria

Proteases comprise about 60% of commercially utilized industrial enzymes. Earlier, the industries utilized alkaline proteases obtained from alkaliphilic and mesophilic eubacteria, but the isolation of halophilic proteases from halophilic bacteria paved a new reservoir for industrial application.

Proteases produced by halophilic bacteria are salt-tolerant extremophilic enzymes that are known for their tolerance to NaCl up to 3.0 M for their activity. These are a more sought-after class of enzymes for use in industrial applications as they display extraordinary properties and functions over a wider range of pH, temperature, and salt. In addition, they showed increased efficiency and enhanced substrate specificity with low immunogenicity under extreme conditions compared to non-halophilic counterparts (Nwankwo et al. 2025; Kour et al. 2025). Figure 4.1 illustrates the unique features of halophilic proteases.

Halophilic microbes and proteases, which are employed in high salt seafood processing industries, improve the flavor, taste, and food quality by catalytic hydrolysis of the protein substrate (Gao et al. 2019; Naveed et al. 2021). They are reported to degrade complex organic compounds present in industrial wastes and water into simpler nontoxic forms (Anh et al. 2021) and improve the quality. In addition, halophilic proteases act as antifungal agents in agriculture and food production, along with ingredients in detergent formulations for effective stain removal (Mokashe et al. 2017). In silico simulation techniques are advancing and known to modify proteases to enhance salt tolerance for novel biotechnological applications (Senba et al. 2023; Fu et al. 2008; Hao et al. 2022; Zhang et al. 2019). Table 4.1 comprises the summary list of some halophilic proteases from halophilic eubacteria isolated from various ecological environments over the last two decades.

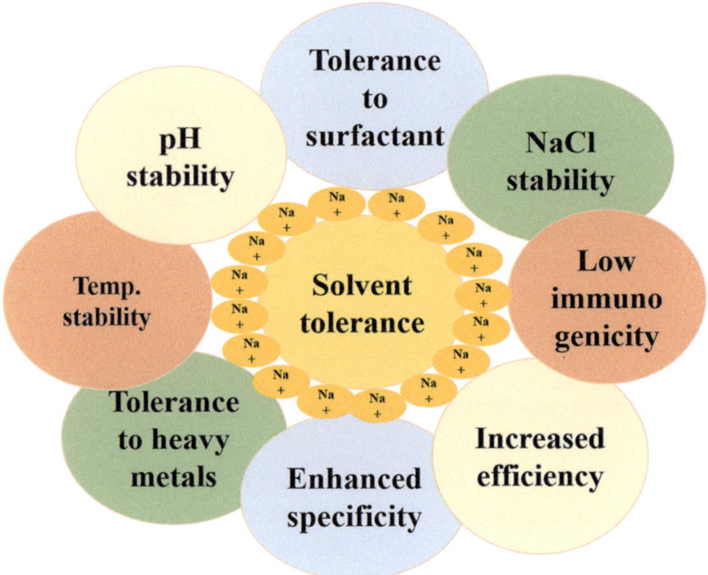

Fig. 4.1 Characteristic features of halophilic proteases

4.3 Production of Salt-Stable Proteases

Production of protease can be constitutive or inducible according to Hanlon and Norman (1981). Protease production by halophilic microorganisms depends on physicochemical conditions. The primary conditions include temperature, aeration, pH, moisture, agitation, and incubation period, while nutrient conditions like nitrogen source, carbon source, minerals, growth promoters, and vitamins also showed an effect on halophilic protease production.

Two different systems, submerged (SmB) and solid-state fermentation (SSF), are used for enzyme production by microorganisms. In SmB, protease production depends on the medium composition and overall pH of the medium. Ease of control of physicochemical parameters, like temperature, pH, and aeration, along with medium components, are some of the important advantages of the SmB (da Silva 2017). On the other hand, the advantages of using SSF of microbial enzyme production are immense; it uses less energy and generates less water, and the usage of different agro-waste residues for the growth of microorganisms is possible (Sharma et al. 2022).

Recently, immobilization cell technology has been a preferred and most utilized methods employed to increase protease production. This method increases enzyme stability and required shorter reaction time. Also, the separation of the enzyme, which is attached to various matrices used for the entrapment, is easy and can be reused for several cycles. Various cell entrapment matrices, such as alginate, gelatin, collagen, agarose, carrageenan, cellulose, and polyacrylamide, are routinely used, of

Table 4.1 The summary list of some halophilic proteases from halophilic eubacteria

Strain	Source and place	NaCl optimum	NaCl range	References
Bacillus cereus TS2	Salt, Kohat	10%	5–30%	Saeed et al. (2023)
Halobacillus sp. HAL1	Sediment, Hurghada City	3%	NS	Fahmy and El-Deeb (2023)
Bacillus subtilis sp.	Soil, Saudi Arabia	0.5%	NS	Mahmoud et al. (2021)
Salicola sp. SBJ9	Salt-lake, Tunisia	3.2 M	5–20%	Daoud et al. (2019)
Bacillus luteus H11	Saline soda lime, Poland	4%	0–9%	Kalwasińska et al. (2018)
Salicola marasensis	Salty areas, Iran	3 M	1-5 M	Khoshnevis et al. (2018)
Idiomarina	Saline and mineral springs, Iran	NS	NS	Satari Faghihi et al. (2018)
Halobacillus sp. CJ4	Soil, hypersaline lake, Tunisia	0.87 M	0–4.15 M	Daoud et al. (2017)
Bacillus iranensis strain X5B	Saline mud, hypersaline lake, Iran	0.87–1.3 M	0.44–2.6	Ghafoori et al. (2016)
Bacillus sp. strain NPST-AK15	Water and sediment, hypersaline soda lake, Egypt	0.87 M	0.3–4.5 M	Ibrahim et al. (2015)
Alkalibacillus sp. NM-Fa4	Alkaline and hypersaline lakes, Egypt	1.7 M	0.9–3.4 M	Mesbah and Wiegel (2014)
Actinopolyspora sp. VITSDK2	Marine sediments India	NS	1.035–3.11 M	Suthindhiran et al. (2014)
Salinivibrio sp. strain MS-7	Salt-lake, Maharlu, Iran	1.75 M	0.87–3.45 M	Shahbazi and Karbalaei-heidari (2012)
Salimicrobium halophilum strain LY20	Saline soil, China	1.75 M	0.087–4.35	Li and Yu (2012)
Geomicrobium sp. EMB2	Sambhar Salt Lake, India	NS	0.87–3.45 M	Karan et al. (2011)
Oceanobacillus iheyensis O.MA18	Soil from salt pans, India	2.6 M	0.87–3.45 M	Purohit and Singh (2011)
Halobacillus sp. LY6	Saline soil, China	2.07 M	0–3.45 M	Xin et al. (2011)
Virgibacillus sp. SK33	1-month-old fish sauce	0.87 M	0–4.35 M	Sinsuwan et al. (2010)
Salicola sp. IC10	Hypersaline habitats, Spain	2.6–3.45 M	2.6–5.2 M	De Lourdes Moreno et al. (2009)
Chromohalobacter sp. TVSP101	Solar salterns, India	5.8–25%	23.2	Vidyasagar et al. (2009)
Halobacillus sp. *SR5–3*	Fish sauce, Japan	1–30%	10	Namwong et al. (2006)

NS not specified

which alginate and polyacrylamide are preferred and most used for protease production. The use of gelatin and radiated gelatin is also reported to give better results (Lapponi et al. 2022; Maghraby et al. 2023). *Micrococcus* sp. VKMM 037, *Bacillus*, *Salicola marasensis*, and *Halobacillus* sp. CJ4 produce protease in NaCl-containing medium (Daoud et al. 2017; Khoshnevis et al. 2018; Manikandan et al. 2011). Metal ions such as K^+, Ca^{2+}, and Mg^{2+} in media contribute to enhanced secretion of the enzyme by *Virgibacillus halodenitrificans* RSK CAS1 (Sathishkumar et al. 2015a, b). Table 4.2 represents the optimal parameters reported for some halophilic protease production from halophilic microbes.

4.4 Purification and Characterization of Salt-Stable Protease

Enzyme purification can be carried out by either a single-step to multistep procedure. While it depends on different characteristics of enzymes, strategies for enzyme purification depend on the level of purity required, enzyme source, size, charge, solubility, binding property with ligands, and application of targeted enzymes. Excess NaCl in media is a challenging and difficult task for purification, while the presence of salt is important to maintain the structural integrity of the protein during purification. Further, the presence of salt is required to overcome the repulsive force created by the presence of amino acids, which are acidic in nature (Sivakumar et al. 2006). Literature documented for the purification of non-halophilic enzymes cannot be generally implemented for halophilic proteins. For instance, ammonium sulfate precipitation results in low yield with considerable loss of protease activity, and protocols using traditional gel-based electrophoresis and chromatographic techniques will also hamper the yield and activity of the enzyme (Paši et al. 2008). To overcome this, optimal NaCl should be used in buffers to avoid loss of enzyme activity for the success of purification of halophilic proteins. Most proteases are extracellular and are usually separated from cells by filtration or centrifugation, followed by purification using different methods (Hou et al. 2019; Tanasupawat et al. 2011). For instance, commonly used methods for protease purification are listed in Table 4.3.

4.5 Biochemical Characterization of Protease

Industries usually offer enzymes with unique biochemical properties. Specifically, halophilic proteases are serine protease, metallo-protease, serine-metallo-protease, aspartic protease, threonine protease, and cysteine protease. Enzyme characterization assigns unique properties necessary to create an environment for different applications without any interference or inhibition of the enzyme reaction. The

Table 4.2 Optimal parameters reported for some halophilic proteases from halophilic bacteria

Halophilic strains	Salt	Optimum NaCl	Carbon	Nitrogen	pH	Temperature (°C)	Inoculum	References
Halobacillus sp. HAL1	Mg_2SO_4. K_2HPO_4, NaCl, Na_2CO_3	9%	Glucose	Peptone Yeast extract Skim milk	9	35	**NS**	Fahmy and EL-Deeb (2023)
Bacillus cereus TS2	$MgSO_4$, Na_2CO_3, KH_2 PO_4	10%	Glucose	Casein Yeast extract Peptone	7	37	0.1	Saeed et al. (2023)
Bacillus licheniformis KB111	NaCl	0.5 M	Glucose	Skimmed milk agar	NS	37	**NS**	Foophow et al. (2022)
Bacillus subtilis sp.	K_2HPO_4, KH_2PO_4 $CaCl_2$, MgSO4.7H2O	NS	Lactose	Soybean powder	9.5	40	1	Mahmoud et al. (2021)
Bacillus licheniformis sp.	NaCl	0.5%	Glucose	Polypeptone	7.2–7.4	**NS**	**NS**	Ji et al. (2020)
Gracilibacillus boraciitolerans Strain LO15	KH_2PO_4, K_2HPO_4, $CaCl_2$, $MgSO_4$, NaCl, $ZnCl_2$, g/L $FeSO_4$ H_3BO_3, and $MoNa_2O_4$	3.5%	**NS**	Casein Yeast extract Tryptone	7.4	28	NS	Ouelhadj et al. (2020)
Virgibacillus halodenitrificans ST-1	NaCl	10%	**NS**	Beef powder Peptone Casein	8.0.	25	NS	Liu et al. (2020)

(continued)

Table 4.2 (continued)

Halophilic strains	Salt	Optimum NaCl	Carbon	Nitrogen	pH	Temperature (°C)	Inoculum	References
Barrientosiimonas sp. V9	K_2HPO_4, $MgSO_4$. $FeSO_4$, KCl, KNO_3	NS	Glucose Sucrose Glycerol Fructose Maltose Lactose Galactose Cellobiose Starch Cane molasses	Gelatin, peptone, casein, skim milk, yeast extract, NH_4Cl, $(NH_4)_2SO_4$, casein hydrolysate, beef extract, urea	7.0	37	2	Flores-Santos et al. (2020)
Salipaludibacillus agaradhaerens strain AK-R	K_2HPO_4 Mg_2SO_4 NaCl Na_2CO_3	NS	Glucose	Yeast extract Peptone Skim milk	10.5	35	2	Ibrahim et al. (2019)
Bacillus luteus H11	NaCl $MgSO_4$ Na_2CO_3 K_2HPO_4	5–6%	Fructose Glucose	Shrimp shell powder and)	9	30	**NS**	Kalwasińska et al. (2018)
Salicola marasensis	NaCl	20%	Glucose	Yeast extract KNO_3 Peptone	7	40	2	Khoshnevis et al. (2018)
Salinicoccus sp. UN-12	NaCl, sodium citrate; $CaCl_2$	6%	Glucose	Soybean meal	p7.5	30	6.	Mokashe et al. (2017)
Bacillus agaradhaerens AK-R	K_2HPO_4 (1 Mg_2SO_4 NaCl Na_2CO_3	2.5–5%	Glucose	Yeast extract (5 g) Peptone	11	35	5	Ibrahim et al. (2016)

Bacillus sp. APCMST-RS3	NaCl, $MgSO_4$ KH_2PO_4	8%	Shrimp & Oyster shell powder	None	NS	50	NS	Maruthiah et al. (2015)
Virgibacillus halodenitrificans RSK CAS1	NaCl, $MgSO_4$	5.5%	Shrimp shell powder	Casein	7	40	1	Sathishkumar et al. (2015a, b)
Virgibacillus sp. EMB13	NaCl, KCl, $CaCl_2$, $MgCl_2$	0.5%	Mannitol	Yeast extract Peptone	8	30	4	Sinha and Khare (2012)
Salinivibrio sp. strain MS-7	NaCl, $MgSO_4$, $CaCl_2$, $MgCl_2$	3.1%	Maltose	Peptone Yeast extract	8	30	NS	Shahbazi and Karbalaei-Heidari (2012)
Bacillus halodurans CAS6	NaCl, K_2HPO_4, $MgSO_4$,	30%	Shrimp shell powder	NS	9	50	1	Annamalai et al. (2013)
Micrococcus sp. VKMM 037	NaCl	0.5%	Glycerol	Peptone	10	40	NS	Manikandan et al. (2011)
Chromohalobacter sp. TVSP101	NaCl, $MgCl_2$, KCl	23%	Trisodium citrate	Peptone Skim milk	7.2	40	NS	Vidyasagar et al. (2009)

NS not specified

Table 4.3 Purification of proteases from some halophilic bacteria

Microbial source	Purification steps	Mw (kDa)	References
Bacillus licheniformis KB111	$(NH_4)_2SO_4$ precipitation	70	Foophow et al. (2022)
	Phenyl Sepharose 6 fast flow column		
Alkalihalobacillus okhensis Kh10–101 T	Ethanol precipitation	27.1	Falkenberg et al. (2022)
	Desalting column (HiPrep 26/10) and S-Sepharose FF column		
Bacillus subtilis sp.	Cell-free supernatant	48	Mahmoud et al. (2021)
	$(NH_4)_2SO_4$ precipitation		
	DEAE-Sepharose (CL-6B column)		
	Sephadex G-100 (FF column)		
Bacillus licheniformis sp.	Cell medium filtrate	48.8	Ji et al. (2020)
	$(NH_4)_2SO_4$ precipitation		
Virgibacillus natechei sp.	Cell-free supernatants	31	Mechri et al. (2019a, b)
	$(NH_4)_2SO_4$ precipitation		
	Dialysis		
	HPLC system (ZORBAX PSM 300 HPSEC column)		
Citricoccus sp.	Crude enzyme	NS	Verma and Pandey (2019)
	$(NH_4)_2SO_4$ precipitation		
	DEAE cellulose		
Alkalibacillus sp. NM-Fa4	Cell-free supernatant	19.7	Mesbahi-Nowrouzi and Mollania (2018)
	Ethanol precipitation		
	Q-Sepharose		
Bacillus sp. APCMST-RS7	Culture filtrate	32	Maruthiah et al. (2017)
	$(NH_4)_2SO_4$ precipitation		
	DEAE Sepharose		
Bacillus iranensis (X5B)	Cell-free supernatant	45–50	Ghafoori et al. (2016)
	Acetone concentration		
	Ultrafiltration		
	Carboxymethyl cellulose		
Virgibacillus halodenitrificans RSK CAS1	Cell-free supernatant	21	Sathishkumar et al. (2015a, b)
	$(NH_4)_2SO_4$		
	DEAE cellulose		
	Sephadex G-75		
Bacillus sp. APCMST-RS3	Culture filtrate	40	Maruthiah et al. (2015)
	$(NH_4)_2SO_4$ precipitation		
	Sephadex G-75		
Bacillus firmus CAS 7	Cell-free supernatant	21	Annamalai et al. (2014a)
	$(NH_4)_2SO_4$ precipitation		
	DEAE cellulose		
	Sephadex G-50		

(continued)

Table 4.3 (continued)

Microbial source	Purification steps	Mw (kDa)	References
Bacillus halodurans CAS6	Cell-free supernatant	28	Annamalai et al. (2013)
	$(NH_4)_2SO_4$ precipitation		
	DEAE cellulose		
	Sephadex G-50		
Virgibacillus sp. EMB13	Crude enzyme	49 & 54	Sinha and Khare (2012)
	Ultrafiltration		
	DEAE -cellulose		
	Sephadex-G75		
Micrococcus sp. VKMM 037	Cell-free supernatant	41	Manikandan et al. (2011)
	$(NH_4)_2SO_4$ precipitation		
	Q-Sepharose		
Halobacillus sp. LY6	Culture supernatant	69	Xin et al. (2011)
	$(NH_4)_2SO_4$ precipitation		
	Sephacryl S-100		
Chromohalobacter sp. strain TVSP101	Ultrafiltration	66	Vidyasagar et al. (2009)
	Ethanol precipitation		
	Hydrophobic interaction column chromatography		
	Sephadex G-100		
	Q-Sepharose		

NS not specified

requirement of NaCl/KCl for efficient catalytic activity and for the enhancement of the thermostability of extremely halophilic enzymes is a known fact. Salts maintain ionic strength and are required to protect the active conformation of the protein. Halophilic proteins are active from 0.5 to 6 M NaCl. Table 4.4 summarizes the requirement of salt for the proteases produced by some halophilic bacteria.*NS not specified*

Proteases active under alkaline conditions obtained from halophilic microorganisms are reported (Ali et al. 2016; Dammak et al. 2016). Setyorini et al. (2006) documented halophilic protease maximally active at pH 6.5. Generally, the optimum pH range for alkaline proteases ranged from pH 9 to pH 11. Serine proteases from haloalkaliphilic bacterium have enzyme activity in a pH range 8–12 (Raval et al. 2014). Alkaliphilic *Salipaludibacillus agaradhaerens* strains AKR (formerly *Bacillus agaradhaerens*) produced AK-R protease optimally active at pH 10 (Ibrahim et al. 2019). Reports on the thermodynamic catalytic properties of salt-stable protease are rare. *Oceanobacillus iheyensis* protease shows maximum activity at 50 °C (Purohit and Singh 2011). The presence of salt, polyethelene glycol (PEG), and Ca^{2+} is reported to enhance the stability of protease at higher temperatures (Dodia et al. 2008).

Table 4.4 Salt requirement for the proteases from halophilic microorganisms

Halophilic bacteria	NaCl range	NaCl optimum	References
Bacillus subtilis sp.	NS	0.5%	Mahmoud et al. (2021)
Bacillus licheniformis sp.	NS	10%	Ji et al. (2020)
Bacillus velezensis sp.	NS	2.0 M	Yang et al. (2020)
Marinirhabdus sp.	NS	10–30%	Anh et al. (2021)
Pseudoalteromonas sp.	NS	0.3 M	Zhao et al. (2012)
Virgibacillus natechei sp.	NS	3.1 M	Mechri et al. (2019a, b)
Virgibacillus sp. CD6	0–12.5%	10%	Lam et al. (2018)
Virgibacillus halodenitrificans RSK CAS1	0–6.9 M	4.31 M	Sathishkumar et al. (2015a, b)
Bacillus firmus CAS 7	0–6.05 M	5.2 M	Annamalai et al. (2014a)
Bacillus alveayuensis CAS 5	0–6.05 M	6.05 M	Annamalai et al. (2014b)
Bacillus halodurans CAS6	0–5.2 M	5.2 M	Annamalai et al. (2013)
Virgibacillus sp. EMB13	0–2.6 M	0.18 M	Sinha and Khare (2012)
Salimicrobium halophilum strain LY20	0–3.4 M	2.16 M	Li and Yu (2012)
Micrococcus sp. VKMM 037	0–3.42 M	0.85 M	Manikandan et al. (2011)
Halobacillus sp. LY6	0–4.31 M	1.73 M	Xin et al. (2011)
Chromohalobacter sp. strain TVSP101	1–4 M	4 M	Vidyasagar et al. (2007)

NS not specified

The activity of protease from *Bacillus pseudofirmus* increased at 80 °C when shifted from 50 °C in the presence of NaCl concentration from 0 to 3 M. Recently Mechri et al. (2019a, b) isolated thermostable and halotolerant serine protease from *Melghiribacillus thermohalophilus* Nari2A^T active at 75 °C. Table 4.5 summarizes the type, pH, and temperature optima with unique features of proteases from some representative genera belonging to halophilic bacteria.

Catalytic efficiency of halophilic proteases is known to be influenced by Ca^{2+} and Mg^{2+}. In addition, studies are available on the improvement of thermostability of halophilic extracellular proteases in the presence of Cu^{2+}, Ba^{2+}, Ca^{2+}, Mg^{2+}, Fe^{3+}, Co^{2+}, Zn^{2+} and Mn^{2+} (Nwankwo et al. 2023; Daoud et al. 2017; Maruthiah et al. 2015).

4.5.1 Substrate Specificity

Substrate specificity and kinetic parameters studies are very important for any enzyme study in way of knowing its application in industries, where enzyme properties differ from their physiological properties. Proteases from halophilic microbes are known to hydrolyze a large range of natural substrates, like elastin, gelatin, casein, bovine serum albumin, insulin, yeast extract, tryptone, peptone, hemoglobin, and wheat gluten. In addition, proteases also hydrolyze modified

Table 4.5 Properties of halophilic proteases from some halophilic bacteria

Halophilic bacteria	Optimal pH	Optimal temperature (°C)	Protease type	Unique activity	References
Bacillus subtilis sp.	9.5	50	Alkaline serine protease	Alkali-halo-thermostable property, promising binding affinities toward the casein	Mahmoud et al. (2021)
Bacillus licheniformis sp.	NS	NS	Serine proteases	NS	Ji et al. (2020)
Bacillus velezensis sp.	8.0	40	Alkaline serine protease	NS	Yang et al. (2020)
Marinirrhabdus sp. and *Marinobacter hydrocarbonoclasticus*	4.0–12.0	25–60	NS	Industrial wastewater treatment: aquaculture effluents	Anh et al. (2021)
Pseudoalteromonas sp.	9.5	25	Metallo-protease	Important role in elastin degradation	Zhao et al. (2012)
Virgibacillus natechei sp.	3.1	60	NS	Moderate tolerance to detergents	Mechri et al. (2019a, b)
Pseudoalteromonas sp. 129–1	8.0	50	Alkaline protease	Activity in the presence of detergents	Wu et al. (2015)
Bacillus sp. NPST-AK$_{15}$	10.5	60	Alkaline serine protease	Application in pharmaceutical industries and laundry	Ibrahim et al. (2015)
Salinicoccus sp. UN-12	8.5	55	Alkaline-protease	Activity in detergents	Mokashe et al. (2017)
Bacillus luteus H11	10.5	45	Halo- alkaline protease	Application in industries: detergent, food, pharmaceutical and in bioremediation.	Kalwasinska et al. (2018)
Vibrio sp. LA-05	6.0–10.0	25–40	Metallo-protease	Used in detergent formulations	Zhang et al. (2018)
Virgibacillus sp. CD6	NS	NS	Protease	Used in detergent formulations	Lam et al. (2018)
Bacillus halodurans strain US193	10.0	70	Halo-alkaline protease	Used in detergent formulations	Daoud et al. (2018)

(continued)

Table 4.5 (continued)

Halophilic bacteria	Optimal pH	Optimal temperature (°C)	Protease type	Unique activity	References
Salipaludibacillus agaradhaerens strain AK-R	10.0	60	Alkaline serine protease	Application in detergent industry	Ibrahim et al. (2019)
Barrientosiimonas sp. V9	9.0	60	Extremophilic proteases	Stability in the presence of chemical additives: Triton X-100, urea, EDTA, and laundry detergents	Flores-Santos et al. (2020)
Virgibacillus halodenitrificans ST-1	8.0	30	Protease	Antioxidant activity and shrimp paste fermentation	Liu et al. (2020)
Gracilibacillus boraciitolerans strain LO15	10.0	65	Alkaline serine protease	Detergent industry	Ouelhadj et al. (2020)
Bacillus licheniformis KB111	7.0	50	Halo- alkaline protease	NS	Foophow et al. (2022)
Alkalihalobacillus okhensis Kh10-101T	9.0–9.5	55	Alkaline-subtilisin	Biotechnology industry	Falkenberg et al. (2022)
Haloalkaliphilic bacterium D-15-9	10.5–10.0	50	Alkaline proteases	Industries: detergent, leather, pharmaceutical, food, peptide synthesis, waste treatment	Raval et al. (2022)
Planococcus sp.	10.0	35	NS	Tolerance to detergents	Zhang et al. (2016)
Stenotrophomonas maltophilia sp.	8.6	45	NS	Tolerant to detergents	Ribitsch et al. (2012)
Salicola sp. strain SBJ9	8.0.	40	Serine protease	Stability in organic solvents and detergents	Daoud et al. (2019)
Alkalibacillus sp. NM-Fa4	9.5	45	Cysteine protease	Stability toward reducing agents and surfactant (Tween 80)	Mesbahi-Nowrouzi and Mollania (2018)
Halobacillus sp. CJ4	9.0	45	Serine protease	Stability in solvents and detergents	Daoud et al. (2017)
Bacillus sp. APCMST-RS7	8	50	Serine protease	Deproteinization of shrimp shell waste	Maruthiah et al. (2017)

Bacillus iranensis (X5B)	9.5	35	Serine protease	Activated by Ca^{2+}	Ghafoori et al. (2016)
Virgibacillus halodenitrificans RSK CAS1	9	50	Metallo-protease	Stability in salts, solvents, metal ions, and detergents	Sathishkumar et al. (2015a, b)
Bacillus sp. APCMST-RS3	9	60	Serine protease	Stability in the presence of detergents and enzyme showed antioxidant potential	Maruthiah et al. (2015)
Actinopolyspora sp. VITSDK2	10	60	Serine protease	Retains 96% activity at RT after 10 days of incubation	Suthindhiran et al. (2014)
Bacillus firmus CAS 7	11	70	Metallo-protease	Solvent and surfactant stability	Annamalai et al. (2014a)
Bacillus alveayuensis CAS 5	9	50	Metallo-protease	Surfactant and detergent stability	Annamalai et al. (2014b)
Bacillus halodurans CAS6	10	70	Metallo-protease	Application as additive in detergent formulation and antioxidant synthesis	Annamalai et al. (2013)
Virgibacillus sp. EMB13	7.5	50	Metallo-protease	Solvent stability	Sinha and Khare (2012)
Salimicrobium halophilum strain LY20	10	80	Serine protease	Activity in solvents and surfactants	Li and Yu (2012)
Micrococcus sp. VKMM 037	10	40	Serine protease	Stability in the presence of reducing agents, organic solvents, and surfactants	Manikandan et al. (2011)
Oceanobacillus iheyensis O.M.A18	11	50	NS	NS	Purohit and Singh (2011)
Halobacillus sp. LY6	10	40	Metallo-protease	Metal ions stability	Xin et al. (2011)
Chromohalobacter sp. strain TVSP101	8	75	Serine protease	Stability and activity in the presence of solvents (DMSO, DMF, ethanol, and acetone)	Vidyasagar et al. (2007)

NS not specified, *DMSO* dimethyl sulfoxide, *DMF* dimethyl fluoride, *RT* room temperature

Table 4.6 Substrate specificity and kinetic parameters evaluation for some halophilic proteases

Halophilic bacteria	Substrate specificity	Km and Vmax	References
Virgibacillus halodenitrificans ST-1	Maximum for casein, least for azo-casein	NS	Liu et al. (2020)
Vibrio sp. LA-05	Maximum for elastin-orcein, least for collagen fiber	NS	Zhang et al. (2018)
Salinicoccus sp. UN-12	Maximum for casein, least for bovine serum albumin	NS	Mokashe et al. (2017)
Bacillus sp. APCMST-RS7	Maximum for casein, least for bovine serum albumin	Km is 0.0002 g/l and Vmax is 1428.57 U/ml	Maruthiah et al. (2017)
Bacillus sp. APCMST-RS3	Maximum for casein, least for bovine serum albumin	Km is 0.66 g/l and Vmax is 1111.11 U/ml	Maruthiah et al. (2015)
Virgibacillus halodenitrificans RSK CAS1	Maximum for casein, least for bovine serum albumin	NS	Sathishkumar et al. 2015a, b)
Alkalibacillus sp. NM-Fa4	Maximum for casein, least for gelatin	Km is 1.3 mg/ml, and Vmax is 1111 mg/min/ml	Mesbah and Wiegel (2014)
Actinopolyspora sp. VITSDK2	Maximum for casein, least for bovine serum albumin	NS	Suthindhiran et al. (2014)
Bacillus firmus CAS 7	Maximum for casein, least for bovine serum albumin	NS	Annamalai et al. (2014a)
Bacillus alveayuensis CAS 5	Maximum for casein, least for bovine serum albumin	NS	Annamalai et al. (2014b)
Bacillus halodurans CAS6	Maximum for casein, least for bovine serum albumin	NS	Annamalai et al. (2013)
Virgibacillus sp. EMB13	Maximum for casein, least for gelatin	Km is 7.5 mg/ml and Vmax is 156.25 μ/min/ml	Sinha and Khare (2012)

NS not specified

substrates, like azocasein, casamino acids, and skim milk. Specificity has also been checked with N-succinyl-LTyr-L-Leu-L-Val-P-nitroanilide or N-succinyl-L-Ala-L-Ala-L-Pro-L-Phe-P-nitroanilide. Kinetic parameters, such as Km, Vmax, and Kcat, reveal information important for scale-up of the biotechnological process. Km and Vmax studies on purified enzymes are determined at various concentrations of substrates and the calculation of kinetic parameters from the Lineweaver-Burk plot. Kinetic properties of proteases from halophilic microorganisms are elucidated using various synthetic chromogenic and natural protein substrates (Sarwa et al. 2024). The substrate specificity and kinetic properties (Km and Vmax) reported from halophilic microbial strains are summarized in Table 4.6.

4.5.2 Effect of Chemical Additives and Detergents on Halophilic Proteases

The detergent industry holds one of the top positions in using protease to remove the protein stains of coffee, grass, blood, sauces, grease, milk, keratin, and spinach (Sinha and Khare 2013). The enzyme-detergent formulations basically consist of enzymes (lipase, protease, cellulase, and amylase), surfactants (ionic and nonionic), bleaching agents, and oxidizing agents, along with other chemical ingredients. Table 4.7 summarizes the stability of halophilic proteases in several detergent formulations. Further, the use of halophilic proteases in detergent formulations depends on the stability of protease under alkaline conditions, stability at higher temperatures (30–60 °C), tolerance to total dissolved solids and salts, and efficiency to remove stains (Mokashe et al. 2015; Patil et al. 2016).

Proteases from halophilic microbes are reported for their use in detergent formulation (Eichler 2001; Oren 2010). Commercial detergents are used in formulations to determine the compatibility of proteases derived from halophilic microbes. Halophilic proteases from *Bacillus* sp. APCMST-RS3, *Bacillus alveayuensis* CAS 5, *Bacillus halodurans* CAS6, *Virgibacillus* sp. EMB13, and *Virgibacillus halodenitrificans* RSK CAS1 are compatible in different detergents (Mokashe et al. 2018). For instance, *Bacillus* sp. APCMST-RS7 halophilic protease remained active in the formulations containing Henko, Technobright, Surf Excel, Rin, Ariel, and Tide, attaining more than 100% residual activity (Maruthiah et al. 2015).

4.5.3 Protease Activity in the Presence of Solvents

Generally, halophilic enzymes are efficiently used as biocatalysts for various chemical reactions containing organic solvents (Gupta and Khare 2009). However, enzymes from non-halophiles face denaturation and loss in catalytic activity (Torres and Castro 2004). Salt-stable proteases remain active in nonaqueous conditions that play a key role in (i) improvement in the catalytic properties, (ii) catalysis of reversible and irreversible reactions involved in peptide synthesis, (iii) alteration of the regio- and enantio substrate selectivity, (iv) improvement in solubility of hydrophobic substrates, and (v) reduction of microbial contamination (Karan et al. 2011). In the last two decades, limited reports have been available on solvent-stable halophilic proteases. For instance, proteases from *Bacillus iranensis* strain X5B (Ghafoori et al. 2016), *Micrococcus* sp. VKMM 037 (Manikandan et al. 2011), *Virgibacillus* sp. EMB13RSK CAS1 (Sinha and Khare 2012), *Halobacillus* sp. CJ4 (Daoud et al. 2017), *Alkalibacillus* sp. NM-Fa4 (Mesbah and Wiegel 2014), and *Salimicrobium halophilum* strain LY20 (Li and Yu 2012) showed stability in the presence of solvents. Table 4.8 summarizes the effect of hydrophilic and hydrophobic organic solvents on halophilic proteases from halophilic microbes.

Table 4.7 Stability of halophilic proteases in several detergent formulations

Halophilic microbes	Ingredients in detergent formulations		Detergents	References
	Ionic surfactants	Nonionic surfactants		
Bacillus sp. APCMST-RS7	5 mM SDS – 28.75%	Tween 20 (5 mM) – 95.04% Tween 40 (5 mM) – 114.52% Tween 60 (5 mM) – 95.61% Tween 80 (5 mM) – 90.69% Triton X-100 (5 mM) –29.78%	Technobright – 117.05% Surf Excel – 126.24% Henko – 119% Ariel – 38.03% Rin – 111.13% Tide – 31.48%	Maruthiah et al. (2017)
Bacillus sp. APCMST-RS3	5 mM SDS – 121%	Tween 20 (5 mM) – 135.65% Tween 40 (5 mM) – 129.67% Tween 60 (5 mM) – 125.21% Tween 80 (5 mM) – 119.65% Triton X-100 (5 mM) – 63.38%	Technobright – 94.87% Surf Excel – 111.11% Henko – 8.65% Ariel – 29.66% Rin – 99.65% Tide – 24.58%,	Maruthiah et al. (2015)
Virgibacillus Halodenitrificans RSK CAS1	1% SDS – 59.21% 10% SDS – 51.10%	1% triton X-100 – 65.17% 1% Tween 80 – 70.41%	Rin – 74.20% Tide – 67% Henko – 71.12% Ariel – 76.41%	Sathishkumar et al. (2015a, b)
Bacillus firmus CAS 7	1% SDS – 78.33% 10% SDS – 70.33%	1% Tween 80 – 88% 10%Tween 80 – 72.33% 1% Triton X-100 – 75.67% 10% Triton X-100 – 70.33%	NS	Annamalai et al. (2014a)
Bacillus alveayuensis CAS 5	1% SDS – 66%	1% Tween 80 – 90.33% 1% Triton X-100 – 78.33% 1% sodium deoxycholate – 69.33%	Ariel – 2.48% Rin – 80.33% Tide – 79.38% Henko – 0.26%	Annamalai et al. (2014b)
Bacillus halodurans CAS6	1% SDS – 82.34%	1% Tween 80 – 94.33%, 1% Triton X-100 – 72.67% 1% sodium deoxycholate – 85.67%	Ariel – 79.00% Tide – 76.67% Henko – 85.67% Rin – 88.34%	Annamalai et al. (2013)

(continued)

Table 4.7 (continued)

Halophilic microbes	Ingredients in detergent formulations		Detergents	References
	Ionic surfactants	Nonionic surfactants		
Chromohalobacter sp. TVSP101	0.1% SDS – 00%	NS	NS	Vidyasagar et al. (2009)

NS not specified, *SDS* sodium dodecyl sulfate

Table 4.8 Halophilic proteases from halophilic microbes and their organic solvent tolerance properties

Halophilic microbes	Solvents	References
Halobacillus sp. HAL1	Methanol, ethanol, acetone, hexane, benzene, propanol, butanol, xylene	Fahmy and El-Deeb (2023)
Bacillus cereus TS2	Methanol, ethanol, ethyl acetate, acetone, hexane	Saeed et al. (2023)
Gracilibacillus boraciitolerans strain LO15	Chloroform, dimethyl sulfoxide, n-hexane, toluene, isopropanol	Ouelhadj et al. (2020)
Virgibacillus halodenitrificans ST-1	Ethanol, ethyl acetate, acetonitrile, methanol, acetone, dimethyl sulfoxide, diethyl ether, PMSF, DTT, Triton X-100, Tween 80, SDS, EDTA, EGTA	Liu et al. (2020)
Salipaludibacillus agaradhaerens strain AK-R	Methanol, dimethyl ether, butanol, toluene, amyl alcohol, diethyl ether	Ibrahim et al. (2019)
Bacillus luteus H11	Butanol, methanol, ethanol acetone	Kalwasinska et al. (2018)
Salinicoccus sp. UN-12	Toluene, hexane, methanol, iso-octane, acetonitrile, dimethyl sulfoxide, ethanol, glycerol.	Mokashe et al. (2017)
Bacillus sp. APCMST-RS7	Acetone, chloroform, benzene, methanol, ethanol, dimethyl sulfoxide, hexane	Maruthiah et al. (2017)
Halobacillus sp. CJ4	Acetonitrile, methanol, cyclohexane, dimethyl sulfoxide, isopropanol, benzene, toluene, ether, chloroform	Daoud et al. (2017)
Bacillus iranensis strain X5B	Ethanol, hexane, methanol, butanol, chloroform	Ghafoori et al. (2016)
Bacillus sp. APCMST-RS3	Ethanol, acetone, methanol, 2-propanol, butanol, hexane, chloroform, benzene, xylene, petroleum ether	Maruthiah et al. (2015)
Virgibacillus halodenitrificans RSK CAS1	Ethanol, ethyl ether, acetonitrile, methanol, isopropanol, benzene, hexane	Sathishkumar et al. (2015a, b)
Alkalibacillus sp. NM-Fa4	Ethanol, methanol, isopropanol, butanol, hexane, benzene, chloroform	Mesbah and Wiegel (2014)
Bacillus firmus CAS 7	Methanol, ethanol, acetonitrile, isopropanol, benzene, ethyl ether hexane, benzene	Annamalai et al. (2014a)

(continued)

Table 4.8 (continued)

Halophilic microbes	Solvents	References
Bacillus alveayuensis CAS 5	Ethanol, methanol, acetonitrile, isopropanol, hexane, benzene, ethyl ether	Annamalai et al. (2014b)
Bacillus halodurans CAS6	Isopropanol, hexane, xylene and benzene	Annamalai et al. (2013)
Virgibacillus sp. EMB13RSK CAS1	Acetonitrile, methanol, propanol, ethanol, butanol, benzyl alcohol, cyclohexane, heptane, octane, toluene, decane, dodecane	Sinha and Khare (2012)
Salimicrobium halophilum strain LY20	Acetone, acetonitrile, ethanol, methanol, dimethyl sulfoxide, dodecane, isooctane, dimethyl fluoride, heptane, benzene	Li and Yu (2012)
Micrococcus sp. VKMM 037	Butanol, ethylene glycol, butanol, ethanol, acetone, dimethyl sulfoxide	Manikandan et al. (2011)
Chromohalobacter sp. TVSP101	Dimethyl sulfoxide, ethanol, DMF, acetone	Vidyasagar et al. (2009)

4.6 Domain Organization of Halophilic Proteases

Domain organization of halophilic protease is a significant factor affecting protein function. Such protein domains are essential for the mechanism of action shown by different proteins. Domains are self-stabilizing regions of a polypeptide chain, which are able to fold independently to produce proteins with diverse functions (Du et al. 2015).

Halophilic proteases are categorized into three domains consisting of N-terminal, catalytic, and C-terminal domains (Zhao et al. 2019). The first is involved in the regulation of pathways of the secretory process and is known as the PepSY domain or fungalysintThermolysin propeptide (FTP) domain (Ban et al. 2020). The reported N-terminal domains in halophilic enzymes include FTP/PEPSY, M4 peptidases, and PepSY_TM-associated peptidase with unknown functions (Yang et al. 2020). The second is responsible for the proteolytic activity, while the C-terminal extension has multifaceted functions, which vary across different organisms. For instance, Mei et al. (2022) documented that C-terminal domains are involved in maintaining the structural flexibility and stability of the enzyme, facilitating processes like substrate binding and translocation. Likewise, C-terminal domains can also contribute to other functions in halophilic proteases. This includes the chitin-binding domain (ChBD), as mentioned by Zhang et al. (2014); the collagen-binding domain (CBD) and C-ter 100, as per Huang et al. (2017); the LPXTG anchor according to Plavec et al. (2019); the polycystic kidney disease domain (PKdD) and P-protein (P), as described by Jha et al. (2017); and prepeptidase C-terminal domain (PPC), according to Huang et al. (2017).

4.7 Discovery of Halophilic Protease: Recent Advancements, Challenges, and Opportunities

Genetic engineering is of utmost importance in the enzyme industry. Genetic modifications in the enzyme have now become easy with the use and advancement of recombinant technology. This results in the enhancement of efficiency and yield of the products by utilizing a few materials. In addition, recombinant technology has revolutionized knowledge on the structural-functional relationship of the genetic systems. Genetic modifications allow experts to isolate and manipulate control genes involved in enzyme overproduction, provide a platform for locating active site residues while studying protein crystallographic structure, and thus are able to alter enzyme properties. The use of advanced culture-independent methods has resulted in an entirely new scenario of protease production by halophiles. The problem of culturing challenges is overcome by heterologous expression overcomes by transferring the desired halophilic protease gene into an easily cultivable host organism, such as *Escherichia coli* and *Bacillus subtilis*, which can be grown quickly and cost-effectively and enables production in high yields (Hou et al. 2021, 2022). A report of cloning and sequencing of the protease gene is from the halophile *Geomicrobium* sp. EMB2 to achieve better expression is available (Karan et al. 2011). Mechri et al. (2019a, b) studied heterologous overexpression and purification of the sapV gene encoding protease from *Virgibacillus*. Li and colleagues successfully expressed a thermophilic serine protease (P3862) from *Ornithinibacillus caprae* L9T in *Bacillus subtilis* SCK6 (Li et al. 2022).

Also, studies are highlighted in the combined use of genetic engineering along with multi-omics techniques that identify the coding and regulatory genes, thus revealing the mechanism for salt tolerance protease. The use of this strategy has helped in increasing the isolation of new enzymes with special functions. Recombination of genes and their subsequent expression in vectors, such as *E. coli* and others, has resulted in maximum enzyme production with enhanced activity. For instance, the Krill metagenomic library is screened for coding genes of peptidase S8A subfamily and successfully expressed in *E. coli*. This has resulted in the production of a salt-tolerant protease with more than 40% residual activity that is retained in the reaction containing 3 M NaCl (Purohit and Singh 2013; Sun et al. 2020).

Hou et al. (2019) expressed a serine protease gene from halophilic *Bacillus* sp. LCB10 in *E. coli*. This has led to the production of salt-tolerant protease maximally active at 7% NaCl, retaining 86% residual activity. Likewise, Sun et al. (2020) obtained a thermostable serine protease from marine sediments of the East China Sea, showing remarkably strong salt tolerance, retaining 40% residual activity in a reaction containing 3 M NaCl.

Other methods, such as affinity chromatography, are used to purify recombinant proteases, which are often produced with an affinity tag, such as a polyhistidine tag. This method utilizes the specific binding between the tag and a complementary ligand (e.g., Ni-NTA for polyhistidine tags) immobilized on a matrix, allowing for the selective capture and elution of the tagged protein. Such an approach yields a

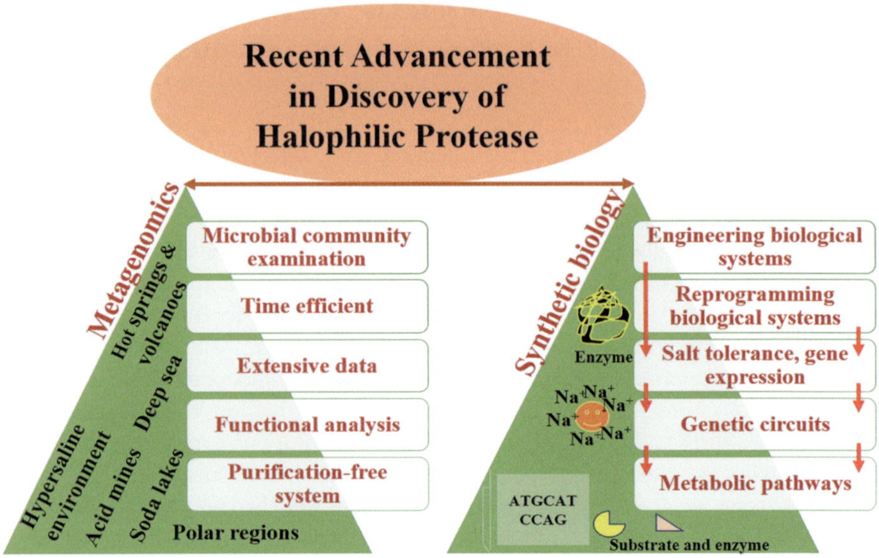

Fig. 4.2 Recent advancement in the discovery of halophilic enzymes

highly pure recombinant protease, overcoming the complex and often multistep purification procedures associated with extracting proteases from native halophilic cultures (Han et al. 2023).

Recently, with the advancement of next-generation sequencing and its use to study biotechnology and microbial ecology, various metagenomics approaches have been developed for the discovery of enzymes produced by genetically diverse microbes. These approaches enable easy identification of enzymes possessing unique functions and their potential biotechnological applications. Likewise, bioinformatics and computational tools help in enzyme-encoding genes, enzyme prediction, and their functions and identification. This allows the researcher to gain insight into enzyme properties, which allows them to retain their stability under different conditions. Recently, the synthetic biology approach has been used in the research of halophilic enzymes. This method involves engineering and reprogramming biological systems that enable the researcher to study different cell mechanisms, the pathways involved, and product synthesis (Yavari-Bafghi and Amoozegar 2025). Both metagenomics and synthetic biology offer various advantages as compared to the traditional approach to discovering halophilic enzymes (Fig. 4.2).

The use of culture-dependent methods of halophilic enzyme production often poses challenges in the purification of extracellular proteases. This includes slow growth, which significantly limits the amount of biomass and, consequently, the protease yield (Promchai et al. 2018). The high salt requirements for growth and subsequent purification steps, which necessitate high concentrations, are another challenge. The high salt concentrations complicate the salting-out process, a

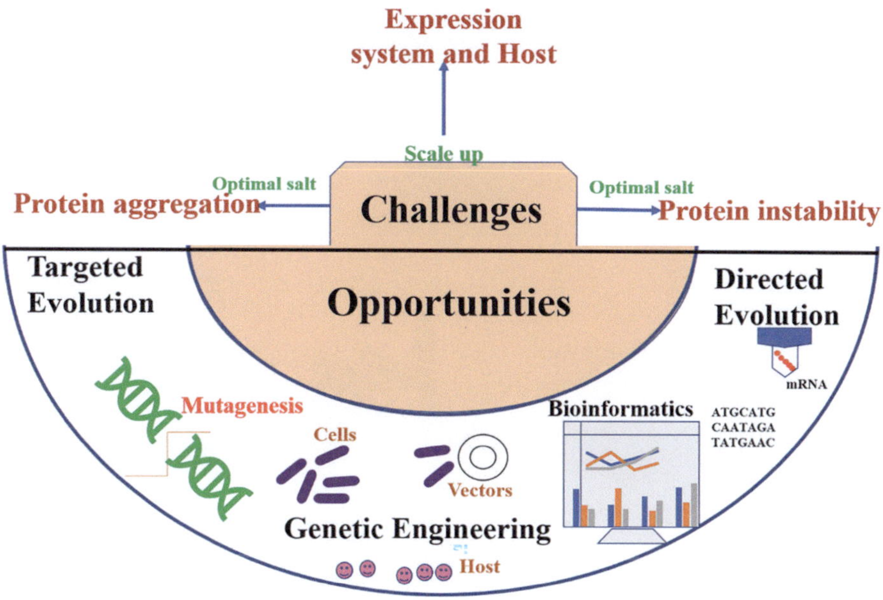

Fig. 4.3 Challenges and research opportunities concerning the discovery of halophilic proteases

common step in protein purification. In other cases, extracellular halophilic proteases can exhibit generic activity, leading to self-degradation. Further, multiple chromatographic steps involved in purification result in further reduction in the final yield of enzyme (Promchai et al. 2018). Considering this, both challenges and research opportunities go hand in hand, which allows further investigation and the study of unique properties of halophilic proteases (Fig. 4.3).

4.8 Biotechnological Applications of Halophilic Protease

The use of chemicals in industrial applications has increased tremendously, giving rise to health issues and other physiological implications. Hence, a search for environmentally friendly products is always a priority for scientific research. Microbial proteases possess unique characteristics applicable for desirable industrial applications (Sathishkumar et al. 2015a, b), such as in the food or drugs industry or in textiles and detergents (Singhal et al. 2012). Genetic manipulation of bacterial genes enhanced the productivity of the enzymes. The yield is also not hindered by climate change (Solanki et al. 2021). Within the protease group of enzymes, the salt-stable proteases are more promising for industrial applications due to their higher tolerance to extreme conditions (salt, temperature, low water activity, etc.) than proteases from non-halophilic microbes, as highlighted in Fig. 4.4. Genetic manipulation has changed the way researchers approach bacterial productivity, particularly

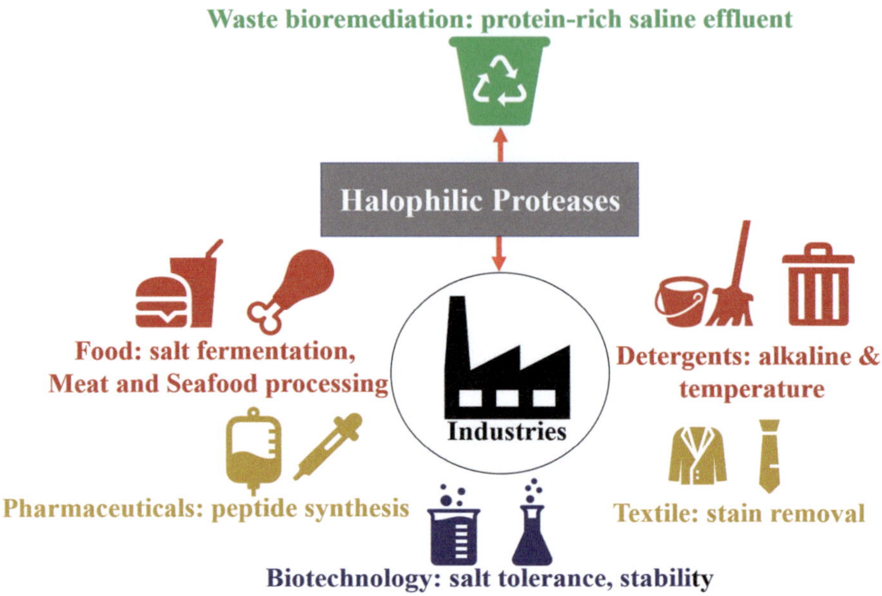

Fig. 4.4 Biotechnological applications of halophilic proteases

by allowing precise modification of genes, resulting in optimizing metabolic pathways. As noted by Solanki et al. (2021), microorganisms are ideal for enzyme production due to ease and rapid growth, minimal requirement of space, and high adaptability to gene editing. A significant advantage of using these microbial systems is that their yield is relatively shielded even under extreme conditions, compared to mesophilic proteases as highlighted in Fig. 4.4.

4.8.1 Food Processing Industry

Salt-tolerant proteases find application in high NaCl-containing food products, such as sausages, curd, soy sauce, and shrimp paste (Gao et al. 2019, 2018; Naveed et al. 2021). These enzymes hydrolyze proteins in the food to produce peptides and free amino acids, enhancing the overall quality of fermented foods (Zhou et al. 2021). In soy sauce brewing, immobilized enzymes boost glutaminase levels by nearly 46%, enhancing product flavor (Zhang et al. 2022). The work of Heng et al. (2022) emphasizes that using a "consortium" of microbes rather than a single strain creates a more robust process. When multiple proteases work together on substrates like bean dregs and soybean meals, the results include improved strain growth, higher organic acid production, and waste valorization. In Korean *doenjang*, specific salt-tolerant proteases of *Lactobacillus brevis* GABA 100 and *A. oryzae* KACC 40250 are combined, which has overall shortened the production cycles of the produced food product (Li et al. 2017) Marine-derived aminopeptidases (like those from *B. licheniformis*) are particularly effective at debittering soy proteins, making the final product more palatable (Lei et al. 2017). He et al. (2022) reported the complexity of Sufu (fermented tofu) in terms of flavor due to the metabolic activity of different salt-tolerant enzymes. These have influenced the unique flavor and nutritional value of Sufu. According to Tien et al. (2022), proteases can replace the addition of external nitrogen sources. By breaking down the internal protein structure of grains (like rice), the enzymes release enough nitrogen to increase ethanol yield, decrease total fermentation time, and eliminate the need for costly chemical nitrogen supplements.

4.8.2 Detergent Formulations

Applications of halophilic and alkaline proteases in the detergent industry represent one of the most successful applications because of their ability, stability, and compatibility in the presence of harsh conditions of modern detergent industry (Mokashe et al. 2015). In powder formulation, protease remains highly stable in granulated form, ensuring they don't lose potency before reaching the wash cycle (Saxena 2015). Jaouadi et al. (2008) stated that the role of protease in detergents is due to its enhanced catalytic activity in the presence of high salt concentration. As documented by Ibrahim et al. (2015), enzymes from *Bacillus* sp. *NPST-AK15* can retain up to 100% activity even when mixed with aggressive commercial brands like Persil, Tide, and Ariel at 40 °C. Also, the proteases of *Bacillus* sp. EMB9 (Sinha and Khare 2013), *Geomicrobium* sp. EMB2 (Karan et al. 2011), and *Virgibacillus* sp. EMB13 (Singhal et al. 2012) are confirmed as detergent compatible enzymes. Proteases from *Bacillus circulans* MTCC 7942 (Patil et al. 2016) and *Salinicoccus* sp. UN-12 (Mokashe et al. 2017) are specifically adapted to function in high-ion environments present in hard water. Mokashe et al. (2017) demonstrated the effectiveness of salt-tolerant protease across a wide range of common household stains. Enzymes hydrolyze hemoglobin and albumin present in blood and egg.

4.8.3 Leather Industry

Proteases from halophilic/halotolerant microbes are capable of replacing toxic and polluting chemicals. *Bacillus alkalitelluris* TWI3 (Anandharaj et al. 2016) produces a thermostable metallo-protease with dehairing capability targeting the hair follicle without damaging the structural collagen of the hide, while *Bacillus tequilensis* P15 produces an organic solvent-tolerant protease effective for gelatin hydrolysis and stain removal, making it a versatile "green" biocatalyst for diverse tannery operations. (Bose et al. 2014).

4.8.4 Other Roles of Protease

The use of salt-tolerant proteases extends far beyond their industrial and food application. For example, the natural antifungal defense of protease is proved by its action on fungal cell walls, ultimately killing the fungus (Essghaier et al. 2009). Park et al. (2020) demonstrated that supplementing diets with *Bacillus licheniformis,* a producer of salt-tolerant protease, enhances protein absorption and gut health in livestock. Salt-tolerant proteases have shown potential in agriculture as antifungal agents, in food production, and in biotechnology (Essghaier et al. 2009). It has been shown that inoculation of *Bacillus licheniformis* to low-protein-containing nursery diets can improve the overall digestibility of the nutrients, growth, and intestinal morphology of weaned pigs (Park et al., 2020). Proteases from *Bacillus pumilus* MTCC 7514 (Gupta 2012) and *Vibrio* sp. V26 (Manjusha et al. 2013) are proposed for their use in tissue culturing and maintaining cell lines. Microbial infections are cured using enzybiotics, which consist of enzyme-antibiotic conjugates (Sellami-Kamoun et al. 2008). Concerning the environmental perspective, salt-tolerant proteases improve the water quality by degradation of complex organic compounds in industrial wastewater (Anh et al. 2021). Proteases are employed in finishing and treatment of effluent in the textile industry (Kumar et al. 2021). Halophilic protease from *Bacillus* sp. strain SP II-4 efficiently removed protein deposits from contact lenses and tear film. These proteases are advantageous because they are less likely to cause eye irritation or allergic reactions (Rejisha and Murugan 2021).

4.9 Conclusion

In conclusion, proteases from halophilic bacteria prove to be a better alternative compared to those from mesophilic bacteria concerning their stability and activity under extreme processing conditions. These conditions were high salt concentrations, high temperature, low water activity, etc., that form the usual basis in any type of industrial settings, involving the production of biomolecules and products. In addition, the applications cited in this chapter hold unique use in different fields of biotechnology involving medicine, pharmaceutical industries, textiles, detergents, food industry, agriculture, and environmental bioremediation.

References

Ali N, Ullah N, Qasim M, Rahman H, Khan SN, Sadiq A, Adnan M (2016) Molecular characterization and growth optimization of halo-tolerant protease producing *Bacillus Subtilis* strain BLK-1.5 isolated from salt mines of Karak, Pakistan. Extremophiles 20(4):395–402. https://doi.org/10.1007/s00792-016-0830-1

Amoozegar MA, Siroosi M, Atashgahi S, Smidt H, Ventosa A (2017) Systematics of haloarchaea and biotechnological potential of their hydrolytic enzymes. Microbiol (U K) 163(5):623–645. https://doi.org/10.1099/mic.0.000463

Anandharaj M, Sivasankari B, Siddharthan N, Rani RP, Sivakumar S (2016) Production, purification, and biochemical characterization of thermostable metallo-protease from novel *Bacillus alkalitelluris* TWI3 isolated from tannery waste. Appl Biochem Biotechnol 178(8):1666–1686. https://doi.org/10.1007/s12010-015-1974-7

Anh HTH, Shahsavari E, Bott NJ, Ball AS (2021) Application of co-culture technology to enhance protease production by two halophilic bacteria, Marinirhabdus sp. and Marinobacter hydrocarbonoclasticus. Molecules 26:3141. https://doi.org/10.3390/molecules26113141

Annamalai N, Rajeswari MV, Thavasi R, Vijayalakshmi S, Balasubramanian T (2013) Optimization, purification and characterization of novel thermostable, haloalkaline, solvent stable protease from *Bacillus halodurans* CAS6 using marine shellfish wastes: a potential additive for detergent and antioxidant synthesis. Bioprocess Biosyst Eng 36(7):873–883. https://doi.org/10.1007/s00449-012-0820-3

Annamalai N, Rajeswari MV, Sahu SK, Balasubramanian T (2014a) Purification and characterization of solvent stable, alkaline protease from *Bacillus firmus* CAS 7 by microbial conversion of marine wastes and molecular mechanism underlying solvent stability. Process Biochem 49(6): 1012–1019. https://doi.org/10.1016/j.procbio.2014.03.007

Annamalai N, Rajeswari MV, Balasubramanian T (2014b) Extraction, purification and application of thermostable andhalostable alkaline protease from *Bacillus alveayuensis* CAS 5using marine wastes. Food Bioprod Process 92(4):335–342. https://doi.org/10.1016/j.fbp.2013.08.009

Ban X, Li C, Zhang Y, Gu Z, Cheng L, Hong Y, Li Z (2020) Importance of C-terminal extension in thermophilic 1,4-α-Glucan branching enzyme from *Geobacillus thermoglucosidans* STB02. Appl Biochem Biotechnol 190(3):1010–1022. https://doi.org/10.1007/s12010-019-03150-7

Bose A, Chawdhary V, Keharia H, Subramanian RB (2014) Production and characterization of a solvent-tolerant protease from a novel marine isolate *Bacillus tequilensis* P15. Ann Microbiol 64(1):343–354. https://doi.org/10.1007/s13213-013-0669-y

da Silva RR (2017) Bacterial and fungal proteolytic enzymes: production, catalysis and potential applications. Appl Biochem Biotechnol 183(1). https://doi.org/10.1007/s12010-017-2427-2

Dammak DF, Smaoui SM, Ghanmi F, Boujelben I, Maalej S (2016) Characterization of halo-alkaline and thermostable protease from *Halorubrum ezzemoulense* strain ETR14 isolated from Sfax solar saltern in Tunisia. J Basic Microbiol 56(4):337–346. https://doi.org/10.1002/jobm.201500475

Daoud L, Jlidi M, Hmani H, Hadj Brahim A, El Arbi M, Ben Ali M (2017) Characterization of thermo-solvent stable protease from *Halobacillus* sp. CJ4 isolated from Chott Eldjerid hypersaline lake in Tunisia. J Basic Microbiol 57(2):104–113. https://doi.org/10.1002/jobm.201600391

Daoud L, Hadj brahim A, Hmani H, Akremi A, Jlidi M, Ben Ali M, Bejar S, Aliye Feto N, Ben Ali M (2019) *Salicola sp* Strain SBJ9: A novel extremely halophilic bacterium with an interesting protease activity 01:6115. https://doi.org/10.3390/mol2net-04-06115

De Lourdes Moreno M, García MT, Ventosa A, Mellado E (2009) Characterization of *Salicola* sp. IC10, a lipase- and protease-producing extreme halophile. FEMS Microbiol Ecol 68(1): 59–71. https://doi.org/10.1111/j.1574-6941.2009.00651.x

Dodia MS, Rawal CM, Bhimani HG, Joshi RH, Khare SK, Singh SP (2008) Purification and stability characteristics of an alkaline serine protease from a newly isolated Haloalkaliphilic bacterium sp. AH-6. J Ind Microbiol Biotechnol 35(2):121–131. https://doi.org/10.1007/s10295-007-0273-x

Du X, Li M, Tang W, Zhang Y, Zhang L, Wang J, Li T, Tang B, Tang X (2015) Secretion of T at-dependent halolysin S pt A capable of autocatalytic activation and its relation to haloarchaeal growth. Mol Microbiol 96(3):548–565. https://doi.org/10.1111/mmi.12955

Eichler J (2001) Biotechnological uses of archaeal extremozymes. Biotechnol Adv 19(4):261–278. https://doi.org/10.1016/S0734-9750(01)00061-1

Essghaier B, Bejji M, Jijakli H, Boudabous A, Sadfi-Zouaoui N (2009) High salt-tolerant protease from a potential biocontrol agent *Bacillus pumilus* M3-16. Ann Microbiol 59(3):553–558. https://doi.org/10.1007/BF03175145

Fahmy NM, El-Deeb B (2023) Optimization, partial purification, and characterization of a novel high molecular weight alkaline protease produced by *Halobacillus* sp. HAL1 using fish wastes as a substrate. J Genet Eng Biotechnol 21(1):48. https://doi.org/10.1186/s43141-023-00509-6

Falkenberg F, Rahba J, Fischer D et al (2022) Biochemical characterization of a novel oxidatively stable, halotolerant, and high-alkaline subtilisin from Alkalihalobacillus okhensis Kh10-101T. FEBS Open Bio 12(10):1729–1746. https://doi.org/10.1002/2211-5463.13457

Flores-Santos JC, Zavaleta AI, Flores-Fernandez CN et al (2020) Production and characterization of extremophilic proteinases from a new enzyme source, Barrientosiimonas sp. V9. Appl Biochem Biotechnol 190(3):1060–1073. https://doi.org/10.1007/s12010-019-03140-9

Foophow T, Sittipol D, Rukying N et al (2022) Purification and characterization of a novel extracellular haloprotease Vpr from Bacillus licheniformis strain KB111. Food Technol Biotechnol 60(2):225–236. https://hrcak.srce.hr/280358

Fu XT, You SG, Kim SM et al (2008) Characterization of a salt-tolerant acid protease produced by Bacillus megaterium KLP-98 and its potential as a fermentation starter for the manufacture of fish sauce. J Food Biochem 32(3):279–298. https://doi.org/10.1111/j.1745-4514.2008.00155.x

Gao XL, Yin YY, Yan JK et al (2019) Separation, biochemical characterization and salt-tolerant mechanisms of alkaline protease from Aspergillus oryzae. J Sci Food Agric 99(7):3359–3366. https://doi.org/10.1002/jsfa.9553

Ghafoori H, Askari M, Sarikhan S (2016) Purification and characterization of an extracellular haloalkaline serine protease from the moderately halophilic bacterium, *Bacillus iranensis* (X5B). Extremophiles 20(2):115–123. https://doi.org/10.1007/s00792-015-0804-8

Gupta RK (2012) Scale-up of an alkaline protease from *Bacillus pumilus* MTCC 7514 utilizing fish meal as a sole source of nutrients. J Microbiol Biotechnol 22(9):1230–1236. https://doi.org/10.4014/jmb.1203.03021

Gupta A, Khare SK (2009) Enzymes from solvent-tolerant microbes: useful biocatalysts for non-aqueous enzymology. Crit Rev Biotechnol 29(November 2008):44–54. https://doi.org/10.1080/07388550802688797

Han D, Hou J, Cui HL (2023) An extracellular protease containing a novel C-terminal extension produced by a marine-originated haloarchaeon. Front Mar Sci 10:1166287. https://doi.org/10.3389/fmars.2023.1166287

Hanlon GW, Norman A (1981) Bacitracin and protease production in relation to sporulation during exponential growth of *Bacillus licheniformis* poorly utilized carbon and nitrogen sources. J Bacteriol 147(2):427–431. https://doi.org/10.1128/jb.147.2.427-431.1981

Hao JY, Zhang ZC, Yang M et al (2022) Micronization using combined alkaline protease hydrolysis and high-speed shearing homogenization for improving the functional properties of soy protein isolates. Bioresour Bioprocess 9(1):77. https://doi.org/10.1186/s40643-022-00565-9

He W, Chen Z, Chung HY (2022) Dynamic correlations between major enzymatic activities, physicochemical properties and targeted volatile compounds in naturally fermented plain sufu during production. Food Chem 378:131988. https://doi.org/10.1016/j.foodchem.2021.131988

Heng X, Chen H, Lu C, Feng T, Li K, Gao E (2022) Study on synergistic fermentation of bean dregs and soybean meal by multiple strains and proteases. LWT 154:112626. https://doi.org/10.1016/j.lwt.2021.112626

Hou Y, Lu F, Tian J et al (2019) Cloning, heterologous expression and characterization of an intracellular serine protease from Bacillus sp. LCB10. Appl Biochem Microbiol 55(5):482–488. https://doi.org/10.1134/S0003683819050168

Hou J, Yin X-M, Li Y, Han D, Lü B, Zhang J-Y, Cui H-L (2021) Biochemical characterization of a low salt-adapted extracellular protease from the extremely halophilic archaeon *Halococcus salifodinae*. Int J Biol Macromol 176:253–259. https://doi.org/10.1016/j.ijbiomac.2021.02.081

Hou J, Li S-Y, Zhao Y-J, Cui H-L (2022) A novel halolysin without C-terminal extension from an extremely halophilic archaeon. Appl Microbiol Biotechnol 106(8):3009–3019. https://doi.org/10.1007/s00253-022-11903-4

Huang J, Wu C, Liu D, Yang X, Wu R, Zhang J, Ma C, He H (2017) C-terminal domains of bacterial proteases: structure, function and the biotechnological applications. J Appl Microbiol 122(1):12–22. https://doi.org/10.1111/jam.13317

Ibrahim ASS, Al-Salamah AA, Elbadawi YB, El-Tayeb MA, Shebl Ibrahim SS (2015) Production of extracellular alkaline protease by new halotolerant alkaliphilic *Bacillus* sp. NPST-AK15 isolated from hyper saline soda lakes. Electron J Biotechnol 18(3):236–243. https://doi.org/10.1016/j.ejbt.2015.04.001

Ibrahim ASS, Al-Salamah AA, Elbadawi YB et al (2016) Alkaline protease from a new halotolerant alkaliphilic Bacillus agaradhaerens strain AK-R isolated from Egyptian soda lakes. Biosci J 32(6):1604–1618

Ibrahim ASS, Elbadawi YB, El-Tayeb MA, Al-maary KS, Maany DAF, Ibrahim SSS, Elagib AA (2019) Alkaline serine protease from the new halotolerant alkaliphilic *Salipaludibacillus agaradhaerens* strain AK-R: purification and properties. *3* Biotech 9(11). https://doi.org/10.1007/s13205-019-1928-9

Jaouadi B, Ellouz-Chaabouni S, Rhimi M, Bejar S (2008) Biochemical and molecular characterization of a detergent-stable serine alkaline protease from *Bacillus pumilus* CBS with high catalytic efficiency. Biochimie 90(9):1291–1305. https://doi.org/10.1016/j.biochi.2008.03.004

Jha SK, Rauniyar K, Karpanen T, Leppänen V-M, Brouillard P, Vikkula M, Alitalo K, Jeltsch M (2017) Efficient activation of the lymphangiogenic growth factor VEGF-C requires the C-terminal domain of VEGF-C and the N-terminal domain of CCBE1. Sci Rep 7(1):4916. https://doi.org/10.1038/s41598-017-04982-1

Ji Z-L, Peng S, Chen L-L, Liu Y, Yan C, Zhu F (2020) Identification and characterization of a serine protease from Bacillus licheniformis W10: a potential antifungal agent. Int J Biol Macromol 145:594–603. https://doi.org/10.1016/j.ijbiomac.2019.12.216

Kalwasińska A, Jankiewicz U, Felföldi T, Burkowska-But A, Brzezinska MS (2018) Alkaline and halophilic protease production by Bacillus luteus H11 and its potential industrial applications. Food Technol Biotechnol 56(4):553. https://doi.org/10.17113/ftb.56.04.18.5553

Karan R, Singh SP, Kapoor S, Khare SK (2011) A novel organic solvent tolerant protease from a newly isolated *Geomicrobium* sp. EMB2 (MTCC 10310): production optimization by response surface methodology. New Biotechnol 28(2):136–145. https://doi.org/10.1016/j.nbt.2010.10.007

Khoshnevis N, Rezaei S, Samaei-Nouroozi A, Amin M, Moshfegh M, Khoshayand MR, Faramarzi MA (2018) Enhanced production and characterization of a highly stable extracellular protease from an extreme halophilic isolate *Salicola marasensis*. Iran J Pharm Res 17(4):1392–1412. https://doi.org/10.22037/ijpr.2018.2311

Kour D, Sharma B, Kaur T, Kaur S, Alqahtani AMS, Khan SS, Jan T, Kadasah SF, Singh S, Maithini D (2025) Extremozymes: unlocking potential of extreme environments for sustainable biotechnology. Syst Microbiol Biomanuf 5(3):915–947. https://doi.org/10.1007/s43393-025-00352-5

Kumar D, Bhardwaj R, Jassal S, Goyal T, Khullar A, Gupta N (2021) Application of enzymes for an eco-friendly approach to textile processing. Environ Sci Pollut Res 30(28):71838–71848. https://doi.org/10.1007/s11356-021-16764-4

Lam MQ, Nik Mut NN, Thevarajoo S, Chen SJ, Selvaratnam C, Hussin H et al (2018) Characterization of detergent compatible protease from halophilic *Virgibacillus* sp. CD6. 3 Biotech 8(2): 104. https://doi.org/10.1007/s13205-018-1133-2

Lanyi JK (1974) Salt dependent properties of proteins from extremely halophilic bacteria. Bacteriol Rev 38(3):272–290. https://doi.org/10.1128/mmbr.38.3.272-290.1974

Lapponi MJ, Méndez MB, Trelles JA, Rivero CW (2022) Cell immobilization strategies for biotransformations. Curr Opin Green Sustain Chem 33:100565. https://doi.org/10.1016/j.cogsc.2021.100565

Lei F, Zhao Q, Lin L, Sun B, Zhao M (2017) Evaluation of the hydrolysis specificity of an aminopeptidase from *Bacillus licheniformis* SWJS33 using synthetic peptides and soybean protein isolate. J Agric Food Chem 65(1):167–173. https://doi.org/10.1021/acs.jafc.6b04426

Li X, Yu HY (2012) Purification and characterization of novel organic-solvent-tolerant β-amylase and serine protease from a newly isolated *Salimicrobium halophilum* strain LY20. FEMS Microbiol Lett 329(2):204–211. https://doi.org/10.1111/j.1574-6968.2012.02522.x

Li Y, Ku S, Park MS, Li Z, Ji GE (2017) Acceleration of aglycone isoflavone and γ-aminobutyric acid production from doenjang using whole-cell biocatalysis accompanied by protease treatment. J Microbiol Biotechnol 27(11):1952–1960

Li X, Zhang Q, Xu Z, Jiang G, Gan L, Tian Y, Shi B (2022) High-expression and characterization of a novel serine protease from *Ornithinibacillus caprae* L9T with eco-friendly applications. Environ Sci Pollut Res 29(24):35996–36012. https://doi.org/10.1007/s11356-021-17495-2

Liu X, Feng Y, Lai X, Deng T, Liu X, Lyu M, Wang S (2020) *Virgibacillus halodenitrificans* ST-1 for fermentation of shrimp paste and hydrolysates of its protease. Food Sci Nutr 8(10): 5352–5361. https://doi.org/10.1002/fsn3.1777

Maghraby YR, El-Shabasy RM, Ibrahim AH, Azzazy HME-S (2023) Enzyme immobilization technologies and industrial applications. ACS Omega 8(6):5184–5196. https://doi.org/10.1021/acsomega.2c07560

Mahmoud A, Kotb E, Alqosaibi AI, Al-Karmalawy AA, Al-Dhuayan IS, Alabkari H (2021) In vitro and in silico characterization of alkaline serine protease from Bacillus subtilis D9 recovered from Saudi Arabia. Heliyon:e08148. https://doi.org/10.1016/j.heliyon.2021

Manikandan M, Kannan V, Paŝić L (2011) Extraction, purification and characterization of a protease from *Micrococcus* sp. VKMM 037. Environ Technol 32(13):1487–1495. https://doi.org/10.1080/09593330.2010.540718

Manjusha K, Jayesh P, Jose D, Sreelakshmi B, Priyaja P, Gopinath P, Saramma AV, Bright Singh IS (2013) Alkaline protease from a non-toxigenic mangrove isolate of *Vibrio* sp. V26 with potential application in animal cell culture. Cytotechnology 65(2):199–212. https://doi.org/10.1007/s10616-012-9472-z

Maruthiah T, Somanath B, Immanuel G, Palavesam A (2015) Deproteinization potential and antioxidant property of haloalkalophilic organic solvent tolerant protease from marine *Bacillus* sp. APCMST-RS3 using marine shell wastes. Biotechnol Reports 8:124–132. https://doi.org/10.1016/j.btre.2015.10.009

Maruthiah T, Immanuel G, Palavesam A (2017) Purification and characterization of halophilic organic solvent tolerant protease from marine *Bacillus* sp. APCMST-RS7 and its antioxidant potentials. Proc Nat Acad Sci India Sect B Biol Sci 87(1):207–216. https://doi.org/10.1007/s40011-015-0603-0

Mechri S, Bouacem K, Amziane M, Dab A, Nateche F, Jaouadi B (2019a) Identification of a new serine alkaline peptidase from the moderately halophilic *Virgibacillus natechei* sp. Nov., strain FarD^{T} and its application as bioadditive for peptide synthesis and laundry detergent formulations. Biomed Res Int 2019:1–17. https://doi.org/10.1155/2019/6470897

Mechri S, Bouacem K, Jabeur F, Mohamed S, Addou NA, Dab A, Bouraoui A, Bouanane-Darenfed A, Bejar S, Hacène H, Baciou L, Lederer F, Jaouadi B (2019b) Purification and biochemical characterization of a novel thermostable and halotolerant subtilisin SAPN, a serine protease from *Melghiribacillus thermohalophilus* Nari2AT for chitin extraction from crab and shrimp shell by-products. Extremophiles 23(5):529–547. https://doi.org/10.1007/s00792-019-01105-8

Mei S, Li M, Sun Y, Deng X, Chen N, Liu Y, Yin J, Luo H, Wu Y, He D, Gan F, Tang B, Tang X-F (2022) Sec-dependent secretion of subtilase SptE in *Haloarchaea* facilitates its proper folding and heterocatalytic processing by Halolysin SptA extracellularly. Appl Environ Microbiol 88(8):e00246–e00222. https://doi.org/10.1128/aem.00246-22

Mesbah NM, Wiegel J (2014) Purification and biochemical characterization of halophilic, alkalithermophilic protease AbCP from *Alkalibacillus* sp. NM-Fa4. J Mol Catal B Enzym 105:74–81. https://doi.org/10.1016/j.molcatb.2014.03.023

Mesbahi-Nowrouzi M, Mollania N (2018) Purification of selenate reductase from *Alcaligenes* sp. CKCr-6A with the ability to biosynthesis of selenium nanoparticle: enzymatic behavior study in imidazolium based ionic liquids and organic solvent. J Mol Liq 249:1254–1262. https://doi.org/10.1016/j.molliq.2017.10.117

Mokashe N, Chaudhari A, Patil U (2015) Optimal production and characterization of alkaline protease from newly isolated halotolerant *Jeotgalicoccus* sp. Biocatal Agric Biotechnol 4(2): 235–243. https://doi.org/10.1016/j.bcab.2015.01.003

Mokashe N, Chaudhari B, Patil U (2017) Detergent-compatible robust alkaline protease from newly isolated halotolerant *Salinicoccus* sp. UN-12. J Surfactant Deterg 20(6):1377–1393. https://doi.org/10.1007/s11743-017-2024-y

Mokashe N, Chaudhari B, Patil U (2018) Operative utility of salt-stable proteases of halophilic and halotolerant bacteria in the biotechnology sector. Int J Biol Macromol 117(2018):493–522. https://doi.org/10.1016/j.ijbiomac.2018.05.217

Namwong S, Hiraga K, Takada K, Tsunemi M, Tanasupawat S, Oda K (2006) A halophilic serine proteinase from *Halobacillus* sp. SR5-3 isolated from fish sauce: purification and characterization. Biosci Biotechnol Biochem 70(6):1395–1401. https://doi.org/10.1271/bbb.50658

Naveed M, Nadeem F, Mehmood T et al (2021) Protease-A versatile and ecofriendly biocatalyst with multi-industrial applications: an updated review. Catal Lett 151(2):307–323. https://doi.org/10.1007/s10562-020-03316-7

Nwankwo C, Hou J, Cui H-L (2023) Extracellular proteases from halophiles: diversity and application challenges. Appl Microbiol Biotechnol 107(19):5923–5934. https://doi.org/10.1007/s00253-023-12721-y

Nwankwo C, Mao YL, Hou J, Cui HL (2025) From salt to gut: the health benefits of halophiles in fermented foods and gut microbiota. Food Rev Intl:1–18. https://doi.org/10.1080/87559129.2025.2503476

Oren A (2010) Industrial and environmental applications of halophilic microorganisms. Environ Technol 31(8–9):825–834. https://doi.org/10.1080/09593330903370026

Ouelhadj A, Bouacem K, Asmani K-L, Allala F, Mechri S, Yahiaoui M, Jaouadi B (2020) Identification and homology modeling of a new biotechnologically compatible serine alkaline protease from moderately halotolerant *Gracilibacillus boraciitolerans* strain LO15. Int J Biol Macromol 161:1456–1469. https://doi.org/10.1016/j.ijbiomac.2020.07.266

Paši L, Velikonja BH, Ulrih NP (2008) Preparative biochemistry and biotechnology optimization of the culture conditions for the production of a bacteriocin from halophilic archaeon Sech7a. Prep Biochem Biotechnol 34(September 2013):37–41. https://doi.org/10.1080/10826060802164637

Patil U, Mokashe N, Chaudhari A (2016) Detergent-compatible, organic solvent-tolerant alkaline protease from *Bacillus circulans* MTCC 7942: purification and characterization. Prep Biochem Biotechnol 46(1):56–61. https://doi.org/10.1080/10826068.2014.979205

Plavec TV, Štrukelj B, Berlec A (2019) Screening for new surface anchoring domains for Lactococcus lactis. Front Microbiol 10:1879. https://doi.org/10.3389/fmicb.2019.01879

Promchai R, Boonchalearn A, Visessanguan W, Luxananil P (2018) Rapid production of extracellular thermostable alkaline halophilic protease originating from an extreme haloarchaeon, Halobacterium salinarum by recombinant Bacillus subtilis. Biocatal Agric Biotechnol 15: 192–198. https://doi.org/10.1016/j.bcab.2018.06.017

Purohit MK, Singh SP (2011) Comparative analysis of enzymatic stability and amino acid sequences of thermostable alkaline proteases from two haloalkaliphilic bacteria isolated from coastal region of Gujarat, India. Int J Biol Macromol 49(1):103–112. https://doi.org/10.1016/j.ijbiomac.2011.04.001

Purohit MK, Singh SP (2013) A metagenomic alkaline protease from saline habitat: cloning, over-expression and functional attributes. Int J Biol Macromol 53:138–143. https://doi.org/10.1016/j.ijbiomac.2012.10.032

Raval VH, Pillai S, Rawal CM, Singh SP (2014) Biochemical and structural characterization of a detergent-stable serine alkaline protease from seawater haloalkaliphilic bacteria. Process Biochem 49(6):955–962. https://doi.org/10.1016/j.procbio.2014.03.014

Raval VH, Rathore DS, Singh SP et al (2022) Comparative studies of the characteristics of two alkaline proteases from Haloalkaliphilic bacterium D-15-9 and Oceanobacillus onchorynchii Mi-10-54. Appl Biochem Microbiol 58(5):551–563. https://doi.org/10.1134/S0003683822050131

Rejisha RP, Murugan M (2021) Alkaline protease production by halophilic *Bacillus* SP. strain SP II-4 and characterization with special reference to contact lens cleansing. Mater Today Proc 45: 1757–1760. https://doi.org/10.1016/j.matpr.2020.08.624

Ribitsch D, Heumann S, Karl W, Gerlach J, Leber R, Birner-Gruen berger R, Gruber K, Eiteljoerg I, Remler P, Siegert P, Lange J, Maurer KH, Berg G, Guebitz GM, Schwab H (2012) Extracellular serine proteases from Stenotrophomonas maltophilia: screening, isolation and heterologous expression in E. coli. J Biotechnol 157:140–147. https://doi.org/10.1016/j.jbiotec.2011.09.025

Saeed K, Riaz S, Adil A, Nawaz I, Naqvi SK-U-H, Baig A, Ali M, Zeb I, Ahmed R, Naqvi TA (2023) Characterization of alkaline metalloprotease isolated from halophilic bacterium *Bacillus cereus* and its applications in various industrial processes. An Acad Bras Cienc 95(3): e20230014. https://doi.org/10.1590/0001-3765202320230014

Sarwa N, Kumari P, Meena D, Udawat P, Chaudhary NS (2024) Alkaline proteases from Haloalkaliphiles: unveiling nature's catalysts for diverse applications. Appl Biochem Microbiol 60(5):855–870. https://doi.org/10.1134/S0003683824603676

Satari Faghihi L, Seyedalipour B, Riazi G, Ahmady-Asbchin S (2018) Introduction of two halo-alkali-thermo-stable biocatalysts: purification and characterization. Catal Lett 148(3):831–842. https://doi.org/10.1007/s10562-018-2295-6

Sathishkumar R, Ananthan G, Raghunathan C (2015a) Production and characterization of haloalkaline protease from ascidian-associated *Virgibacillus halodenitrificans* RSK CAS1 using marine wastes. Ann Microbiol 65(3):1481–1493. https://doi.org/10.1007/s13213-014-0987-8

Sathishkumar R, Ananthan G, Arun J (2015b) Production, purification and characterization of alkaline protease by ascidian associated Bacillus subtilis GA CAS8 using agricultural wastes. Biocatal Agric Biotechnol 4:214–220. https://doi.org/10.1016/j.bcab.2014.12.003

Saxena S (2015) Microbial enzymes and their industrial applications. In: Applied microbiology. Springer India, New Delhi, pp 121–154. https://doi.org/10.1007/978-81-322-2259-0_9

Sellami-Kamoun A, Haddar A, Ali NEH, Ghorbel-Frikha B, Kanoun S, Nasri M (2008) Stability of thermostable alkaline protease from *Bacillus licheniformis* RP1 in commercial solid laundry detergent formulations. Microbiol Res 163(3):299–306. https://doi.org/10.1016/j.micres.2006.06.001

Senba H, Nishikawa A, Kimura Y et al (2023) Improvement in salt-tolerance of Aspergillus oryzae gamma-glutamyl transpeptidase via protein chimerization with Aspergillus sydowii homolog. Enzym Microb Technol 167:110240. https://doi.org/10.1016/j.enzmictec.2023.110240

Setyorini E, Kim YJ, Takenaka S, Murakami S, Aoki K (2006) Purification and characterization of a halotolerant intracellular protease from *Bacillus subtilis* strain FP-133. J Basic Microbiol 46(4): 294–304. https://doi.org/10.1002/jobm.200510086

Shahbazi M, Karbalaei-heidari HR (2012) A novel low molecular weight extracellular protease from a moderately halophilic bacterium *Salinivibrio* sp. strain MS-7: production and biochemical properties. Mol Biol Res Commun 1(2):45–56

Sharma V, Tsai M-L, Nargotra P, Chen C-W, Kuo C-H, Sun P-P, Dong C-D (2022) Agro-industrial food waste as a low-cost substrate for sustainable production of industrial enzymes: a critical review. Catalysts 12(11):1373. https://doi.org/10.3390/catal12111373

Singhal P, Nigam V, Vidyarthi A (2012) Studies on production, characterization and applications of microbial alkaline proteases. Int J Adv Biotechnol Res 3:653–669

Sinha R, Khare SK (2012) Isolation of a halophilic *Virgibacillus* sp. EMB13: characterization of its protease for detergent application. Indian J Biotechnol 11(4):416–426

Sinha R, Khare SK (2013) Characterization of detergent compatible protease of a halophilic *Bacillus* sp. EMB9: differential role of metal ions in stability and activity. Bioresour Technol 145:357–361. https://doi.org/10.1016/j.biortech.2012.11.024

Sinsuwan S, Rodtong S, Yongsawatdigul J (2010) A NaCl-stable serine proteinase from *Virgibacillus* sp. SK33 isolated from Thai fish sauce. Food Chem 119(2):573–579. https://doi.org/10.1016/j.foodchem.2009.06.064

Sivakumar N, Li N, Tang JW, Patel BKC, Swaminathan K (2006) Crystal structure of AmyA lacks acidic surface and provide insights into protein stability at poly-extreme condition. FEBS Lett 580:2646–2652. https://doi.org/10.1016/j.febslet.2006.04.017

Solanki P, Putatunda C, Kumar A, Bhatia R, Walia A (2021) Microbial proteases: ubiquitous enzymes with innumerable uses. 3 Biotech 11(10):428. https://doi.org/10.1007/s13205-021-02928-z. Epub 2021 Sep 8. PMID: 34513551; PMCID: PMC8425024

Sun JA, Li P, Liu Z et al (2020) A novel thermostable serine protease from a metagenomic library derived from marine sediments in the east China sea. Appl Microbiol Biotechnol 104(21):9229–9238. https://doi.org/10.1007/s00253-020-10879-3

Suthindhiran K, Jayasri MA, Dipali D, Prasar A (2014) Screening and characterization of protease producing actinomycetes from marine saltern. J Basic Microbiol 54(10):1098–1109. https://doi.org/10.1002/jobm.201300563

Tanasupawat S, Taprig T, Akaracharanya A et al (2011) Characterization of Virgibacillus strain TKNR13-3 from fermented shrimp paste (ka-pi) and its protease production. Afr J Microbiol Res 5(26):4714–4721. https://doi.org/10.5897/AJMR10.520

Tien TN, Nguyen TC, Nguyen CN, Nguyen TT, Pham TA, Pham NH, Chu-Ky S (2022) Protease increases ethanol yield and decreases fermentation time in no-cook process during very-high-gravity ethanol production from rice. Process Biochem 117:10–18. https://doi.org/10.1016/j.procbio.2022.03.005

Torres S, Castro GR (2004) Non-aqueous biocatalysis in homogeneous solvent systems. Food Technol Biotechnol 42(4):271–277

Verma J, Pandey S (2019) Biocatalysis and agricultural biotechnology characterization of partially purified alkaline protease secreted by halophilic bacterium *Citricoccus* sp. Isolated from agricultural soil of northern India. Biocatal Agric Biotechnol 17:605–612. https://doi.org/10.1016/j.bcab.2019.01.020

Vidyasagar M, Prakash S, Jayalakshmi SK, Sreeramulu K (2007) Optimization of culture conditions for the production of halothermophilic protease from halophilic bacterium *Chromohalobacter* sp. TVSP101. World J Microbiol Biotechnol 23(5):655–662. https://doi.org/10.1007/s11274-006-9279-1

Vidyasagar M, Prakash S, Mahajan V, Shouche YS, Sreeramulu K (2009) Purification and characterization of an extreme halothermophilic protease from a halophilic bacterium *Chromohalobacter* sp. Tvsp101. Braz J Microbiol 40(1):12–19. https://doi.org/10.1590/S1517-83822009000100002

Wu S, Liu G, Zhang D, Li C, Sun C (2015) Purification and biochemical characterization of an alkaline protease from marine bacteria *Pseudoalteromonas* sp. 129-1. J Basic Microbiol 55(12):1427–1434. https://doi.org/10.1002/jobm.201500327

Xin L, Hui-Ying Y, Xiao-Xue L, Xiao S (2011) Production and characterization of a novel extracellular metalloproteinase by a newly isolated moderate halophile, *Halobacillus* sp. LY6. Folia Microbiol 56(4):329–334. https://doi.org/10.1007/s12223-011-0046-9

Yang H, Liu Y, Ning Y, Wang C, Zhang X, Weng P, Wu Z (2020) Characterization of an intracellular alkaline serine protease from Bacillus velezensis SW5 with fibrinolytic activity. Curr Microbiol 77:1610–1621. https://doi.org/10.1007/s00284-020-01977-6

Yavari-Bafghi M, Amoozegar MA (2025) Pharmaceutical applications of halophilic enzymes. Heliyon 11(4):e42754. https://doi.org/10.1016/j.heliyon.2025.e42754

Zambare VP, Nilegaonkar SS (2016) Proteases in leather processing. In: Thangadurai D, Sangeetha J (eds) Industrial biotechnology, sustainable production and bioresource utilization. Apple Academic Press/CRC Press., (e-Book ISBN-13: 978-1-77188-262-0), pp 209–238

Zhang Y, Wang M, Du X, Tang W, Zhang L, Li M, Wang J, Tang B, Tang X-F (2014) Chitin accelerates activation of a novel Haloarchaeal serine protease that Deproteinizes chitin-containing biomass. Appl Environ Microbiol 80(18):5698–5708. https://doi.org/10.1128/AEM.01196-14

Zhang H, Mu H, Mo Q, Sun T, Liu Y, Xu M, Wang H, Dai Y, Lu F (2016) Gene cloning, expression and characterization of a novel cold-adapted protease from Planococcus sp. J Mol Catal B Enzym 130:1–8. https://doi.org/10.1016/j.molcatb.2016.04.002

Zhang H, Li H, Lang DA et al (2018) Purification, characterization and potential applications of a novel halotolerant metalloprotease from marine bacterium vibrio sp LA-05. J Chem Technol Biotechnol 93(12):3627–3637. https://doi.org/10.1002/jctb.5744

Zhang H, Li H, Liu HR et al (2019) The application of a halotolerant metal loprotease from marine bacterium vibrio sp. LA-05 in liquid detergent formulations. Int Biodeterior Biodegrad 142:18–25. https://doi.org/10.1016/j.ibiod.2019.04.006

Zhang H, Hu M, Wang Q, Liu F, Xu M, Zhang X, Rao Z (2022) A novel salt-tolerant l-glutaminase: efficient functional expression, computer-aided design, and application. Fermentation 8(9):444. https://doi.org/10.3390/fermentation8090444

Zhao H-L, Chen X-L, Xie B-B, Zhou M-Y, Gao X, Zhang X-Y, Zhou B-C, Weiss AS, Zhang Y-Z (2012) Elastolytic mechanism of a novel M23 metalloprotease pseudoalterin from deep-sea Pseudoalteromonas sp. CF6-2: cleaving not only glycyl bonds in the hydrophobic regions but also peptide bonds in the hydrophilic regions involved in cross-linking. J Biol Chem 287: 39710–39720. https://doi.org/10.1074/jbc.M112.405076

Zhao W, Li G-P, Wang J, Zhou Y-K, Gao Y, Du P-F (2019) Predicting protein sub-Golgi locations by combining functional domain enrichment scores with pseudo-amino acid compositions. J Theo ret Biol 473:38–43. https://doi.org/10.1016/j.jtbi.2019.04.025

Chapter 5
Improving the Production and Properties of Microbial Proteases by Genetic Engineering

Shashank Vasanthkumar, Chandan Kumar Sahu, and Ravi-Kumar Kadeppagari

Abstract Genes of different microbial proteases were cloned for the overproduction of enzymes with food, biomedical, and detergent applications. These microbial proteases include different subtilisin enzymes, neutral protease, different alkaline proteases, aspartic protease, aspergillopepsin, acidic protease, and proteinase K. Proteases were also genetically modified to obtain more pH-resistant, more substrate-specific, thermostable, less autocatalytic, and halotolerant enzymes. Proteases from *Bacillus, Aspergillus, Acremonium, Tritirachium* species were cloned into *Escherichia coli, Bacillus subtilis,* and *Saccharomyces cerevisiae*. Aminopeptidases are widely applied in the food industry for controlling bitterness in protein hydrolysates and improving flavor in fermented foods, and advancements like codon optimization in genetic engineering have focused on improving the stability of aminopeptidases. Serine proteases like Carlsberg protease, derived from *B. licheniformis,* and subtilisin, from *B. amyloliquefaciens*, have been genetically modified for better performance in detergent applications. Site-directed mutagenesis was used to improve the catalytic efficiency and thermal stability of cysteine proteases, and these proteases are commercially used for meat tenderization, baking, leather processing, wound treatment, anticancer therapy, and treating inflammatory diseases. Recent developments on the improvement of production and properties of microbial proteases by genetic engineering will be discussed here.

Keywords Microbial proteases · Subtilisin · Proteinase K · *Bacillus* · *Aspergillus* · *Saccharomyces*

S. Vasanthkumar · C. K. Sahu · R.-K. Kadeppagari (✉)
Centre for Incubation Innovation Research and Consultancy, Food Technology Department, Jyothy Institute of Technology, Bengaluru, Karnataka, India
e-mail: ravikumar.k@ciirc.jyothyit.ac.in

V. Zambare et al. (eds.), *Biotechnological Advances and Applications in Microbial Proteases*, Interdisciplinary Biotechnological Advances,
https://doi.org/10.1007/978-981-95-9011-7_5

5.1 Introduction

Proteases, also called proteinases or proteolytic enzymes, break down proteins into smaller peptides by catalyzing the hydrolysis of peptide bonds. They play a major role in industrial applications (Fig. 5.1) (Mótyán et al. 2013). Proteases occupy 60% of the global enzyme market by volume and account for nearly 40% of total enzyme sales worldwide. Traditionally, they are sourced from animals (e.g., calf rennet), plants (e.g., papaya), and microbes (e.g., *Bacillus* spp.) (Mamo and Assefa 2018). However, enzymes derived from animal and plant sources come with several limitations, like low yield, ethical concerns, seasonal availability, and environmental impact. Hence, industries are focusing on microbes as an alternative for producing proteases with higher stability, activity, and yield. Microbes, especially bacteria and fungi, have numerous advantages, like shorter life cycles and being easy for genetic manipulation, and they can be produced economically at a large scale with higher yields by fermentation in bioreactors (Mirsalami and Mirsalami 2025). Consequently, microbial proteases have become the primary choice for industrial application, now accounting for nearly two-thirds of global enzyme production.

Among microbial producers, bacterial strains like *P. aeruginosa*, *B. cereus*, and *B. subtilis* and fungal strains like *Penicillium* and *Aspergillus* species have been widely explored (Nilegaonkar et al. 2007; Zambare et al. 2011; Razzaq et al. 2019). According to industrial needs, these organisms not only produce high levels of extracellular proteases but also display a broad spectrum of enzymatic activity. But the major challenge for using this wild type of microbes lies in the complexity of their protease profile. For example, *A. oryzae*, known for its significant protein secretion system, contains over 130 protease genes. These genes simultaneously

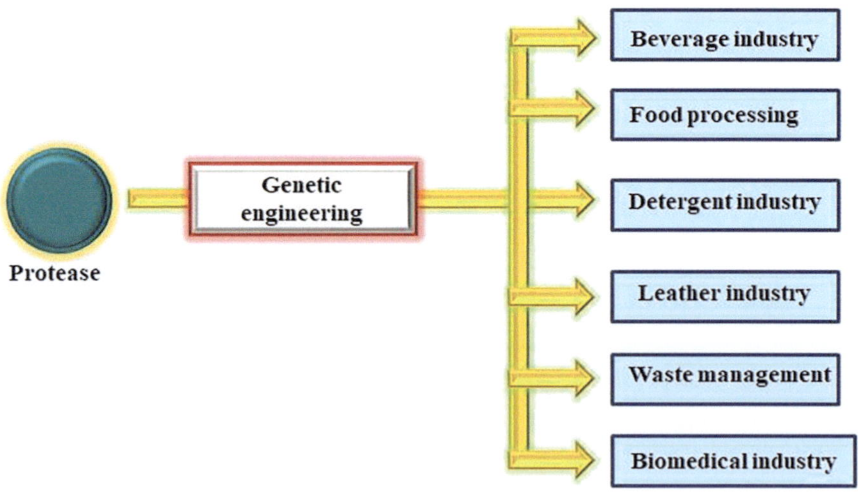

Fig. 5.1 Industrial applications of recombinant proteases

express and produce multiple proteases that hinder downstream processing and degrade the desired product (Yoon et al. 2008).

In addition, genetic engineering of filamentous fungi like *Aspergillus* spp. poses a significant challenge in achieving product-specific expression due to factors like gene silencing, low transformation efficiency, and complex genomic structure (Li et al. 2022). These limitations have driven researchers to explore alternative hosts for producing recombinant proteases. Model organisms like *Bacillus*, yeast, *Escherichia coli,* and *Pichia pastoris* have gained importance due to their simple genome, easy for genetic manipulation, and ability to be grown in large scale for higher yields of product. The right host for recombinant expression plays an important role in increasing the activity, structure, and stability of protease (Komarevtsev et al. 2021). Hence, prokaryotic organisms like *E. coli* and *B. subtilis* are advantageous for their short life cycle and genomic properties, but they may complicate with complex posttranslational modification process. On the other hand, eukaryotic systems like *Saccharomyces cerevisiae* are more capable of correct protein folding and glycosylation, making them more suitable for producing functional protease enzymes (Ke et al. 2018). Among eukaryotes, yeast-based expression systems have emerged as a practical and efficient choice for protein production. It offers simple genome, cloning benefits, and expression of target molecule, making it an increasingly popular alternative to filamentous fungi, which are often more difficult to work with due to their complex genome (Kjeldsen et al. 2023). This book chapter will focus on different types of proteases and microbes and their genetic manipulation for obtaining higher yield, good stability, and enhanced activity.

5.2 Improvement of Production and Properties of Microbial Proteases by Genetic Engineering

Microbial proteases are of different types (Fig. 5.2) based on their site of action (endopeptidases that act on the inner regions of the polypeptide chain and exopeptidases- which act on N-terminal or C-terminal of peptide chain), based on the pH range in which they are active (acidic, neutral, or alkaline), or the nature of the catalytic residue in their active site (serine, cysteine, aspartic, etc.). On the other hand, the Nomenclature Committee of the International Union of Biochemistry and Molecular Biology (IUBMB) has placed the proteases as part of group 3, which includes hydrolases and is further classified into subgroup 4, which includes mode of action on peptides (Song et al. 2023). This classification was based on the type of reaction it carries, the characteristics of the active site, and the evolutionary relationship with its structure. Proteases are further classified based on evolutionary traits and amino acid sequences into families and subdivided into "clans." The clans are groups of proteases that do not have similar sequences but have similar

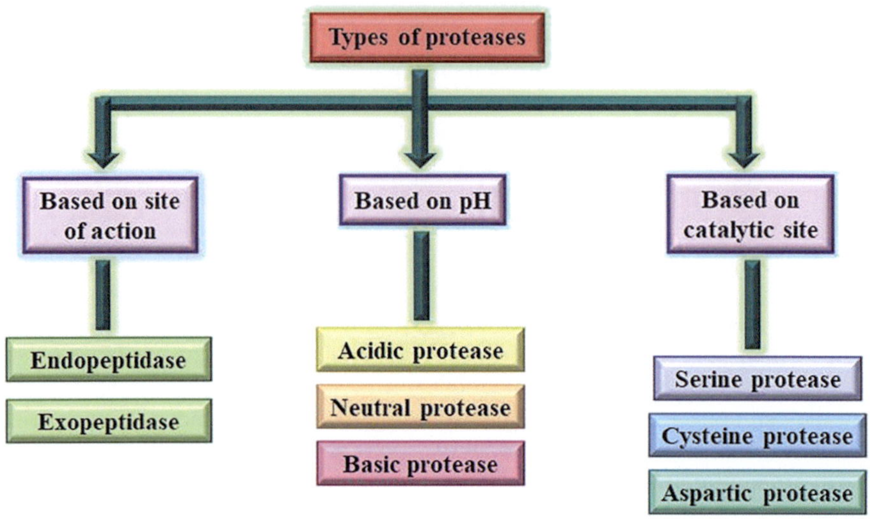

Fig. 5.2 Classification of proteases

evolutionary traits. Additionally, it also includes different types of catalytic classes based on catalytic site with similar tertiary folds (Mótyán et al. 2013).

Over the past few decades, microbes contribute 50% of industrial enzymes in the market. Selection of a proper host and gene plays an important role in the production of recombinant protease (Liu and Kokare 2016). These recombinant proteases are produced by isolating and cloning the protease-producing gene into a model organism for higher yields (Table 5.1) and also enhance physicochemical properties, like stability, specificity, and resistance against broad pH and temperature (Table 5.2). Generally, protease genes are cloned into bacteria to overproduce enzymes that are commercially used in food, detergent industries, and medical applications (Pawar et al. 2009; Zambare et al. 2014). They were also cloned to study regulatory systems with their biological roles (Song et al. 2023). Different recombinant proteases are discussed below.

5.2.1 *Aminopeptidases*

Aminopeptidases are exopeptidases that remove N-terminal amino acids from the peptide chain by using a zinc ion. They act on the peptide chain by hydrolysis reaction and help in protein digestion, peptide processing, and regulation of biological functions (Nandan and Nampoothiri 2020). They are produced from *A. oryzae*, *B. subtilis*, *S. griseus*, and lactic acid bacteria by genetic modifications. These enzymes are widely applied in the food industry for controlling bitterness in protein hydrolysates and improving flavor in fermented foods (Nandan and Nampoothiri

Table 5.1 Improving the production of proteases in heterologous hosts

Protease name	Gene	Source of gene	Expression host	Yield of protease	References
Alkaline protease	Apr	*B. licheniformis 2709*	*B. subtilis WB600*	1.6 fivefold increase	Tang et al. (2004)
Alkaline serine protease	aprE	*B. subtilis WB600*	*B. clausii*	Threefold increase	Mohsen et al. (2011)
Alkaline protease	BLAP	*B. lehensis*	*E. coli BL21 (DE3)*	4.1-fold increase	Joshi and Satyanarayana (2012)
Alkaline protease	Apr	*A. sojae GIM3.33*	*P. pastoris KM71*	Twofold increase	Ke et al. (2018)
Alkaline serine protease	GsProS8	*G. stearothermophilus*	*B. subtilis.*	2.1twofold increase	Chang et al. (2021)
Alkaline protease	Bsp-1	*B. subtilis*	*B. amyloliquefaciens BAX-9*	4.87-fold increase	Jiang et al. (2022)
Alkaline proteases	Rsep and tpa	*Exiguobacterium*	*E. coli BL21(DE3)*	1 twofold increase	Srivastava et al. (2024)
Aspartic protease	apa1	*A. niger*	*P. pastoris*	9.36-fold increase	Wei et al. (2023)
Thermostable alkaline protease	F1	*B. stearothermophilus*	*P. pastoris*	Twofold increased	Latiffi et al. (2013)
Thermostable alkaline serine protease	EU181368	*G. stearothermophilus B-1172*	*E. coli BL21 (DE3)*	1.32-fold increase	Iqbal et al. (2014)
Serine protease	Isp	*Salt-tolerant Bacillus* sp. *LCB10*	*E. coli BL21 (DE3)*	1.8-fold increase	Hou et al. (2019)
Trypsin	SGT	*S. griseus*	*P. pastoris*	3.75-fold increase	Zhang et al. (2018)
Neutral protease	nprS	*B. stearothermophilus TELNE*	*B. subtilis*	Fivefold increase	Nishiya and Imanaka (1990)
Neutral protease	rNpI	*A. oryzae*	*P. pastoris*	Highest 43,101 U/mL yield	Ke et al. (2012)
Serine protease	aprE	*B. safensis lab 418*	*E. coli BL21 (DE3)*	3.87-fold increase	El-Sayed et al. (2024)
Subtilisin Carlsberg	aprE, subtilisin	*B. licheniformis*	*B. subtilis* ISW 1214	20-fold increase	Razzaq et al. (2019)
Keratinase	pPZK3	*B. licheniformis MKU3*	*P. pastoris X33*	Threefold increase	Radha and Gunasekaran (2009)

Table 5.2 Improving properties of proteases by heterologous expression systems

Protease name	Gene	Source of gene	Expression host	Improved properties	References
Serine alkaline protease	sapN	*M. thermohalophilus*	*P. pastoris* SMD1168 and X33	Resistant to high temperature and alkaline pH.	Mechri et al. (2021)
Subtilisin-like protease	TTHA0724	*T. thermophilus HB8*	*E. coli Transetta (DE3)*	Broad substrate specificity	Xie et al. (2019)
Heat-sensitive serine protease	NHSSP	*O. corvina*	*P. pastoris*	Resistant to a wide range of alkaline pH	Skowron et al. (2020)
Aspartic protease	PepAb and PepAc	*A. niger F0215*	*P. pastoris GS115*	Increased the specific activity	Song et al. (2020)
Thermostable serine protease	TfpA	*T. fusca*	*P. pastoris*	Resistant to high temperature (80 °C) and pH (8.5).	Kim and Lei (2005)
Serine protease	Ap-subtilisin	*A. pernix*	*E. coli BL21*	Active at broad range of pH (5.0–12.0) and temperature (60–120 °C)	Catara et al. (2003)
Alkaline protease	Apr	*B. licheniformis 2709*	*E. coli BL21*	Molecular weight increased	Lin et al. (2014)
Alkaline protease	R5	*B. subtilis RD7*	*E. coli BL21*	Higher specific activity and stability at pH 10 and temperature of 40 °C	Suberu et al. (2019)
Keratinase	Tcsp	*T. sporacurvata*	*E. coli BL21 (DE3)*	Gained resistance to nonionic surfactants	Sittipol et al. (2019)
Alkaline serine protease	ASP	*B. halodurans C-125*	*B. subtilis WB800*	Activity at higher temperature (60 °C) and alkaline pH (12).	Tekin et al. (2020)
Alkaline protease	AprV	*Vibrio* sp. *DA1–1*	*E. coli BL21 (DE3)*	Increased specific activity	Chen et al. (2018)

(continued)

Table 5.2 (continued)

Protease name	Gene	Source of gene	Expression host	Improved properties	References
Subtilisin-like serine protease	VlPr1	*Verticillium lecanii*	*E. coli DH5α*	Increased specific activity and showed inhibitory activity against several plant pathogens.	Yu et al. (2012)
Thermostable keratinase	kerT1	*Thermoactinomyces* sp. *YT06*	*E. coli BL21(DE3)*	Gained stability at high temperature (60 °C) and resist wide range of pH (9–11).	Wang et al. (2019)
Aspartic protease	TaAsp	*Trichoderma asperellum*	*P. pastoris*	Stable at wide range of temperature (25–60 C) and pH (3–9).	Yang et al. (2013)
Alkaline protease	Ap	*Aspergillus sojae GIM3.33*	*P. pastoris KM71*	Resistant to a wide range of pH (7–10).	Ke et al. (2018)

2017). Recent advancements in genetic engineering have focused on improving the stability of aminopeptidases. For example, Savijoki et al. (2006) produced aminopeptidase N from *Lactobacillus delbrueckii* by plasmid-based overexpression technique for higher yield and activity. This enzyme played a crucial role in cheese ripening and flavor development by hydrolyzing peptides during fermentation. Yao et al. (2023) used a strong promoter aprE in *B. subtilis* to produce aminopeptidase T, which was salt-tolerant and thermostable. It was also used for peptide processing during industrial biocatalysis.

Song et al. (2023) explained the importance of aminopeptidase Y, which was a zinc-dependent exopeptidase, by isolating the gene from *Streptomyces griseus* and expressing it in *E. coli* to provide broad substrate specificity character, and used in peptide synthesis. Alternatively, aminopeptidase N and B were produced from *E. coli* by using pET-26b, and PepN plasmids to enhance the production yield and simplify the downstream processing for industrial applications (Golich et al. 2005). On further research, Kang et al. (2014) used codon optimization technique on prolyl aminopeptidase (PAP) gene from *A. oryzae* and cloned it in *P. pastoris*, where α-mating factor and glycosylation enhanced the enzyme secretion and thermostability (Fig. 5.3). Similarly, *B. subtilis*-derived aminopeptidases have been modified

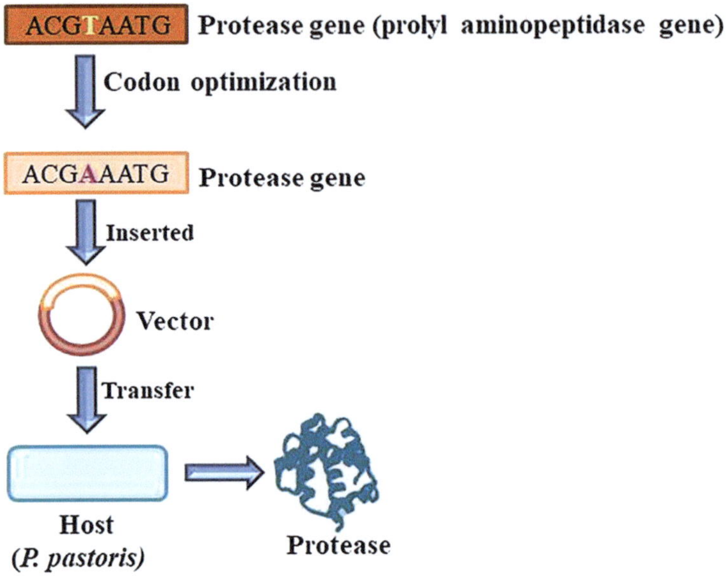

Fig. 5.3 Schematic representation of the codon optimization technique for increasing the protease yield

using a plasmid vector under strong promoters, resulting in higher yields of thermostable enzymes for industrial use (Tang et al. 2016).

5.2.2 Carboxypeptidases

Carboxypeptidases are exopeptidases that cleave the peptide chain at the C-terminal, either as zinc-dependent metalloproteases or as serine-type peptidases. They are produced from *S. cerevisiae*, *A. oryzae*, *B. subtilis*, and *T. aquaticus*. They help in peptide maturation, protein scavenging, flavor development, and protein sequencing (Rao et al. 1998). Their thermostability, specificity, and resistance characteristics make them valuable in food processing, pharmaceutical purity analysis, and biocatalysis. Commercially, they are used in cheese ripening, controlling bitterness in protein hydrolysate, and detoxification of food toxins. Human procarboxypeptidase B was heterologously expressed in *P. pastoris* using the α-factor secretion signal and AOX1 promoter, and it increased the enzyme activity in fed-batch culture (Kim et al. 2008).

5.2.3 Serine Proteases

Serine proteases belonging to a family of alkaline proteases are one of the most explored industrial enzymes. They have serine residues on their active site. They break down proteins in two steps: first, they form a temporary bond with the peptide by acylation, and water breaks this bond by deacylation, which leads to protein breakdown. This group of proteases is active at alkaline or neutral condition showing optimal activity at pH 7 and 11 (Pawar et al. 2023). In addition, serine proteases have shown broad substrate specificity. They are usually of low molecular weight (18–35 kDa), whereas some studies have found serine proteases of high molecular mass (90 kDa) from *B. subtilis* (natto) (Yamagata et al. 1995). Serine proteases are further classified by the specificity of their lateral chains and their structural homology, which includes 20 families, where the "clans" contain proteases with common ancestors. Among those "clans" of serine proteases, some of the most important are the chymotrypsin-like, the subtilisin-like, and the carboxypeptidase-like enzymes (Krem 2001).

In addition to animals, chymotrypsin is also present in microbes such as *Streptomyces erythraeus*, *S. fradiae*, and *S. griseus*. Generally, they are known as digestive enzymes that share structural similarities with other enzymes like trypsin, elastase, and thrombin. It has optimum pH activity around 8 (Parekh et al. 2011). On the other hand, subtilisin-like proteases, also known as subtilases, are produced by *Bacillus* spp. They are active in alkaline conditions above pH 10 (Figueiredo et al. 2017). Wild subtilisin-producing *Bacillus* strains cannot be used for industrial applications because they are susceptible to oxidation and sensitive to pH and temperature. To overcome these problems, genetic engineering approaches like r-DNA technology, site-directed mutagenesis, promoter modification, and codon optimization are used to modify *Bacillus* strains. These recombinant proteases showed higher stability, yield, and activity. For example, the Carlsberg protease derived from *B. licheniformis* and subtilisin BPN′ from *B. amyloliquefaciens* have been genetically modified for better performance in detergent applications. It was achieved by mutating methionine-222 and other oxidation-sensitive amino acid that resulted in an oxidation-resistant variety suitable for formulation in detergents that can resist hydrogen peroxide and other bleaching agents (Rao et al. 1998). Similarly, Thakrar et al. (2023) used a heterologous expression system for producing a thermostable serine protease from an alkaliphilic *Nocardiopsis* for resisting higher temperature suitable for detergent application. Takenaka et al. (2021) incorporated charged amino acids into *B. subtilis* to produce serine protease with enhanced stability in alkaline conditions (pH 5–11) and resistance to high salt concentrations. Alternatively, Miao et al. (2023) used site-directed mutagenesis on the PB92 gene from *B. alcalophilus* to increase the catalytic activity and thermostability of serine proteases. Chen et al. (2007) improved the stability of nattokinase, a serine protease, by modifying the plasmid origin of expression, and this also resulted in a twofold increase in the production of enzymes. Similarly, Zhao et al. (2020) used *B. subtilis* as a host and optimized the insertion site of the plasmid, which resulted in a higher yield of nattokinase and alkaline proteases.

One more example of serine endopeptidase is proteinase K, derived from the fungus *Tritirachium album*. It can degrade a wide range of proteins by acting on aliphatic and aromatic amino acids on peptides (Betzel et al. 2001). The enzyme has shown optimum activity at wide range of pH (7–12) and resist temperature up to 70 ° C. This enzyme depends on Ca^{2+} ions for stability and activity (Bajorath et al. 1988). Hence, it is used in industrial research like nucleic acid purification, peptide mapping, prion inactivation, tissue damage, and clinical diagnosis. In order to satisfy the industrial need, researchers are improving its expression levels, stability, and cost-effectiveness through genetic strategies. For example, the codon optimization strategy is used on a strong promoter like AOX1 (alcohol oxidase promoter) in *P. pastoris* and T7 in *E. coli* for getting a higher yield of proteinase K (Gustafsson et al. 2004).

Among different types of serine proteases, Gong et al. (2020) increased keratinase activity in *B. subtilis* WB600 by using aprE promoter. This promoter, also known as a growth phase promoter, has increased the yield up to twofold and controlled the toxic level of keratinase in the microbe. Alternatively, Miao et al. (2020) studied the transcriptional analysis of *B. pumilus* BA06 and developed a new promoter named P2069 for expressing alkaline protease gene aprE in *B. subtilis* WB600. This promoter increased the activity of protease by 3.65-fold compared to the control strain. Wu et al. (2010) modified the promoter sequence P_{aprN} in the wild type, which increased the nattokinase yield up to 136% in *B. subtilis.*

Among the recombinant strategies, promoter tandem repeats also increased the production, efficiency, and characteristics of recombinant proteases. For example, Liu et al. (2019) used two different tandem promoters P_{Bsamy}-P_{Baamy} to exhibit higher alkaline serine protease activity and yield. Similarly, Cui et al. (2018) deleted the Cod Y binding sequence in the 5’UTR of the promoter to increase the yield up to 1.2-fold. Fang et al. (2022) optimized and replaced spacer sequences at 5’UTR to increase the keratinase activity by 19.7-fold. This strategy also helped Degering et al. (2010) to increase the activity of subtilisin BPN’ by about sevenfold compared to its native signal peptide. Similarly, Tian et al. (2019) explained the highest extracellular keratinase activity and increased the yield by 1.95-fold by following Tat pathway in *B. subtilis*. Molecular techniques also helped *B. licheniformis* to increase serene protease activity and production by optimizing protein secretion pathways. Zhou et al. (2020) directly integrated protease genes into genomic loci for increasing alkaline protease activity in *B. licheniformis.*

On the other hand, Cai et al. (2017) increased the yield of nattokinase and α-amylase by overexpressing the signal peptide peptidase (SppA) in *B. licheniformis* BL10. Mo et al. (2019) increased the nattokinase activity by deleting three genes associated with cell wall components. One more subtilisin from *Bacillus amyloliquefaciens* was produced by changing the optimum pH and mutating Asp99 gene (Rao et al. 1998). On further research, the serine protease was also expressed in other prokaryotic expression systems like *E. coli, Lactococcus lactis*, *B. megaterium*, *B. methanolicus*, and *B. pumilus*. Increase in recombinant protease production was also achieved by modifying transcription factors, signal peptides, and optimizing fermentation conditions. Zhu et al. (2016) increased the extracellular protease activity of *E. coli* by replacing the single peptide with pel B (pectate lyase B) of *Erwinia carotovora*. Srivastava et al. (2024) developed a halotolerant serine protease by heterologous expression in *E. coli* to resist organic solvents, surfactants, and a broad pH range without losing the enzyme activity.

5.2.4 Cysteine Proteases

Cysteine proteases are a group of endopeptidases that belong to family of neutral protease. They contain cysteine in the active site and break down peptide bonds from protein. The mode of action is by following nucleophilic attack where the thiol (-SH) group of cysteine is cleaved with the help of histidine residue (Buttle and Mort 2013). It is active at acidic to neutral pH. Cysteine proteases are found in microbes, plants, and animals. Cysteine proteases are naturally produced by bacterial and fungal species. They help in pathogenicity, nutrient uptake, and host tissue degradation (Yang et al. 2023). Cysteine proteases are commercially used for meat tenderization, baking, and leather processing. They are also used in biomedical applications like wound treatment, anticancer therapy, and to treat inflammatory diseases (Haque et al. 2023). Since wild-type strains cannot satisfy the enzyme market, recombinant strains are created for higher yield and activity. These industrial strains have to produce a cysteine protease that can resist a harsh environment during enzyme activity. For example, papain is a cysteine protease found expressed in *E. coli* and *P. pastoris* that has structural similarities with papain from papaya. It was achieved by codon optimization and fusion with solubility-enhancing tags, like GST or MBP. The resulting strain had a higher solubility rate and enzymatic activity (Hoque et al. 2024). Alternatively, the site-directed mutagenesis technique also helped in increasing the protease stability and yield of protease (Fig. 5.4). For example, Dutta et al. (2013) developed a plant cysteine protease called Ervatamin C by site-directed mutagenesis at S32T, A67Y gene that increased catalytic efficiency and thermal stability.

Similarly, Karbalaei et al. (2020) improved the yield and activity of cysteine proteases in *P. pastoris* by using codon optimization. Additionally, De Marco et al. (2007) significantly improved the yield and solubility of cysteine proteases by prodomain deletion strategies and co-expression of chaperones such as GroEL-GroES in *E. coli*, which controlled the inclusion body formation and increased protein recovery. Dağlioğlu (2017) expressed cathepsin S, a mammalian lysosomal cysteine protease, in *P. pastoris* by using fed-batch fermentation and codon optimization, which increased the enzyme activity levels and its rate of secretion. In addition, Zhang et al. (2020) expressed viral cysteine proteases like SARS-CoV-2 3CLpro in *E. coli* by using self-cleaving linker peptides and fusion tags like MBP or His-tags for higher yield and used them in biomedical applications.

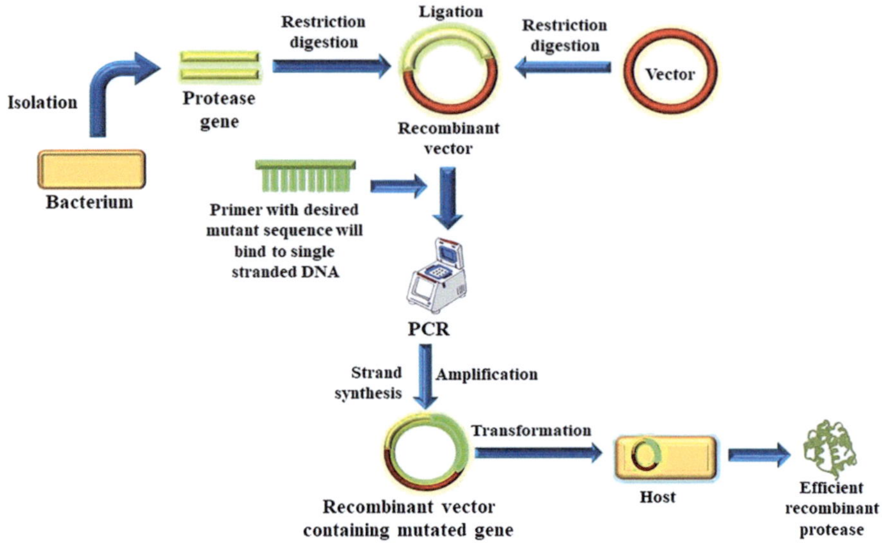

Fig. 5.4 Increasing the protease efficiency by the site-directed mutagenesis technique

5.2.5 Aspartic Proteases

Aspartic proteases, or aspergillopepsin, also known as acidic proteases, use aspartic acid in their active site to cleave protein. During acidic conditions (pH 3.0–5.0), they follow an acid-base catalytic mechanism, and water acts as the nucleophile, activated by aspartate to cleave the protein substrate into smaller peptides (Sussman et al. 2013). Aspartic proteases are found in fungi, bacteria, plants, and animals. Among them, industries use aspartic proteases from filamentous fungi like *Aspergillus niger*, *Aspergillus oryzae*, and *Rhizomucor miehei*. This enzyme helps particularly with cheese processing, beverage clarification, and the production of protein hydrolysate (Wei et al. 2023).

Advancements in recombinant DNA technology have provided an opportunity to isolate, modify, or synthesize aspartic protease genes and clone them into heterologous hosts like *P. pastoris*, *S. cerevisiae*, and *Aspergillus* spp., which can provide significant yields, improved stability, enhanced activity, and shorten downstream processing. For example, the pepsin coding gene from *Rhizomucor miehei* was isolated and cloned into *P. pastoris* to express and produce recombinant aspartic proteases that can resist higher temperature (Sun et al. 2017). Similarly, Wang et al. (2023) cloned the apa1 gene from *A. niger* by codon optimization into *P. pastoris* that resulted in enhanced enzyme production. In another study, Souza et al. (2017) optimized fermentation conditions in stirred-tank bioreactors for *A. foetidus* to produce a novel aspartic protease that can withstand pH (5.0) and temperatures (up to 55 °C), with enhanced enzyme activity.

Alternatively, Wang et al. (2021) developed a recombinant aspartic protease by cloning the RmproB gene from *Rhizomucor miehei* CAU432 and expressed in *Aspergillus niger*, which was stable at a broad pH range and resist temperature up to 45 °C. Takyu et al. (2022) cloned the pep E gene from *A. oryzae* RIB40 and *A. luchuensis* under a strong promoter into *A. oryzae* to produce recombinant proteases that can cleave κ-casein at the Phe105-Met106 bond, similar to chymosin used in rennet. In addition, this enzyme had substrate specificity, acidic pH activity, and low proteolytic activity suitable for cheese processing.

Since the eukaryotic expression system in yeast provides posttranslational modifications, it offers unique benefits to express genes for recombinant proteases from other organisms. Heterologous proteases were expressed in *S. cerevisiae* by removing intron sequences for higher yield and activity. Additionally, optimizing promoter and terminator sequences will also enhance proteinase activity. For example, Horiuchi et al. (1990) used *S. cerevisiae* strain R277C-1C to express aspartic proteinase I (RNAP-P) by optimizing the promoter sequence. This resulted in a 4.8-fold increase in enzyme activity.

5.2.6 Metalloproteases

Metalloproteases or metalloendopeptidase belongs to a group of endopeptidases that depend on divalent metal ions like zinc (Zn^{2+}) for catalytic activity. They have an active site containing a histidine, glutamate, or aspartate residue (Cerdà-Costa and Gomis-Rüth 2013). They are active at broad pH and temperature. Metalloproteases are classified into subtypes, like thermolysin-like proteases, matrix metalloproteinases (MMPs), astacins, and alkaline metalloproteases. They play a major role in cellular signaling, tissue remodeling, and protein degradation (Klein and Bischoff 2010). Hence, they are applied in biotechnology applications, medicine, food processing, and environmental engineering by modifying the wild strain through genetic engineering techniques, which provides a significant benefit to increase the yield, stability, and enzyme activity.

For example, Li et al. (2022) developed thermostable zinc metalloprotease by cloning the thermolysin (npr) gene from *B. thermoproteolyticus* into *B. subtilis* by using a strong promoter aprE that significantly increased the extracellular enzyme yield and secretion efficiency. Similarly, Lin-Cereghino et al. (2013) used the alkaline metalloprotease gene from *B. licheniformis* and cloned it into *P. pastoris* by fusing with *S. cerevisiae* α-mating factor signal peptide and applying the codon optimization technique, which enhanced the secretion and enzyme activity at broad pH and temperatures. In addition, Liu et al. (2022c) could increase the enzyme activity of a metalloprotease, pseudolysin (lasB) by 2.41-fold in *S. cerevisiae* CEN. PK 530-1C by supplementing metal ions in the media. Alternatively, signal peptide and propeptide were also used to produce recombinant proteases. This strategy has benefited different types of organisms, like *Hathewaya histolytica*, *P. aeruginosa,* and *A. oryzae*, to express their genes in *S. cerevisiae*. Xiao et al. (2022) used

S. cerevisiae CEN.PK 530-1C to express collagenase genes Col G and Col H by optimizing the signal peptide and supplementing metal ions. This could increase the extracellular protease activity.

5.3 Conclusions

Conventionally, microbial proteases will show a decrease in their activity under harsh industrial conditions like extreme pH, temperature, or the presence of inhibitory chemicals. Even though genetic and protein engineering techniques such as site-directed mutagenesis, codon optimization, and rational design have offered significant benefits, recombination efficiency is still very low and time-consuming. Alternative approaches like introducing disulfide bonds, modifying metal-binding sites, or altering amino acids in the active site are gaining importance since they increase the thermal stability and gain resistance to chemical inhibitors. On further advancements, high-throughput screening techniques, like fluorescence-activated cell sorting (FACS), phage-assisted continuous evolution (PACE), and bead-based enzyme screening methods, have offered the selection of highly resistant and active protease strains. Computational tools, like Rosetta, FoldX, and FireProt, will now support the design of stable enzyme mutants by predicting favorable structural changes. Additionally, newer approaches like light-activated protease systems and microfluidic screening have expanded the scope of protease functions in diagnostics and cell signaling. Collectively, these modern strategies will significantly improve the yield and qualities of recombinant proteases required for food processing, biopharmaceutical applications, environmental applications, and disease diagnostics.

References

Bajorath J, Hinrichs W, Saenger W (1988) The enzymatic activity of proteinase K is controlled by calcium. Eur J Biochem 176(2):441–447. https://doi.org/10.1111/j.1432-1033.1988.tb14301.x

Betzel C, Gourinath S, Kumar P, Kaur P, Perbandt M, Eschenburg S, Singh TP (2001) Structure of a serine protease proteinase K from *Tritirachium album* limber at 0.98 Å resolution. Biochemistry 40(10):3080–3088. https://doi.org/10.1021/bi002538n

Buttle D, Mort J (2013) Cysteine proteases. In: Elsevier eBooks, pp 589–592. https://doi.org/10.1016/b978-0-12-378630-2.00009-8

Cai D, Wang H, He P, Zhu C, Wang Q, Wei X, Nomura CT, Chen S (2017) A novel strategy to improve protein secretion via overexpression of the SppA signal peptide peptidase in *Bacillus licheniformis*. Microb Cell Factories 16(1). https://doi.org/10.1186/s12934-017-0688-7

Catara G, Ruggiero G, La Cara F, Digilio FA, Capasso A, Rossi M (2003) A novel extracellular subtilisin-like protease from the hyperthermophile *Aeropyrum pernix* K1: biochemical properties, cloning, and expression. Extremophiles 7(5):391–399. https://doi.org/10.1007/s00792-003-0337-4

Cerdà-Costa N, Gomis-Rüth FX (2013) Architecture and function of metallopeptidase catalytic domains. Protein Sci 23(2):123–144. https://doi.org/10.1002/pro.2400

Chang C, Gong S, Liu Z, Yan Q, Jiang Z (2021) High level expression and biochemical characterization of an alkaline serine protease from *Geobacillus stearothermophilus* to prepare anti-hypertensive whey protein hydrolysate. BMC Biotechnol 21(1). https://doi.org/10.1186/s12896-021-00678-7

Chen P, Chiang C, Chao Y (2007) Strategy to approach stable production of recombinant nattokinase in *Bacillus subtilis*. Biotechnol Prog 23(4):808–813. https://doi.org/10.1021/bp070108j

Chen X, Zhou C, Xue Y, Shi J, Ma Y (2018) Cloning, expression, and characterization of an alkaline protease, AprV, from vibrio sp. DA1-1. Bioprocess Biosyst Eng 41(10):1437–1447. https://doi.org/10.1007/s00449-018-1972-6

Cui W, Suo F, Cheng J, Han L, Hao W, Guo J, Zhou Z (2018) Stepwise modifications of genetic parts reinforce the secretory production of nattokinase in *Bacillus subtilis*. Microb Biotechnol 11(5):930–942. https://doi.org/10.1111/1751-7915.13298

Dağlioğlu C (2017) Cloning, expression, and activity analysis of human cathepsin C in the yeast *Pichia pastoris*. Turk J Biol 41:746–753. https://doi.org/10.3906/biy-1704-4

De Marco A, Deuerling E, Mogk A, Tomoyasu T, Bukau B (2007) Chaperone-based procedure to increase yields of soluble recombinant proteins produced in *E. coli*. BMC Biotechnol 7(1). https://doi.org/10.1186/1472-6750-7-32

Degering C, Eggert T, Puls M, Bongaerts J, Evers S, Maurer K, Jaeger K (2010) Optimization of protease secretion in *Bacillus subtilis* and *Bacillus licheniformis* by screening of homologous and heterologous signal peptides. Appl Environ Microbiol 76(19):6370–6376. https://doi.org/10.1128/aem.01146-10

Dutta S, Dattagupta JK, Biswas S (2013) Enhancement of proteolytic activity of a thermostable papain-like protease by structure-based rational design. PLoS One 8(5):e62619. https://doi.org/10.1371/journal.pone.0062619

El-Sayed GM, Agwa MM, Emam MT, Kandil H, Abdelhamid AE, Nour SA (2024) Utilizing immobilized recombinant serine alkaline protease from *bacillus safensis* lab418 in wound healing: gene cloning, heterologous expression, optimization, and characterization. Int J Biol Macromol 270:132286. https://doi.org/10.1016/j.ijbiomac.2024.132286

Fang J, Zhou G, Ji X, Zhang G, Peng Z, Zhang J (2022) Design of 5′-UTR to enhance keratinase activity in *Bacillus subtilis*. Fermentation 8(9):426. https://doi.org/10.3390/fermentation8090426

Figueiredo J, Silva MS, Figueiredo A (2017) Subtilisin-like proteases in plant defence: the past, the present and beyond. Mol Plant Pathol 19(4):1017–1028. https://doi.org/10.1111/mpp.12567

Golich FC, Han M, Crowder MW (2005) Over-expression, purification, and characterization of aminopeptidase N from *Escherichia coli*. Protein Expr Purif 47(2):634–639. https://doi.org/10.1016/j.pep.2005.11.012

Gong J, Ye J, Tao L, Su C, Qin J, Zhang Y, Li H, Li H, Xu Z, Shi J (2020) Efficient keratinase expression via promoter engineering strategies for degradation of feather wastes. Enzym Microb Technol 137:109550. https://doi.org/10.1016/j.enzmictec.2020.109550

Gustafsson C, Govindarajan S, Minshull J (2004) Codon bias and heterologous protein expression. Trends Biotechnol 22(7):346–353. https://doi.org/10.1016/j.tibtech.2004.04.006

Haque MM, Khatun M, Mosharaf MK, Rahman A, Haque MA, Nahar K (2023) Biofilm producing probiotic bacteria enhance productivity and bioactive compounds in tomato. Biocatal Agric Biotechnol 50:102673. https://doi.org/10.1016/j.bcab.2023.102673

Hoque MA, Gross RA, Koffas M a G (2024) Papain expression in the *Escherichia coli* cytoplasm by T7-promoter engineering and co-expression with human protein disulfide isomerase (PDI) and thiol peroxidase (GPx7) genes. Appl Environ Microbiol 91(1). https://doi.org/10.1128/aem.02119-24

Horiuchi H, Ashikari T, Amachi T, Yoshizumi H, Takagi M, Yano K (1990) High-level secretion of a *Rhizoupus niveus* aspartic proteinase in *Saccharomyces cerevisiae*. Agric Biol Chem 54(7): 1771–1779. https://doi.org/10.1271/bbb1961.54.1771

Hou Y, Lu F, Tian J, Tian Y (2019) Cloning, heterologous expression and characterization of an intracellular serine protease from *bacillus* sp. LCB10. Appl Biochem Microbiol 55(5):482–488. https://doi.org/10.1134/s0003683819050168

Iqbal I, Aftab MN, Afzal M, Ur-Rehman A, Aftab S, Zafar A, Ud-Din Z, Khuharo AR, Iqbal J, Ul-Haq I (2014) Purification and characterization of cloned alkaline protease gene of *Geobacillus stearothermophilus*. J Basic Microbiol 55(2):160–171. https://doi.org/10.1002/jobm.201400190

Jiang C, Ye C, Liu Y, Huang K, Jiang X, Zou D, Li L, Han W, Wei X (2022) Genetic engineering for enhanced production of a novel alkaline protease BSP-1 in *Bacillus amyloliquefaciens*. Front Bioeng Biotechnol 10:977215. https://doi.org/10.3389/fbioe.2022.977215

Joshi S, Satyanarayana T (2012) Characteristics and applications of a recombinant alkaline serine protease from a novel bacterium *bacillus lehensis*. Bioresour Technol 131:76–85. https://doi.org/10.1016/j.biortech.2012.12.124

Kang C, Yu X, Xu Y (2014) Cloning and expression of a novel prolyl endopeptidase from *aspergillus oryzae* and its application in beer stabilization. J Ind Microbiol Biotechnol 42(2): 263–272. https://doi.org/10.1007/s10295-014-1571-8

Karbalaei M, Rezaee SA, Farsiani H (2020) *Pichia pastoris*: a highly successful expression system for optimal synthesis of heterologous proteins. J Cell Physiol 235(9):5867–5881. https://doi.org/10.1002/jcp.29583

Ke Y, Huang W, Li J, Xie M, Luo X (2012) Enzymatic characteristics of a recombinant neutral protease I (rNpI) from *aspergillus oryzae* expressed in *Pichia pastoris*. J Agric Food Chem 60(49):12164–12169. https://doi.org/10.1021/jf303167r

Ke Y, Yuan X, Li J, Zhou W, Huang X, Wang T (2018) High-level expression, purification, and enzymatic characterization of a recombinant *Aspergillus sojae* alkaline protease in *Pichia pastoris*. Protein Expr Purif 148:24–29. https://doi.org/10.1016/j.pep.2018.03.009

Kim T, Lei XG (2005) Expression and characterization of a thermostable serine protease (TfpA) from *Thermomonospora fusca* YX in *Pichia pastoris*. Appl Microbiol Biotechnol 68(3): 355–359. https://doi.org/10.1007/s00253-005-1911-8

Kim M, Kim S, Lee JH, Seo J, Lee J, Kim J, Kim Y, Nam S (2008) High-level secretory expression of human procarboxypeptidase B by fed-batch cultivation of *Pichia pastoris* and its partial characterization. J Microbiol Biotechnol 18(12):1938–1944. https://doi.org/10.4014/jmb.0800.078

Kjeldsen T, Andersen AS, Hubálek F, Johansson E, Kreiner FF, Schluckebier G, Kurtzhals P (2023) Molecular engineering of insulin for recombinant expression in yeast. Trends Biotechnol 42(4): 464–478. https://doi.org/10.1016/j.tibtech.2023.09.012

Klein T, Bischoff R (2010) Physiology and pathophysiology of matrix metalloproteases. Amino Acids 41(2):271–290. https://doi.org/10.1007/s00726-010-0689-x

Komarevtsev SK, Evseev PV, Shneider MM, Popova EA, Tupikin AE, Stepanenko VN, Kabilov MR, Shabunin SV, Osmolovskiy AA, Miroshnikov KA (2021) Gene analysis, cloning, and heterologous expression of protease from a Micromycete *Aspergillus ochraceus* capable of activating protein C of blood plasma. Microorganisms 9(9):1936. https://doi.org/10.3390/microorganisms9091936

Krem MM (2001) Molecular markers of serine protease evolution. EMBO J 20(12):3036–3045. https://doi.org/10.1093/emboj/20.12.3036

Latiffi AA, Salleh AB, Rahman RN, Oslan SN, Basri M (2013) Secretory expression of thermostable alkaline protease from Bacillus stearothermophilus FI by using native signal peptide and α-factor secretion signal in *Pichia pastoris*. Genes Genet Syst 88(2):85–91. https://doi.org/10.1266/ggs.88.85

Li Q, Lu J, Zhang G, Liu S, Zhou J, Du G, Chen J (2022) Recent advances in the development of *Aspergillus* for protein production. Bioresour Technol 348:126768. https://doi.org/10.1016/j.biortech.2022.126768

Lin S, Zhang M, Liu J, Jones GS (2014) Construction and application of recombinant strain for the production of an alkaline protease from *Bacillus licheniformis*. J Biosci Bioeng 119(3): 284–288. https://doi.org/10.1016/j.jbiosc.2014.08.002

Lin-Cereghino GP, Stark CM, Kim D, Chang J, Shaheen N, Poerwanto H, Agari K, Moua P, Low LK, Tran N, Huang AD, Nattestad M, Oshiro KT, Chang JW, Chavan A, Tsai JW, Lin-Cereghino J (2013) The effect of α-mating factor secretion signal mutations on recombinant protein expression in *Pichia pastoris*. Gene 519(2):311–317. https://doi.org/10.1016/j.gene.2013.01.062

Liu X, Kokare C (2016) Microbial enzymes of use in industry. In: Elsevier eBooks, pp 267–298. https://doi.org/10.1016/b978-0-12-803725-6.00011-x

Liu Y, Shi C, Li D, Chen X, Li J, Zhang Y, Yuan H, Li Y, Lu F (2019) Engineering a highly efficient expression system to produce BcaPRO protease in *Bacillus subtilis* by an optimized promoter and signal peptide. Int J Biol Macromol 138:903–911. https://doi.org/10.1016/j.ijbiomac.2019.07.175

Mamo J, Assefa F (2018) The role of microbial aspartic protease enzyme in food and beverage industries. J Food Qual 2018:1–15. https://doi.org/10.1155/2018/7957269

Mechri S, Jaouadi NZ, Bouacem K, Allala F, Bouraoui A, Ferard C, Rekik H, Noiriel A, Abousalham A, Bouanane-Darenfed A, Hacène H, Lederer F, Baciou L, Jaouadi B (2021) Cloning and heterologous expression of subtilisin SAPN, a serine alkaline protease from *Melghiribacillus thermohalophilus* Nari2AT in *Escherichia coli* and *Pichia pastoris*. Process Biochem 105:27–41. https://doi.org/10.1016/j.procbio.2021.03.020

Miao C, Han L, Lu Y, Feng H (2020) Construction of a high-expression system in *Bacillus* through transcriptomic profiling and promoter engineering. Microorganisms 8(7):1030. https://doi.org/10.3390/microorganisms8071030

Miao H, Xiang X, Han N, Wu Q, Huang Z (2023) Improving the Thermostability of serine protease PB92 from *bacillus alcalophilus* via site-directed mutagenesis based on semi-rational design. Foods 12(16):3081. https://doi.org/10.3390/foods12163081

Mirsalami SM, Mirsalami M (2025) Advances in genetically engineered microorganisms: transforming food production through precision fermentation and synthetic biology. Future Foods:100601. https://doi.org/10.1016/j.fufo.2025.100601

Mo F, Cai D, He P, Yang F, Chen Y, Ma X, Chen S (2019) Enhanced production of heterologous proteins via engineering the cell surface of *bacillus licheniformis*. J Ind Microbiol Biotechnol 46(12):1745–1755. https://doi.org/10.1007/s10295-019-02229-8

Mohsen AHS, Fereshteh E, Bagher Y, Tehrani DM (2011) Cloning and enhanced expression of an extracellular alkaline protease from a soil isolate of *Bacillus clausii* in *Bacillus subtilis*. Iran J Biotechnol 9(4):275–280. https://www.sid.ir/en/VEWSSID/J_pdf/92820110404.pdf

Mótyán J, Tóth F, Tőzsér J (2013) Research applications of proteolytic enzymes in molecular biology. Biomolecules 3(4):923–942. https://doi.org/10.3390/biom3040923

Nandan A, Nampoothiri KM (2017) Molecular advances in microbial aminopeptidases. Bioresour Technol 245:1757–1765. https://doi.org/10.1016/j.biortech.2017.05.103

Nandan A, Nampoothiri KM (2020) Therapeutic and biotechnological applications of substrate specific microbial aminopeptidases. Appl Microbiol Biotechnol 104(12):5243–5257. https://doi.org/10.1007/s00253-020-10641-9

Nilegaonkar SS, Zambare VP, Kanekar PP, Dhakephalkar PK, Sarnaik SS (2007) Production of dehairing protease by *Bacillus cereus* MCM B 326. Bioresour Technol 98(6):1238–1245. https://doi.org/10.1016/j.biortech.2006.05.003

Nishiya Y, Imanaka T (1990) Cloning and nucleotide sequences of the *Bacillus stearothermophilus* neutral protease gene and its transcriptional activator gene. J Bacteriol 172(9):4861–4869. https://doi.org/10.1128/jb.172.9.4861-4869.1990

Parekh V, Rathod V, Pandit A (2011) Substrate hydrolysis. In: Elsevier eBooks, pp 103–118. https://doi.org/10.1016/b978-0-08-088504-9.00094-5

Pawar R, Zambare V, Barve S, Paratkar G (2009) Application of protease isolated from *Bacillus* sp.-158 in enzymatic cleaning of contact lenses. Biotechnology 8(2):276–280. https://doi.org/10.3923/biotech.2009.276.280

Pawar KS, Singh PN, Singh SK (2023) Fungal alkaline proteases and their potential applications in different industries. Front Microbiol 14. https://doi.org/10.3389/fmicb.2023.1138401

Radha S, Gunasekaran P (2009) Purification and characterization of keratinase from recombinant *Pichia* and *Bacillus* strains. Protein Expr Purif 64(1):24–31. https://doi.org/10.1016/j.pep.2008.10.008

Rao MB, Tanksale AM, Ghatge MS, Deshpande VV (1998) Molecular and biotechnological aspects of microbial proteases. Microbiol Mol Biol Rev 62(3):597–635. https://doi.org/10.1128/mmbr.62.3.597-635.1998

Razzaq A, Shamsi S, Ali A, Ali Q, Sajjad M, Malik A, Ashraf M (2019) Microbial proteases applications. Front Bioeng Biotechnol 7. https://doi.org/10.3389/fbioe.2019.00110

Savijoki K, Ingmer H, Varmanen P (2006) Proteolytic systems of lactic acid bacteria. Appl Microbiol Biotechnol 71(4):394–406. https://doi.org/10.1007/s00253-006-0427-1

Sittipol D, Saelao P, Lohnoo T, Lerksuthirat T, Kumsang Y, Yingyong W, Khunrae P, Rattanarojpong T, Jongruja N (2019) Cloning, expression, purification and characterization of a thermo- and surfactant-stable protease from *Thermomonospora curvata*. Biocatal Agric Biotechnol 19:101111. https://doi.org/10.1016/j.bcab.2019.101111

Skowron PM, Krefft D, Brodzik R, Kasperkiewicz P, Drag M, Koller K (2020) An alternative for proteinase K-heat-sensitive protease from fungus *Onygena corvina* for biotechnology: cloning, engineering, expression, characterization and special application for protein sequencing. Microb Cell Factories 19(1). https://doi.org/10.1186/s12934-020-01392-3

Song P, Cheng L, Tian K, Zhang M, Mchunu NP, Niu D, Singh S, Prior B, Wang Z (2020) Biochemical characterization of two new *Aspergillus niger* aspartic proteases. *3* Biotech 10(7). https://doi.org/10.1007/s13205-020-02292-4

Song P, Zhang X, Wang S, Xu W, Wang F, Fu R, Wei F (2023) Microbial proteases and their applications. Front Microbiol 14. https://doi.org/10.3389/fmicb.2023.1236368

Souza PM, Werneck G, Aliakbarian B, Siqueira F, Filho EXF, Perego P, Converti A, Magalhães PO, Pessoa A, Junior (2017) Production, purification and characterization of an aspartic protease from *Aspergillus foetidus*. Food Chem Toxicol 109:1103–1110. https://doi.org/10.1016/j.fct.2017.03.055

Srivastava N, Shiburaj S, Khare SK (2024) Improved production of alkaline and solvent-stable proteases from a halotolerant *Exiguobacterium* isolate through heterologous expression. Int J Biol Macromol 260:129507. https://doi.org/10.1016/j.ijbiomac.2024.129507

Suberu Y, Akande I, Samuel T, Lawal A, Olaniran A (2019) Cloning, expression, purification and characterisation of serine alkaline protease from *Bacillus subtilis* RD7. Biocatal Agric Biotechnol 20:101264. https://doi.org/10.1016/j.bcab.2019.101264

Sun Q, Chen F, Geng F, Luo Y, Gong S, Jiang Z (2017) A novel aspartic protease from *Rhizomucor miehei* expressed in *Pichia pastoris* and its application on meat tenderization and preparation of turtle peptides. Food Chem 245:570–577. https://doi.org/10.1016/j.foodchem.2017.10.113

Sussman F, Villaverde MC, Dominguez JL, Danielson UH (2013) On the active site protonation state in aspartic proteases: implications for drug design. Curr Pharm Des 19(23):4257–4275. https://doi.org/10.2174/13816128113199230009

Takenaka S, Takada A, Kimura Y, Watanabe M, Kuntiya A (2021) Improvement of the halotolerance of a *Bacillus* serine protease by protein surface engineering. J Basic Microbiol 62(2):174–184. https://doi.org/10.1002/jobm.202100335

Takyu Y, Asamura T, Okamoto A, Maeda H, Takeuchi M, Kusumoto K, Katase T, Ishida H, Tanaka M, Yamagata Y (2022) A novel milk-clotting enzyme from *Aspergillus oryzae* and *A. luchuensis* is an aspartic endopeptidase PepE presumed to be a vacuolar enzyme. Biosci Biotech Bioch 86(3):413–422. https://doi.org/10.1093/bbb/zbac005

Tang X, Shen W, Lakay F, Shao W, Wang Z, Prior B, Zhuge J (2004) Cloning and over-expression of an alkaline protease from *Bacillus licheniformis*. Biotechnol Lett 26(12):975–979. https://doi.org/10.1023/b:bile.0000030042.91094.38

Tang W, Li Z, Li C, Yu X, Wang F, Wan X, Wang Y, Ma L (2016) High-level expression and characterization of the *Bacillus subtilis* subsp. subtilis str. BSP1 YwaD aminopeptidase in *Pichia pastoris*. Protein Expr Purif 122:23–30. https://doi.org/10.1016/j.pep.2016.02.009

Tekin A, Uzuner U, Sezen K (2020) Homology modeling and heterologous expression of highly alkaline subtilisin-like serine protease from *Bacillus halodurans* C-125. Biotechnol Lett 43(2): 479–494. https://doi.org/10.1007/s10529-020-03025-6

Thakrar FJ, Koladiya GA, Singh SP (2023) Heterologous expression and structural elucidation of a highly thermostable alkaline serine protease from Haloalkaliphilic *Actinobacterium, Nocardiopsis* sp. Mit-7. Appl Biochem Biotechnol 195(12):7583–7602. https://doi.org/10.1007/s12010-023-04472-3

Tian J, Long X, Tian Y, Shi B (2019) Enhanced extracellular recombinant keratinase activity in *Bacillus subtilis* SCK6 through signal peptide optimization and site-directed mutagenesis. RSC Adv 9(57):33337–33344. https://doi.org/10.1039/c9ra07866e

Wang L, Zhou Y, Huang Y, Wei Q, Huang H, Guo C (2019) Cloning and expression of a thermostable keratinase gene from *Thermoactinomyces* sp. YT06 in *Escherichia coli* and characterization of purified recombinant enzymes. World J Microbiol Biotechnol 35(9). https://doi.org/10.1007/s11274-019-2710-1

Wang S, Zhang P, Xue Y, Yan Q, Li X, Jiang Z (2021) Characterization of a novel aspartic protease from *Rhizomucor miehei* expressed in *Aspergillus niger* and its application in production of ACE-inhibitory peptides. Foods 10(12):2949. https://doi.org/10.3390/foods10122949

Wang X, Qin X, Tong L, Zheng J, Dong T, Wang X, Wang Y, Huang H, Yao B, Zhang H, Luo H (2023) Improving the catalytic activity of a detergent-compatible serine protease by rational design. Microb Biotechnol 16(5):947–960. https://doi.org/10.1111/1751-7915.14218

Wei M, Chen P, Zheng P, Tao X, Yu X, Wu D (2023) Purification and characterization of aspartic protease from *Aspergillus niger* and its efficient hydrolysis applications in soy protein degradation. Microb Cell Factories 22(1):42. https://doi.org/10.1186/s12934-023-02047-9

Wu S, Feng C, Zhong J, Huan L (2010) Enhanced production of recombinant nattokinase in *Bacillus subtilis* by promoter optimization. World J Microbiol Biotechnol 27(1):99–106. https://doi.org/10.1007/s11274-010-0432-5

Xiao H, Liu X, Feng Y, Zheng L, Zhao M, Huang M (2022) Secretion of collagenases by *Saccharomyces cerevisiae* for collagen degradation. Biotechnol Biofuels Bioprod 15(1). https://doi.org/10.1186/s13068-022-02186-y

Xie G, Shao Z, Zong L, Li X, Cong D, Huo R (2019) Heterologous expression and characterization of a novel subtilisin-like protease from a thermophilic *Thermus thermophilus* HB8. Int J Biol Macromol 138:528–535. https://doi.org/10.1016/j.ijbiomac.2019.07.101

Yamagata Y, Abe R, Fujita Y, Ichishima E (1995) Molecular cloning and nucleotide sequence of the 90k serine protease gene, hspK, from *Bacillus subtilis* (natto) no. 16. Curr Microbiol 31(6): 340–344. https://doi.org/10.1007/bf00294696

Yang X, Cong H, Song J, Zhang J (2013) Heterologous expression of an aspartic protease gene from biocontrol fungus *Trichoderma asperellum* in *Pichia pastoris*. World J Microbiol Biotechnol 29(11):2087–2094. https://doi.org/10.1007/s11274-013-1373-6

Yang N, Matthew MA, Yao C (2023) Roles of cysteine proteases in biology and pathogenesis of parasites. Microorganisms 11(6):1397. https://doi.org/10.3390/microorganisms11061397

Yao H, Liu S, Liu T, Ren D, Zhou Z, Yang Q, Mao J (2023) Microbial-derived salt-tolerant proteases and their applications in high-salt traditional soybean fermented foods: a review. Bioresour Bioprocess 10(1). https://doi.org/10.1186/s40643-023-00704-w

Yoon J, Kimura S, Maruyama J, Kitamoto K (2008) Construction of quintuple protease gene disruptant for heterologous protein production in *Aspergillus oryzae*. Appl Microbiol Biotechnol 82(4):691–701. https://doi.org/10.1007/s00253-008-1815-5

Yu G, Liu J, Xie L, Wang X, Zhang S, Pan H (2012) Characterization, cloning, and heterologous expression of a subtilisin-like serine protease gene VlPr1 from *Verticillium lecanii*. J Microbiol 50(6):939–946. https://doi.org/10.1007/s12275-012-2199-x

Zambare VP, Nilegaonkar SS, Kanekar PP (2011) Production optimization and purification of a novel extracellular protease from *Pseudomonas aeruginosa* MCM B-327. New Biotechnol 28(2):173–181. https://doi.org/10.1016/j.nbt.2010.10.002

Zambare VP, Nilegaonkar SS, Kanekar PP (2014) Scale up production of protease using *Pseudomonas aeruginosa* MCM B-327 and its detergent compatibility. J Biochem Technol 5(2): 698–707

Zhang Y, Huang H, Yao X, Du G, Chen J, Kang Z (2018) High-yield secretory production of stable, active trypsin through engineering of the N-terminal peptide and self-degradation sites in *Pichia pastoris*. Bioresour Technol 247:81–87. https://doi.org/10.1016/j.biortech.2017.08.006

Zhang L, Lin D, Sun X, Curth U, Drosten C, Sauerhering L, Becker S, Rox K, Hilgenfeld R (2020) Crystal structure of SARS-CoV-2 main protease provides a basis for design of improved α-ketoamide inhibitors. Science 368(6489):409–412. https://doi.org/10.1126/science.abb3405

Zhao X, Xu J, Tan M, Zhen J, Shu W, Yang S, Ma Y, Zheng H, Song H (2020) High copy number and highly stable *Escherichia coli–Bacillus subtilis* shuttle plasmids based on pWB980. Microb Cell Factories 19(1). https://doi.org/10.1186/s12934-020-1296-5

Zhou C, Zhou H, Li D, Zhang H, Wang H, Lu F (2020) Optimized expression and enhanced production of alkaline protease by genetically modified *Bacillus licheniformis* 2709. Microb Cell Factories 19(1). https://doi.org/10.1186/s12934-020-01307-2

Zhu F, Liu F, Wu B, He B (2016) Efficient extracellular expression of metalloprotease for Z-aspartame synthesis. J Agric Food Chem 64(51):9631–9638. https://doi.org/10.1021/acs.jafc.6b04164

Chapter 6
Partitioning Techniques for Protease Purification: Exploring the Potential of Two-Phase and Three-Phase Systems

Endang Ismail, Asep Awaludin Prihanto, and Sukoso Sukoso

Abstract Purification of proteases is an important step in obtaining high-purity enzymes for industrial and research applications. One promising technique to achieve this goal is phase separation, which utilizes differences in molecular affinity to two or more separate phases. This chapter looks at how well two-phase and three-phase systems work for purifying proteases. It focuses on the different ways they separate, how to choose the right solvent, and the pros and cons of each system. Two-phase solutions, such as polymer-based systems and organic solvents, have been shown to work well for separating proteases based on their hydrophobic properties. Three-phase techniques also have the extra benefit of splitting the mixture into three parts, which makes the purification process more efficient and allows for more exact enzyme separation. We look at a lot of experimental studies that use two-phase and three-phase systems to see how well these methods work for purifying proteases from both native and recombinant sources. The results of these studies show that two-phase and three-phase systems could be very useful in industry, but there are problems with stability and operational costs that need to be worked out before they can be used. For the future, this chapter gives useful information that will help make protease purification procedures more efficient and less expensive.

Keywords Aqueous · Enzyme stability · Phase system separation · Protease · Purification

E. Ismail
Department of Fishery Product Technology, Faculty of Fisheries and Marine Sciences, Brawijaya University, Malang, East Java, Indonesia

Secondary Fisheries Business School, Kotaagung, Lampung, Indonesia

A. A. Prihanto (✉) · S. Sukoso
Department of Fishery Product Technology, Faculty of Fisheries and Marine Sciences, Brawijaya University, Malang, East Java, Indonesia
e-mail: asep_awa@ub.ac.id

V. Zambare et al. (eds.), *Biotechnological Advances and Applications in Microbial Proteases*, Interdisciplinary Biotechnological Advances,
https://doi.org/10.1007/978-981-95-9011-7_6

6.1 Introduction

Proteases or proteolysis enzymes are a group of enzymes that play a role in the hydrolysis of peptide bonds in proteins, making them one of the most important enzymes in the hydrolysis of peptide bonds in proteins and in various biotechnological, industrial, and medical fields. Proteases are widely used in the food industry, detergents, leather, textiles, and in the pharmaceutical field, especially in the therapy of diseases such as cancer, inflammation, and metabolic disorders (Zambare and Nilegaonkar 2016; Ojo Omoniyi et al. 2024). With the increasing demand for these enzymes in various sectors, the need for efficient, economical, and environmentally friendly purification methods is becoming increasingly important.

Conventional protease purification processes involve multiple stages of chromatography and precipitation techniques, which, although effective, are often expensive, time-consuming, and difficult to implement on an industrial scale. In recent decades, aqueous two-phase partitioning systems (ATPS) and three-phase systems (TPP) have developed as viable alternatives for the purification of biomolecules, such as proteases. These approaches are recognized for their simplicity, scalability, and operation under mild settings that maintain the biological activity of the enzymes (Pereira and Coutinho 2020).

ATPS typically have two primary components that do not fully amalgamate, such as polymers (e.g., polyethylene glycol and dextran) or polymer-salt combinations in an aqueous solution. These systems facilitate selective separation by exploiting variations in solubility, size, and surface interactions among biomolecules and phase components. This technology has advanced swiftly with the advent of eco-friendly solvents such as ionic liquids (ILs) and deep eutectic solvents (DES), providing sustainable and tailored options for many biotechnological applications (Buarque et al. 2023). Conversely, TPP is an integrated precipitation and extraction technique that employs an organic solvent (typically t-butane) and a salt, such as ammonium sulphate, to enable protein partitioning into three distinct phases: the upper phase (organic solvent), the middle phase (precipitated protein), and the lower phase (salt solution). This technique has demonstrated significant efficacy in the purification of proteases from diverse sources, including microbes and plants, yielding large outputs and satisfactory purity (Gimenes et al. 2021). The benefits of TPP encompass the integration of extraction and concentration into a singular phase, hence enhancing overall process efficiency.

The efficacy of partitioning systems is influenced by various elements, including the characteristics of the phases, the specific protease employed, and the operational parameters. Two-phase and three-phase partitioning systems provide numerous benefits, including simplicity, ease of scalability, and the capacity to function at mild circumstances, hence maintaining protease activity and stability. Moreover, these methods can frequently be tailored for the targeted separation of proteases possessing distinct characteristics, hence enhancing the overall efficacy of the purifying process. (Hao et al. 2022).

Numerous studies indicate that hybrid methodologies integrating ATPS and TPP demonstrate enhanced yield and purity relative to the exclusive application of either technique. This method highlights the significance of comprehending the synergy among purification approaches to formulate more efficient and cost-effective downstream processing solutions. Moreover, the use of partitioning systems has been expanded to the purification of recombinant proteases, which are essential to the contemporary biotechnology sector (Akkaya et al. 2025). Due to the escalating intricacy of biological systems and industrial requirements for efficiency and sustainability, study into two-phase and three-phase partitioning systems has emerged as a primary emphasis in the advancement of enzyme purification technologies. This review is to investigate the efficacy of two-phase and three-phase partitioning systems in protease purification, emphasizing their latest advancements and uses in both laboratory and industrial settings. The chapter examines the obstacles and future prospects regarding the incorporation of partitioning strategies into protease purification operations.

6.2 The Important Role of Proteases in Industry and Medicine

Proteases, or proteinases and peptidases, are enzymes that catalyze the breakdown of peptide bonds in proteins. These enzymes are indispensable in the industrial and medical fields due to their diverse functions and broad substrate specificity. Its capacity to decompose proteins into peptides and amino acids renders it significant in numerous biological processes and industrial uses (David Troncoso et al. 2022). Due to their characteristic of being able to recognize and break peptide bonds specifically, proteases are becoming invaluable biological tools in various industrial and medical sectors. In the industrial field, proteases account for more than 60% of the global market for industrial enzymes, making them the enzyme with the largest production volume (Pawar et al. 2023). Proteases contribute significantly to improving efficiency and sustainability and reducing the use of hazardous chemicals in industrial processes. Some of the major applications of proteases in various industrial sectors include food processing, detergents, leather and textile industries, and pharmaceuticals, as presented in Fig. 6.1. In the detergent business, proteases are extensively utilized for the elimination of protein-based stains, including blood and milk. Alkaline proteases (Ragab et al. 2022), Especially favored for their stability and efficacy under elevated pH and temperature conditions are a major component in modern detergent formulations (Zambare et al. 2014; Yang et al. 2024), These enzymes not only improve cleaning performance but are also environmentally friendly as they reduce the need for phosphates and synthetic chemicals, making them effective for household and industrial cleaning activities (Nilegaonkar et al. 2006, 2007a, b). Food processing: Proteases play an important role in the production of various processed food products, such as tenderizing meat (Patil et al. 2024);

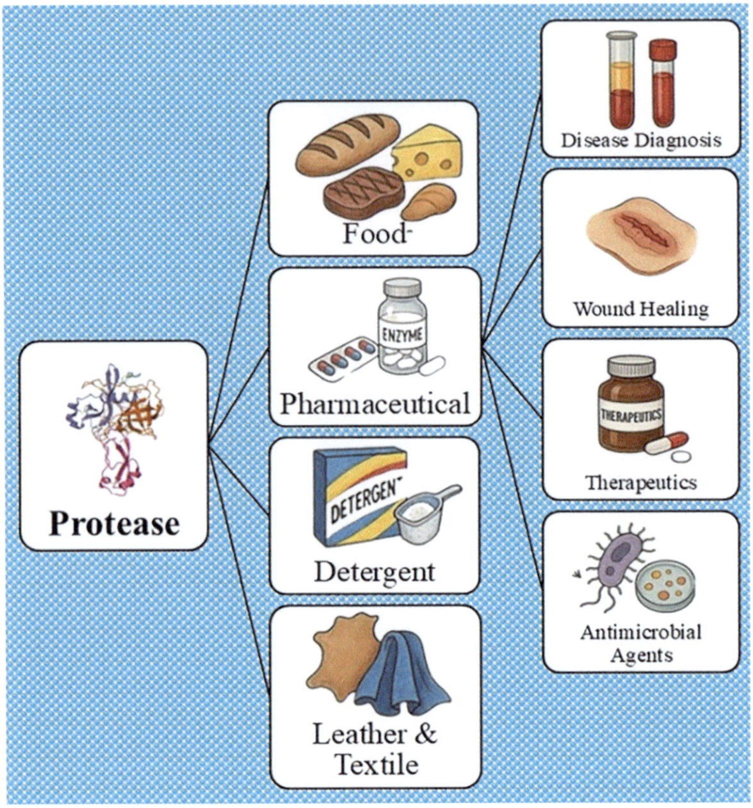

Fig. 6.1 Application of proteases in several industries, especially in the field of health

(Handayani et al. 2025), cheese production, clarifying juices and beer (Santos et al. 2024). Moreover, proteases contribute to the synthesis of protein hydrolyses utilized as dietary supplements or infant nutrition due to their enhanced digestibility (Sargazi et al. 2024); (Honrado et al. 2024), Leather and textile industry: In the tanning industry, protease is a substitute for harsh chemicals for dehairing and bating processes, reducing the environmental impact of toxic waste, and improving work safety (Nilegaonkar et al. 2007; Zambare et al. 2011; Becker da Silva et al. 2025). In the textile industry, proteases are used to remove sericin (degumming) in order to improve the quality of silk fibers and optimize the dyeing process, Protease also applied in the refining of cotton fabrics to enhance fabric quality (Sarwa et al. 2024); (Pan et al. 2025). Pharmaceutical and biomedical industry: Proteases have important value in the production of bioactive peptides, drug development, and enzyme-based therapies. Some microbial proteases are used for contact lens cleaning (Pawar et al. 2009). Some proteases, such as trypsin and pepsin, are used on an industrial scale to produce vaccines and recombinant antibodies (Kurnianto et al. 2024). In the clinical field, proteases are used in enzyme therapy, wound treatment, and disease detection, such as treating lysosomal storage disorders (Manzoli et al. 2024).

Proteases also play a role in disease treatment, such as enzyme therapy and the development of protease-based anti-cancer drugs (Bege and Borbás 2024) (Hussain et al. 2025). In medicine, proteases are essential in understanding disease mechanisms and developing therapeutic interventions. Proteases are important for breaking down proteins, which is necessary for cells to work properly and stay in balance. Proteases can be used as biomarkers to help doctors figure out what is wrong with a person, such as cancer, inflammation, or an infection. Matrix metalloproteinase (MMPs) is often overexpressed in cancerous tissues and may be used as markers to find cancer early (Vizovisek et al. 2021). Therapy: Protease inhibitors are widely used as drugs to treat illnesses, including HIV/AIDS (e.g., HIV protease inhibitors) and high blood pressure (e.g., ACE inhibitors). Also, tissue plasminogen activator (tPA) is an enzyme that is used to break up blood clots in persons who have had a stroke (Ćwilichowska et al. 2022).

1. Wound healing: Surgeons use proteolysis enzymes as collagen Channelize to remove dead tissue and speed up the healing process. These enzymes are a less invasive alternative to surgery (Baydogan et al. 2025).
2. Antimicrobial agents: Some proteases have antimicrobial properties and are being looked at as possible replacements for traditional antibiotics. Their ability to break down bacterial biofilms is a good sign for the treatment of diseases that do not respond to antibiotics (Aini et al. 2024, (Debbarma et al. 2024).

6.3 Drawbacks in Protease Purification

Genetic engineering and recombinant DNA technologies have made it easier to make proteases utilizing biotechnology. These changes have made it easier to make tailored proteases that are more stable, specialized, and active for certain tasks. However, there are still problems in making proteases on a large scale and keeping them stable in harsh conditions. Protein engineering and the search for new sources, such as extremophile microorganisms, are two ways that people are trying to get around these problems. Proteases are useful tools in both industry and medicine. They help encourage new ideas and make many processes more efficient. Their varied uses and capacity to adapt make them a major area of research and development. As technology improves, proteases are projected to have more uses, leading to long-lasting solutions and better treatments for problems around the world.

Proteases are a type of enzyme that breaks down peptide links in proteins. They are important for many biological and commercial processes. Despite their many uses, maintaining proteases as they are throughout the purification process is very difficult. These problems come from the way proteases work, how they interact with other proteins, and how complicated the purifying process is. This chapter talks about the main problems that come up while trying to purify proteases.

Protease unstable: Proteases often break down or denature easily, which makes them unstable during purification (Sujitha and Shanthi 2023). Changes in temperature, pH, or ionic strength during the purification process could cause this instability. To avoid autolysis, researchers must carefully tune conditions and often use protease inhibitors, which may make it harder to use them later.

Low yield: Protease purification often results in low yields due to losses during several purification steps. These losses can be caused by nonspecific binding to the chromatography resin, degradation by other proteases in the sample, or precipitation during buffer exchange (Medina et al. 2024).

Source material complexity: Proteases are often extracted from complex biological matrices such as microbial cultures, animal tissues, or plant materials. These sources contain diverse proteins and other biomolecules that can interfere with purification, thus requiring multiple steps to achieve adequate purity (Moreno et al. 2024).

Protein co-purification: Many proteases share physical and chemical properties with other proteins, making selective separation difficult. Co-purification of unwanted proteins can lead to contamination and reduce the overall activity and specificity of the target protease (Contreras et al. 2024).

Optimization of chromatographic techniques: Although techniques such as ion exchange, size exclusion, and affinity chromatography are widely used, each protease requires a unique combination of methods. Developing a robust and scalable purification protocol involves extensive trial and error, which can be time-consuming and resource-intensive (Oliveira et al. 2025).

Cost and scalability: Purification methods that work efficiently on a laboratory scale may not translate well into industrial-scale processes due to increased costs and equipment limitations. Affinity tags or resins, for example, can be very expensive for large-scale applications (Melchor-Moncada et al. 2024).

Accordingly, this chapter delineates the principles, mechanisms, and practical applications of two-phase and three-phase separation systems as robust and efficient strategies for the partial purification of enzymes.

6.4 Aqueous Two-Phase System (ATPS) and Three Phase Partitioning (TPP)

Conventional methods such as column chromatography, membrane filtration, and chemical precipitation often require long time, high cost, and low yield, especially when applied on a large scale (Du et al. 2022). To tackle these issues, phase-based separation systems, specifically ATPS and TPP, are being developed as efficient and cost-effective alternative methods.

ATPS segregates proteases between two immiscible liquid phases, often created by combining polymers such as polyethylene glycol (PEG) (Zhang et al. 2025) or dextran with salts. The partitioning of proteases according to their affinity for each phase is influenced by parameters such as hydrophobicity, charge, and molecule size. ATPS is extensively utilized owing to its simplicity, affordability, and compatibility with biological substances. Enhancing system parameters, including pH, polymer concentration, and salt type, can markedly elevate protease recovery and purity (Jiang et al. 2021).

TPP is an approach for the separation and purification of proteins (or other biomolecules) that integrates the concepts of precipitation and extraction. This technique was initially identified by Dennison and Lovrien (1997) and entails the incorporation of a salt, typically ammonium sulfate, and an organic solvent, commonly t-butanol, into the sample solution. This mix creates three phases: The upper phase has a lot of t-butane in it, the lower phase is a salt-saturated aqueous solution, and the middle phase (interphase) is where the target protein or biomolecule precipitates. The way TPP works is by having sulfate ions (Cosmo tropic) and t-butane work together to make proteins less soluble and cause them to precipitate at the interface of the two liquid phases. TPP is often used to refold denatured proteins or proteins that come from inclusion bodies (Gupta and Roy 2021). It also often increases enzyme activity.

6.4.1 Separation Efficiency and Selectivity

The main benefit of ATPS and TPP is that they can separate and clean proteases in one step or with very few further steps. This method takes advantage of the fact that proteases have different solubility and affinity for two or three phase components. These can be polymers such as PEG-dextran or a mix of salts and mild organic solvents. This method makes it possible to increase the amount of protease in one phase (either the top or bottom), while other proteins or nucleic acids are moved to different phases. This makes selective fractionation better (Zhang et al. 2025). ATPS based on PEG and sodium citrate has been shown to be able to purify proteases from *Streptococcus zooepidemicus*. The recovery rate is above 79.4%, and the purity level is 3.5 times higher than that of standard precipitation methods (Srivastav. 2022). TPP has also been widely used to purify proteases from plants and microbes, obtaining very high separation efficiency in a much shorter amount of time (Lemes et al. 2023).

6.4.2 Simple Process and Nondestructive

Affinity or size-exclusion chromatography needs a lot of processing and buffers, while ATPS and TPP are gentler and do not hurt the enzyme structure, so the protease's biological activity can stay the same. This is very crucial for protease uses in medicine or food, where the enzyme activity needs to stay the same (Cui et al. 2025). This technique also lets the separation happen at room temperature and a neutral pH, which lowers the chance of denaturation or autolysis. Also, using compounds such as ammonium sulfate salts or t-butane with TPP makes the enzyme precipitate in a selective way without changing its three-dimensional structure (Vicente et al. 2024).

6.4.3 Lower Production Costs

Two-phase and three-phase systems are much cheaper than traditional purification methods since they do not need expensive resins, high-pressure chromatography systems, or complicated lab setups. Also, the materials utilized in ATPS and TPP are usually recyclable and easy to find (Bekavac et al. 2024; Oliveira et al. 2025). This efficiency is particularly important for producing large quantities of proteases, for example, for detergent or food applications, where the final product price needs to be competitive.

6.4.4 Scalability and Potential for Process Integration

Another advantage of ATPS and TPP is the ease with which they can be integrated into large-scale processes, both in the form of batch and continuous systems. The technique is not only compatible with fermentation bioreactors but can also be used as part of in-line processing, thus reducing time and labor (Li et al. 2022). This purification technique can also be combined with other purification methods, such as ultrafiltration or short chromatography, to achieve higher purity without losing efficiency. This combination can increase the overall yield while reducing the number of purification steps (Cai et al. 2024; Nemani et al. 2024).

6.5 Basic Principles of Separation Systems

Phase separation is an important biophysical process that contributes to the formation of biomolecular condensates, which are membrane-less compartments within cells (O'Flynn and Mittag 2021). These structures regulate certain biochemical reactions, regulating intracellular signaling (Banani et al. 2017), and mediate stress responses (Wen and Ma 2022). Proteases are key enzymes that help break down peptide bonds. They can use the idea of phase separation to make their activity, control, and substrate specificity better in a cellular setting.

Figure 6.2 schematic representation of the process of isolating protease from fish waste using both aqueous two-phase system (ATPS) and three-phase partitioning (TPP) techniques. In the ATPS approach, the crude enzyme extract is distributed between two immiscible aqueous phases, typically a polymer-rich top phase (e.g., PEG) and a salt-rich bottom phase. Simultaneously, the TPP approach recommends

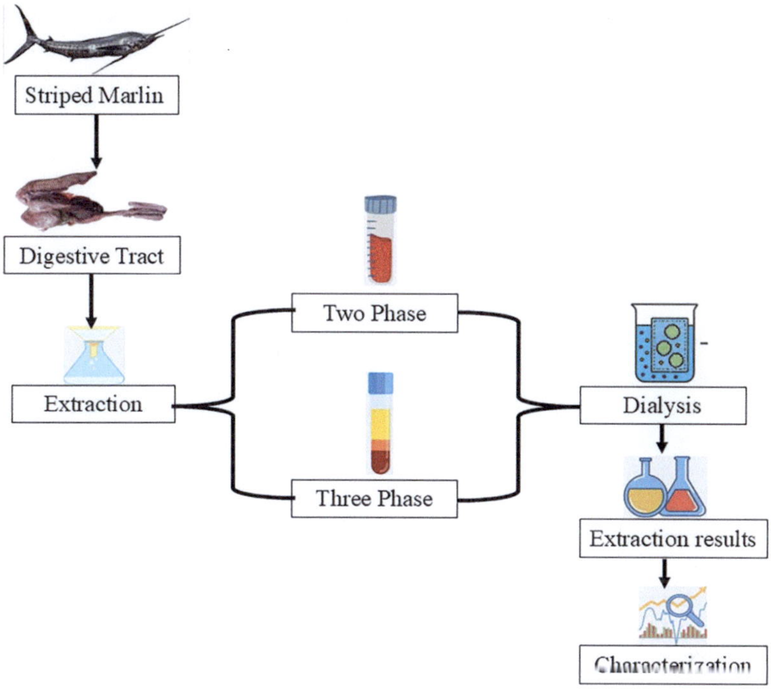

Fig. 6.2 Illustration mechanism separating protease from fish discard with two aqueous two-phase system and three-phase partitioning method

incorporating ammonium sulfate and t-butanol to create three distinct layers: an organic upper phase, an interphase layer rich in protein where protease precipitates, and a lower aqueous phase that holds salts and other hydrophilic impurities

The idea behind phase separation is to mix biomolecules such as proteins and nucleic acids into separate liquid phases. Weak multivalent interactions between intrinsically disordered regions (IDRs) (Borcherds et al. 2020), low complexity domains (LCDs), or certain protein motifs (Banani et al. 2017); (Mittag and Pappu 2022) often cause this to happen. Phase separation can affect several important activities for proteases: Substrate concentration: To speed up the reaction, protease activity needs a spatial concentration of substrate. Phase separation makes a small area where the substrate is concentrated, which makes proteolysis more efficient (Patel et al. 2017). Activity regulation: Proteases can be kept out of or put into certain condensates, which controls how easily they can get to substrates and inhibitors. This dynamic adjustment stops undesired proteolysis and cell injury (Alberti and Dormann 2019). Structural stability: Phase-separated environments can stabilize the structure of proteases or their zymogen precursors, maintaining activity under stress conditions (Franzmann et al. 2018) (Huang et al. 2025).

6.5.1 Two-Phase System Mechanism

The concept of a two-phase system (TPS) refers to a unique liquid–liquid separation process in which two distinct water-rich phases are generated when certain combinations of polymers and/or salts are mixed at concentrations exceeding a critical threshold. The phase diagram shows how the system responds to different combinations of phase-forming components, with their concentrations plotted on the horizontal axis (bottom phase component) and vertical axis (top phase component). The binodal curve serves as the boundary between two conditions: Below this line, the components remain fully mixed and form a homogeneous solution known as the monophasic region, whereas compositions above the curve lead to spontaneous separation into two aqueous phases due to thermodynamic incompatibility, defining the biphasic region. When a mixture is prepared with an overall composition located within this biphasic area, it resolves along a tie line into two equilibrium layers, typically a polymer-rich top phase (such as PEG-rich) and a salt-rich bottom phase.

Target solutes distribute unevenly between these phases based on their chemical affinity and molecular interactions, a phenomenon quantitatively represented by the partition (Fig. 6.3).

Liquid two-phase systems are formed through spontaneous liquid–liquid phase separation in which two immiscible solutions form two separate phases. The mechanism of ATPS formation involves interactions between molecules in the solution, leading to phase separation based on their interaction preferences (Fick et al. 2023). ATPS can be categorized as segregate or associative systems, depending on the thermodynamic drivers of phase separation (Martin 2019); in segregate systems, the incompatibility between solutions of solute (macro) molecules results in a liquid phase enriched in one solute (macro) molecule (Prabhu 2021). Solutes such as proteins or enzymes will be distributed between the upper and lower phases

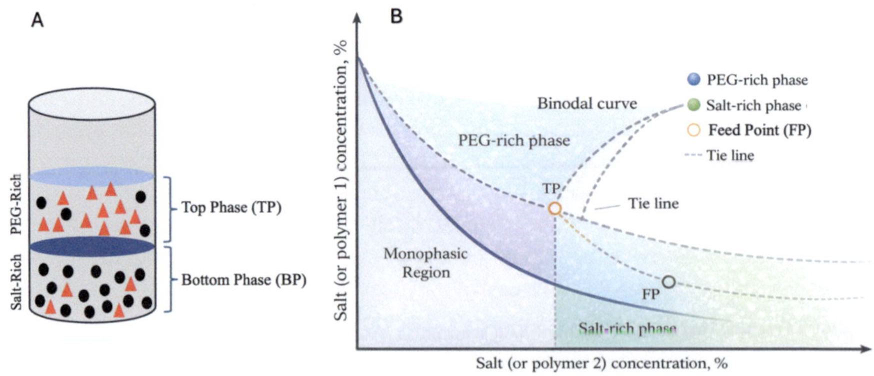

Fig. 6.3 Phase separation behavior and partitioning mechanism in ATPS. (**a**) Solute partitioning between PEG-rich top phase (TP) and salt-rich bottom phase (BP) in a polymer–salt ATPS. (**b**) Phase diagram showing binodal curve, monophasic and biphasic regions, tie line, and positions of feed (FP), TP, and BP

depending on solubility (Urge et al. 2023), charge (Anupa et al. 2024), hydrophobicity (Liu et al. 2022) (Nouri and Khayati 2024), and molecular size (Singla and Sit 2023).

Common components: The materials used in the separation of protein molecules with ATPS are generally polymers, salts, charged solvents, and complex polyelectrolytes (Fick et al. 2023). The two phases formed consist of two main components, namely, polymers (such as polyethylene glycol/PEG) and salts (ammonium sulfate or other salts). Furthermore, Yusree et al. (2021) revealed that variations in PEG and salt concentrations can increase or decrease the selectivity of the system for protease purification, depending on the type of protease being processed.

Polymers and salts can bind water and form separate aqueous phases. This allows biomolecules to be separated based on key physicochemical traits, such as size and polarity. Beyond conventional systems, alternative approaches have been reported, including organic solvent–salt mixtures and deep eutectic solvent (DES)- salt systems (Nguyen et al. 2024), Similar strategies using DES-derived salts have also emerged (Ortizo et al. 2024). The material combinations used in these systems are summarized in Table 6.1.

6.5.1.1 Separation Principle

The separation of proteases in two-phase systems (TPS) is influenced by various parameters that can significantly affect the purification process. Because it can selectively separate biomolecules depending on their physicochemical properties, TPS is often used to separate and purify proteins. The way proteases are separated in TPS is complicated and depends on the characteristics of their molecules. The protease's size, charge, and hydrophobicity decide if it will separate into the upper or lower phase. Proteases that stick better to one phase will build up in that phase, making it easy to separate them from other biomolecules. One essential element of enhancing protease purification is the capacity to customize the partition mechanism (Norouzi and Rahimpour 2024) (Chen et al. 2024).

6.5.1.2 Influencing Factors

Comprehending the factors influencing protease separation is crucial for enhancing the efficacy of the separation process. The following factors influence the purity level of two-phase protease separation systems:

(a) Composition and Concentration

One of the most essential things about TPS is the concentration of the aqueous and organic phases. The choice of phase formation agents, such as polyethylene glycol (PEG) and dextran, is based on the concentration, volume, and phase characteristics during protease separation. Proteases usually break apart into phases that match their charge or hydrophobicity. For instance, proteins that do not like water

Table 6.1 General components of materials used in ATPS

Combination of materials	Ingredients composition	Result	References
Polymer–polymer	Dextran-PEG +25% silica particle	PEG-rich and dextran particle silica particle in secondary phase, adjustment strength, and stress on the network	Ruiz-martinez et al. (2025)
Polymer–salt	PEG 8000-Na_2SO_4	Small particle size, fast phase formation, ideal for biomolecule purification	Titus et al. (2022)
	PEG-$Na_3C_6H_5O_7$	Aspartyl proteinase secreted with ATPS shows effective cytotoxicity in HeLa cell line (cancer cells)	Kokhdani et al. (2024)
Organic solvent–salt	Ethanol-$(NH_4)_2SO_4$	Concentration of 10% $(NH_4)_2SO_4$ and 35% ethanol during ATPS extraction produced the highest hyperin and quercitrin content	Nguyen et al. (2024)
Salt–salt	DES–DES	Produces peptides that inhibit DPP-IV	Ortizo et al. (2024)
	DES–dipotassium hydrogen phosphate (K_2HPO4)	Separation of drug ingredients	Mashhaditafreshi and Haghtalab (2024)
Polyelectrolyte complex	poly(diallyldimethylammonium chloride)/PDADMAC, poly (ionic liquids)	Efficient desalination of seawater	Zhang et al. (2023)

tend to stick to the PEG phase, while proteins that do like water tend to stick to the dextran phase (Mittag and Pappu 2022).

(b) pH and Ionic Strength

The pH of the solution has a big effect on how soluble and charged proteases are. When building TPS for purification, it's important to keep in mind that proteases usually work best and stay stable in a certain pH range. The ionic strength of the system also changes the electrostatic interactions between the protease and the phase-forming parts. Higher ionic strength can influence the partition behavior and break up hydrophobic contacts, which makes the purification process less successful (Sarwa et al. 2024).

(c) Temperature

Temperature substantially affects the stability and functionality of proteases. Some proteases have increased activity at elevated temperatures, whereas others may experience a loss of activity or denaturation. The partitioning of proteases in

TPS is temperature-dependent, as phase separation and protease solubility fluctuate with temperature changes. The appropriate temperature for partitioning is a crucial factor in protease purification (Teymoornejad and Jabbari 2023).

(d) Size and Charge of Proteases

Proteases differ in size, charge, and structural characteristics, which directly influence their partitioning behavior in TPS. Larger proteases may segregate differently than smaller proteases due to steric effects, whereas the charge of proteases influences electrostatic interactions with phase components. A comprehensive study of these parameters facilitates the customization of TPS settings for a specific protease, hence ensuring elevated purity and yield (Buarque et al. 2023) (Minagawa et al. 2025).

(e) Phase-Forming Agent

The selection of the phase-forming agent is essential in TPS design. PEG and dextran are frequently utilized for their capacity to create clear and stable biphasic systems. The molecular weight of PEG can be adjusted to regulate protease partitioning; larger PEG molecules often enhance selectivity and improve separation efficiency for proteins with varying hydrophobicities (Moteshafi et al. 2022).

6.5.2 Three-Phase System Mechanism

The three-phase system in protease separation is evident in the mechanism or structure via which the protease enzyme functions across distinct phases to perform its catalytic role. Protease is an enzyme that facilitates the degradation of proteins into smaller polypeptides or amino acids through the hydrolysis of peptide bonds. Comprehending the operational mechanism of protease activity, particularly in systems where enzymes function throughout three distinct phases, can facilitate several biotechnological and industrial applications. The three-phase system consists of two liquid phases and one solid or gel phase. The first liquid phase usually consists of ethanol or salt, while the second phase contains proteins and surfactants. When used together, the intermediate layer produced between the liquid and solid phases can separate proteins more effectively, especially for hydrophobic proteins (Liu et al. 2025).

The TPP mechanism (Gupta and Roy 2021) is as follows: First, adding salt: The solution containing the protein is mixed with salt, usually ammonium sulfate $(NH_4)_2SO_4$, which acts as a salting agent. This entails diminishing the protein's solubility in the aqueous phase, resulting in precipitation. In addition to $(NH_4)_2SO_4$, other salts such as dipotassium hydrogen phosphate (K_2HPO_4) and trisodium citrate ($Na_3C_6H_5O_7$) may also be employed (Kuepethkaew and Klomklao 2023). Second, adding organic solvents: Organic solvents such as t-butanol (tert-butanol) are then added. These solvents, especially at lower temperatures, interact with the precipitated protein and can cause it to further aggregate at the interface between the

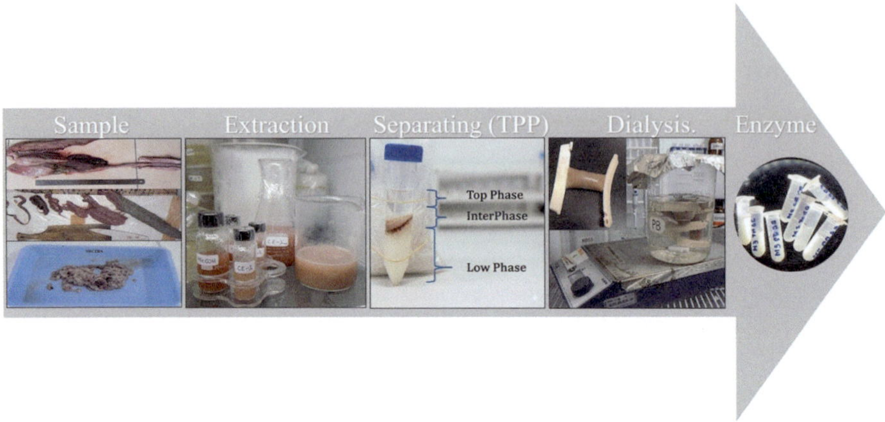

Fig. 6.4 Separation of protease from fish stomach contents using the TPP method

aqueous and organic phases. In addition, the use of T-butanol in high concentrations can reduce protease activity, so for certain purposes in producing target enzymes, DES can also be used (Liu et al. 2025). Third, phase separation: After mixing and centrifugation, the solution separates into three phases: Upper organic phase: This phase, rich in t-butanol, contains lipids, pigments, and other nonpolar compounds. Lower aqueous phase: This phase contains polar compounds such as carbohydrates and smaller molecules. Interphase (protein-rich layer): The desired protein, along with some contaminants, precipitates at the interface between the aqueous and organic phases (Fig. 6.4).

The key to the above principle is salting: Ammonium sulfate reduces the solubility of proteins, causing them to precipitate. Co-solubility: t-Butanol acts as a co-solvent, changing the solubility of proteins and promoting precipitation and aggregation. Phase affinity: Different components of the mixture have different affinities for the three phases, allowing their separation. Selective separation: By optimizing the salt and t-butane concentrations, the desired protein can be selectively separated into the interfacial precipitate phase, while other contaminants are concentrated in the other phases. Advantages of TPP: Simplicity and speed: TPP is a simple, one-step process that is relatively fast compared to other separation techniques. Nonchromatography: TPP does not require a chromatography column, making it scalable and cost-effective. Versatile: TPP has been used to purify a wide range of proteins and enzymes (Qiu et al. 2024).

6.5.2.1 General Principles (Salting Out, Use of Surfactants)

TPP is a nonchromatographic technique for separating and purifying proteins from intricate mixtures. The method involves using a combination of a salting agent (such as ammonium sulfate) and an organic solvent (usually t-butanol) to create three distinct phases. The desired protein is selectively separated into one phase, while contaminants are concentrated in the other phase.

Three-phase systems, such as ammonium sulfate, are employed to segregate proteins into distinct phases according to their solubility. By altering the salt content, an initial two-phase system is established; subsequent modifications result in a third phase where proteins, including proteases, are concentrated. This method is frequently employed in protein purification and stabilization, providing a straightforward and economical means of isolating proteases while preserving their activity. Ammonium sulfate precipitation can isolate protease enzymes from other cellular proteins by utilizing variations in their solubility characteristics. The method involves incrementally elevating the concentration of ammonium sulfate in the solution until protein precipitation occurs, allowing for subsequent analysis or purification (Zhao et al. 2024). In a three-phase system, surfactants are used to modulate the interactions between proteins and the formed phases, thereby directing protein partitioning to specific phases. This approach significantly improves the selectivity and efficiency of separating protease enzymes from other proteins and contaminants. Surfactants such as Triton X-100 (Rony et al. 2024) and polyethylene glycol (PEG) (Azad et al. 2025) are frequently employed to formulate these systems. These surfactants create micelles in the organic phase that can interact with proteins, resulting in their distribution throughout various phases based on their hydrophobic or hydrophilic characteristics. This partitioning is especially advantageous for the purification of membrane-associated proteases, which are frequently challenging to isolate using traditional techniques. Salt-induced three-phase systems and surfactant-mediated systems are essential methodologies for the investigation and separation of proteases. Salt-induced systems offer a straightforward and economical technique for enzyme separation, whereas surfactant-mediated systems present a more advanced strategy for addressing membrane-bound or hydrophobic proteins. Collectively, these methodologies enhance the potential uses of proteases in biotechnology, medicines, and industrial applications.

Additional benefit: TPP intermediate-layer separation is a partial purification technique for protein molecules that combines the principles of selective precipitation and liquid-liquid extraction. This technique generates three primary layers: a superior layer abundant in t-butane, an intermediate layer typically housing the target protein, and an inferior layer including an aqueous phase with contaminants and salts. In the context of protease enzyme purification, intermediate layer separation has several significant additional advantages in terms of enzyme efficiency, selectivity, and stability (Belchior and Freire 2021). One of the main advantages of intermediate layer separation is simultaneous concentration and purification. As the protease partitions into the intermediate layer, other unwanted proteins remain in the aqueous phase or are dissolved in t-butanol. This allows for highly efficient initial purification, reducing the need for expensive and time-consuming further chromatographic steps (Patil & Rathod 2022). Another additional advantage is protection against enzyme denaturation. Unlike other purification methods involving high amounts of organic solvents or extreme pH conditions, the intermediate layer in the TPP system creates a relatively stable semi-hydrophobic environment for proteases. This environment can help maintain the biological activity of proteases that are sensitive to environmental conditions (Torabizadeh and Khaki 2025). The separation of the intermediate layer also facilitates the selective isolation of target enzymes, especially in crude extract mixtures. Numerous studies indicate that proteases derived from microbial sources, such as *Bacillus subtilis* or *Aspergillus niger*, demonstrate preferential distribution to the intermediate layer, achieving high yields and specific activity that increase by almost twofold following a single TPP step (Rajendran et al. 2023). This makes the TPP method very efficient as an initial step in downstream processes.

Furthermore, the system's value comes from its scalability and minimal operational expenses. The TPP technique can be scaled for industrial use without requiring specialized equipment. Reagents such as t-butane and ammonium sulfate utilized in TPP are also comparatively inexpensive and readily accessible. This combination renders this approach appealing for commercial applications, particularly in the manufacture of protease enzymes for the food, detergent, and pharmaceutical sectors (Ismail et al. 2021). The intermediate layer can be altered, or its efficacy enhanced by modifying parameters such as solvent ratio, salt content, and temperature. These alterations influence both yield and enhance selectivity for specific protease isoforms, which is advantageous in therapeutic and diagnostic applications (Babagil and Nadaroglu 2022). The segregation of the intermediate layer in the TPP system functions as a purification method and an effective strategy to enhance both the quality and yield of proteases in a cost-efficient and eco-friendly manner. Consequently, this approach represents a viable option for protease separation and purification at both laboratory and industrial sizes.

6.5.2.2 Factors Affecting Separation Efficiency

The efficiency of protease separation in the three phase partitioning (TPP) system is influenced by various physicochemical factors that determine the distribution and stability of the enzyme during the process. The following are the main factors that affect the efficiency of the process:

(a) pH and ionic strength

The pH of the medium is crucial as it influences the net charge of the protease molecule, hence affecting the electrostatic interactions between the protease and its surroundings. Generally, the optimal pH for TPP is close to the isoelectric point (pI) of the enzyme (Chen et al. 2024), where protein solubility is lowest, thus increasing precipitation to the interface phase. In addition, the ionic strength of the solution can modulate protein solubility and its interaction with the added ammonium sulfate salt, which acts as a precipitating agent (salting out) in the TPP system (Antunes et al. 2024).

(b) Protease Hydrophobicity

The surface hydrophobicity of the protease affects its ability to interact with the organic phase in the TPP system, but in this case, t-butanol is substituted with DES (Chen et al. 2023). According to Dennison and Lovrien (1997), proteins that have high hydrophobic properties tend to be more easily pushed to the interface between the organic phase and the salt phase, which facilitates selective precipitation. Changes in hydrophobicity can affect protein distribution and separation selectivity.

(c) Phase Composition and Concentration

The concentration of ammonium sulfate and the volume ratio of t-butanol to crude extract solution are critical parameters in the TPP system. Elevating the salt concentration enhances the salting-out effect, whereas the t-butanol ratio influences the system's viscosity and protein distribution. The selection of alcohol type and its saturation level dictates the efficiency of the separation process (Maskey et al. 2024).

(d) Temperature and Stabilizer Agents

The temperature during the TPP process influences the enzyme's structure and activity. The ideal temperature is often used to preserve the stability of the protease during separation. An excessively high temperature can lead to denaturation, whilst an excessively low temperature can diminish separation efficiency (Gagaoua et al. 2021). Moreover, including additives like polyethylene glycol (PEG), protectants such as trehalose, or certain buffers can preserve the enzyme's structure and activity throughout the TPP process (Antunes et al. 2024).

6.6 Advantages and Limitations

TPS and TPP are extensively utilized in the purification of proteins and enzymes, particularly proteases, because to their numerous advantages over traditional techniques.

6.6.1 Advantages of Two- and Three-Phase Systems

Some of the advantages of two- and three-phase systems are: high selectivity, environmentally friendly, cost-effective.

(a) High selectivity: A main advantage of this system is its high selectivity in separating target proteins from complex mixtures. In TPS, separation is based on the differential distribution of biomolecules between two immiscible phases (e.g., polymer and salt phases), whereas TPP introduces an intermediate phase as a precipitate that improves the selective separation of target proteins (Robescu and Bavaro 2025). This allows the separation of proteases from other impurities with high efficiency.

(b) Ecologically sustainable: TPS and TPP typically employ comparatively safe and biodegradable substances, such as polyethylene glycol (PEG), phosphate salts, ammonium sulphate, or benign organic solvents such as t-butanol (Ng et al. 2021). This renders this process more environmentally sustainable than alternative techniques such as organic solvent or resin-based chromatography (Du et al. 2022). Some approaches have even developed TPS based on natural solvents or DES, which strengthen the sustainability aspect of the process (Chen et al. 2022; Muniasamy et al. 2022; Zhou et al. 2022).

(c) Cost-effective and fast process: TPS and TPP do not require sophisticated equipment and are scalable, making them suitable for industrial applications, especially in the biotechnology and food sectors. The purification process can be carried out in a single step, combining extraction and concentration, reducing time and operational costs (Yadav et al. 2023). This technique frequently yields elevated specific activity of the isolated enzymes.

6.6.2 Limitations and Challenges

Although promising, the implementation of TPS and TPP on a large scale still faces several technical and economic challenges: phase stability, industrial scalability, and solvent selection.

(a) Phase Stability

Phase stability and reproducibility in two-phase or three-phase systems can be problematic, especially if the material composition, temperature, or pH changes. Certain TPS systems exhibit heightened sensitivity to minor variations that may influence the separation of biomolecules (Zhang et al. 2024). This stability is especially important when the process is run continuously in industry.

(b) Industrial Scalability

One of the major challenges in the implementation of ATPS and TPP at the industrial level is the issue of scalability. Although both systems offer excellent efficiency at the laboratory level, the transition to industrial scale often faces problems related to phase stability, material costs, and control of process conditions (Somader et al. 2025). The scalability of two- and three-phase systems has not been fully established in many industrial applications. While these systems are simple at the laboratory scale, transitioning to industrial scale requires tighter parameter control, including phase recovery and material recycling (García et al. 2025).

(c) Selection of Solvents and System Components

The selection of phase components, such as the type of PEG, salt, or organic solvent, greatly affects the separation efficiency. This optimization process can be time-consuming and is not always universal for various types of enzymes or proteins. In addition, some solvents, such as t-butanol, can affect protein stability if used in high concentrations (Al-Saidi et al. 2024). Therefore, the system design must be adjusted to the characteristics of the target protein.

6.7 Applications and Case Studies

6.7.1 Two-Phase Systems

TPS are extensively utilized in the separation and purification of proteins, including protease enzymes, owing to their capacity to segregate biomolecules based on varying solubility and affinity for the two established phases. Case studies of two-phase PEG/salt systems. A prevalent system is the amalgamation of polyethylene glycol (PEG) with salts such as phosphate or sulphate. PEG generates a more hydrophobic upper phase, whereas salt produces a more hydrophilic lower phase. An illustration of an application is a TPS from an experiment involving PEG35000, sodium carbonate (Na_2CO_3), and potassium carbonate (K_2CO_3), wherein the PEG35000 and Na_2CO_3 system exhibited a superior salting-out strength (Khavidaki et al. 2024).

The localization of the target protein is significantly influenced by the molecular weight of PEG, salt concentration, as well as the pH and temperature of the system (Fakhari et al. 2025). A further illustration of the PEG/Dextran System, the PEG/dextran system is utilized for protein separation according to size and polarity. Research by Mahdi et al. (2023) indicates that big proteins are predominantly found in the dextran phase (bottom), whereas tiny and more hydrophobic proteins are located in the PEG phase (top). This renders the technique highly selective in the separation of complex mixed proteins, including those derived from plant or microbe extracts. The ATPS method is extensively utilized for the purification of proteases in industrial applications, including the extraction of proteases from plant sources for food product manufacturing. Furthermore, ATPS has been employed for the purification of therapeutic proteases utilized in illness treatment (Saha et al. 2022).

6.7.2 Three-Phase System

A TPP is a technique that concurrently facilitates protein precipitation and phase separation through the use of organic solvents and salts, specifically employing polymer-salt combinations such as PEG and potassium citrate tribasic monohydrate ($K_3C_6H_5O_7{\cdot}H_2O$), as demonstrated in the purification of proteases and antibodies, where target proteins precipitate between the upper liquid phase (PEG) and the lower solution ($K_3C_6H_5O_7{\cdot}H_2O$) (Capela *et al.*, 2023). The enzyme retains its activity as concentration and purification transpire concurrently. Furthermore, surfactant-based

TPP systems, such as Triton X-114 or natural biosurfactants, have been formulated for more eco-friendly biotechnological applications. The utilization of nonionic surfactants and ammonium sulphate for the separation of proteases from fermentation waste resulted in enhanced purification and improved separation, attributed to micelle formation and protein partitioning at the interface (Zerizer et al. 2024). A further study demonstrated that the three-phase approach effectively purifies proteases from the hydrophobic proteins of Chia seeds (*Salvia hispanica L.*), which were previously challenging to isolate using traditional methods (Jiang and Zheng 2024).

6.8 Prospects and Future Direction

Recent studies increasingly link protease purification with a range of enabling technologies. These include membrane-based processes, chromatographic techniques, and even artificial intelligence. At the same time, more specialized systems are being explored. Aqueous two-phase systems (ATPS) and three-phase partitioning (TPP), for instance, are now being adapted to work alongside microfluidic platforms and other modern tools. This integration allows better control of separation conditions and can significantly improve purification performance at the microscale. Microfluidics technology facilitates enhanced precision in fluid flow and temperature regulation, alongside improved separation efficiency utilizing significantly reduced solvent quantities relative to traditional approaches (Yuan et al. 2025). Artificial intelligence (AI) and machine learning: Recent research has demonstrated the capability of AI technology to enhance experimental settings in protease purifying operations. Through the application of AI and machine learning, we can more efficiently analyze experimental data to identify optimal settings for protease purification and anticipate the impact of variables such as temperature, pH, and concentration on protein yield and purity (Alloun and Calvio 2024).

The combined use of nanotechnology with ATPS and TPP creates novel opportunities in enzyme purification. Nanoparticles, including metal nanoparticles and carbon nanotubes, enhance protein separation by leveraging the electrostatic interactions and hydrophobicity of proteases (Ariaeenejad and Motamedi 2025), garnering increasing interest for biotechnological applications (Eskandari et al. 2024). Possibility for employing natural and sustainable materials. Growing interest in sustainability has driven the use of renewable materials in ATPS and TPP research. In particular, natural polymers and carbohydrate-based compounds are increasingly explored as greener phase-forming agents. These bio-derived materials can serve as viable substitutes for synthetic chemicals, which are often more costly and may pose environmental or safety concerns. For instance, incorporating gelatin or pectin as phase components in ATPS has been shownto support more eco-friendly purification processes (Patil et al. 2024). Sustainable materials: The focus on renewable materials, such as cellulose or agar, is increasingly important amidst the need to reduce the use of hazardous chemicals and reduce the environmental impact of purification processes. Research related to the use of biodegradable and nontoxic materials provides a promising direction for the pharmaceutical and food industries.

6.9 Conclusion

Overall, aqueous two-phase systems (ATPS) and three-phase partitioning (TPP) offer several notable benefits compared with conventional purification approaches, such as chromatography or centrifugation. Their advantages are significant. These systems provide strong selectivity toward proteases while enabling efficient separation under relatively mild conditions. In addition, they help minimize the reliance on hazardous reagents, making the purification process safer and more environmentally responsible. The application of ATPS and TPP in several industrial sectors, including food, pharmaceuticals, and biotechnology, demonstrates significant promise for processing proteases in substantial quantities, with reduced costs and more eco-friendly methods. Continuous development and research, especially in terms of integration with the latest technologies (such as AI and microfluidics), the use of TPS and TPP for protease purification can be further expanded and optimized to meet larger industrial needs. With potential applications in the laboratory and industrial scale at the laboratory level, ATPS and TPP provide a flexible and efficient approach to purify proteases under a variety of conditions, allowing for faster optimization of experiments. At the industrial scale, the main challenges are solving the issues of phase stability, process scalability, and cost efficiency. However, with the continuous development of technological innovations, there is great potential for the application of these systems in large-scale and commercial protease purification.

Acknowledgments The authors would like to express sincere gratitude to the Faculty of Fisheries and Marine Science, Brawijaya University, for their support and all facilities provided throughout the process of writing this book chapter.

Competing Interests The authors declare that they have no competing interests.

References

Aini N, Wahyuningsih SPA, Uzakky FN, Betzy C, Fatimah, & Andriyono, S. (2024) Analysis of two strain probiotics on digestive enzymes, liver function and antimicrobial activity of catfish (*Clarias gariepinus*) treated with *Aeromonas hydrophila*. Biodiversitas 25(11):4333–4339. https://doi.org/10.13057/biodiv/d251132

Akkaya SN, Almansour A, Altintas R, Sisecioglu M, Adiguzel A (2025) Purification, characterization, optimization, and docking simulation of alkaline protease produced by *Brevibacillus agri* SAR25 using fish wastes as a substrate. Food Chem 471(4):142816. https://doi.org/10.1016/j.foodchem.2025.142816

Alberti S, Dormann D (2019) Liquid-liquid phase separation in disease. Annu Rev Genet 53:171–194. https://doi.org/10.1146/annurev-genet-112618-043527

Alloun W, Calvio C (2024) Bio-driven sustainable extraction and AI-optimized recovery of functional compounds from plant waste: a comprehensive review. Fermentation 10(3). https://doi.org/10.3390/fermentation10030126

Al-Saidi S, Mjalli FS, Al-Azzawi M, Al-Wahaibi T, Abutarboosh B, AlSaadi MA (2024) Aqueous two-phase extraction of amoxicillin using miniaturization technology. Korean J Chem Eng 41(5):1399–1413. https://doi.org/10.1007/s11814-024-00061-0

Antunes EC, Temmink H, Schuur B (2024) Polymer and alcohol-based three-phase partitioning systems for separation of polysaccharide and protein. J Chem Technol Biotechnol 99(1): 259–269. https://doi.org/10.1002/jctb.7534

Anupa A, Metya S, Mihooliya KN, Rathore AS (2024) Development of continuous processing platform utilizing aqueous two-phase extraction for purification of monoclonal antibodies. J Chromatogr A 1715(11):464605. https://doi.org/10.1016/j.chroma.2023.464605

Ariaeenejad S, Motamedi E (2025) Carboxylated nanocellulose from quinoa husk for enhanced protease immobilization and stability of protease in biotechnological applications. Sci Rep 15(1):1–14. https://doi.org/10.1038/s41598-024-77292-y

Azad MAK, Paul RK, Khan JM, Banjare MK, Hasan T, Anis-Ul-Haque KM, Rana S, Hoque MA (2025) The influence of anionic and nonionic hydrotropes on the phase separation of Triton-X 100 and polyethylene glycol mixtures: assessment of several physico-chemical parameters. J Solut Chem 54(5):641–659. https://doi.org/10.1007/s10953-025-01438-3

Babagil A, Nadaroglu H (2022) Purification of pectin lyase enzyme from *Bacillus pumilus* bacteria by three-phase partitioning method (TPP), nanoflower preparation and investigation of fruit juice clarification. Biointerf Res Appl Chem 12(3):3938–3955. https://doi.org/10.33263/BRIAC123.39383955

Banani SF, Lee HO, Hyman AA, Rosen MK (2017) Biomolecular condensates: organizers of cellular biochemistry. Nat Rev Mol Cell Biol 18(5):285–298. https://doi.org/10.1038/nrm.2017.7

Baydogan B, Kucuk A, Kozan B, Erdal M, Abas BI, Cevik O (2025) Hydrogels made with Tilapia fish skin increase collagen production and have an effect on MMP-2/MMP-9 enzymes in burn treatment. Biochemist 5(2):1–16. https://doi.org/10.3390/biochem5020008

Becker da Silva C, Poll Moraes G, Godoy Hoffmann R, Lenz Goettems T, Mendonça Azambuja A, Daroit DJ (2025) Production of bacterial protease via solid-state cultivation with agro-industrial by-products and eco-friendly perspectives of enzyme application. Clean Techn Environ Policy 27(5):2029–2042. https://doi.org/10.1007/s10098-024-02943-7

Bege M, Borbás A (2024) The design, synthesis and mechanism of action of Paxlovid, a protease inhibitor drug combination for the treatment of COVID-19. Pharmaceutics 16(2). https://doi.org/10.3390/pharmaceutics16020217

Bekavac N, Benković M, Jurina T, Valinger D, Gajdoš Kljusurić J, Jurinjak Tušek A, Šalić A (2024) Advancements in aqueous two-phase systems for enzyme extraction, purification, and biotransformation. Molecules 29(16):3776. https://doi.org/10.3390/molecules29163776

Belchior DCV, Freire MG (2021) Simultaneous separation of egg white proteins using aqueous three-phase partitioning systems. J Mol Liq 336:116245. https://doi.org/10.1016/j.molliq.2021.116245

Borcherds, Bremmer, Borgia, Mittag (2020) How do intrinsically disordered protein regions encode a driving force for liquid–liquid phase separation? Curr Opin Struct Biol 67(4):41–50. https://doi.org/10.1016/j.sbi.2020.09.004

Buarque FS, Gautério GV, Coelho MAZ, Lemes AC, Ribeiro BD (2023) Aqueous two-phase systems based on ionic liquids and deep eutectic solvents as a tool for the recovery of non-protein bioactive compounds a review. PRO 11(1). https://doi.org/10.3390/pr11010031

Cai C, Ma S, Li F, Tan Z (2024) Aqueous two-phase system based on pH-responsive polymeric deep eutectic solvent for efficient extraction of aromatic amino acids. Food Chem 430(1): 137029. https://doi.org/10.1016/j.foodchem.2023.137029

Chen X, Cai C, Tan Z (2022) Deep eutectic solvents-based three-phase partitioning for tomato peroxidase purification: a promising method for substituting t-butanol. Food Chem 393(11): 133379. https://doi.org/10.1016/j.foodchem.2022.133379

Chen X, Hu Y, Tan Z (2023) Innovative three-phase partitioning based on deep-eutectic solvents and sugars (sugaring-out effect) for cucumber peroxidase purification. Microchem J 190(7): 108702. https://doi.org/10.1016/j.microc.2023.108702

Chen X, He Y, Liu Z, Huang Z, Xu C, Liu Y, Haran Y, Nisar W, Yan S, Li J (2024) Ultrasound-assisted extraction of polyphenols from lotus rhizome epidermis by alcohol / salt-based aqueous two-phase system : Optimization, extraction mechanism and antioxidant activities. Food Chem 453(9):139620. https://doi.org/10.1016/j.foodchem.2024.139620

Contreras W, Groenendyk J, Gentzel M, Schönberg PY, Buchholz F, Michalak M, Schröder B, Mentrup T (2024) Selective regulation of aspartyl intramembrane protease activity by calnexin. Cell Mol Life Sci 26;81(1):441. https://doi.org/10.1007/s00018-024-05478-8

Cui Z, Wang X, Zhang Q, Li L, Liu F (2025) Liquid–liquid phase separation technologies of α-synuclein governed by its C-terminus via electrostatic interactions: unveiling potential therapeutic targets for Parkinson's disease. Adv Technol Neurosci 2(1):1–8. https://journals.lww.com/10.4103/ATN.ATN-D-24-00027

Ćwilichowska N, Świderska KW, Dobrzyń A, Drąg M, Poręba M (2022) Diagnostic and therapeutic potential of protease inhibition. Mol Asp Med 88(12):101144. https://doi.org/10.1016/j.mam.2022.101144

David Troncoso F, Alberto Sánchez D, Luján Ferreira M (2022) Production of plant proteases and new biotechnological applications: qn updated review. Chemistry Open 11(3). https://doi.org/10.1002/open.202200017

Debbarma S, Narsale SA, Acharya A, Singh SK, Mocherla BP, Debbarma R, Yirang Y (2024) Drawing immune-capacity of fish-derived antimicrobial peptides for aquaculture industry: a comprehensive review. Comparat Immunol Reports 6(6):200150. https://doi.org/10.1016/j.cirep.2024.200150

Dennison C, Lovrien R (1997) Three phase partitioning: concentration and purification of proteins. Protein Expr Purif 11(2):149–161. https://doi.org/10.1006/prep.1997.0779

Du M, Hou Z, Liu L, Xuan Y, Chen X, Fan L, Li Z, Xu B (2022) Progress, applications, challenges and prospects of protein purification technology. Front Bioeng Biotechnol 6(10):1028691. https://doi.org/10.3389/fbioe.2022.1028691

Eskandari A, Leow TC, Rahman MBA, Oslan SN (2024) Utilization and prospect of purification technologies in natural proteins, peptides and recombinant proteins. J Proteins Proteomics 15(2):233–257. https://doi.org/10.1007/s42485-024-00139-7

Fakhari M, Momeni Z, Zare A (2025) Liquid-liquid equilibria of an affinity aqueous two-phase system containing polyethylene glycol and sodium sulfate salt at various ligand concentrations and pH: experimental determination, correlation, and thermodynamic modeling. J Chem Eng Data 70(2):1051–1062. https://doi.org/10.1021/acs.jced.4c00430

Fick C, Khan Z, Srivastava S (2023) Interfacial stabilization of aqueous two-phase systems: a review. Mater Adv 4(20):4665–4678. https://doi.org/10.1039/D3MA00307H

Franzmann TM, Jahnel M, Pozniakovsky A, Mahamid J, Holehouse AS, Nüske E, Richter D, Baumeister W, Grill SW, Pappu RV, Hyman AA, Alberti S (2018) Phase separation of a yeast prion protein promotes cellular fitness. Science 359(6371). https://doi.org/10.1126/science.aao5654

Gagaoua M, Dib AL, Lakhdara N, Lamri M, Botineştean C, Lorenzo JM (2021) Artificial meat tenderization using plant cysteine proteases. Curr Opin Food Sci 359(6371):177–188. https://doi.org/10.1126/science.aao5654

García Y, del Campo M, Solís MF, Gómez-Angulo JH, Martínez Preciado AH, Silva-Jara JM, Padilla de la Rosa JD, Escalante-Garcia ZY (2025) Industrial applications, principal sources, and extraction of galactomannans: a review. Foods 14(9):1–30. https://doi.org/10.3390/foods14091587

Gimenes NC, Silveira E, Tambourgi EB (2021) An overview of proteases: production, downstream processes and industrial applications. Sep Purif Rev 50(3):223–243. https://doi.org/10.1080/15422119.2019.1677249

Gupta MN, Roy I (2021) Three phase partitioning: applications in separation and purification of biological molecules and natural products. In: Three phase partitioning: applications in separation and purification of biological molecules and natural products. Elsevier, pp 1–2. https://doi.org/10.1016/B978-0-12-824418-0.00031-X

Handayani R, Boing Sitanggang A, Sulistiani, Hasanah U, Dewantari Kusumaningrum H (2025) Screening of protease-producing bacteria from fermented shrimp products (terasi) for meat tenderization. BIO Web Conf 169(2025):02005. https://doi.org/10.1051/bioconf/202516902005

Hao X, Chen J, Xu M, Zheng H, Li X, Wang M, Liu T (2022) Separation and purification of enveloped and non-enveloped viruses from water samples using an aqueous two-phase system. Process Biochem 119(8):58–67. https://doi.org/10.1016/j.procbio.2022.05.014

Honrado A, Miguel M, Ardila P, Beltrán JA, Calanche JB (2024) From waste to value: fish protein hydrolysates as a technological and functional ingredient in human nutrition. Foods 13(19): 3120. https://doi.org/10.3390/foods13193120

Huang Z, Liu Z, Chen L, Liu Y, Yan G, Ni Y, Yan Q, He W, Liu J, Luo S, Xie J (2025) Liquid-liquid phase separation in cell physiology and cancer biology: recent advances and therapeutic implications. Front Oncol 15(3):1–16. https://doi.org/10.3389/fonc.2025.1540427

Hussain A, Kumar M, Mukhopadhyay K, Dutta A, Sachan SG (2025) Antimicrobial peptides in *Clarias batrachus* epidermal mucus: characterization and therapeutic potential. Fish Shellfish Immunol 159(4):110191. https://doi.org/10.1016/j.fsi.2025.110191

Ismail E, Prihanto AA, Sukoso K, Nurul Huda H, Al Zamzami IM (2021) Isolation and characterization of a novel acid protease from striped Marlin (*Kajikia audax*) stomach with potential as a rennet substitute in dairy processing. Egypt J Aquat Biol Fish 29(4):1593–1613. https://doi.org/10.21608/ejabf.2025.443720

Jiang, Zheng K (2024) Extraction of mucilage polysaccharides from chia seed by hydrophobic deep eutectic solvents-based three-phase partitioning system: a phase behavior-driven approach. Int J Biol Macromol 280(3):135913. https://doi.org/10.1016/j.ijbiomac.2024.135913

Jiang B, Wang M, Wang X, Wu S, Li D, Liu C, Feng Z, Li J (2021) Effective separation of prolyl endopeptidase from *Aspergillus niger* by aqueous two phase system and its characterization and application. Int J Biol Macromol 169:384–395. https://doi.org/10.1016/j.ijbiomac.2020.12.120

Khavidaki HD, Asadi Z, Salabat A (2024) Effect of salt mixtures on aqueous two-phase system composed of PEG 35000: experiment and correlation. J Solut Chem 53(10):1409–1418. https://doi.org/10.1007/s10953-024-01381-9

Kokhdani E, Khodavandi P, Hasan M (2024) Toxicon anti-cancer activity of secreted aspartyl proteinase protein from *Candida tropicalis* on human cervical cancer HeLa cells. Toxicon 249(10):108073. https://doi.org/10.1016/j.toxicon.2024.108073

Kuepethkaew S, Klomklao S (2023) RSC advances partitioning and recovery of proteases from lizard fish (*Saurida micropectoralis*) stomach using combined phase partitioning systems and their. RSC Adv 13:14931–14942. https://doi.org/10.1039/D3RA01069D

Kurnianto MA, Defri I, Syahbanu F, Aulia SS (2024) Fish-derived bioactive peptide: bioactivity potency, structural characteristics, and conventional and bioinformatics approaches for identification. Future Foods 9(6):100386. https://doi.org/10.1016/j.fufo.2024.100386

Lemes AC, Bioconversion M, Bacillus P, Lemes AC, Victoria G, Augusto C, Brandelli A, Kalil SJ (2023) Two-step purification and partial characterization of Keratinolytic proteases from feather meal Bioconversion by *Bacillus sp.* P45 Processes 11(3):803. https://doi.org/10.3390/pr11030803

Li S, Li T, Hu X, Yang Y, Huang Y, He K (2022) A neglected issue: stationary phase retention determination of classic high-speed counter-current chromatography solvent systems. Separations 9(11):357. https://doi.org/10.3390/separations9110357

Liu Y, Yu J, Zhang J, Wang K, Hu H, Wang X (2022) Liquid liquid triboelectric nanogenerator based on the immiscible interface of an aqueous two-phase system. 9;13(1):5316. https://doi.org/10.1038/s41467-022-33086-2

Liu Y, Wang H, Xu Y, Wang D, Zhu X, Wang M, Teng J (2025) Menthol-based hydrophobic deep eutectic solvents for the purification of tea peroxidase using a three-phase partition method. LWT 218(2):117528. https://doi.org/10.1016/j.lwt.2025.117528

Mahdi HA, Hameed KW, Ali A-JA (2023) Application of aqueous two-phase systems in the extraction of bovine serum albumin. J Polymer Composit 11(1):55–65. https://doi.org/10.37591/jopc.v11i1.97677

Manzoli R, Badenetti L, Bruzzone M, Macario MC, Rubin M, Dal Maschio M, Roveri A, Moro E (2024) Mucopolysaccharidosis type II zebrafish model exhibits early impaired proteasomal-mediated degradation of the axon guidance receptor dcc. Cell Death Dis 15(4):1–11. https://doi.org/10.1038/s41419-024-06661-2

Martin N (2019) Dynamic synthetic cells based on liquid–liquid phase separation. Chembiochem 20(20):2553–2568. https://doi.org/10.1002/cbic.201900183

Mashhaditafreshi S, Haghtalab A (2024) Study on drug separation in two-phase aqueous systems using deep eutectic solvent consisting of choline chloride and 1,2 propanediol. J Mol Liq 393(7):123603. https://doi.org/10.1016/j.molliq.2023.123603

Maskey B, Karki DB, Shrestha NK (2024) Optimization of three-phase partitioning system for the purification of protease from *Calotropis gigantea* latex using response surface methodology. Chem Pap 78(4):2549–2559. https://doi.org/10.1007/s11696-023-03259-4

Medina DM, Acevedo-Gomez AV, Pellegrini Malpiedi L, Leiva LC (2024) Biochemical characterization of acid proteases from the stomach of palometa (*Pygocentrus nattereri*, Kner 1858) with potential industrial application. Int J Biol Macromol 264(1):130548. https://doi.org/10.1016/j.ijbiomac.2024.130548

Melchor-Moncada JJ, García-Barco A, Zuluaga-Vélez A, Veloza LA, Sepúlveda-Arias JC (2024) Scale-up of the fermentation process for the production and purification of Serratiopeptidase using silkworm pupae as a substrate. Methods Protocols 7(2):1–17. https://doi.org/10.3390/mps7020019

Minagawa Y, Yabuta M, Su M, Noji H (2025) Self-growing protocell models in aqueous two-phase system induced by internal DNA replication reaction. Nat Commun 16(1):1–11. https://doi.org/10.1038/s41467-025-56172-7

Mittag T, Pappu RV (2022) A conceptual framework for understanding phase separation and addressing open questions and challenges. Mol Cell 82(12):2201–2214. https://doi.org/10.1016/j.molcel.2022.05.018

Moreno JF, Oulego P, Collado S, Díaz M (2024) Valorisation of waste activated sludge for protease production by *Bacillus licheniformis*. J Clean Prod 445(1):141282. https://doi.org/10.1016/j.jclepro.2024.141282

Moteshafi H, Jabbari L, Hashemi M (2022) Performance of *Bacillus subtilis* D3d xylanase separated through optimized aqueous two-phase system in bio-bleaching of sugar beet pulp. Process Saf Environ Prot 159:749–756. https://doi.org/10.1016/j.psep.2022.01.030

Muniasamy R, Balamurugan BS, Rajamahendran D, Rathnasamy S (2022) Switchable deep eutectic solvent driven micellar extractive fermentation of ultrapure fibrin digesting enzyme from *Bacillus subtilis*. Sci Rep 12(1):1–14. https://doi.org/10.1038/s41598-022-04788-w

Nemani N, Dehnavi SM, Pazuki G (2024) Extraction and separation of astaxanthin with the help of pre-treatment of *Haematococcus pluvialis* microalgae biomass using aqueous two-phase systems based on deep eutectic solvents. Sci Rep 14(1):1–13. https://doi.org/10.1038/s41598-024-55630-4

Ng HS, Kee PE, Yim HS, Tan JS, Chow YH, Lan JCW (2021) Characterization of alcohol/salt aqueous two-phase system for optimal separation of gallic acids. J Biosci Bioeng 131(5):537–542. https://doi.org/10.1016/j.jbiosc.2021.01.004

Nguyen M, Thi L, Han T, Le N (2024) Heliyon response surface methodology for aqueous two-phase system extraction : An unprecedented approach for the specific flavonoid-rich extraction of *Houttuynia cordata* Thunb leaves towards acne treatment. Heliyon 10(4):e25245. https://doi.org/10.1016/j.heliyon.2024.e25245

Nilegaonkar SS, Zambare VP, Kanekar PP, Dhakephalkar PK, Sarnaik SS, Chandrababu NK, Rajaram R, Ramanaiah B, Ramasami T, Saikumari YK, Balaram P (2006) A novel protease for industrial application. Indian Patent 2471DEL2006

Nilegaonkar SS, Zambare VP, Kanekar PP, Dhakephalkar PK, Sarnaik SS (2007) Production of dehairing protease by *Bacillus cereus* MCM B- 326. Bioresour Technol 98(6):1238–1245. https://doi.org/10.1016/j.biortech.2006.05.003

Nilegaonkar SS, Zambare VP, Kanekar PP, Dhakephalkar PK, Sarnaik SS, Chandrababu NK, Rajaram R, Ramanaiah B, Ramasami T, Saikumari YK, Balaram P (2007a) A novel protease for industrial application. German Patent 102007013950:2

Nilegaonkar SS, Zambare VP, Kanekar PP, Dhakephalkar PK, Sarnaik SS, Chandrababu NK, Rajaram R, Ramanaiah B, Ramasami T, Saikumari YK, Balaram P (2007b) A novel protease for industrial application, US Patent:11/931537

Norouzi GS, Rahimpour F (2024) Investigating and optimizing insulin partitioning with conjugated Au nanoparticles in aqueous two-phase systems using response surface methodology. 9(8): 9676–9685. https://doi.org/10.1021/acsomega.3c09664

Nouri E, Khayati G (2024) A review of background and application of ATPSs in protein and enzyme extraction. J Solut Chem 53(10):1374–1408. https://doi.org/10.1007/s10953-024-01380-w

O'Flynn BG, Mittag T (2021) The role of liquid–liquid phase separation in regulating enzyme activity. Curr Opin Cell Biol 69:70–79. https://doi.org/10.1016/j.ceb.2020.12.012

Ojo Omoniyi OA, Dighitoghi Moro D, Bridget Afolabi O (2024) Microbial proteases: sources, significance and industrial applications. Int J Curr Microbiol App Sci 13(6):1–23. https://doi.org/10.20546/ijcmas.2024.1306.001

Oliveira AS, Ferreira C, Pereira JO, Pintado ME, Carvalho AP (2025) Valorisation of protein-rich extracts from spent brewer's yeast (*Saccharomyces cerevisiae*): an overview. Biomass Convers Biorefinery 15(2):1771–1793. https://doi.org/10.1007/s13399-022-02636-5

Ortizo RGG, Sharma V, Tsai ML, Nargotra P, Sun PP, Chen CW, Dong CD (2024) A novel deep eutectic solvent-based green extraction and purification of DPP-IV inhibitory peptides from tilapia (*Oreochromis niloticus*) viscera hydrolysate. Food Biosci 61(4):104658. https://doi.org/10.1016/j.fbio.2024.104658

Pan Y, He W, Gao J, Luo Y (2025) Microencapsulated alkaline protease with enhanced stability and temperature resistance for silk degumming. Colloids Surf A Physicochem Eng Asp 707(9): 135894. https://doi.org/10.1016/j.colsurfa.2024.135894

Patel A, Malinovska L, Saha S, Wang J, Alberti S, Krishnan Y, Hyman AA (2017) Biochemistry: ATP as a biological hydrotrope. Science 356(6339):753–756. https://doi.org/10.1126/science.aaf6846

Patil U, Palamae S, Zhang B, Hong H, Ma L, Benjakul S (2024) Tenderization of giant squid (*Dosidicus gigas*) meat pretreated with pulsed electric field and Asian seabass trypsin with the aid of vacuum impregnation. Food Biosci 61(10):104822. https://doi.org/10.1016/j.fbio.2024.104822

Pawar R, Zambare V, Barve S, Paratkar G (2009) Application of protease isolated from *Bacillus* sp.-158 in enzymatic cleaning of contact lenses. Biotechnology 8(2):276–280. https://doi.org/10.3923/biotech.2009.276.280

Pawar KS, Singh PN, Singh SK (2023) Fungal alkaline proteases and their potential applications in different industries. Front Microbiol 30(14):1138401. https://doi.org/10.3389/fmicb.2023.1138401

Pereira JFB, Coutinho JAP (2020) Aqueous two-phase systems: towards novel and more disruptive applications. Fluid Phase Equilib 505(2):112341. https://doi.org/10.1016/j.fluid.2019.112341

Prabhu VM (2021) Interfacial tension in polyelectrolyte systems exhibiting associative liquid–liquid phase separation. Curr Opin Coll Interf Sci 53:101422. https://doi.org/10.1016/j.cocis.2021.101422

Qiu C, Chen Z, Hu F, Wu H, Wang L, Liu M, Zhou K (2024) Simultaneous extraction of oil, protein and polysaccharide from residual seed cake of *Camellia oleifera* Abel. Using three phase partitioning. Ind Crop Prod 209(12):117994. https://doi.org/10.1016/j.indcrop.2023.117994

Ragab MM, Othman HA, Hassabo AG (2022) Various extraction methods of different enzymes and their potential applications in various industrial sector (a review). Egypt J Chem 65(10): 495–508. https://doi.org/10.21608/EJCHEM.2022.115484.5239

Rajendran DS, Chidambaram A, Kumar PS, Venkataraman S, Muthusamy S, Vo DVN, Rangasamy G, Vaithyanathan VK, Vaidyanathan VK (2023) Three-phase partitioning for the separation of proteins, enzymes, biopolymers, oils and pigments: a review. Environ Chem Lett 21(2):911–934. https://doi.org/10.1007/s10311-022-01540-8

Patil, Rathod VK (2022) Simultaneous extraction and partial purification of proteins from spent turmeric powder using ultrasound intensified three phase partitioning and its potential as antidiabetic agent. Chem Eng Process Process Intensif 172(9):108788. https://doi.org/10.1016/j.cep.2022.108788

Robescu MS, Bavaro T (2025) A comprehensive guide to enzyme immobilization: all you need to know. Molecules 30(4):39. https://doi.org/10.3390/molecules30040939

Rony MRI, Khan JM, Islam MR, Alam KMK, Kumar D, Ahmad A, Rana S, Hoque MA (2024) Exploration of phase separation and several physicochemical parameters of the mixture of triton X-100 and ceftriaxone sodium salt: influences of the composition of sodium salts. Chem Pap 78(1):307–319. https://doi.org/10.1007/s11696-023-03085-8

Ruiz-martinez L, Leermakers F, Stoyanov S, Van Gucht J, Der. (2025) Tunable capillary suspensions from aqueous two-phase systems. Langmuir 41(18):11604–11613. https://doi.org/10.1021/acs.langmuir.5c00749

Saha N, Sarkar B, Sen K (2022) Aqueous biphasic systems: a robust platform for green extraction of biomolecules. J Mol Liq 363:119882. https://doi.org/10.1016/j.molliq.2022.119882

Santos MPF, de Souza Junior EC, Villadóniga C, Vallés D, Castro-Sowinski S, Bonomo RCF, Veloso CM (2024) Proteases: importance, immobilization protocols, potential of activated carbon as support, and the importance of modifying supports for immobilization. Biotech 13(2). https://doi.org/10.3390/biotech13020013

Sargazi Z, Miri MA, Najafi M (2024) Evaluation the effect of hydrolysis conditions and type of protease on the degree of hydrolysis and antioxidant properties of the protein hydrolysate from the skipjack fish (*Katsuwonus pelamis*) viscera by the response surface methodology. J Food Sci Technol (Iran) 1-Introduction 20(143):91–107. https://doi.org/10.22034/FSCT.20.144.131

Sarwa N, Kumari P, Meena D, Udawat P, Chaudhary NS (2024) Alkaline proteases from Haloalkaliphiles: unveiling nature's catalysts for diverse applications. Appl Biochem Microbiol 60(5):855–870. https://doi.org/10.1134/S0003683824603676

Singla M, Sit N (2023) Theoretical aspects and applications of aqueous two - phase systems. ChemBioEng Rev 10(1):65–80. https://doi.org/10.1002/cben.202200026

Somader PD, Trzcinski A, Chen G, Chow Y, Abdul Manan M (2025) Fermentation of sorghum with *Aspergillus* strains: a promising and sustainable pathway to enzyme production- comprehensive review. Renew Sust Energ Rev 213(5):115456. https://doi.org/10.1016/j.rser.2025.115456

Srivastav. (2022) Hyaluronic acid generated by *Streptococcus zooepidemicus* is recovered Utilising PEG citrate aqueous two phase systems. J Pharmaceut Negat Results 13(1):100–108. https://doi.org/10.47750/pnra.2022.13.01.018

Sujitha P, Shanthi C (2023) Importance of enzyme specificity and stability for the application of proteases in greener industrial processing- a review. J Clean Prod 425(7):138915. https://doi.org/10.1016/j.jclepro.2023.138915

Teymoornejad G, Jabbari M (2023) Thermodynamic insights into phase behavior a new aqueous two-phase system at different temperatures : Experimental equilibria, data correlation and modeling. J Mol Liq 375:121282. https://doi.org/10.1016/j.molliq.2023.121282

Titus AR, Madeira PP, Ferreira LA, Chernyak VY, Uversky VN, Zaslavsky BY (2022) Mechanism of phase separation in aqueous two-phase systems. Int J Mol Sci 23(22). https://doi.org/10.3390/ijms232214366

Torabizadeh H, Khaki L (2025) Sesame oil and protein hydrolysates extraction by combined Carbohydrases and proteases simultaneously in a single unit aqueous phase enzymatic bioreactor. Food Bioprocess Technol 18(8):7330–7348. https://doi.org/10.1007/s11947-025-03872-z

Urge AY, Pampanin DM, Martino ME, Knudsen DL (2023) Salting-out assisted liquid-liquid extraction for UPLC-MS / MS determination of thyroxine and steroid hormones in human serum and fish plasma. Separations 10(4):240. https://doi.org/10.3390/separations10040240

Vicente FA, Tkalec N, Likozar B (2024) Responsive deep eutectic solvents: mechanisms, applications and their role in sustainable chemistry. Chem Commun 61(6):1002–1013. https://doi.org/10.1039/d4cc05157b

Vizovisek M, Ristanovic D, Menghini S, Christiansen MG, Schuerle S (2021) The tumor proteolytic landscape: a challenging frontier in cancer diagnosis and therapy. Int J Mol Sci 22(5):1–30. https://doi.org/10.3390/ijms22052514

Wen Y, Ma J (2022) Phase separation drives the formation of biomolecular condensates in the immune system. Front Immunol 13(11):1–15. https://doi.org/10.3389/fimmu.2022.986589

Yadav M, Taliyan SK, Kumar A, Singh P (2023) Extensive studies on physiochemical properties, purification strategies, application of microbial l-amino acid oxidase. Vegetos 36(1):38–51. https://doi.org/10.1007/s42535-023-00569-6

Yang H, Ren X, Zhao Y, Xu T, Xiao J, Chen H (2024) Enhancing alkaline protease stability through enzyme-catalyzed crosslinking and its application in detergents. PRO 12(3). https://doi.org/10.3390/pr12030624

Yuan X, Xu H, Liu X, Zhang J, Li J, Liang Q, An B,釘Paternò GM, Zhang M, Tang Y, Zhang C, Xu D, Zhong C, Li K, Wang X (2025) Engineered living energy materials. Interdiscipl Mater 4(3):412–455. https://doi.org/10.1002/idm2.12245

Yusree F, Mohd F, Peter AP, Zuhair M, Nor M, Show PL, Mokhtar MN (2021) Latest advances in protein-recovery technologies from agricultural waste. Foods 10(11):2748. https://doi.org/10.3390/foods10112748

Zambare VP, Nilegaonkar SS (2016) Proteases in leather processing. In: Thangadurai D, Sangeetha J (eds) Industrial biotechnology, sustainable production and bioresource utilization. Apple Academic Press/CRC Press., (e-Book ISBN-13: 978-1-77188-262-0), pp 209–238

Zambare VP, Nilegaonkar SS, Kanekar PP (2011) Production optimization and purification of a novel extracellular protease from *Pseudomonas aeruginosa* MCM B-327. New Biotechnol 28(2):173–181. https://doi.org/10.1016/j.nbt.2010.10.002

Zambare VP, Nilegaonkar SS, Kanekar PP (2014) Scale up production of protease using *Pseudomonas aeruginosa* MCM B-327 and its detergent compatibility. J Biochem Technol 5(2): 698–707

Zerizer H, Boughachiche F, Mebarki A, Sinacer O, Rachedi K, Ait Kaki A (2024) Partitioning purification, biochemical characterization, and milk coagulation efficiency of protease from a newly *Streptomyces* sp. isolate. Braz J Microbiol 55(3):2179–2187. https://doi.org/10.1007/s42770-024-01386-y

Zhang C, Liu X, Gong J, Zhao Q (2023) Liquid sculpture and curing of bio-inspired polyelectrolyte aqueous two-phase systems. Nature Commun 14(1). https://doi.org/10.1038/s41467-023-38236-8

Zhang, Cai Z, Wang L, Xie S, Zong W (2024) Unlocking liquid-liquid separation: exploring the marvels of aqueous two-phase systems. Microchem J 200(2):110445. https://doi.org/10.1016/j.microc.2024.110445

Zhang, Han M, Han S, Zong W (2025) Aqueous two-phase system (ATPS): from basic science to applications. RSC Adv 15(12):9041–9054. https://doi.org/10.1039/d4ra08232j

Zhao ZC, Huang LX, Dong XL, Wu JY (2024) Evaluation of three-phase partitioning for efficient and simultaneous isolation of immunomodulatory polysaccharides and proteins from *Lentinula Edodes* mushroom. Food Bioprocess Technol 17(8):2277–2291. https://doi.org/10.1007/s11947-023-03262-3

Chapter 7
Metagenomics and Genome Mining for Novel Protease Discovery

Alexander Machado Cardoso and P. Saranraj

Abstract Proteases are key enzymes with vast biological and industrial relevance. Recent developments in metagenomics and genome mining have significantly expanded the landscape of protease discovery by enabling access to genetic resources from uncultivable microorganisms. This chapter reviews functional and sequence-based strategies for metagenomic exploration, highlights major discoveries from diverse ecosystems, discusses bioinformatic tools and databases for genome mining, and outlines current challenges and emerging technologies. The convergence of bioinformatics, omics integration, synthetic biology, and artificial intelligence is critical to accelerate the identification and application of novel proteases for use in biotechnology, medicine, and sustainable industries.

Keywords Bioinformatics · Biotechnology · Functional screening · Enzyme discovery · Microbial enzymes

7.1 Introduction

Proteases are among the most often used enzymes in biotechnology, accounting for over 60% of the entire worldwide enzyme market due to their diverse catalytic capabilities and wide range of industrial uses, including detergents, medicines, and food processing (Zambare and Nilegaonkar 2016; Song et al. 2023). Microbial proteases have gained popularity due to their simplicity of manufacture, stability, and capacity to work in a variety of environmental settings (Razzaq et al. 2019). The identification of new proteases with desired properties, such as activity at extreme pH, temperature, or salinity, has become a top goal in enzyme biotechnology.

A. M. Cardoso (✉)
Department of Biology, Rio de Janeiro State University (UERJ), Rio de Janeiro, RJ, Brazil

P. Saranraj
Department of Microbiology, Sacred Heart College (Autonomous), Tirupattur, Tamil Nadu, India

V. Zambare et al. (eds.), *Biotechnological Advances and Applications in Microbial Proteases*, Interdisciplinary Biotechnological Advances,
https://doi.org/10.1007/978-981-95-9011-7_7

However, established approaches for isolating novel proteases are heavily culture-dependent, restricting access to the microbial diversity that cannot be cultured in laboratory conditions (Uchiyama and Miyazaki 2009).

The discovery of metagenomics, the culture-independent examination of genetic material extracted directly from environmental samples, has transformed the search for new enzymes, including proteases (Robinson et al. 2021). By eliminating the necessity for microbial cultivation, metagenomic techniques allow researchers to access the vast pool of functional genes carried by uncultured microorganisms in a variety of settings, including soil, ocean, sediments, compost, and even the gut microbiota (Cardoso et al. 2012; Popovic et al. 2015). Functional metagenomics enables direct screening of metagenomic libraries for enzymatic activity, making it an effective method for identifying novel enzymes with no known homologs (Rabausch et al. 2013).

Genome mining has emerged as a complementary strategy, allowing in silico identification of biosynthetic gene clusters and enzymatic domains from sequenced microbial genomes and metagenomes using bioinformatic pipelines (Meesil et al. 2023). The integration of metagenomics with genome mining approaches has created new opportunities for enzyme discovery by enabling high-throughput identification and annotation of putative protease genes across large datasets, often guided by sequence motifs or conserved catalytic domains such as those found in the MEROPS peptidase database (Neil et al. 2018). These integrative approaches have significantly expanded the landscape of known proteases and hold promises of unlocking previously hidden catalytic potential.

With an increasing need for enzymes designed for industrial, medicinal, and environmental applications, it is critical to create strong, high-throughput enzyme discovery methodologies. Metagenomic and genome mining techniques are especially adapted to this endeavor, offering both breadth and depth in the investigation of microbial proteolytic potential. Recent developments in sequencing technology, bioinformatics tools, and machine learning have improved our ability to uncover, predict, and describe new proteases in complex microbial communities. This chapter presents a summary of recent breakthroughs in metagenomics and genome mining for the discovery of new microbial proteases. We begin with a review of microbial protease diversity and its industrial applications, followed by a thorough explanation of metagenomic methods and genome mining approaches. By combining traditional enzymology with cutting-edge omics and computational tools, this chapter attempts to showcase the transformational potential of metagenomics and genome mining in enzyme biotechnology.

7.2 Microbial Proteases: Classification and Applications

Proteases (sometimes called peptidases or proteinases) are enzymes that break down peptide bonds in proteins and peptides. They are abundant in nature and required for many physiological activities in both prokaryotes and eukaryotes, such as protein

Table 7.1 Classification of major microbial protease families based on catalytic mechanisms and key features according to the MEROPS database

Classification	Features
Serine proteases	Use a serine residue in their active site to mediate catalysis. These include subtilisins, trypsin-like enzymes, and are widely produced by *Bacillus* spp.
Cysteine proteases	They include enzymes such as papain, are common in both prokaryotes and eukaryotes, and use nucleophilic cysteine residue
Aspartic proteases	Contain two aspartic acid residues in their catalytic site. These are active at acidic pH and are produced by fungi such as *Aspergillus* and *Rhizopus* spp.
Metalloproteases	Require divalent metal ions (usually Zn^{2+}) for activity. Examples include thermolysin and pseudolysin
Threonine and glutamic proteases	Less common, but increasingly recognized with the advancement of genome mining tools

turnover, signal transduction, and cellular control (López-Otín and Bond 2008). Proteases are regarded essential biocatalysts in biotechnology because of their flexibility and broad substrate specificity (Song et al. 2023). Microbial proteases generated from bacteria, fungi, and archaea are the most important in industry, owing to their fast growth, simplicity of genetic modification, and adaptation to large-scale fermentation (López-Otín and Bond 2008; Song et al. 2023). They can be classified based on several criteria, such as site of action, catalytic mechanism, and pH preference. The MEROPS database serves as an authoritative resource for classification and nomenclature of peptidases and their inhibitors, dividing them into clans and families based on evolutionary and functional properties (Neil et al. 2018). The most widely accepted classification system is based on the catalytic mechanism (Table 7.1).

Proteases dominate the worldwide industrial enzyme landscape, accounting for over two-thirds of the entire enzyme market, owing to their catalytic flexibility, broad substrate range, and use in a variety of industries (Song et al. 2023). These enzymes are important in many sectors, including home and industrial detergents, food and dairy processing, leather and textile treatment, medicines, and environmental biotechnology. Alkaline serine proteases, notably those from the subtilisin family, are among the most important economically. These enzymes are widely used in laundry and dishwashing detergents, facilitating the breakdown of proteinaceous stains under hard washing conditions. To be effective in such formulations, these proteases must be stable at high temperatures (usually 40–60°C), alkaline pH (8–11), and in the presence of surfactants, oxidizing agents, and chelators. Subtilisin variations designed for oxidant resistance and extended shelf life have become industry standards, notably those generated from *Bacillus licheniformis* and *Bacillus subtilis* (Song et al. 2023; Allison et al. 2024).

In the dairy and food industries, proteases such as chymosin (rennin) are essential to cheese production, where they facilitate the specific splitting of -casein, facilitating milk coagulation and formation of curls. Traditional animal-derived chymosins are increasingly replaced by recombinant microbial versions (e.g., *Aspergillus niger* var. *awamori* or *Kluyveromyces lactis* genetically modified) because of improved safety, scale, and regulatory acceptance (Kumar et al. 2010). In addition to milk clotting, proteins are used for meat tenderization, protein hydrolysis, production of bioactive peptides, flavor enhancement, and texture modification, contributing to product quality and consumer appeal (MSI et al. 2023).

Microbial proteases have shown to be quite useful in therapeutic formulations in the pharmaceutical and medical industries. Streptokinase and urokinase, for example, are used as thrombolytics to dissolve blood clots in myocardial infarction and stroke patients. Proteases are also utilized in wound debridement treatments, where enzymes such as papain, bromelain, and microbial proteases preferentially break down dead tissue to promote healing in chronic ulcers and burns (Avila-Rodríguez et al. 2020). Furthermore, protease inhibitors are a key component of antiviral medication development, with HIV-1 protease inhibitors (e.g., ritonavir, darunavir) being an important class of antiretroviral therapy (Liu and Kong 2016).

Enzyme-based processing has replaced traditional chemical treatments in the leather and textile sectors in an effort to reduce their environmental impact. In order to effectively remove keratin and other structural proteins without the use of harsh alkali treatments or poisonous sulfide-lime, proteases are now frequently employed in the dehairing, batting, and degumming processes. While preserving product quality, this enzymatic method lowers water pollution and improves worker safety (Hasan et al. 2022). Microbial proteases aid in the bioremediation of protein-rich agro-industrial effluents and the breakdown of organic waste in the context of environmental sustainability. Extremophilic microorganism-derived proteases, frequently obtained by metagenomic mining, are becoming increasingly popular because of their solvent stability, halotolerance, and thermostability, which make them appropriate for high-load, variable-condition waste processing systems (Kikani et al. 2023). The biotechnological applicability of these enzymes has been expanded by recent studies that have reported the discovery of novel proteases from metagenomes of compost, oil-contaminated soils, and marine sediments that exhibit activity across broad pH and temperature ranges and remain functional in the presence of heavy metals or organic solvents (Alma'abadi et al. 2015).

The combination of metagenomic discovery, directed evolution, and synthetic biology promises to speed up the synthesis of next-generation proteases with specialized functions for particular commercial applications. These activities will help to address worldwide demand for sustainable, efficient, and environmentally friendly biocatalysts (Radley et al. 2023). Aside from their traditional activities in detergents, food processing, and medicines, proteases have lately emerged as multifunctional biocatalysts with new applications in antifouling surfaces, antimicrobial coatings, and biosensor technologies. Proteases that may degrade microbial biofilms or cleave quorum-sensing molecules have shown promise in interrupting microbial colonization on medical implants and marine structures (Rather et al. 2021). By leveraging their substrate selectivity and catalytic amplification characteristics, proteases may now be used to identify disease biomarkers such as cancer-associated peptides and bacterial virulence factors (Buss et al. 2018).

The increased interest in proteases that work under severe physicochemical conditions has pushed bioprospecting to extreme habitats like deep-sea hydrothermal vents, hypersaline lakes, and acidic mines drainage. These niches are home to microbial populations that have evolved to severe temperatures, pH, salinity, and pressure, and their enzymes, known as extremozymes, frequently exhibit extraordinary thermostability, halotolerance, acid-alkali resistance, and solvent stability (Kikani et al. 2023). Comparing these enzymes to their mesophilic counterparts reveals several advantages, such as a reduced risk of contamination, a longer shelf life, and compatibility with nonaqueous industrial environments. A significant obstacle to obtaining this enzymatic variety, nevertheless, is the inability to cultivate most extremophiles in a laboratory setting. In such environments, conventional culture methods recover less than 1% of the overall microbial diversity, resulting in a large discrepancy between exploitable and potential enzymatic resources. This "great plate count anomaly" highlights how important metagenomics and genome mining are to the current process of discovering new enzymes. Direct access to functional genes from complex microbial consortia is made possible by these methods, which do not need microbial isolation. Metagenomics has become essential for finding novel proteases with distinct structural characteristics, catalytic processes, and industrial importance, especially when combined with developments in high-throughput sequencing, bioinformatic annotation, and functional screening in heterologous hosts (Liu et al. 2022).

7.3 Metagenomics: Concepts and Techniques

Metagenomics is the study of entire genetic material obtained directly from environmental samples that allows the exploration of microbial diversity and function without the need for cultivation. This approach revolutionized microbial ecology and biotechnology, revealing the enormous genomic potential of microorganisms, accounting for more than 99% of all microbial species on Earth (Liu et al. 2022). Metagenomics provides a powerful platform for the discovery of new enzymes, such as proteases, from complex microbial communities that live in a variety of environments, including soils, marine ecosystems, compost, hot springs, and extreme habitats. Metagenomics can be broadly divided into two methodological frameworks:

7.3.1 Sequence-Based Metagenomics

This method involves sequencing the collected DNA and evaluating it using bioinformatic tools to estimate gene function and phylogenetic identity. DNA extraction and quality control are critical procedures, followed by shotgun sequencing (Illumina, Nanopore, PacBio), assembly and binning of metagenome-assembled genomes (MAGs), gene prediction and functional annotation, and identification of potential enzymes based on conserved domains. Sequence-based metagenomics is especially useful for genome mining because it allows the discovery of genes producing proteases based on conserved motifs, such as those found in the MEROPS database (Neil et al. 2018).

7.3.2 Functional Metagenomics

Functional metagenomics entails creating libraries of DNA fragments and expressing them in a heterologous host (e.g., *Escherichia coli*), then screening for enzymatic activity. This method does not require prior sequence information and can identify unique proteins with no known homologs "ORFans" (Liu et al. 2022). Functional screening for proteases frequently uses indicator medium containing substrates such as casein, skim milk, or gelatin, where zones of clearance indicate proteolytic activity. While function-based approaches can be time-consuming, they provide direct evidence of enzyme activity and are invaluable for discovering novel biocatalysts. High-throughput sequencing platforms have revolutionized metagenomics by enabling comprehensive analysis of environmental DNA, but environmental sampling and DNA extraction are critical. Samples are collected from targeted environments, with strategies adapted to specific goals (e.g., sampling saline sediments for halophilic proteases). DNA extraction methods must ensure high molecular weight and minimal contaminants such as humic acids, especially from soil samples. Sampling strategies must be adapted based on the environment, such as saline sediments for halophilic proteases or mangrove sediments for an efficient DNA extraction (Yeates et al. 1998).

Assembling the short reads into longer contiguous sequences (contigs) is essential for reconstructing genes and pathways. Assemblers such as MEGAHIT, MetaSPAdes, and IDBA-UD are commonly used (Nurk et al. 2017). Binning algorithms (e.g., MetaBAT, MaxBin, CONCOCT) cluster contigs into metagenome-assembled genomes (MAGs), allowing taxonomic assignment and metabolic reconstruction. Also, gene prediction and annotation are complex steps; the open reading frames (ORFs) may be predicted using tools such as Prodigal, MetaGeneMark, or GeneMarkS-2. Functional annotation is carried out using BLAST/DIAMOND against databases such as NCBI NR, SwissProt, Pfam, InterPro, and eggNOG for domain-based annotation, and MEROPS for protease classification and family identification (Yeates et al. 1998; Nurk et al. 2017; Hera et al. 2024).

Putative protease genes can be filtered based on catalytic triads, sequence homology, or conserved domains (e.g., peptidase S8 domain in subtilisins). Enzymes of interest may also be prioritized using machine learning models that predict protein function or activity under specific conditions (Denesyuk et al. 2024). Several public repositories and tools have become essential resources for researchers working with metagenomic datasets, including IMG/M (Integrated Microbial Genomes &

Microbiomes) and KBASE (The Department of Energy Systems Biology Knowledgebase), which offer tools for comparative genomics and genome mining (Chen et al. 2019); MEROPS Database, a specialized resource for classification and annotation of peptidases and their inhibitors (Neil et al. 2018); and antiSMASH, which, although primarily used for secondary metabolites, can identify biosynthetic clusters encoding protease-related enzymes when co-localized with biosynthetic gene clusters (Blin et al. 2025).

7.4 Functional Metagenomics for Protease Discovery

The effective recovery and characterization of new proteases depend on each of the multiple well-defined steps that make up the functional metagenomic process. Environmental matrices, including soil, rhizosphere, compost, marine sediments, wastewater sludge, or hypersaline brines, are used to extract high-molecular-weight (HMW) DNA. To reduce inhibitors that obstruct downstream cloning and expression, such as humic acids, phenolic compounds, or polysaccharides, the extraction procedure must be improved. Producing DNA fragments appropriate for large-insert libraries, commercial kits, or CTAB-based techniques is frequently changed. The extracted eDNA is broken manually or by enzymes and then ligated into appropriate vectors, such as bacterial artificial chromosomes (BACs), cosmids, fosmids, or plasmids (Ji et al. 2024).

The target screening throughput, library complexity, and desired insert size are taken into consideration while choosing the vector system. When targeting protease complexes or accessory genes involved in secretion, fosmids and BACs are useful for maintaining operon architecture and co-regulatory elements. Recombinant libraries are converted into appropriate hosts for expression. Because of its genetic tractability and ease of transformation, *Escherichia coli* continues to be the most widely used host; however, genes that require particular folding environments or posttranslational modifications are increasingly being accommodated by alternative hosts such as *Streptomyces lividans*, *Bacillus subtilis*, or *Pseudomonas putida*. Expression fidelity can be further enhanced by using expression methods optimized with synthetic promoters or broad-host-range shuttle vectors (Ji et al. 2024).

Proteinaceous substrates are added to a solid or liquid medium to screen transformants for protease activity. Hemoglobin, gelatin, casein, and skim milk are examples of common screening substrates. Clearing zones, or halos, on agar plates or colorimetric or fluorometric tests in microtiter plate formats are commonly used to observe enzymatic activity. Proteases with certain industrial qualities can be enriched by applying selective circumstances (such as high salt, an acidic pH, or the addition of detergent). To determine the open reading frames (ORFs) causing activity, positive clones are sequenced. Tools such as Prodigal or GeneMark are used for gene prediction, while domain databases (MEROPS, Pfam, and TIGRFAMs) are used for functional annotation. Also, 3D structure predictions using programs such as AlphaFold2 or SWISS-MODEL may offer additional insights on substrate

Table 7.2 Detection of protease activity. The choice of assay should be guided by the expected enzyme properties (e.g., pH, temperature, salt tolerance) and the intended industrial application

Test/assay	Methodology
Casein agar plates	Clones secreting extracellular proteases create clear halos on casein-containing media. This assay is simple and broadly applicable
Skim Milk agar	Milk proteins serve as substrates. Useful for detecting both broad-spectrum and specific proteases
Azocasein or Azocoll assay	Colorimetric substrates in liquid cultures offer quantification of proteolytic activity
Fluorescent probes	Substrates such as FITC-labeled casein allow sensitive detection of protease activity, enabling high-throughput screening in microtiter plates
Zymography	Gel-based assay using SDS-PAGE with substrate-embedded gels (e.g., gelatin or casein) to identify proteases based on molecular weight and activity profiles

selectivity, structural stability, and possible industrial robustness. catalytic domains and active site motifs may be examined to verify protease class (Jumper et al. 2021).

The choice of substrate and assay sensitivity is crucial for the successful finding of proteases in functional metagenomics. Proteases with limited or noncanonical substrate specificities may not be detected by conventional substrates, such as casein or gelatin, despite their wide application. Numerous test formats have been created in an effort to increase functional coverage (García-Moyano et al. 2021). In order to improve throughput and decrease false negatives, particularly in large metagenomic libraries, sophisticated techniques, including microfluidics, droplet-based screening, and activity-based probes, are also being developed. The substrate and detection technique selection have a significant impact on functional screening for proteases. Table 7.2 lists a few popular methods for checking for protease activity.

Bypassing the constraints of culture, functional metagenomics offers a potent platform for direct identification of new biocatalysts from complex microbiomes. Researchers have discovered proteases with remarkable biochemical characteristics that are frequently absent from reference databases by expressing environmental DNA in surrogate hosts and checking for enzymatic activity. The effectiveness of this method to find proteolytic enzymes that have adapted to harsh or specialized ecological environments has been demonstrated by a number of seminal researches (Hogg et al. 2024).

A large-insert fosmid metagenomic library made from soil was examined for proteolytic activity on casein agar in a pioneering work. An unidentified serine protease with a limited sequence similarity (less than 30%) to known enzymes was encoded by one positive clone. Its broad pH tolerance (pH 5.0–10.0), thermostability up to 60°C, and compatibility with existing detergents were discovered by biochemical evaluation. These characteristics highlight their potential for use in laundry formulations and industrial cleaning agents. From Antarctic marine sediments, a metagenomic library was screened using azocasein as a substrate to identify cold-active enzymes. Over 80% of the maximum activity of a cold-adapted metalloprotease (MPA) was maintained at 10 °C. The characteristics of psychrophilic enzyme adaptation, such as decreased core hydrophobicity and increased surface loop flexibility, were demonstrated by structural modeling (Kennedy et al. 2011; Falkenberg et al. 2023; Hogg et al. 2024). Such cold-active proteases are particularly attractive for food processing and pharmaceutical applications, where low-temperature operations help preserve product quality and bioactivity. The enzyme showed high thermal stability and maintained its activity in the presence of organic solvents and ionic detergents, indicating its usefulness in bioindustrial processes that require solvent resistance and thermal resistance, highlighting the potential of functional genomics to identify proteins that are sensitive to environmental conditions, from the polar ecosystem to geothermal habitats, which are not well explored in traditional microbiology (Kennedy et al. 2011; Falkenberg et al. 2023; Hogg et al. 2024).

Despite its promises, functional metagenomics faces several methodological challenges, especially regarding heterogeneous expression and the sensitivity of tests. Expression is often inefficient. *E. coli* is the most common host, lacking the regulatory elements, chaperones, or posttranslation machinery needed to express and fold many foreign proteins correctly. Compound usage bias, incompatible promoters, improper folding, or absence of basic cofactors may cause nonfunctional expression of metagenomic genes (Kennedy et al. 2011; García-Moyano et al. 2021; Jumper et al. 2021; Falkenberg et al. 2023; Hogg et al. 2024; Ji et al. 2024). Clone screening is labor-intensive and low-throughput, especially when libraries consist of millions of environmental DNA fragments. While agar plate–based assays (e.g., skim milk or casein hydrolysis zones) are simple and widely used, they suffer from limited sensitivity and may miss enzymes with narrow substrate specificity. Substrate generality is another constraint; common substrates such as casein or gelatin enable broad-spectrum screening but fail to detect proteases with specificity for nonproteinaceous targets or regulatory substrates. This leads to the underrepresentation of functionally novel enzymes (García-Moyano et al. 2021). Library design trade-offs also influence success. While large-insert libraries (fosmids, BACs) preserve gene clusters and regulatory elements, they are more complex to handle and manipulate. Small-insert plasmid libraries offer higher transformant efficiency but may truncate functional genes or omit upstream regulatory sequences.

7.5 Genome Mining and Bioinformatics for Novel Protease Discovery

The field of enzyme discovery has seen a significant transformation due to the rapid growth of microbial genome and metagenome databases, which is fueled by advancements in next-generation sequencing (NGS) technology and bioinformatics. A potent in silico technique for the methodical discovery of protease-encoding genes from both cultivated and uncultured microbes is genome mining. By using curated domain models, structural predictions, and annotation workflows to investigate the proteolytic potential embedded within large-scale sequencing data, this method gets beyond the drawbacks of conventional culture and biochemical screening. The foundation of genome mining is the idea that sequence encodes function and that conserved domains, catalytic motifs, and protein structures may be used as stand-ins for biochemical characteristics, substrate selectivity, and enzymatic activity (Gerlt 2017).

Proteases are classified according to their catalytic mechanism, mostly into classes that are characterized by conserved motifs: serine, cysteine, aspartic, threonine, and metalloprotease. For example, metalloproteases usually contain Zn^{2+}-coordinating motifs such as HEXXH, whereas serine proteases are characterized by the catalytic triad His–Asp–Ser. Profile Hidden Markov Models (HMMs), which are used in programs such as HMMER, InterProScan, and PfamScan, make it easier to detect these domains. Even among new or distantly related sequences, these methods allow for the sensitive and reliable detection of catalytic fingerprints (Sonnhammer et al. 1998; Quevillon et al. 2005; Eddy 2011). Protease categorization relies heavily on databases such as Pfam, MEROPS, and SMART, which curate hundreds of functionally annotated protein domains. Remarkably, the MEROPS database, which groups proteases into families and clans according to catalytic type, domain organization, and phylogenetic distribution, continues to be the gold standard for peptidase taxonomy (Neil et al. 2018). Using this method, very diversified metagenomes from soil, marine sediments, thermal springs, and host-associated habitats have yielded previously unidentified or cryptic protease families, including multidomain fusions and divergent homologs.

Workflows involving genome mining depend on the precise identification of protein-coding regions. Tools such as Prodigal, MetaGeneMark, and GeneMarkS are often used in bacterial genomes and metagenomes for open reading frame (ORF) prediction, particularly from assembled contigs or metagenome-assembled genomes (MAGs). Sequence similarity searches (e.g., BLASTp, DIAMOND) are used to annotate predicted ORFs against curated databases, including eggNOG (evolutionary context), COG (Clusters of Orthologous Groups), and MEROPS (proteolytic function). Multiple information sources are integrated via functional annotation pipelines such as RAST, Prokka, and DRAM to provide consensus predictions that boost functional assignment certainty (Neil et al. 2018; Hogg et al. 2024). In addition to traditional proteases, this layered annotation approach makes it possible to identify potential regulatory proteases, hybrid enzymes, and new variations of proteases that are frequently overlooked by single-source annotations.

There is increasing evidence that several proteases are co-localized inside operons involved in host interactions, nutrient scavenging, or quorum-sensing networks, despite the fact that they are not normally found in traditional secondary metabolite biosynthesis gene clusters (BGCs). Protease genes embedded in BGCs or close to mobile genetic elements, toxin–antitoxin systems, or secretion modules can be found using genome mining techniques such as antiSMASH, PRISM, and BAGEL (Table 7.3). In pathogenic or symbiotic bacteria, for instance, several secreted proteases are co-transcribed with genes involved in siderophore production or virulence regulators, suggesting ecological specialization and coordinated expression (Sharon

Table 7.3 Some bioinformatics tools for protease discovery

Tool	Features
MEROPS	The primary database for proteases, inhibitors, and regulatory proteins. It classifies proteases into families/clans based on catalytic mechanism (serine, cysteine, aspartic, metallo) and provides structural, substrate, and specificity data
Pfam/InterPro	Identify conserved proteolytic domains (e.g., Pfam's "Peptidase_S8" (PF00082) for subtilisins)
antiSMASH/BAGEL	Detect BGCs potentially encoding extracellular proteases involved in microbial defense or competition
Anvi'o/IMG/M	Enable advanced metagenomic/pangenomic analyses, including metagenome-assembled genome (MAG) reconstruction, gene mapping, and ecological correlation
AlphaFold/I-TASSER/ SWISSModel	Predict 3D protease structures to reveal active-site topology and stability. AlphaFold2, in particular, achieves near-crystallographic accuracy

et al. 2009). Thus, assessing the genomic neighborhood offers crucial hints regarding the control of proteases, their ecological role, and their potential for expression, especially in species that have not been cultured or functionally defined.

A protease's subcellular location and secretion profile are essential to its usefulness in both industrial and medicinal settings. Proteases that are intended for extracellular use, as in food processing, detergents, or wound debridement, need to be effectively secreted and stable in challenging conditions. While PSORTb, CELLO, and TMHMM infer ultimate subcellular localization and transmembrane domain design, SignalP and Phobius predict N-terminal signal peptides guiding export via Sec or Tat pathways. By selecting the proteases that are most likely to be efficiently produced and secreted, as well as those that are compatible with host expression platforms, these predictions help guide the selection of enzymes. Extracellular proteases found in halophilic or thermophilic metagenomes, for example, could be naturally stable in industrial settings (Mendes et al. 2017).

The capacity to predict protease tertiary structures from sequence alone, even for orphan or low-homology proteins, has been transformed by recent developments in structural bioinformatics, especially the emergence of AlphaFold2 (Jumper et al. 2021). Resolving active sites, locating substrate-binding pockets, and differentiating between catalytic and noncatalytic homologs are all possible with structural models. This structural resolution enables logical predictions of metal or cofactor dependencies through ligand docking or pocket geometry, temperature, or pH stability by surface charge distribution and hydrophobic core analysis, and enzymatic mechanism (e.g., confirmation of catalytic triads or dyads). These predictions are further supported by tools such as I-TASSER, DeepEC, and Protscale, which estimate biochemical characteristics including thermostability, solvent tolerance, and isoelectric point. Mechanistic and physicochemical criteria can replace identity-based heuristics in candidate ranking as these structural tools become increasingly incorporated into genome mining pipelines. This will accelerate the process of moving from in silico prediction to laboratory validation.

The application of genome mining to metagenomic and environmental datasets has yielded a growing number of successful case studies in the discovery of novel proteases. These studies highlight the power of in silico tools to uncover enzymes with desirable industrial properties from previously inaccessible microbial communities, particularly those inhabiting extreme environments.

Hypersaline environments such as marine salterns harbor extremophilic microbial consortia adapted to high salt concentrations, making them fertile grounds for the discovery of salt-tolerant enzymes. A notable genome mining effort involving metagenomic datasets from hypersaline brines identified over 50 candidate protease genes bearing Peptidase M4 and S8 domains, families known for metalloprotease and serine protease activity, respectively (Mendes et al. 2017). The existence of unique lineages may be indicated by the intriguing fact that a number of these genes have less than 35% sequence identity with any described proteases in the MEROPS database (Neil et al. 2018). The authors discovered that several of the candidate enzymes had surface-exposed acidic residues, a characteristic of halophilic adaptation that preserves protein stability and solubility under hyperosmotic conditions, using structural modeling. These results highlight the ability of structure-guided mining to forecast methods for environmental adaptation prior to functional characterization (Mak et al. 2015; Mevarech et al. 2000).

For low-temperature industrial operations, psychrophilic enzymes provide clear benefits such as reduced substrate denaturation and energy savings. Researchers identified 18 potential protease genes, some of which were connected to metallopeptidase families, by screening metagenomic data from Arctic maritime sediments using the Integrated Microbial Genomes & Microbiomes (IMG/M) platform (Kumari et al. 2021). The presence of flexible surface loops, a known adaptation for sustaining enzymatic activity under low-temperature conditions by improving catalytic site dynamics, was suggested by structural modeling of one protease that had an atypical HEXXH motif, which deviated from canonical Zn^{2+}-binding configurations. This example shows how functional innovation may be indicated by departures from conserved motifs, which are frequently missed in rigorous domain-based searches (Bauvois et al. 2008).

7.6 Challenges and Future Directions

There are still several obstacles that prevent metagenomics and genome mining from reaching their full potential in protease discovery and application, despite the remarkable advancements they have made. These restrictions cover the fields of methodology, computing, biology, and translation. Nonetheless, new developments in synthetic biology, machine learning, multiomics integration, and expression system design present encouraging answers and might usher in a new age of functional protease bioprospecting. The heterologous expression of environmental genes, which is frequently carried out in *Escherichia coli* and has a number of drawbacks, is a significant problem in activity-based metagenomics. These include biased codon use, promoters or ribosome binding sites that are unsuitable, and incorrect protein folding brought on by the lack of natural chaperones or secretion mechanisms. Furthermore, certain proteases are generated as inactive zymogens or need particular cofactors or posttranslational modifications (such as glycosylation or disulfide bond formation) that are not accurately replicated in bacterial hosts. As a result, a significant portion of protease genes in metagenomic libraries is not detectable using conventional activity tests (Craig et al. 2010).

Sequence-based mining avoids expression problems, but it is mostly dependent on curated databases and computational tools, which are not perfect. Many potential proteases are below the cutoff point for accurate functional prediction, sharing less than 30% of their identity with known enzymes. Through public repositories, annotation mistakes and domain misclassification spread even further, resulting in false negatives or incorrectly assigned functions. It is still very difficult to distinguish active enzymes from inactive homologs (such as pseudogenes or noncatalytic paralogs) based on sequence. Furthermore, full-length gene prediction and functional inference are frequently made more difficult by the short and fragmented structure of many metagenomic contigs. DNA extraction and sample collection are still crucial processes that may skew the representation and diversity of metagenomic libraries. Because of low biomass, the existence of PCR inhibitors (such as humic acids), or resistant cell walls that prevent lysis, extreme environments, including deep-sea sediments, acidic soils, or hypersaline brines, present challenges. Rare or ecologically significant species may be left out by these restrictions, which might result in a lack of coverage of possibly new enzymatic lineages. Furthermore, the existence of mobile genetic elements, such as transposons and plasmids, might make functional annotation and genome assembly more difficult. For proteases discovered by metagenomic processes to satisfy industrial or biological performance standards, considerable tweaking is sometimes necessary. Enhancing stability, substrate selectivity, temperature and pH tolerance, and resistance to inhibitors or denaturing agents are all part of this. However, protein engineering is time-consuming and resource-intensive, regardless of whether it is done by directed evolution or rational design. Incomplete structural knowledge limits rational methods, and high-throughput screening platforms are necessary for directed evolution, although these platforms are not always available. Furthermore, especially when balancing promiscuity and specificity, improving one characteristic (like activity) may undermine another (like stability). In order to turn a promising protease into a product that is ready for the market, scalable fermentation, purification, and formulation techniques must be developed.

There are ethical and legal issues with using environmental DNA and gene editing technologies, especially when the samples come from delicate or protected ecosystems (such as human microbiota, forest soils, or plants). For food-grade, medicinal, or environmental uses, regulatory permission is required; the specific criteria vary according to the jurisdiction. Commercialization may be made more difficult by intellectual property issues, such as the patentability of enzymes that are created by synthetic biology or obtained from natural sources. Additionally, heterologous expression methods frequently produce inadequate amounts of enzymes, have batch variability, and pose difficulties with scaling up, all of which might postpone or impede commercial use. Countries' biosafety laws pertaining to genetically modified organisms (GMOs) vary greatly, which may limit the application of modified enzymes in the food processing or agricultural industries. The acceptability of such technologies by society will also depend on public opinion and openness. Combining metagenomic data with multiomics datasets, such as metabolomics, metatranscriptomics, and metaproteomics, is a potential area in functional metagenomics. Finding actively transcribed genes and searching for proteases with immediate functional significance are made possible by metatranscriptomics.

Metaproteomics may also allow the identification of posttranslational changes and offers direct proof of protein expression. Stable isotope probing can also be used to provide ecological context for enzyme function by connecting proteolytic activity to taxonomic identification. This comprehensive approach increases ecological validation, candidate ranking, and decreases false positives. Sequence-function prediction is being quickly transformed by artificial intelligence and machine learning techniques. Primary sequences may be used to predict substrate specificity, activity, and stability using deep learning models that have been trained on experimentally verified datasets. Structure-guided design and prioritization are now possible thanks to tools such as AlphaFold2, which enable high-accuracy structural modeling of potential proteases. Additionally, machine learning is being utilized to optimize expression vectors for heterologous hosts and create synthetic promoters. These techniques can significantly shorten the time between sequence identification and functional characterization when used in conjunction with metagenomic mining.

The modular construction of genes for regulated or inducible protease production, including the co-expression of auxiliary proteins such as chaperones, foldases, or secretion machinery, is made possible by synthetic biology. Complex proteolytic tasks, such as trash degradation or bioremediation, can be carried out in situ by engineered microbial consortia. In the meantime, many of the obstacles caused by cellular toxicity or expression incompatibility are removed by cell-free protein synthesis platforms, which provide a host-independent substitute for the quick production and functional evaluation of metagenome-derived enzymes. Researchers are increasingly using alternative expression methods to get around *E. coli*'s drawbacks. Extracellular enzymes are commonly secreted by *Bacillus subtilis*, while eukaryotic posttranslational modifications can be carried out by yeasts such as *Pichia pastoris* and *Yarrowia lipolytica*. The host compatibility of metagenomic inserts will continue to be enhanced by developments in promoter engineering, synthetic biology toolkits, and chassis design.

Sustainable bioprospecting methods, open-access data exchange, and multidisciplinary cooperation amongst microbiologists, chemists, engineers, and computational biologists are essential for the future of protease discovery. Massive and unused metagenomic resources are still being produced by international projects such as the Earth Microbiome Project and the Tara Oceans Expedition. To make data integration and replication easier, standardization of gene annotations, environmental conditions, and metadata will be crucial. In order to ensure that innovation is in line with global sustainability goals, it is equally crucial to incorporate life cycle evaluations and green chemistry metrics into the development and use of protease-based biotechnologies.

7.7 Conclusions

The discovery of novel proteases represents a keystone for advancing both biotechnology and industrial applications. Traditional cultivation methods, which inherently limit access to the vast microbial diversity, have been increasingly complemented and surpassed by metagenomics and genome mining approaches. These transformative strategies enable the exploration of genetic resources from uncultivable microorganisms, facilitating the identification of enzymes with unique and highly valuable catalytic properties. Both sequence-based and functional metagenomics, in combination with genome mining, have emerged as powerful tools to reveal hidden proteolytic potential across diverse ecosystems. When integrated with advances in bioinformatics, synthetic biology, and artificial intelligence, these approaches accelerate the discovery, characterization, and application of novel proteases. Often sourced from extremophiles, such enzymes exhibit remarkable stability under harsh physicochemical conditions, making them particularly suitable for applications in detergents, food processing, pharmaceuticals, leather and textile industries, and environmental bioremediation. The future of protease discovery will rely on continued innovation and interdisciplinary synergy. The ability to identify, optimize, and deploy proteases with enhanced specificity and efficiency is essential to meet the growing demand for sustainable and effective biocatalysts. Continued exploration of extreme environments, coupled with the application of advanced computational tools, promises to uncover a new generation of proteases, driving innovation across biotechnology and contributing to greener, more efficient solutions for global challenges.

Acknowledgments We thank the technical support and insightful discussions from FIOCRUZ, UERJ, and SHC staff.

Conflict of Interest The authors declare no conflict of interest.

Funding This research was funded by Fundação Carlos Chagas Filho de Amparo à Pesquisa do Estado do Rio de Janeiro (FAPERJ), Coordenação de Aperfeiçoamento de Pessoal de Nível Superior (CAPES), and Conselho Nacional de Desenvolvimento Científico e Tecnológico (CNPq).

References

Allison SD, AdeelaYasid N, Shariff FM, Rahman AN (2024) Molecular cloning, characterization, and application of organic solvent-stable and detergent-compatible thermostable alkaline protease from Geobacillus thermoglucosidasius SKF4. J Microbiol Biotechnol 34(2):436–456. https://doi.org/10.4014/jmb.2306.06050

Alma'abadi AD, Gojobori T, Mineta K (2015) Marine metagenome as a resource for novel enzymes. Genomics Proteomics Bioinform 13(5):290–295. https://doi.org/10.1016/j.gpb.2015.10.001

Avila-Rodríguez IM, Meléndez-Martínez D, Licona-Cassani C et al (2020) Practical context of enzymatic treatment for wound healing: a secreted protease approach (review). Biomed Rep 13(1):3–14. https://doi.org/10.3892/br.2020.1300

Bauvois C, Jacquamet L, Huston AL et al (2008) Crystal structure of the cold-active aminopeptidase from Colwellia psychrerythraea, a close structural homologue of the human bifunctional leukotriene A4 hydrolase. J Biol Chem 283(34):23315–23325. https://doi.org/10.1074/jbc.M802158200

Blin K, Shaw S, Vader L et al (2025) antiSMASH 8.0: extended gene cluster detection capabilities and analyses of chemistry, enzymology, and regulation. Nucleic Acids Res 53:32–38. https://doi.org/10.1093/nar/gkaf334

Buss CG, Dudani JS, Akana RTK et al (2018) Protease activity sensors noninvasively classify bacterial infections and antibiotic responses. EBioMedicine 38:248–256. https://doi.org/10.1016/j.ebiom.2018.11.031

Cardoso AM, Cavalcante JJV, Cantão ME et al (2012) Metagenomic analysis of the microbiota from the crop of an invasive snail reveals a rich reservoir of novel genes. PLoS One 7(11):e48505. https://doi.org/10.1371/journal.pone.0048505

Chen IA, Chu K, Palaniappan K et al (2019) IMG/M v.5.0: an integrated data management and comparative analysis system for microbial genomes and microbiomes. Nucleic Acids Res 47:666–677. https://doi.org/10.1093/nar/gky901

Craig JW, Chang FY, Kim JH et al (2010) Expanding small-molecule functional metagenomics through parallel screening of broad-host-range cosmid environmental DNA libraries in diverse proteobacteria. Appl Environ Microbiol 76(5):1633–1641. https://doi.org/10.1128/AEM.02169-09

Denesyuk AI, Denessiouk K, Johnson MS, Uversky VN (2024) Structural catalytic core in subtilisin-like proteins and its comparison to trypsin-like serine proteases and alpha/Beta-hydrolases. Int J Mol Sci 25(22):11858. https://doi.org/10.3390/ijms252211858

Eddy SR (2011) Accelerated profile HMM searches. PLoS Comput Biol 7(10):e1002195. https://doi.org/10.1371/journal.pcbi.1002195

Falkenberg F, Kohn S, Bott M et al (2023) Biochemical characterisation of a novel broad pH spectrum subtilisin from Fictibacillus arsenicus DSM 15822T. FEBS Open Bio 13(11):2035–2046. https://doi.org/10.1002/2211-5463.13701

García-Moyano A, Diaz Y, Navarro J et al (2021) Two-step functional screen on multiple proteinaceous substrates reveals temperature-robust proteases with a broad-substrate range. Appl Microbiol Biotechnol 105(8):3195–3209. https://doi.org/10.1007/s00253-021-11235-9

Gerlt JA (2017) Genomic enzymology: web tools for leveraging protein family sequence-function space and genome context to discover novel functions. Biochemistry 56(33):4293–4308. https://doi.org/10.1021/acs.biochem.7b00614

Hasan MJ, Haque P, Rahman MM (2022) Protease enzyme based cleaner leather processing: a review. J Clean Prod 365:132826. https://doi.org/10.1016/j.jclepro.2022.132826

Hera MR, Liu S, Wei W et al (2024) Metagenomic functional profiling: to sketch or not to sketch? Bioinformatics 40:165–173. https://doi.org/10.1093/bioinformatics/btae397

Hogg BN, Schnepel C, Finnigan JD et al (2024) The impact of metagenomics on biocatalysis. Angew Chem Int Ed Engl 63(21):e202402316. https://doi.org/10.1002/anie.202402316

Ji L, Xu S, Zhang Y, Cheng H (2024) Screening of broad-host expression promoters for shuttle expression vectors in non-conventional yeasts and bacteria. Microb Cell Factories 23(1):230. https://doi.org/10.1186/s12934-024-02506-x

Jumper J, Evans R, Pritzel A et al (2021) Highly accurate protein structure prediction with AlphaFold. Nature 596(7873):583–589. https://doi.org/10.1038/s41586-021-03819-2

Kennedy J, O'Leary ND, Kiran GS et al (2011) Functional metagenomic strategies for the discovery of novel enzymes and biosurfactants with biotechnological applications from marine ecosystems. J Appl Microbiol 111(4):787–799. https://doi.org/10.1111/j.1365-2672.2011.05106.x

Kikani B, Patel R, Thumar J et al (2023) Solvent tolerant enzymes in extremophiles: adaptations and applications. Int J Biol Macromol 238:124051. https://doi.org/10.1016/j.ijbiomac.2023.124051

Kumar A, Grover S, Sharma J, Batish VK (2010) Chymosin and other milk coagulants: sources and biotechnological interventions. Crit Rev Biotechnol 30(4):243–258. https://doi.org/10.3109/07388551.2010.483459

Kumari M, Padhi S, Sharma S et al (2021) Biotechnological potential of psychrophilic microorganisms as the source of cold-active enzymes in food processing applications. 3 Biotech 11(11): 479. https://doi.org/10.1007/s13205-021-03008-y

Liu C, Kong X (2016) The current progress and challenge in the development of protease inhibitors of HIV-1. Front Clin Drug Res HIV 3:238. https://doi.org/10.2174/9781681082295116030009

Liu S, Moon CD, Zheng N et al (2022) Opportunities and challenges of using metagenomic data to bring uncultured microbes into cultivation. Microbiome 10:76. https://doi.org/10.1186/s40168-022-01272-5

López-Otín C, Bond JS (2008) Proteases: multifunctional enzymes in life and disease. J Biol Chem 283(45):30433–30437. https://doi.org/10.1074/jbc.R800035200

Mak WS, Tran S, Marcheschi R et al (2015) Integrative genomic mining for enzyme function to enable engineering of a non-natural biosynthetic pathway. Nat Commun 6:10005. https://doi.org/10.1038/ncomms10005

Meesil W, Muangpat P, Sitthisak S, et al. (2023) Rattanarojpong, T., Chantratita, N., Machado, R. A.R., Shi, Y.M., Bode, H.B., Vitta, A., Thanwisai, A.: Genome mining reveals novel biosynthetic gene clusters in entomopathogenic bacteria. Sci Rep 13(1):20764. doi: https://doi.org/10.1038/s41598-023-47121-9

Mendes LW, Braga LPP, Navarrete AA et al (2017) Using metagenomics to connect microbial community biodiversity and functions. Curr Issues Mol Biol 24:103–118. https://doi.org/10.21775/cimb.024.103

Mevarech M, Frolow F, Gloss LM (2000) Halophilic enzymes: proteins with a grain of salt. Biophys Chem 86(2–3):155–164. https://doi.org/10.1016/s0301-4622(00)00126-5.37

MSI A, Kumar P, Sharma N et al (2023) Application of plant proteases in meat tenderization: recent trends and future prospects. Foods 12(6):1336. https://doi.org/10.3390/foods12061336

Neil D, Rawlings AJB, Thomas PD et al (2018) The MEROPS database of proteolytic enzymes, their substrates and inhibitors in 2017 and a comparison with peptidases in the PANTHER database. Nucleic Acids Res 46:624–632. https://doi.org/10.1093/nar/gkx1134

Nurk S, Meleshko D, Korobeynikov A, Pevzner PA (2017) metaSPAdes: a new versatile metagenomic assembler. Genome Res 27(5):824–834. https://doi.org/10.1101/gr.213959.116

Popovic A, Tchigvintsev A, Tran H et al (2015) Metagenomics as a tool for enzyme discovery: hydrolytic enzymes from marine-related metagenomes. Adv Exp Med Biol 883:1–20. https://doi.org/10.1007/978-3-319-23603-2_1

Quevillon E, Silventoinen V, Pillai S et al (2005) InterProScan: protein domains identifier. Nucleic Acids Res 33:116–120. https://doi.org/10.1093/nar/gki442

Rabausch U, Juergensen J, Ilmberger N et al (2013) Functional screening of metagenome and genome libraries for detection of novel flavonoid-modifying enzymes. Appl Environ Microbiol 79(15):4551–4563. https://doi.org/10.1128/AEM.01077-13

Radley E, Davidson J, Foster J et al (2023) Engineering enzymes for environmental sustainability. Angew Chem Int Ed Engl 62(52):e202309305. https://doi.org/10.1002/ange.202309305

Rather MA, Gupta K, Mandal M (2021) Microbial biofilm: formation, architecture, antibiotic resistance, and control strategies. Braz J Microbiol 52(4):1701–1718. https://doi.org/10.1007/s42770-021-00624-x
Razzaq A, Shamsi S, Ali A et al (2019) Microbial proteases applications. Front Bioeng Biotechnol 7:451237. https://doi.org/10.3389/fbioe.2019.00110
Robinson SL, Piel J, Sunagawa S (2021) A roadmap for metagenomic enzyme discovery. Nat Prod Rep 38(11):1994–2023. https://doi.org/10.1039/d1np00006c
Sharon H, Hagag S, Osherov N (2009) Transcription factor PrtT controls expression of multiple secreted proteases in the human pathogenic mold Aspergillus fumigatus. Infect Immun 77(9): 4051–4060. https://doi.org/10.1128/IAI.00426-09
Song P, Zhang X, Wang S et al (2023) Microbial proteases and their applications. Front Microbiol 14:1236368. https://doi.org/10.3389/fmicb.2023.1236368
Sonnhammer EL, Eddy SR, Birney E et al (1998) Pfam: multiple sequence alignments and HMM-profiles of protein domains. Nucleic Acids Res 26(1):320–322. https://doi.org/10.1093/nar/26.1.320
Uchiyama T, Miyazaki K (2009) Functional metagenomics for enzyme discovery: challenges to efficient screening. Curr Opin Biotechnol 20(6):616–622. https://doi.org/10.1016/j.copbio.2009.09.010
Yeates C, Gillings MR, Davison AD et al (1998) Methods for microbial DNA extraction from soil for PCR amplification. Biol Proced Online 1:40–47. https://doi.org/10.1251/bpo6
Zambare VP, Nilegaonkar SS (2016) Proteases in leather processing. In: Thangadurai D, Sangeetha J (eds) Industrial biotechnology, sustainable production and bioresource utilization. Apple Academic Press, CRC Press., (e-book ISBN-13: 978-1-77188-262-0), pp 209–238

Chapter 8
CRISPR-Cas9-Mediated Genome Editing for Protease Engineering

Hasanain A. J. Gharban, Muthanna N. K. Altaee, P. Saranraj, Vasudeo Zambare, K. Gayathri, and U. Subhalakshmi

Abstract Proteases represent one of the most important classes of industrial enzymes, with pivotal roles in detergents, food processing, leather treatment, and therapeutic development. However, naturally occurring proteases often lack the optimal properties, such as specific activity, stability, and substrate specificity, required for harsh industrial processes or precise medical applications. Traditional protease engineering methods, including rational design and directed evolution, have achieved significant success but are often hampered by their labor-intensive nature, limited mutational scope, and reliance on in vitro systems. The emergence of CRISPR-Cas9 as a powerful, precise, and programmable genome-editing technology has revolutionized the field of metabolic engineering and synthetic biology. This review provides a comprehensive examination of the application of CRISPR-Cas9 for protease engineering. We begin by outlining the fundamental biology of proteases and the molecular mechanics of the CRISPR-Cas9 system. We then contrast CRISPR-based strategies with traditional engineering approaches, highlighting the transformative advantages of in vivo, multiplexed genome editing. A detailed analysis of specific applications, including gene knockouts to create clean genetic backgrounds, precise point mutations to refine catalytic properties, and multiplexed editing to generate complex combinatorial libraries, is presented. Supported by key case studies, we demonstrate how CRISPR-Cas9 accelerates the engineering of proteases with enhanced thermostability, altered substrate specificity, and improved

H. A. J. Gharban
Department of Internal and Preventive Veterinary Medicine, College of Veterinary Medicine, University of Wasit, Wasit, Iraq

M. N. K. Altaee
Department of Biology, College of Science, University of Wasit, Wasit, Iraq

P. Saranraj (✉) · K. Gayathri · U. Subhalakshmi
PG and Research Department of Microbiology, Sacred Heart College (Autonomous), Tirupattur, Tamil Nadu, India

V. Zambare
R & D Department, Balaji Enzyme & Chemical Pvt Ltd, Mumbai, Maharashtra, India

V. Zambare et al. (eds.), *Biotechnological Advances and Applications in Microbial Proteases*, Interdisciplinary Biotechnological Advances,
https://doi.org/10.1007/978-981-95-9011-7_8

functional expression. Finally, we discuss the prevailing challenges, such as off-target effects and low HDR efficiency, and explore future perspectives, including the integration of base editing, artificial intelligence, and high-throughput screening to usher in a new era of designer protease creation for advanced biotechnological applications.

Keywords CRISPR-Cas9 · Genome editing · Protease engineering · Directed evolution · Rational design · Homology-directed repair and nonhomologous end joining

8.1 Introduction

Proteases or peptidases are enzymes that catalyze the hydrolysis of peptide bonds in proteins. They are ubiquitous in nature, found in all forms of life from viruses and bacteria to humans, where they are indispensable for critical physiological processes such as digestion, blood coagulation, apoptosis, and immune response. Beyond their biological essentiality, proteases have been harnessed as workhorse biocatalysts across a multitude of industries. They constitute approximately 60% of the global industrial enzyme market, a testament to their economic and practical significance (Zambare and Nilegaonkar 2016; Ibrahim et al. 2025). In the detergent industry, proteases such as subtilisin are added to break down protein-based stains. In food and beverage processing, they are used for cheese production (rennet), meat tenderization, and brewing. Additional applications span leather processing, silk degumming, waste management, and molecular biology (e.g., trypsin for cell culture). In the pharmaceutical sector, proteases are vital therapeutic targets (e.g., HIV-1 protease inhibitors) and themselves are used as drugs, such as thrombolytic agents (t-PA, streptokinase) for dissolving blood clots and digestive aids (papain, pepsin) (Adiego-Pérez et al. 2019).

Despite their abundance and diversity, naturally evolved proteases are often not optimal for industrial applications. Industrial processes frequently involve extreme conditions of temperature, pH, and organic solvents that can denature native enzymes, leading to a loss of activity and economic inefficiency. Furthermore, achieving high specificity for a particular target protein sequence is crucial in therapeutic contexts to avoid off-target cleavage. Natural proteases may also exhibit autoproteolysis (self-digestion), low catalytic efficiency on nonnatural substrates, or poor expression yields in heterologous production hosts. These limitations create a compelling need to engineer proteases with tailor-made properties enhanced thermostability, altered pH activity profiles, increased specific activity, reduced autolysis, and novel substrate specificities. The field of protease engineering aims to bridge this gap between the natural function of proteases and the demanding requirements of modern technology (Amri and Mamboya 2012).

For decades, protease engineering relied on two primary strategies: rational design and directed evolution. Rational design requires detailed structural knowledge and is often limited by our incomplete understanding of protein folding and function. Directed evolution, which mimics natural evolution in the laboratory through iterative rounds of random mutagenesis and screening, can be powerful but is often a "black box" process: It is laborious, time-consuming, and typically limited to generating and screening libraries of 104,104–106,106 variants in vitro, a minuscule fraction of sequence space (Kwon et al. 2023). The discovery and subsequent repurposing of the CRISPR-Cas9 system have fundamentally transformed Genetic engineering. Derived from an adaptive immune system in bacteria, CRISPR-Cas9 functions as a programmable RNA-guided DNA endonuclease. Its simplicity, efficiency, and unparalleled versatility enable precise, targeted modifications to the genomes of virtually any organism. This technology offers a paradigm shift for protease engineering by moving the site of diversification from in vitro plasmids to the native genome in vivo, allowing for direct coupling of genotype and phenotype, enabling complex multiplexed editing, and facilitating the engineering of the cellular host alongside the protease itself (Kraut 1977).

8.2 Type of Protease and Its Function

Mechanistic and structural frameworks form the essential lexicon for understanding this diverse enzyme family. Unlike many enzymes classified solely by their reaction, proteases are most informatively categorized by the chemical mechanism of peptide bond hydrolysis, as the key catalytic residues involved define their biochemical behavior, optimal conditions, and inhibition profiles. The four major classes, serine, cysteine, aspartic, and metalloproteases, represent evolutionary distinct solutions to the same catalytic challenge, though they share common mechanistic principles (Table 8.1) (Jones et al. 2021). This classification, endorsed by the MEROPS database, the premier repository for peptidase information, provides a systematic framework that transcends phylogenetic origins, meaning a serine protease from a bacterium and one from a human can be mechanistically more similar to each other than to a cysteine protease from the same bacterium. Beyond these four, there also exist threonine proteases (e.g., the proteasome) and glutamic acid proteases, but these are less common in industrial contexts. The initial step in the catalytic cycle for most proteases involves the binding of the substrate polypeptide into the active site cleft, positioning the scissile peptide bond for nucleophilic attack. The precise nature of this nucleophile and the machinery that supports it is the primary differentiator between the classes, dictating the enzyme's sensitivity to pH, specific inhibitors, and redox potential, which are critical considerations for both their natural function and their engineered application (Hsu et al. 2014).

Table 8.1 Classification, mechanism, and characteristics of major protease types

Protease type	Catalytic mechanism/ residues	Optimal pH	Industrial/biological roles	References
Serine proteases	Catalytic triad: Ser, His, Asp. Nucleophilic serine attacks carbonyl carbon, forming a covalent acyl-enzyme intermediate.	Neutral to alkaline (7.0–11.0)	Detergent additives, food processing, digestion, blood coagulation, bacterial pathogenesis	Li and Ellington (2025)
Cysteine proteases	Catalytic dyad: Cys, His. Nucleophilic cysteine thiolate attacks carbonyl carbon, forming a thioester intermediate.	Lightly acidic to neutral (4.5–7.0)	Meat tenderizing, brewing, apoptosis, antigen processing, parasitic infection	Rawlings et al. (2018)
Aspartic proteases	Two Asp residues. Activate a water molecule for nucleophilic attack. No covalent intermediate; single-step mechanism.	Acidic (2.0–5.0)	Cheese production (rennet), digestive aids, therapeutic targets for HIV	Liu et al. (2022)
Metalloproteases	Divalent metal ion (usually Zn^{2+}). Metal-bound water acts as nucleophile. Metal ion stabilizes the transition state.	Neutral (5.0–9.0)	Bio-catalysis for aspartame production, therapeutic targets, extracellular matrix remodeling	Hsu et al. (2014)

8.2.1 *Serine Proteases*

Serine proteases are arguably the most well-studied class, renowned for their roles in digestion (trypsin, chymotrypsin), blood coagulation (thrombin), and bacterial extracellular digestion (subtilisin). Their name derives from the uniquely reactive serine residue within the active site that acts as the primary nucleophile. However, this serine does not act alone; it is part of a highly conserved constellation of amino acids known as the catalytic triad. In the classic chymotrypsin-like fold, this triad is ordered as Ser195, His57, and Asp102 (chymotrypsin numbering). The mechanism is an elegant two-step dance of proton transfers: The histidine residue, acting as a general base, abstracts a proton from the serine hydroxyl group, enhancing its nucleophilicity to attack the carbonyl carbon of the substrate's peptide bond. This results in the formation of a transient, covalent acyl-enzyme intermediate and the release of thc first product (thc amine-terminus fragment). The now protonated histidine then acts as a general acid, donating a proton to the departing amine group. In the second step, a water molecule enters the active site. The histidine, now deprotonated again, abstracts a proton from this water, generating a nucleophilic hydroxide ion that attacks the acyl-enzyme intermediate. This hydrolyzes the covalent bond, releasing the second product (the carboxyl-terminus fragment) and regenerating the free enzyme. It is crucial to note that the aspartate residue plays a critical role in orienting the histidine and stabilizing its charged state, fine-tuning the entire process. The bacterial serine protease subtilisin is a convergent evolution marvel; it possesses an identical catalytic triad (Ser221, His64, Asp32) but within a completely different protein fold, underscoring the catalytic efficiency of this triadic arrangement (Fernández et al. 2025).

8.2.2 Cysteine Proteases

Cysteine proteases mirror the catalytic strategy of serine proteases but employ a sulfur atom instead of an oxygen atom as the nucleophile. The mechanism similarly involves the formation of a covalent thioester intermediate but utilizes a catalytic dyad, most commonly composed of Cys25 and His159 (papain numbering). The lower pKa of the cysteine thiol group makes it a potent nucleophile even at neutral pH when deprotonated. The reaction initiates with the histidine residue abstracting a proton from the cysteine thiol, generating a reactive thiolate-imidazolium ion pair (Doudna and Charpentier 2014). The nucleophilic thiolate ion attacks the peptide bond carbonyl carbon, forming the tetrahedral intermediate transition state, which is stabilized by an oxyanion hole, and subsequently collapses into the acyl-enzyme thioester intermediate. The first product is released. Deacylation then occurs by a water molecule, activated by the now protonated histidine, leading to hydrolysis of the thioester and release of the second product. A defining feature of cysteine proteases is their sensitivity to the cellular redox environment; they are often compartmentalized to avoid oxidizing conditions (e.g., in lysosomes) and can be inhibited by oxidizing agents or specific alkylating agents like iodoacetate. This sensitivity must be carefully managed in industrial processes. Prominent examples include papain from papaya, used as a meat tenderizer, caspases that execute programmed cell death (apoptosis), and cathepsins involved in lysosomal protein degradation. Their mechanism makes them particularly useful in applications requiring moderate conditions where their activity can be precisely controlled (Cox et al. 2015).

8.2.3 Aspartic Proteases

Aspartic proteases utilize a novel mechanism that relies on the direct activation of a water molecule and not on the creation of a covalent intermediate. Two strictly conserved aspartic acid residues within the catalytic site define this family, which is exemplified by the digestive enzymes pepsin and rennin (chymosin) and the clinically important HIV-1 protease. For their normal gastric functions, these enzymes function optimally at an acidic pH (pH 3–4). The catalytic activity is symmetric and cooperative. The two aspartate residues, normally stored in close proximity to each other, act in concert. To enhance its nucleophilic attack on the carbonyl carbon of the substrate peptide bond, the protonated aspartate polarizes and activates a strategically located water molecule. By donating a proton to the amide group's nitrogen atom of the peptide bond, the deprotonated second aspartate also serves as a general acid (Clement et al. 2019). This forms a single coordinated mechanism that both releases the two products and breaks the peptide bond without a stable intermedium. Its transition state is stabilized by interaction with both aspartates and by the gem-diol conformation. As a result, transition-state analogs such as pepstatin, which is strongly bound to the active site, can inhibit aspartic proteases with high sensitivity. With one catalytic aspartate contributed by each subunit, the HIV-1 protease is an excellent example of a retroviral aspartic protease. Therefore, it is a perfect candidate for antiviral drugs to hinder its critical symmetry and activity (Bornscheuer et al. 2012).

8.2.4 Metalloproteases

The broad and varied class of metalloproteases is bound together by their complete reliance on a divalent metal ion, most frequently zinc (Zn_2^+), for catalytic activity. The metal ion is typically coordinated in the active site by three amino acid ligands (often two histidines and one glutamate, forming a His-Glu-X-X-His motif) and by water molecules. The catalytic mechanism involves the metal ion polarizing a bound water molecule, effectively generating a potent nucleophilic hydroxide ion at neutral pH. This hydroxide attacks the carbonyl carbon of the scissile peptide bond. Simultaneously, the metal ion acts as an electrophile, stabilizing the developing negative charge on the carbonyl oxygen in the transition state (Li and Ellington 2025). A general acid (often a nearby glutamate or a second water molecule) then protonates the departing amine group. Unlike the previous classes, there is no covalent enzyme-substrate intermediate. Metalloproteases are involved in a wide array of processes, including extracellular matrix remodeling (Matrix metalloproteinases, or MMPs), blood pressure regulation (Angiotensin-converting enzyme, ACE), and are produced by bacteria for nutrient acquisition (e.g., Thermolysin from *Bacillus thermoproteolyticus*). A common characteristic of metalloproteases is their sensitivity to chelation by substances such as EDTA or 1,10-phenanthroline, which eliminate activity by removing the required metal ion (Odoi and Jewett 2025).

8.2.5 Structure–Function Relationships

The overall mechanistic principles are exquisite in their specificity and efficacy because of architectural features such as oxyanion holes, substrate pockets, and catalytic triads. The active site's geometry, aside from the catalytic residues, is also crucial. The oxyanion hole is a feature shared by almost all protease classes. This region of the enzyme helps stabilize the negatively charged, tetrahedral oxyanion transition state created during the nucleophilic attack on the planar peptide bond. It is usually composed of backbone amide groups of conserved residues. By offering strong hydrogen bonds, the oxyanion hole significantly lowers the energy of this transition state, delivering most of the catalytic rate acceleration. More important for engineering use is the structure of the substrate-binding pocket. In a conceptual sense, the active site can be separated into subsites (S1, S2, S3. Sn) that bind the substrate amino acid side chains (labeled P1, P2, P3. Pn) on the N-terminal side of the scissile bond. The S1 pocket usually serves as the main specificity determinant. For example, trypsin's narrow, deep S1 pocket, at the end of which an aspartate makes a salt bridge with the positively charged P1 residue side chains of arginine or lysine, is the origin of its legendary specificity. However, the broad, deep hydrophobic S1 pocket of chymotrypsin can hold bulk aromatic residues such as tryptophan, tyrosine, and phenylalanine. Only small, uncharged side chains such as glycine and alanine can be bound by Elastase's shallow S1 pocket, which is capped by valine and threonine. The main method for altering protease specificity, one of the main objectives of Biotechnology, is to intentionally widen these pockets by adding new charged or hydrophobic residues (Arnold 2018).

8.3 CRISPR-Cas9 Genome Editing of Proteases

The arrival of CRISPR-Cas9 gene editing technology has also revolutionized enzyme engineering. Pioneered by an immune system of bacteria, this gene editing giant offers an unprecedented ability to make precise and accurate alterations to genetic material. Its use in protease engineering is a paradigm shift from the limitations of conventional techniques to achieve multiplexed, direct, and rational genomic alterations that optimize these precious enzymes for application in research, industry, and medicine. By incorporating CRISPR-Cas9 into the metabolic engineering pipeline, scientists are able to address at once yield, stability, and specificity issues by having the microbial host as a high-performance production chassis alongside enhancing the protease gene (Pohl et al. 2016). In the pre-CRISPR-Cas9 era, directed evolution and rational design were the two primary complementary pillars of protease engineering.

Close structural knowledge is the foundation of rational design, which predicts particular amino acid substitutions that will result in a desired activity, such as improved thermostability through the addition of disulfide bridges or changed substrate specificity through the modification of active site pockets (Table 8.2). Although sometimes effective, this method is frequently frustrated by the complicated and capricious character of protein folding and activity, in which individual mutations can have unexpected destabilizing effects. Directed evolution, however, simulates natural selection by creating massive libraries of random mutants and exposing them to repeated cycles of high-throughput screening for better variants. Although this method has yielded remarkable successes, it is inherently labor-intensive, time-consuming, and limited by the throughput of screening methods, often exploring only a minuscule fraction of the possible sequence space.

Table 8.2 Comprehensive overview of CRISPR-Cas9-mediated strategies for protease engineering

Engineering strategy	CRISPR tool	Mechanism	Example application	Advantages	Limitations	References
Gene knockout (host engineering)	Wild-type Cas9 + sgRNA	Exploits error-prone nonhomologous end joining (NHEJ) to create frameshift mutations and gene disruptions	Knocking out extracellular proteases (e.g., *aprE*, *nprE*) in *Bacillus subtilis* to prevent degradation of recombinant subtilisin	Dramatically increases target protease yield Creates a universal production host	Requires complete knowledge of host's protease genome. Multiplexed knockouts require high efficiency	Zhang et al. (2025)
Precision mutagenesis	Wild-type Cas9 + sgRNA + ssODN donor	Utilizes homology-directed repair (HDR) with a single-stranded oligodeoxynucleotide (ssODN) repair template to introduce specific point mutations	Introducing mutations in subtilisin to enhance thermostability or in trypsin to reduce autolysis (K- > Q mutations)	Enables true rational design directly on the chromosome Creates stable, well-defined production strains	HDR efficiency is often very low in industrial microbial strains. Requires design and synthesis of repair templates	Lian et al. (2017)
Multiplexed editing	Wild-type Cas9 + multiple sgRNAs (± donors)	Uses several sgRNAs simultaneously to target multiple genomic loci for knockout (via NHEJ) or editing (via HDR)	Simultaneously knocking out 5+ protease genes in *Aspergillus niger* or introducing multiple stability mutations in a protease gene	Drastically reduces strain generation time Explores genetic space more efficiently than sequential methods	Can be difficult to map genotype to phenotype in complex libraries. Requires high transformation efficiency	Anzalone et al. (2019)
Gene integration and overexpression	Wild-type Cas9 + sgRNA + dsDNA donor	Uses HDR with a double-stranded DNA (dsDNA) donor template to insert a large DNA cassette (e.g., promoter + gene) into a specific genomic locus	Integrating a plant-derived protease (e.g., papain) gene into the genome of *Pichia pastoris* for stable, high-yield production	Stable, consistent expression without antibiotic selection Avoids issues related to plasmid instability	Low HDR efficiency for large fragments. Requires careful selection of genomic integration site	Billon et al. (2017)

Transcriptional tuning (CRISPRi/a)	dCas9 fused to repressor (KRAB) or activator (VP64) + sgRNA	Catalytically dead Cas9 (dCas9) is targeted to promoter regions to block (CRISPRi) or recruit (CRISPRa) transcriptional machinery, modulating gene expression without editing the sequence	Downregulating host proteases that cause degradation or upregulating chaperones to improve folding of a difficult-to-express protease	Reversible and tunable control of gene expression. Avoids permanent genetic changes	Requires continuous expression of dCas9 and guide RNAs. Can have off-target transcriptional effects	Chandra et al. (2025)
Base editing	Cas9-domain fusion (e.g., Cas9-cytidine deaminase)	A catalytically impaired Cas9 (nCas9) fused to a deaminase enzyme directly converts one DNA base into another (C- > T or A- > G) without double-strand breaks or donor templates	Shifting the pH optimum of a fungal aspartic protease by converting specific codons (e.g., changing a histidine codon)	Highly precise editing without DSBs. Dramatically reduces unwanted indels. Does not require a donor template	Limited by the need for a PAM site and a specific editing window. Potential for off-target editing	Chen et al. (2020)
Prime editing	Cas9-reverse transcriptase fusion + pegRNA	A Cas9 nickase fused to a reverse transcriptase is guided by a prime editing guide RNA (pegRNA) to directly write new genetic information into a target DNA site	Installing a specific disulfide bond by inserting two cysteine codons or altering a catalytic triad residue with high fidelity	Unprecedented precision and versatility. Can make all 12 possible base-to-base changes. No DSBs required	Complexity of pegRNA design. Efficiency can be variable and currently lower than other methods	Jinek et al. (2012)
In vivo directed evolution	Cas9 + sgRNA library	A library of sgRNAs targets a specific genomic region with a pool of repair templates, creating a diverse mutant library directly on the chromosome for continuous evolution	Creating a library of protease active site variants and using fluorescence-activated cell sorting (FACS) to select for altered specificity	Links genotype to phenotype directly in the host. Allows for continuous evolution experiments	Library diversity can be limited by transformation efficiency. Screening throughput is critical	Kaur et al. (2022)

Furthermore, both methods typically operate on plasmids in vitro, decoupling the enzyme from its genomic context and ignoring the critical influence of the host organism's physiology on production and function (Ran et al. 2013).

CRISPR-Cas9 circumvents these limitations by performing edits directly on the chromosome of the host organism. This in vivo engineering ensures stable inheritance of traits without the need for antibiotic selection and allows the enzyme to be studied and optimized within its authentic cellular environment. The technology's core components are elegantly simple: the Cas9 endonuclease, which acts as molecular scissors, and a single-guide RNA (sgRNA), which programmably directs Cas9 to a specific DNA sequence adjacent to a protospacer adjacent motif (PAM). The resulting double-strand break (DSB) activates the cell's innate DNA repair machinery, which can be harnessed to achieve different engineering outcomes (Peterson and Liu 2025).

8.3.1 CRISPR-Cas9 for Protease Optimization

8.3.1.1 Nonhomologous End Joining (NHEJ) Repair Pathway

The most initial and simplest use is the utilization of CRISPR-Cas9 to knock out genes using the error-inducing nonhomologous end joining (NHEJ) repair pathway (Fig. 8.1). This is especially useful for developing specialized hosts for microbial

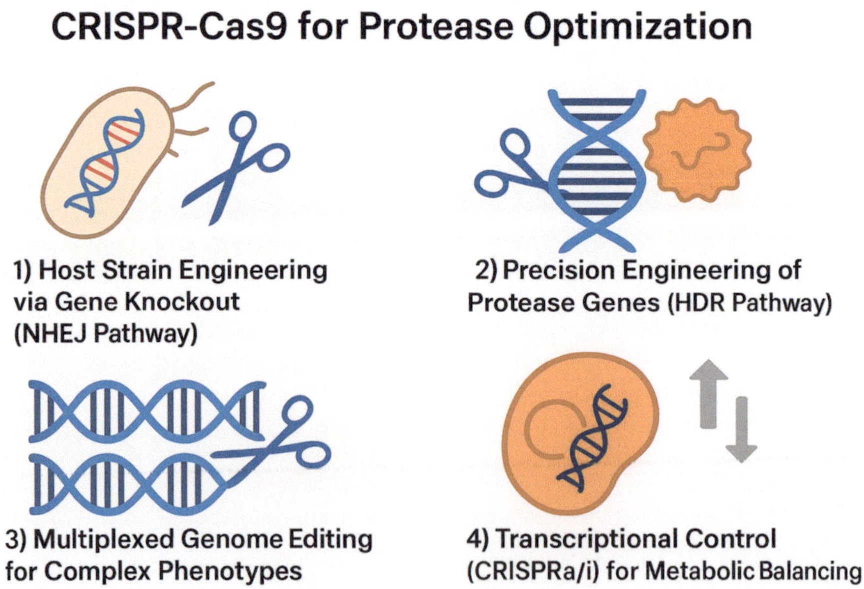

Fig. 8.1 CRISPR Cas9 for protease optimization

production. For example, the native genome of *Bacillus subtilis*, a major industrial host for the production of proteases, contains a large number of extracellular proteases (such as AprE and NprE) that can contaminate and degrade the target recombinant enzyme, reducing yield and purity to unacceptable levels. These undesirable protease genes can be swiftly and concurrently eliminated using CRISPR-Cas9, creating a "clean background" host strain that is unable to produce those contaminants. This makes it possible to significantly increase the target recombinant protease's yield and stability. Any robust production platform should start with this host rationalization strategy (Schwartz et al. 2016).

8.3.1.2 Homology-Directed Repair (HDR) Pathway

The high-fidelity homology-directed repair (HDR) pathway is used to achieve direct protease gene engineering. This entails delivering the target mutation between flanking homology arms along with a repair template single-stranded oligodeoxynucleotide (ssODN) or double-stranded DNA fragment. This enables precise, single-amino-acid alteration of the protease coding sequence. For rational design, this is revolutionary. Scientists today can now specifically target residues in the active site to modify catalytic capacity or change substrate specificity. An example is optimizing subtilisin to make it more effective in organic solvents or at elevated temperatures by the addition of stabilizing mutations in its architecture. Likewise, trypsin, which is susceptible to autolytic degradation, can be genetically engineered by modifying its susceptible autolysis sites (e.g., lysine and arginine residues) to resistant ones, thereby substantially enhancing its functional stability and shelf-life. Although HDR efficiency is challenging, particularly in nondividing cells, optimization of delivery protocols, as well as the use of Cas9 variants, makes it more feasible (Shi et al. 2019).

8.3.1.3 Multiplexed Editing Simultaneous Targeting

The most powerful benefit of CRISPR-Cas9 is multiplexed editing simultaneous targeting of several genomic sites using several sgRNAs. This allows the construction of complex combinatorial libraries in vivo that would be extremely challenging to create by the conventional plasmid-based method. This is especially useful for developing specialized hosts for microbial production. For example, the native genome of *Bacillus subtilis*, a major industrial host for the production of proteases, contains a large number of extracellular proteases (such as AprE and NprE) that can contaminate and degrade the target recombinant enzyme, reducing yield and purity to unacceptable levels. These undesirable protease genes can be swiftly and concurrently eliminated using CRISPR-Cas9, creating a "clean background" host strain that is unable to produce those contaminants. This makes it possible to significantly increase the target recombinant protease's yield and stability. Any robust production platform should start with this host rationalization strategy (Schwartz et al. 2016).

8.3.1.4 High-Fidelity Homology-Directed Repair (HDR)

The high-fidelity homology-directed repair (HDR) pathway is used to achieve direct protease gene engineering. This entails delivering the target mutation between flanking homology arms along with a repair template single-stranded oligodeoxynucleotide (ssODN) or double-stranded DNA fragment. This enables precise, single-amino-acid alteration of the protease coding sequence. For rational design, this is revolutionary.

8.4 Case Studies and Industrial Impact of CRISPR-Cas9 Genome Editing of Proteases

The technical implementation of CRISPR-Cas9 genome editing has proven to be highly influential in both academic and industrial settings, and the following case studies demonstrate how the new technology is being used to overcome particular challenges in protease engineering, resulting in notable improvements in the functional activity, enzymatic stability, and production capacity. These developments are directly improving the quality of final products and commercial manufacturing operations. Together, these instances reflect a move away from the older, time-consuming approaches and toward a more accurate, efficient, and streamlined approach to enzyme engineering (Yang and Zhang 2018).

8.4.1 Enhancing Subtilisin Performance for Detergent Applications

Subtilisin, an alkaline serine protease mainly produced by *Bacillus* species, is a basic building block of the global detergent enzyme market. It is extremely valued for its ability to hydrolyze proteinaceous stains under harsh washing conditions. However, industrial use necessitates better thermal stability after exposure to high temperatures, oxidizing bleach reagents, and nonaqueous environments. Even though standard directed evolution methods used on subtilisin in *B. licheniformis* had yielded gradual improvements, the process was limited by slow and iterative rounds of random genetic variation and selection. To transcend such limitations, researchers employed a multiplex CRISPR-Cas9 strategy. They constructed a set of sgRNAs and corresponding single-stranded oligodeoxynucleotide (ssODN) repair templates, which allowed them to target several genetic loci simultaneously. This allowed direct introduction of successful stabilizing mutations, for example, N218S and Q271K, into the native subtilisin gene. A diverse population of *B. licheniformis* strains was created using this one-step, in vivo method, each of which had different combinations of these beneficial genetic alterations. Utilizing high-throughput screening on heat-treated or organic solvent-treated agar plates, such as dimethyl sulfoxide (DMSO), researchers were able to quickly distinguish among the strongest variants. The resulting modified proteases possessed an extremely extended half-life at 60 °C with a gain of more than 50% over the wild-type enzyme. Furthermore, these designed enzymes were operative in liquid detergent formulations comprising oxidizing bleaching agents with satisfactory effectiveness in fulfilling a significant industrial requirement for more robust and efficient biocatalysts (Liu et al. 2022).

8.4.2 Combating Autolysis in Trypsin for Biopharmaceutical Manufacturing

Trypsin, a serine protease with a specificity for cleavage of peptide bonds following lysine and arginine residues, plays a crucial role in biopharmaceutical manufacture, primarily for propagation of cell cultures and manufacture of insulin. One of the serious problems of trypsin is its tendency to self-digestion, where trypsin molecules digest other trypsin molecules enzymatically, resulting in loss of activity very quickly, lack of consistency between lots from production to production, and excessive cost of manufacture. Earlier attempts at trying to engineer trypsin proteins involved production of recombinant versions in *Escherichia coli* but were marred by protein misfolding and poor yields of production. A new approach utilized CRISPR-Cas9-mediated homology-directed repair (HDR) in *Pichia pastoris* yeast, an organism renowned for its unparalleled capability for protein secretion and expertise in proper protein folding. The approach utilized specially designed single-stranded oligonucleotide templates (ssODNs) to introduce specific point mutations at critical sites of self-cleavage in the trypsin genetic material. The changes substituted noncleavable substitutes, such as glutamine, for sensitive lysine and arginine residues (like those in K/S and R/S patterns). The results of the targeted genetic editing were trypsin versions having significantly lowered self-degradation characteristics. The altered enzyme maintained its full catalytic capability but exhibited improved stability throughout the fermentation process and in stored products. The results contributed to a significant increase in functional enzyme yield per production lot and improved uniformity of products, providing the biopharmaceutical industry with a more reliable and cost-effective means of producing this critical biochemical device (Yang and Zhang 2018).

8.4.3 Optimizing Fungal Hosts for Recombinant Protease Production

Filamentous fungi such as *Aspergillus niger* and *Trichoderma reesei* are ideal systems for industrial enzyme production due to their better ability to secrete high levels of protein. However, one major challenge associated with them is their intrinsic extracellular proteases, which can extensively degrade heterologously expressed proteins, including important recombinant proteases, thereby significantly reducing yields. Traditional methods for generating protease-deficient strains involved multiple rounds of successive gene inactivation with cumbersome marker systems and repeated rounds of selection, a process that took several years. The introduction of CRISPR-Cas9 technology has dramatically streamlined this host optimization process. In an early study, researchers used a CRISPR-Cas9 system in *A. niger* to simultaneously inactivate multiple genes responsible for major extracellular proteases. Employing a single transformation event with one plasmid co-expressing Cas9 and several sgRNAs, they were able to effectively achieve five simultaneous gene knockouts using this method. This modified host, having been stripped of its initial protease capacity, was then employed for the production of a recombinant aspartic protease. The outcome was more than twentyfold higher production of the target protease compared to the wild-type strain because of the prevention of enzymatic degradation after secretion. This example highlights how CRISPR-Cas9 shifts research activity away from the enzyme itself to the entire production organism, allowing engineering of tailored cellular systems that maximize both the efficiency and purity of industrial enzymes (Nodvig et al. 2015).

8.4.4 Engineering Plant-Derived Proteases for Heterologous Expression

Microbial proteases such as papain, derived from papaya, and bromelain, derived from pineapple, have widespread usage in food, cosmetic, and pharmaceutical industries. Their supply remains inconsistent owing to agricultural uncertainty, seasonal variation, and contamination hazards, highlighting the requirement for stable microbial synthesis systems. Traditional efforts to produce these enzymes in bacterial or yeast systems often face challenges, including inappropriate protein maturation, low secretion efficiency, and host's own protease-mediated enzymatic degradation, resulting in unacceptable production levels. In order to overcome these limitations, an integrated approach using CRISPR technology and synthetic biology was adopted. A gene for a commercially valuable plant cysteine protease was initially codon-optimized and then targeted into a defined genomic locus of *Pichia pastoris* using CRISPR-Cas9-mediated homology-directed repair. At the same time, the same CRISPR system was used to inactivate multiple endogenous genes that encode vacuolar proteases in *P. pastoris* to protect the heterologously expressed plant enzyme from being degraded. CRISPR activation (CRISPRa) was used to moderately increase expression of genes involved with the unfolded protein response, enhancing the yeast's natural capacity to fold and export this complex protein. The outcome was a genetically fortified and stable *P. pastoris* strain, proficient in generating elevated quantities of active, uncontaminated plant protease within a regulated fermentation environment. This innovation guarantees a steady and expandable production stream, effectively circumventing the limitations imposed by traditional agricultural sourcing (Vasquez Lopez et al. 2021).

8.4.5 Expanding the Host Range for High-Value Protease Production in Escherichia coli

Escherichia coli continues to be a predominant host system for recombinant protein manufacture, favored for its extensively mapped genetics, swift proliferation, and considerable production capabilities. Nevertheless, synthesizing specific proteases, especially those necessitating particular posttranslational modifications or exhibiting host toxicity, has remained problematic. A significant advancement was achieved through the application of CRISPR-Cas9 to tailor an *E. coli* strain for optimized synthesis of a valuable mitochondrial protease. Investigators initially executed multiplexed CRISPR modifications to deactivate three genes, DegP, Lon, and ClpP, responsible for periplasmic and cytoplasmic proteases known to significantly contribute to the breakdown of recombinant proteins. They subsequently employed homology-directed repair (HDR) to incorporate a genetic module featuring a codon-optimized variant of the mitochondrial protease, controlled by a stringent promoter. To combat toxicity problems, the scientists utilized CRISPR interference (CRISPRi) to silence temporarily the expression of essential genes adversely impacted by protease build-up. The engineered strain that resulted from this had a fortyfold increase in active protease yield from that of traditional expression methods. This is evidence in itself that CRISPR-Cas9 is efficient enough to design engineered *E. coli* systems with the potential to overcome previous limitations in the expression of difficult enzymatic products, making new opportunities available for large-scale production of bespoke proteases (Jones et al. 2021).

8.4.6 Engineering Protease Specificity for Therapeutic Applications Using Phage-Assisted Continuous Evolution

A pioneering strategy, while not entirely CRISPR technology reliant, effectively merged CRISPR-Cas9 with phage-assisted continuous evolution (PACE) to evolve proteases with novel specific characteristics at high speed for therapeutic uses. Scientists aimed at a viral protease indispensable for reproducing the virus that originally showed too low specificity to be of any therapeutic utility. Scientists initially utilized CRISPR-Cas9 to create a vast library of protease mutants in *E. coli* by providing precisely defined alterations at substrate-contacting residues. This set of variants was subsequently screened using PACE, a model in which bacteriophage replication was coupled directly to protease activity against a particular human protein target to connect it with disease processes. The continuous process of evolution, augmented by intermittent CRISPR-mediated variation of advantageous mutations, enabled rapid discovery of genetic potential. Within 2 weeks, the research team purified protease variants with dramatically reengineered specificity patterns having more than a 1000-fold stronger preference for the therapeutic target compared to native substrates. These bioengineered proteases had robust biological activity in disease model organisms, validating the combined CRISPR-PACE method as an effective platform for developing new protease-based therapies. This research is an example of how CRISPR-Cas9 can be synergistically blended with other advanced technologies to accelerate the evolution of proteases with designed properties for medical applications (Chen et al. 2020).

8.4.7 Engineering pH Stability in a Fungal Aspartic Protease for Food Processing

The food and beverage industry often requires enzymes that are active under specific and sometimes narrow pH conditions. A key fungal aspartic protease used in cheese manufacturing exhibited optimal activity at pH 4.5 but lost significant function at the pH of milk (pH ~6.7), requiring process adjustments and increasing costs. Scientists used a CRISPR-Cas9 base-editing technique in Aspergillus oryzae to reengineer the pH activity profile. In order to alter the pKa of the catalytic aspartates, they employed a cytidine base editor (CBE) that was matured with a nickase Cas9 (nCas9) and knocked at residues on the lining of the enzyme's active site cleft. With little introduction of random indels, this technique allowed precise CG to TA mutations without causing double-strand breaks or requiring donor templates. To find a variant with two important mutations that moved the pH optimum more than 1.0 units toward neutrality, library screening was used. This engineered protease showed a threefold increase in activity at pH 6.5, significantly improving its efficiency in the cheese curding process without altering its specificity for κ-casein. This case demonstrates the application of advanced CRISPR tools such as base editing for fine-tuning biochemical properties such as pH optimum with high precision (Sato et al. 2022).

8.4.8 Development of a Protease-Null **Lactobacillus** *Chassis for Probiotic Delivery*

Lactobacillus species are attractive candidates for delivering therapeutic proteins to the gut. However, their native extracellular and intracellular proteolytic systems can degrade recombinant biotherapeutics before release. A project aimed to create a "universal" protease-deficient *Lactobacillus casei* chassis to serve as a platform for oral delivery of protease-sensitive drugs, including insulin and other hormones. Researchers used a iterative CRISPR-Cas9 NHEJ strategy to systematically disrupt eight genes encoding key proteases, including the cell envelope protease (PrtP) and various intracellular peptidases (PepO, PepT). The engineering process was accelerated by using a CRISPR plasmid with a temperature-sensitive origin of replication, which was easily cured after editing. The resulting octuple mutant strain showed a $> 99\%$ reduction in extracellular proteolytic activity. When engineered to express human insulin, the protease-null strain protected the hormone from degradation and delivered up to 50% more active insulin to gut epithelial cells in vitro compared to the wild-type strain. This engineered chassis has significant implications for the development of next-generation oral biologic drugs (Kwon et al. 2023).

8.5 Application of CRISPR-Cas9 for Protease Engineering

8.5.1 Strain Development Through Targeted Gene Disruption

Using CRISPR-Cas9 technology has completely changed how microbial strains are created for the production of proteases. In this instance, gene-targeting technologies use the system's targeting capability to create optimal production hosts through systematic gene knockout. To direct the Cas9 nuclease to cause double-strand breaks in the DNA of production strains used in industrial processes, specific single-guide RNAs (sgRNAs) complementary to the genes coding for proteases are engineered. The resulting frameshift mutations caused by the clumsy nonhomologous end joining (NHEJ) repair actually silence target genes. The method has been particularly convenient for the removal of interfering proteolytic activities in production hosts such as *Bacillus subtilis* and *Aspergillus niger*, in which native protease genes can significantly reduce yields of recombinant target proteases by degradation and competition. Because of the strategy's effectiveness, multiple gene knockouts can be produced in a single transformation event, significantly speeding up the development of clean background strains that are better hosts for heterologous protease expression. In terms of accuracy and effectiveness in host strain engineering, the precise gene disruption approach is superior to the conventional random mutagenesis techniques (Gao et al. 2018).

8.5.2 Improvement of Enzyme Properties Through Precision Genome Editing

Protease-encoding genes can be directly and accurately edited by host engineering and CRISPR-Cas9 to enhance their functional properties. It is done by creating intended nucleotide changes in the genomic DNA by the mechanism of homology-directed repair (HDR). Scientists employ single-stranded oligodeoxynucleotide (ssODN) repair templates with perfect mutations that bear flanking homology arms that complement sequences on both sides of the Cas9-break point. Through the application of this method, amino acid substitutions may be incorporated to enhance some of the most important protease properties, such as resistance to pH, specificity for substrate, thermal stability, and protection against inactivation by oxidation. Site-directed trypsin gene alteration, for example, has generated mutants with decreased autoproteolysis and better working stability, and site-directed point mutagenesis of subtilisin genes has generated mutants with significantly enhanced performance in commercial detergents. Protease variants with predetermined characteristics that are optimally suited for a specific industrial use can be designed due to the accuracy of this genome editing technology, allowing rational design from structure information and computer simulations. Researchers are now able to directly implement design-led optimization with unprecedented precision, which is an advancement over conventional directed evolution strategies (Gasiumas et al. 2012).

8.5.3 Accelerating Directed Evolution by Multiplexed Genomic Alterations

By accomplishing multiplexed genomic corrections to create heterogeneous protease variant libraries, CRISPR-Cas9 has revolutionized directed evolution techniques. Multiple sgRNAs that guide to different regions of a protease gene and pools of repair templates having various combinations of mutations are co-transfected in this method. Without requiring assembly of plasmid-based libraries and attendant limits on transformation efficiency and library size, the system yields high-variant libraries directly on the chromosome. By generating combinatorial sets of mutations at multiple positions that are thought to affect protease activity, it allows scientists to probe enormous regions of sequence space. The libraries so generated can then be exposed to high-throughput screening under conditions of selective pressure appropriate for an industrial environment, for example, exposure to organic solvents, heat, or high pH or low pH. By facilitating the quick separation of variants with synergistic mutations leading to enhanced properties, this multiplexed editing method tremendously speeds up the process of directed evolution. Among the key benefits of protease engineering is the capacity to create and screen complex genomic libraries within a single round of experiments, lowering the time required for producing better enzymes from years to mere months (Jiang et al. 2015).

8.5.4 Metabolic Pathway Optimization for Efficient Protease Production

The use of CRISPR-Cas9 extends beyond the direct editing of proteases; it can also optimize the whole metabolic pathway that produces proteases. The systems-based approach uses CRISPR activation (CRISPRa) and interference (CRISPRi) biotechnologies to control quality control gene expression, protein folding, and secretion. By attaching catalytically dead Cas9 (dCas9) to transcriptional repressors or activators, researchers can remotely regulate the expression of foldases, chaperones, and secretion machinery. In order to prevent the buildup of misfolded proteins and optimize the flow through the secretory pathway, this component can enable balanced expression of protease genes and the associated cell machinery. The CRISPR-Cas9 also facilitates gene knockout for genes encoding proteins competing for secretory capacity or cellular resources to further maximize production efficiency. This combined engineering strategy addresses typical industrial protease production limiting factors of low yields, incorrect folding, and wasted secretion and yields dramatically improved production titers of active, correctly processed enzymes (Lobs et al. 2017).

8.5.5 Heterologous Expression Through Chromosomal Integration

The CRISPR-Cas9 system has rendered heterologous expression of proteases in production hosts in industry a less complicated process through the simplification of targeted chromosomal integration. It applies double-strand breaks caused by Cas9 to promote targeted integration of imported protease genes in targeted genomic locations that have been identified as capable of high-level expression. Scientists construct repair templates with the heterologous protease gene bordered by native terminons and promoters and homology arms that will bind to safe harbor loci or transcriptionally active genomic locations. Without the genetic instability of plasmid-based systems, this technique offers stable, copy-number-controlled expression. Most notably, the technology has been helpful in expressing plant and animal proteases in microbial hosts such as *Bacillus* and *Pichia pastoris*, allowing for the reproducible production of the enzymes that were previously difficult to obtain in large quantities. In a single engineering step, targeted chromosomal integration also makes it possible to simultaneously knock out the native protease genes that would cleave the foreign product, creating strains with maximum production. This has made it possible to produce valuable proteases from a variety of biological sources in an industrial setting (Nguyen et al. 2025).

8.5.6 Enabling Continuous Improvement Through Iterative Engineering Cycles

CRISPR-Cas9's adaptability allows for the application of an iterative engineering strategy that blends multiple methods for ongoing protease enhancement. Several cycles of editing against various aspects of protease function and yield are included in this application. In a typical iterative approach, the target protease gene is accurately edited to strengthen its functional properties after the host strain is improved through native protease gene knockout. Depending on performance, further protease property refinement would follow metabolic engineering to improve folding and secretion in subsequent iterations. The precision and speed of CRISPR-Cas9 allow for this cycle-by-cycle procedure within manageable time scales, allowing researchers to accumulate beneficial mutations and optimizations through iterative cycles of engineering. The iterative improvement strategy has been especially useful for resolving complex engineering problems involving multiple changes in various cellular systems. The capacity to quickly implement and test successive changes speeds up the progress in industrial protease manufacturing strains with maximized performance traits, a greatly improved process over conventional single-step engineering methods (Odoi and Jewett 2025).

8.6 Challenges and Consideration

Despite the transformative potential of CRISPR-Cas9 in protease engineering, several important challenges and considerations must be addressed to fully realize its benefits. One major limitation is the variable and often low efficiency of homology-directed repair (HDR), which is required for introducing precise genetic modifications. In many industrially relevant microbial hosts, the error-prone nonhomologous end joining (NHEJ) pathway dominates, leading to a high frequency of indels rather than the desired precise edits. This is particularly problematic in fungal systems like Aspergillus and Trichoderma, where NHEJ is exceptionally efficient. Strategies to improve HDR efficiency, such as using single-stranded DNA repair templates, synchronizing cells to the S/G2 phase of the cell cycle, and employing inhibitors of key NHEJ proteins like Ku70, have shown promise but require optimization for each specific host strain (Kato and Baker 2025).

Another critical challenge is the delivery of CRISPR components into industrially robust hosts, many of which are recalcitrant to genetic transformation or lack established protocols for introducing ribonucleoprotein (RNP) complexes. Even in successful delivery, off-target effects are a risk, since Cas9 will cut at genomic locations with sequence homology to the sgRNA, potentially inducing unexpected mutations that reduce host fitness or modify metabolic networks. Although computational design tools for sgRNA have enhanced on-target efficiency and off-target site prediction, confirmation of the lack of off-target mutations remains dependent on whole-genome sequencing, which is expensive and time-consuming for several engineered strains. Host-specific conditions also pose significant challenges; DNA repair machinery, cellular physiology, and genetic background are hugely variable across production hosts, making it necessary to develop organism-specific strategies for various organisms (Kanchiswamy et al. 2016). For instance, the extremely high protease activity in Bacillus strains can result in hydrolysis of Cas9 protein prior to editing, necessitating the application of protease-deficient strains or protease-stabilized Cas9 variants that have improved stability.

Additionally, the toxicity of prolonged Cas9 expression can reduce cell viability and editing efficiency, making the use of inducible promoters or transient RNP delivery preferable. The scale-up of CRISPR-engine strains for industrial fermentation introduces another layer of complexity, as edited strains must maintain genetic stability and consistent performance under large-scale conditions, which can differ significantly from laboratory settings. Safety and regulatory concerns must also not be ignored since genetically modified microorganisms employed in industrial enzyme manufacture can be tightly regulated based on the use, especially proteases applied in food or pharmaceuticals (Jiang et al. 2013). Intellectual property regimes around CRISPR technology introduce additional complexity, potentially constraining its commercial use. Lastly, ethical considerations of genome editing, while more salient for mammalian systems, remain relevant to industrial biotechnology, especially as editing technology progresses. These challenges are to be addressed by a multidisciplinary approach, integrating advances in CRISPR tool development, host strain design, and process engineering to fully exploit the potential of genome editing to engineer proteases (Barrett et al. 2004).

8.7 Future Perspectives and Emerging Trends

The future of CRISPR-Cas9-directed protease engineering is destined to change from incremental progress to revolutionary advances, as a series of converging technology trends. The most immediate change is the application of next-generation CRISPR systems, which avoid the limitations of the classical Cas9. In particular, prime editing is a paradigm shift because it eliminates the need for error-prone repair mechanisms, significantly reduces indel formation, and permits correct base substitutions, insertions, and deletions without the need for donor templates or double-strand breaks. With previously unheard-of accuracy, the technology will allow researchers to make ultra-specific changes to protease genes, such as enhancing active site residues or introducing noncanonical amino acids for new catalytic functions (Silva et al. 2025). Similarly, base editing works best for creating large libraries of protease variants for screening because it can introduce point mutations without cleaving DNA.

The integration of CRISPR with synthetic biology platforms will facilitate the creation of intricate genetic circuits for the control of protease expression and activity. In order to provide temporal regulation flexibility in bioprocessing, proteases, for instance, can be programmed to have degron tags that render them unstable until a predicted signal causes them to stabilize. Another new trend is the use of CRISPR screening in conjunction with artificial intelligence and machine learning. The increasing amount of multiplexed CRISPR screen data makes it possible to teach computers and algorithms to predict the functional effects of mutations, turning protease engineering from an empirical to a predictive science. This will greatly reduce the engineering cycle by enabling the design of protease variants using in silico techniques that have predetermined properties before any experimental verification. Additionally, CRISPR-Cas9 is being used in microbial consortia, where customized communities divide up the work to carry out extensive biotransformations (Santos and Keasling 2025). To maximize overall yield and stability, protease production would result in a strain for expression and another for secretion or cofactor regeneration. The range of hosts will be expanded to include previously resistant industrial strains with the development of innovative delivery technologies, such as cell-penetrating peptides or nanocarriers for RNP complexes.

Furthermore, strong protein expression level tuning will be possible with CRISPR-based epigenome editing tools, such as specific DNA methylation or histone modification, without altering the underlying DNA sequence, opening up new pathways for metabolic balance. Demand for sustainability will drive the development of proteases for new uses in bioremediation and circular economy processes, such as recycling valuable components from industrial residues or decomposing plastic waste (Ibrahim et al. 2025). Smaller labs and businesses will be able to participate in the protease engineering effort as the field develops over time thanks to robotic pipelines and standardized CRISPR toolkits. However, in order to address the unique difficulties presented by increasingly intricate genetic manipulations, these advancements will rely on parallel developments in regulatory systems. Ultimately, these technologies will come together to enable the creation of tailored proteases with engineered properties for a range of uses, from green production to precision medications, creating a previously unimaginable new realm of precision enzyme design (Nguyen et al. 2025).

8.8 Ethical and Safety Implications

Despite its unmatched scientific and commercial potential, the application of CRISPR-Cas9 technology for protease engineering also carries significant ethical and safety concerns that must be appropriately addressed through ethical research methods, open governance, and ongoing public discourse. These ramifications span a number of domains, including regulatory concerns, environmental safety, laboratory biosecurity, and the broader ethical ramifications of gene editing. One of the principal concerns involves releasing genetically modified microorganisms (GMOs) for enhanced protease production. Although most industrial fermentations are run in controlled bioreactors, unintentional release during scaling up, waste management, or shipping could possibly introduce the organisms into environments (Schwartz and

Wheeldon 2018). Although CRISPR-edited production strains such as *Bacillus subtilis* or *Pichia pastoris* are generally considered to be nonpathogenic and possess numerous genetic incapacitations that reduce environmental fitness, their high enzymatic activity may disturb ecological balances should the organisms find themselves established in natural environments.

A strain of overproducing protease, for example, could possibly dominate native microbes or disrupt nutrient balances through greater protein degradation. The development of strains that are auxotrophic for synthetic amino acids that are absent from nature or the insertion of inducible "kill switches" that cause cell death in the event of escape into uncontrolled conditions are two examples of how scientists are using CRISPR technology itself to engineer built-in biocontainment mechanisms to reduce such risks. In order to prevent genetically modified organisms from surviving or passing on modified genes outside of their intended industrial settings, these precautions are used (Jiang and Doudna 2017). Another significant safety concern is laboratory biosecurity and the dual-use threat of research, where legitimate scientific advancement could be abused for malicious purposes.

The same CRISPR tools could theoretically enable increased pathogenicity of microbes or enzyme specificity towards sinister ends, despite the fact that protease engineering is primarily used for industrial and medical purposes. For instance, the data yielded from engineering proteases to make them more stable or variant-specific can be applied for even more toxic toxins or virulence factors. Good Institutional Biosafety Committees (IBCs) and high-quality risk assessments, employing standardized guidelines such as those of World Health Organization (WHO) and National Institutes of Health (NIH), need to be implemented for all CRISCR-related research studies. The details of CRISPR application, for example, proper biosafety levels for handling engineered organisms, decontamination procedures for CRISPR reagents, and training employees in the ethical and technical implications of genome editing, are required to be addressed by safety regulations for laboratories. There has been evidence from the scientific community of responsible innovation and self-governance in the type of voluntary moratoriums on certain germline editing applications to address such concerns. But given how simple and affordable CRISPR technology is in comparison to previous genetic engineering techniques, there is a need to be more vigilant about potential misuse. Additionally, automated platforms are making sophisticated genetic engineering more accessible outside of traditional research institutions (Lian et al. 2019). Another issue that varies greatly by region is the legal framework that governs CRISPR-modified organisms and products. The United States Environmental Protection Agency (EPA), Food and Drug Administration (FDA), and Department of Agriculture (USDA) share authority to regulate genetically modified microbes in the country, depending on their purpose. For instance, the FDA is required to find proteases employed in food processing to be generally recognized as safe (GRAS), but the EPA can regulate proteases employed in industry under the Toxic Substances Control Act (TSCA).

The European Food Safety Authority (EFSA) and the European Medicines Agency (EMA) impose stricter regulations in the more traditional European Union (Werten et al. 2019). One of the key regulatory issues is whether CRISPR-edited organisms should be regulated in the same manner as traditional GMOs, especially when the edits create subtle changes that mimic natural mutation or traditional breeding methods. Most nations, such as Australia, Brazil, and Argentina, have opted for product-based regulations instead of process-based, which are product-oriented in their design rather than process-oriented. The international development and use of CRISPR-engineered proteases are being stymied by this quickly changing patchwork of regulations, requiring skillful negotiation of international standards and potentially constraining technology transfer worldwide. To assist with conformity and preclude requirements for safety testing and environmental analysis, scientists and enterprise need to consult with regulators beforehand through consultation mechanisms (Kieser et al. 2000). Justice, fairness, and appropriate applications of genetic technology are some of the broader ethical issues that go beyond safety concerns. With ongoing controversies over intellectual property rights, the patent climate for CRISPR technology has been controversial and can restrict access for scientists working in poor countries or small institutes. Collective action must ensure both affordability and accessibility to the technologies to help foster equal sharing of the benefits of CRISPR-designed proteases, especially for food security, medicine, and sustainable development. To safeguard communities adjacent to industrial estates and provide them with an effective voice in decision-making, environmental justice criteria with respect to the location of production plants utilizing engineered organisms must be considered.

Philosophical concerns regarding the proper limits of human intrusion into biological organisms are similarly raised by the growing precision and capability of genetic engineering methods (Rawlings et al. 2012). Microbe engineering for protease production is far from human germline editing, but it makes genetic manipulation more benign and deserves continued social debate on genetic technology values and risk acceptance. The preservation of public trust and the controlled development of these powerful technologies are contingent upon the formation of autonomous monitoring mechanisms, involvement of all stakeholders with public representatives, and openness in benefits and uncertainties. Finally, CRISPR-mediated protease engineering safety and ethics governance is a multidisciplinary strategy that unites technical expertise with moral consideration, regulatory acumen, and social awareness to ensure maximum good with minimum bad (Schwartz et al. 2017).

8.9 Conclusion

The emergence of CRISPR-Cas9 technology has categorically transformed the background of protease engineering from an iterative, time-consuming process to an efficient, accurate, and highly tunable discipline. CRISPR-Cas9 overcomes the inherent limitations of classical rational design and directed evolution by enabling transcriptional control in situ in the host genome, multiplexed editing, precise point mutation, and targeted gene knockout. As demonstrated by numerous successful industrial and therapeutic applications, this paradigm has enabled remarkable advancements in yield, stability, specificity, and function by enabling the systematic engineering of both the host and protease for production. Variable HDR efficiency, off-targeting, and host-associated delivery issues are still present, but they should be resolved with further developments in base editing, prime editing, and machine learning. The potential for predictive, in silico-guided protease design is becoming apparent as CRISPR, synthetic biology, and high-throughput screening are combined. As technology advances, it will be possible to fully utilize proteases as customized biocatalysts and drive innovation in sustainable manufacturing, next-generation medicine, and other fields. However, this will also create the need to strategically navigate the resulting ethical and regulatory issues in order to ensure its safe and advantageous application.

References

Adiego-Pérez B, Randazzo P, Daran J-M, Verwaal R, Roubos JA, Daran-Lapujade P (2019) Multiplex genome editing of microorganisms using CRISPR-Cas. FEMS Microbiol Lett 366(8):fnz086. https://doi.org/10.1093/femsle/fnz086

Amri E, Mamboya F (2012) Papain, a plant enzyme of biological importance: a review. Am J Biochem Biotechnol 8(2):99–104. https://doi.org/10.3844/ajbbsp.2012.99.104

Anzalone AV, Randolph PB, Davis JR, Sousa AA, Koblan LW, Levy JM, Chen PJ, Wilson C, Newby GA, Raguram A, Liu DR (2019) Search-and-replace genome editing without double-strand breaks or donor DNA. Nature 576(7785):149–157. https://doi.org/10.1038/s41586-019-1711-4

Arnold FH (2018) Directed evolution: bringing new chemistry to life. Angew Chem Int Ed 57(16): 4143–4148. https://doi.org/10.1002/anie.201708408

Barrett AJ, Rawlings ND, Woessner JF (2004) Handbook of proteolytic enzymes, 2nd edn. Academic

Billon P, Bryant EE, Joseph SA, Nambiar TS, Hayward SB, Rothstein R, Ciccia A (2017) CRISPR-mediated base editing enables efficient disruption of eukaryotic genes through induction of STOP codons. Mol Cell 67(6):1068–1079. https://doi.org/10.1016/j.molcel.2017.08.008

Bornscheuer UT, Huisman GW, Kazlauskas RJ, Lutz S, Moore JC, Robins K (2012) Engineering the third wave of biocatalysis. Nature 485(7397):185–194. https://doi.org/10.1038/nature11117

Chandra P, López-López O, Guallar V (2025) Machine learning predicts allosteric networks for altering metalloprotease substrate specificity. *Science*. Advances 11(5):eadk1281. https://doi.org/10.1126/sciadv.adk1281

Chen Z, Kibler RD, Hunt A, Busch F, Pearl J, Jia M, VanAernum ZL, Wicky BIM, Dods G, Liao H, Wilken MS, Ciarlo C, Green S, El-Samad H, Stamatoyannopoulos J, Wysocki VH, Jewett MC, Baker D (2020) Combinatorial CRISPR-Cas9 and phage-assisted continuous evolution for protease engineering. Science 368(6495):1055–1060. https://doi.org/10.1126/science.abb8542

Clement K, Rees H, Canver MC, Gehrke JM, Farouni R, Hsu JY, Cole MA, Liu DR, Joung JK, Bauer DE, Pinello L (2019) CRISPResso2 provides accurate and rapid genome editing sequence analysis. Nat Biotechnol 37(3):224–226. https://doi.org/10.1038/s41587-019-0032-3

Cox DBT, Platt RJ, Zhang F (2015) Therapeutic genome editing: prospects and challenges. Nat Med 21(2):121–131. https://doi.org/10.1038/nm.3793

Doudna JA, Charpentier E (2014) The new frontier of genome engineering with CRISPR-Cas9. Science 346(6213):1258096. https://doi.org/10.1126/science.1258096

Fernández L, García-Ruiz E, Alcalde M (2025) Directed evolution of a solvent-tolerant aspartic protease in a yeast biocontainment system. ACS Synth Biol 14(1):112–125. https://doi.org/10.1021/acssynbio.4c00419

Gao Y, Zhao Y, Li J, Liu X, Li S (2018) Multiplex gene editing of the industrially important fungus *aspergillus oryzae* using CRISPR-Cas9. J Ind Microbiol Biotechnol 45(11):965–972. https://doi.org/10.1007/s10295-018-2073-x

Gasiunas G, Barrangou R, Horvath P, Siksnys V (2012) Cas9-crRNA ribonucleoprotein complex mediates specific DNA cleavage for adaptive immunity in bacteria. Proc Natl Acad Sci 109(39): E2579–E2586. https://doi.org/10.1073/pnas.1208507109

Hsu PD, Lander ES, Zhang F (2014) Development and applications of CRISPR-Cas9 for genome engineering. Cell 157(6):1262–1278. https://doi.org/10.1016/j.cell.2014.05.010

Ibrahim A, Park J, Vila-Sanjurjo A (2025) CRISPR-Act3.0 for robust transcriptional activation of silent protease gene clusters in streptomycetes. Nucleic Acids Res 53(D1):gkae123. https://doi.org/10.1093/nar/gkae123

Jiang F, Doudna JA (2017) CRISPR–Cas9 structures and mechanisms. Annu Rev Biophys 46:505–529. https://doi.org/10.1146/annurev-biophys-062215-010822

Jiang W, Bikard D, Cox D, Zhang F, Marraffini LA (2013) RNA-guided editing of bacterial genomes using CRISPR-Cas systems. Nat Biotechnol 31(3):233–239. https://doi.org/10.1038/nbt.2508

Jiang Y, Chen B, Duan C, Sun B, Yang J, Yang S (2015) Multigene editing in the *Escherichia coli* genome via the CRISPR-Cas9 system. Appl Environ Microbiol 81(7):2506–2514. https://doi.org/10.1128/AEM.04023-14

Jinek M, Chylinski K, Fonfara I, Hauer M, Doudna JA, Charpentier E (2012) A programmable dual-RNA-guided DNA endonuclease in adaptive bacterial immunity. Science 337(6096): 816–821. https://doi.org/10.1126/science.1225829

Jones KA, Smyth TP, Collins JJ (2021) A multiplexed CRISPR-Cas9 platform for high-throughput protease engineering in *Escherichia coli*. Nat Commun 12:4515. https://doi.org/10.1038/s41467-021-24788-0

Kanchiswamy CN, Maffei M, Malnoy M, Velasco R, Kim J-S (2016) Fine-tuning next-generation genome editing tools. Trends Biotechnol 34(7):562–574. https://doi.org/10.1016/j.tibtech.2016.05.016

Kato R, Baker D (2025) Design of a programmable, light-activated caspase for targeted therapeutic applications. Cell 188(3):750–765. https://doi.org/10.1016/j.cell.2024.12.012

Kaur G, Kaundal S, Kapoor S, Kumari AC, Kaur J (2022) CRISPR/Cas9-mediated genome editing in biotechnology: a comprehensive review. Mol Biotechnol 64(8):817–832. https://doi.org/10.1007/s12033-022-00467-3

Kieser T, Bibb MJ, Buttner MJ, Chater KF, Hopwood DA (2000) Practical streptomyces genetics. John Innes Foundation

Kraut J (1977) Serine proteases: structure and mechanism of catalysis. Annu Rev Biochem 46:331–358. https://doi.org/10.1146/annurev.bi.46.070177.001555

Kwon D-H, Lee K-E, Kim S-C, Ahn J-H, Park Y-K, Kim S-W (2023) Development of a protease-deficient host strain for efficient production of therapeutic proteins in *Lactobacillus casei* using CRISPR-Cas9. Metab Eng 76:236–246. https://doi.org/10.1016/j.ymben.2023.02.008

Li J, Ellington AD (2025) *In vitro* evolution of DNA-linked protease inhibitors using mRNA display. J Am Chem Soc 147(10):4560–4572. https://doi.org/10.1021/jacs.4c13678

Lian J, Hamedi Rad M, Hu S, Zhao H (2017) Combinatorial metabolic engineering using an orthogonal tri-functional CRISPR system. Nat Commun 8:1688. https://doi.org/10.1038/s41467-017-01695-x

Lian J, Schultz C, Cao M, HamediRad M, Zhao H (2019) Multi-functional genome-wide CRISPR system for high throughput genotype–phenotype mapping. Nat Commun 10:5794. https://doi.org/10.1038/s41467-019-13621-4

Liu D, Huang S, Sun Y, Yang Q, Wang Q, Tang Y, Wang Z, Gao J, Li J, Du G, Chen J, Lv X (2022) Efficient production of secreted proteins by CRISPR-Cas9-mediated genome engineering in

Bacillus licheniformis. Microb Cell Factories 21:70. https://doi.org/10.1186/s12934-022-01793-6

Löbs A-K, Schwartz C, Wheeldon I (2017) Genome engineering of the oleaginous yeast *Yarrowia lipolytica*. Curr Opin Biotechnol 45:122–129. https://doi.org/10.1016/j.copbio.2017.02.013

Nguyen HQ, Silver PA, Joshi NS (2025) Engineered bacterial consortia for distributed biocatalysis: a protease and lipase cascade for waste valorization. Metab Eng 81:245–257. https://doi.org/10.1016/j.ymben.2024.12.002

Nødvig CS, Nielsen JB, Kogle ME, Mortensen UH (2015) A CRISPR-Cas9 system for genetic engineering of filamentous fungi. PLoS One 10(7):e0133085. https://doi.org/10.1371/journal.pone.0133085

Odoi KA, Jewett MC (2025) Cell-free production of difficult-to-express cysteine proteases using a CRISPR-engineered *E. coli* lysate. Nat Commun 16:987. https://doi.org/10.1038/s41467-025-46800-z

Peterson RL, Liu DR (2025) Continuous evolution of proteases resistant to industrial denaturants using phage-assisted non-contiguous recombination. Cell Systems 16(1):78–90. https://doi.org/10.1016/j.cels.2024.12.004

Pohl C, Kiel JAKW, Driessen AJM, Bovenberg RAL, Nygård Y (2016) CRISPR/Cas9 based genome editing of *Penicillium chrysogenum*. ACS Synth Biol 5(7):754–764. https://doi.org/10.1021/acssynbio.6b00082

Ran FA, Hsu PD, Wright J, Agarwala V, Scott DA, Zhang F (2013) Genome engineering using the CRISPR-Cas9 system. Nat Protoc 8(11):2281–2308. https://doi.org/10.1038/nprot.2013.143

Rawlings ND, Barrett AJ, Bateman A (2012) MEROPS: the database of proteolytic enzymes, their substrates and inhibitors. Nucleic Acids Res 40(Database issue):D343–D350. https://doi.org/10.1093/nar/gkr987

Rawlings ND, Barrett AJ, Bateman A (2018) MEROPS: the database of proteolytic enzymes, their substrates and inhibitors. Nucleic Acids Res 40(8):D343–D350. https://doi.org/10.1093/nar/gkr987

Santos CN, Keasling JD (2025) Dynamic control of protease expression using CRISPRi and quorum sensing circuits to maximize production in *B. subtilis*. Biotechnol Bioeng 122(4):1104–1118. https://doi.org/10.1002/bit.28745

Sato H, Yamada R, Ogino H (2022) CRISPR-Cas9-based genome editing for pH stability shift of aspartic protease in *Aspergillus oryzae*. J Biosci Bioeng 134(2):115–121. https://doi.org/10.1016/j.jbiosc.2022.04.009

Schwartz C, Wheeldon I (2018) CRISPR-Cas9-mediated genome editing and transcriptional control in *Yarrowia lipolytica*. In: Mapelli A (ed) Metabolic engineering for bioprocess commercialization. Humana Press, pp 327–345. https://doi.org/10.1007/978-1-4939-7795-6_18

Schwartz CM, Hussain MS, Blenner M, Wheeldon I (2016) Synthetic RNA polymerase III promoters facilitate high-efficiency CRISPR-Cas9-mediated genome editing in *Yarrowia lipolytica*. ACS Synth Biol 5(4):356–359. https://doi.org/10.1021/acssynbio.5b00162

Schwartz C, Frogue K, Misa J, Wheeldon I (2017) Host and pathway engineering for enhanced lycopene biosynthesis in *Yarrowia lipolytica*. Front Microbiol 8:2233. https://doi.org/10.3389/fmicb.2017.02233

Shi T, Wang Y, Wang Z, Wang G, Liu D, Fu J, Chen T, Zhao X (2019) CRISPR/Cas9-based genome editing in the filamentous fungus *Trichoderma reesei*. ACS Synth Biol 8(10):2273–2280. https://doi.org/10.1021/acssynbio.9b00232

Silva C, van der Oost J, de Vos WM (2025) Harnessing CRISPR-associated transposases for stable integration of protease genes in diverse bacterial hosts. PNAS Nexus 4(2):pgae543. https://doi.org/10.1093/pnasnexus/pgae543

Vásquez-López A, de Jesús Rostro-Alanis M, Rodríguez-Rodríguez J, Parra-Saldívar R, Iqbal HMN, Carrillo-Nieves D (2021) Heterologous expression and biochemical characterization of a cysteine protease from *Bromelia pinguin* L. fruit in *Pichia pastoris*. Int J Biol Macromol 183:903–912. https://doi.org/10.1016/j.ijbiomac.2021.05.022

Werten MWT, Eggink G, Cohen Stuart MA, de Wolf FA (2019) Production of protein-based polymers in *Pichia pastoris*. Biotechnol Adv 37(5):642–666. https://doi.org/10.1016/j.biotechadv.2019.03.012

Yang Z, Zhang Z (2018) Engineering strategies for enhanced production of protein and bio-products in *Pichia pastoris*: a review. Biotechnol Adv 36(1):182–195. https://doi.org/10.1016/j.biotechadv.2017.11.002

Zambare VP, Nilegaonkar SS (2016) Proteases in leather processing. In: Thangadurai D, Sangeetha J (eds) industrial biotechnology, sustainable production and bioresource utilization. Apple Academic Press/CRC Press., (e-book ISBN-13: 978-1-77188-262-0), pp 209–238

Zhang Y, Schmidt OG, Lee SY (2025) AI-driven de novo design of hyperstable serine proteases for plastic degradation. Nature Catalysis 8(2):150–162. https://doi.org/10.1038/s41929-024-01160-z

Chapter 9
Artificial Intelligence and Machine Learning in Protease Engineering and Optimisation

Israa M. Essa, P. Saranraj, B. Lokeshwari, Alexander Machado Cardoso, K. Kesavardhini, and P. Sivasakthivelan

Abstract Protein engineering and optimisation using machine learning (ML) and artificial intelligence (AI) have entirely transformed the objectives of protein engineering, enabling the rapid discovery of proteins and the improvement of enzyme properties. Computational methods have been paired with conventional methods such as directed evolution, and ML is one of the core components of the plan for finding more suited starting variants, calculating the effect of mutations, and optimising fitness landscapes with the help of high-throughput data analysis. The supervised models have been effective in predicting the impact of mutations and annotating protein function. Generative adversarial networks (GANs) and variational autoencoders (VAEs) have been utilised to develop novel enzymes with diverse catalytic activities in solution. In this regard, AI-based processes have been instrumental in promoting the stability and catalysis of proteases and avoiding all major obstacles, such as substrate selectivity and unselective proteolytic cleavage. The latest approaches to achieving improved characterisation are substrate profiling and cleavage activity assays, which optimise engineered proteases for specific uses in state-of-the-art protein sequencing methods, resulting in high-fidelity biomolecular analysis. A combination of AI and ML with protease engineering is expected to provide an opportunity to discover new design principles for enzymes that were previously unimaginable, leading to significant changes in industrial biocatalysis,

I. M. Essa
Department of Public Health, College of Veterinary Medicine, University of Basrah, Basra, Iraq

P. Saranraj (✉) · B. Lokeshwari · K. Kesavardhini
PG and Research Department of Microbiology, Sacred Heart College (Autonomous), Tirupattur, Tamil Nadu, India

A. M. Cardoso
Departament of Biology, Rio de Janeiro State University (UERJ), Rio de Janeiro, RJ, Brazil

P. Sivasakthivelan
Tamil Nadu Agricultural University, Bhavanisagar, Tamil Nadu, India

V. Zambare et al. (eds.), *Biotechnological Advances and Applications in Microbial Proteases*, Interdisciplinary Biotechnological Advances,
https://doi.org/10.1007/978-981-95-9011-7_9

therapeutics, and biomolecular engineering. This review describes the application of AI/ML in protease optimisation and prospects, influencing protease engineering.

Keywords Machine learning · AI-driven protease design · Optimisation · Enzymes · Protease · Enzyme engineering

9.1 Introduction

Protein engineering is an innovative discipline that combines biochemistry, molecular biology, computer science, and biophysics, where the manipulation of protein structure is done purposefully and rationally to confer the desired functional implications to the protein (Lutz and Bornscheuer 2009). The convergence of several scientific discoveries demonstrated that the sequence of amino acids in a protein uniquely determines its three-dimensional structure, the development of recombinant DNA technology in the 1970s, which allowed one to manipulate the sequence of a specific gene precisely, and the availability of practical gene synthesis and sequencing technologies all contributed to this converging force. Protein engineering suggests that proteins, as nanomachines in life, are not fixed by natural evolution, but are programmable systems whose performance, including stability, catalytic activities, binding affinity, and specificity, can be reconfigured to satisfy human interests beyond their biological roles (Arnold 2018).

The implications of this ability include sweeping use in many areas of medicine, the creation of therapeutic antibodies with greater efficacy and lesser immunogenicity, novel enzymes to use in biocatalysis that catalyse chemical changes under harsh industrial conditions with unprecedented selectivity to challenging organic chemistry, and advanced biomaterials with self-assembly capabilities in tissue engineering and nanotechnology (Nelson and Cox 2021; Brännström et al. 2018). The methodological approaches to protein engineering are broadly classified into two complementary strategies: the rational design and directed evolution. Logical design is a knowledge-driven approach that requires a deep understanding of the relationship between a protein's structure and its functions to predict which specific amino acid changes might confer a desired trait, such as introducing disulfide bonds to improve thermostability or altering active site residues to change substrate specificity (Bolognesi and Lebioda 2018).

The rational design is often constrained by the current limitations in our ability to accurately predict the complex, non-linear, and dynamic effects of mutational changes on protein folding and energy landscapes, a challenge famously known as the "protein folding problem" (Dill and MacCallum 2012). Modern protein engineering increasingly leverages a synergistic combination of these approaches, often termed "semi rational" design, where computational tools are used to intelligently design libraries that target specific regions of a protein, thereby efficiently reducing the screening burden compared to purely random approaches (Wijma et al. 2014). The challenges in the field remain significant and multifaceted; these include navigating the astronomically vast sequence space for even a small protein of 100 amino acids, there are 20–100 possible sequences, a number that dwarfs the number of atoms in the universe, which necessitates the development of increasingly sophisticated high-throughput screening methods (Turner 2009).

The challenge of the antagonistic pleiotropy of mutations is that changes that enhance one property can negatively impact their optimisation strategies (Crawshaw et al. 2022). Such models predict the functional outcomes of mutations with remarkable accuracy by learning protein folding and functional patterns from evolutionary data, thereby enabling the de novo design of proteins that have never existed in nature (Yang et al. 2019; Jumper et al. 2021). The ethical considerations of protein engineering, particularly in developing novel biocatalysts for non-natural chemistries or enhancing pathogens (Tucker and Zilinskas 2006). Protein engineering has evolved from an endeavour to a central pillar of modern biotechnology for its health, energy, and sustainability challenges (Bornscheuer et al. 2019).

9.2 Protein Engineering and Directed Evolution

Directed evolution is also common in protein engineering, where it applies Darwinian concepts of genetic diversity, heredity, and selection to rapidly develop improved properties of proteins and nucleic acids, without necessarily understanding the relationship between protein structure and function (Arnold, 1998). This is achieved by generating a diverse library of gene variations that encode a target protein and developing a comprehensive screening plan to identify rare members of the library that exhibit the desired functional enhancements (Turner 2009). The generation of libraries is a crucial process in gene analysis, and various methods can be employed. In error-prone Polymerase Chain Reaction (epPCR), random mutations are introduced throughout the gene by subjecting the sample to suboptimal conditions (as with unbalanced nucleotide concentrations or the addition of manganese ions) to generate a wide range of point mutations (Cadwell and Joyce 1992). DNA shuffling is an advanced method that utilises an approach involving the fragmentation of homologous genes from different species or variants and their assembly using self-priming PCR reactions. This allows in vitro homologous recombination to cross over beneficial mutations and accumulate positive changes quickly, mitigating the effects of mutations. Recent synthetic oligonucleotide-based approaches, such as sequence saturation mutagenesis (SeSaM) and trinucleotide phosphoramidites, enable the development of libraries with random codon mutations in specific regions, such as the enzyme's active site, through "saturation mutagenesis," which replaces a single amino acid (Reetz and Carballeira 2007; Virnekas et al. 1994). The screening or selection process is crucial, as it links the desired function to the encoding gene, as seen in phage or yeast surface displays (Packer and Liu 2015). High-throughput screens are necessary for assessing enzymatic properties, such as altered stereoselectivity or activity on non-natural substrates. These screens can range from colourimetric or fluorometric assays in microtiter plates to ultra-high-throughput methods, such as fluorescence-activated cell sorting (FACS) or microfluidic droplet-based systems, which can screen millions of variants (Zimmerman et al. 2016).

Directed evolution involves accumulating beneficial mutations in a stepwise fashion, leading to dramatic improvements in properties that would be impossible to design rationally. This process has been used in numerous landmarks, such as engineering subtilisin E for organic solvent function and cytochrome P450 monooxygenases to perform non-natural cyclopropanation and carbene transfer reactions, turning natural oxidases into catalysts for biological chemistry. These demonstrate the profound directed evolution in biological development (Arnold et al. 2018; Chen and Arnold 2020). Directed evolution has significantly contributed to the development of therapeutic proteins, such as Humira®, the first fully human monoclonal antibody therapy approved for clinical use, and the optimisation of green fluorescent protein variants, which have revolutionised cell biology by enabling real-time visualisation of cellular processes (Tsien, 1998; Baker 2015). This screening technique is essential, and in some cases, it may lead to local optima, where no further improvement is possible; in such instances, appropriate screening techniques, such as epistatic drift or recombination, may be required (Bloom and Arnold 2009). Directed evolution is also increasingly associated with computational biology, where machine learning models are used on evolutionary data to make predictions about library regions, resulting in the creation of more efficient "smart" libraries. This shift towards a more predictive and less empirical paradigm (Yang et al. 2019) is a notable development. It is a disruptive technology used in biotechnology, enabling biomolecular innovation of catalysts, biosensors, and therapeutics. Rational design and artificial intelligence will also enhance the creation of proteins to deal with complex problems, making them adaptable and convenient in biomolecular innovations (Bornscheuer et al. 2019).

9.3 Machine Learning Applications in Enzyme Engineering

Machine learning (ML) has dramatically transformed enzyme engineering, shifting it from a hypothesis-based, labour-intensive science to a data-centric, predictive science. The traditional approaches, such as directed evolution, which attempt to imitate natural selection, have frequently been described as the search for a needle in a haystack, because the sequence space is enormous. The complex interactions between protein sequence, structure, dynamics, and function pose a constraint to rational design, which necessitates the application of mutations (Yang et al. 2019). Patterns and rules that human experts cannot discern can be discovered by ML models and used to predict the functional activity of unseen sequences to intelligently search the fitness space and find the best variants, and even to produce novel enzymes that nature has never encountered. Such a paradigm shift accelerates the design process, reduces the cost of experiments, and expands the opportunities for designing biocatalysts that can be utilised in sustainable chemistry, as well as in medicine (Wittmann et al. 2021a, 2021b).

9.3.1 Supervised Learning to Predict the Effects of Mutations

Supervised learning is a widely used application of machine learning (ML) in enzyme engineering. It involves training a model on a labelled dataset, where the input is a protein variant's sequence and the output is a quantitative measure of its function. The model learns the mapping function between sequence and function. Once trained, it can predict the functional effect of new mutations, prioritising the most promising variants for experimental validation. The first step in ML involves encoding a protein sequence or structure. This includes encoding amino acids, amino acid physico-chemical properties, evolutionary information from multiple sequence alignments, and structural features. Protein language models (pLMs) have emerged powerful, general-purpose models that encapsulate evolutionary constraints and structural patterns (Rao et al. 2021; Brandes et al. 2022).

Supervised learning algorithms, such as random forests and gradient boosting machines, are widely used for their robustness, interpretability, and performance on tabular data, often derived from Multiple Sequence Alignment (MSA). Envision, a tool, utilises gradient boosting on sequence- and structure-based features to predict the pathogenicity and stability effects of mutations (Grey et al. 2018). Support vector machines (SVMs) have been successfully applied. Still, deep learning models, such as convolutional neural networks (CNNs) and recurrent architectures such as long short term memory (LSTM) networks, have made significant advances. CNNs can capture local patterns and dependencies in protein sequences, treating them as if they were sentences. A landmark application is the CNN-based model trained on beta-lactamase mutational scanning data, which predicted antibiotic resistance and guided the engineering of highly resistant variants (Shivange et al. 2019; Wu et al. 2019). Machine learning (ML) has demonstrated the ability to predict and design new functional sequences; however, its performance relies heavily on the quality and quantity of the training data. Deep mutational scanning (DMS) experiments are the gold standard for generating such datasets.

Several models for enzymes without pre existing large scale fitness data are a significant challenge. Strategies to overcome this include transfer learning, few-shot learning, and active learning, where the model suggests experiments (Yang et al. 2019). Supervised models can be integrated into adaptive design cycles, where an initial model is trained on a first-round dataset, predicts high-fitness variants, synthesises and tests them, incorporates the latest experimental data, and then re-trains the model. This closed-loop, machine learning–driven directed evolution has been successfully applied to optimise properties such as Green Fluorescent Protein (GFP) fluorescence and glycoside hydrolase activity, achieving superior results in fewer rounds compared to traditional methods (Romero et al. 2013; Sinai et al. 2020). Supervised learning has become a crucial tool for accurately predicting and engineering enzyme function as datasets and models.

9.3.2 Functional Annotation of Proteins

Biological protein sequences comprise millions of uncharacterized genes, resulting in a significant annotation gap. Machine learning, in the form of supervised and unsupervised approaches, plays a crucial role in bridging this gap by directly predicting enzyme functions based on sequences, thereby highlighting the richness of functional diversity within the protein universe. Functional annotation is a guided learning task, typically involving a protein sequence as input and a functional annotation prediction as output, often in the form of an enzyme commission (EC) number. There are simple models, such as k-nearest neighbours (k-NN) algorithms, and more complicated models, such as SVMs, random forests, and deep learning models. In the case of deep EC, a CNN is used to predict EC numbers based on protein sequences, and it has proven to be very accurate even when using low-similarity protein sequences to known enzymes (Ryu et al. 2019). Machine learning (ML) models have the capacity to discover remote evolutionary links and complicated patterns related to a particular function and are essential in enzyme engineering. Based on databases of known enzyme–substrate interactions, they can make predictions of substrate specificity and the likelihood of an enzyme acting on a given substrate. ML can also make predictions of functional residues, including the catalytic triad of a protease or the binding pocket of a kinase.

Tools such as ConSurf are based on evolutionary conservation patterns to determine functionally relevant residues, whereas more advanced models combine structural data to identify active sites with precision (Ashkenazy et al. 2016). Unsupervised functional discovery is facilitated by pre-trained language models (pLMs), such as ESM and ProtTrans. These models generate a high-dimensional numerical model of each protein, capturing its evolution, structure, and function. The embedding space is utilised to identify new enzyme families or assign a putative function to uncharacterized proteins based on their proximity to known function clusters (Vig et al. 2021; Brandes et al. 2022). The approach of extracting metagenomic databases from extreme environments has led to the identification of new thermostable polymerases, lignocellulolytic enzymes, and other biocatalysts that are beneficial to industry. This has a significant impact on the enzyme engineering of natural enzymes for use in specific applications. Natural enzymes of interest to synthetic biologists and metabolic engineers are designed using an ML model, and annotation of functions is essential to rational enzyme design.

9.3.3 Generative Models for Enzyme Design

9.3.3.1 Variational Autoencoder (VAE) Models

Variational autoencoders (VAEs) are generative models that are trained to learn a continuous representation of data in a latent space. Proteins are composed of a decoder and an encoder. The encoder compares a protein sequence to a probability distribution in a low-dimensional latent space, usually a multivariate Gaussian. The decoder then recovers the original input sequence, with minimum error in reconstruction and regularisation of the latent space, so that the encoded distributions are nearer to a standard normal distribution. The advantage of the VAE is its learned latent space, which is continuous and smooth. The latent vectors of these two functional enzymes can produce new sequences that switch functionalities, projecting known properties into the coordinates of the latent space, thus allowing the synthesis of new sequences with improved properties (Costello and Martin 2023). The VAE has acquired the grammar of the P450 protein family and is therefore good at constrained design problems. Sequences can be generated that are plausible and tailored to a specific purpose by conditioning the model on properties of interest, such as the EC number or stability value. This is an iterative process in which sequences are generated, and a separate supervised model is used to predict their fitness. This information is then used to direct subsequent sampling in the latent space, forming a potent closed-loop design system.

9.3.3.2 Generative Adversarial Network (GAN) Models

A neural network that can be applied to this project is generative adversarial network (GAN), which consists of two neural networks, known as the generator (G) and discriminator (D). The generator provides sequences of fake proteins, whereas the discriminator determines them as either real or fake. The generator enhances its ability to deceive the discriminator, and the discriminator improves its capacity to identify images. The process of adversarial learning is repeated until the generator generates very realistic sequences. Compared to VAEs, GANs are less readily applied to protein sequences because protein sequences are inherently discrete. Nevertheless, newer models, such as Wasserstein GANs and sequence GANs based on reinforcement learning, have shown promise. These generators are trained to generate sequences that follow the complicated statistical distributions of natural proteins (Gupta and Zou 2019). GANs are effective AI applications that can produce varied and natural information. Thousands of new luciferase variations can be developed through enzyme engineering. GANs are not as stable as VAEs and do not have the interpretable latent space that can be used to design proteins. Hybrid models, including adversarial autoencoders, aim to combine the stable encodings of VAEs with the high-quality generation of GANs. Both VAEs and GANs are based on a transition from predictive to creative AI in enzyme engineering, presenting novel solutions to design problems. As these models develop and are combined with high-quality predictors of structure and function, they are automated to assist in de novo enzyme design pipelines.

9.3.4 Applications of AI-Designed Enzymes

AI is transforming biotechnology by accelerating the development of enzyme-based drugs. The convergence of supervised learning, functional annotation, and generative modelling has led to the introduction of AI-designed enzymes in various industries. One notable achievement is the development of a computationally designed enzyme for treating Phenylketonuria (PKU), a genetic disorder in which patients are unable to break down specific amino acids. The enzyme, designed using protein structure prediction and physical energy calculations, has enhanced stability and reduced immunogenicity and is currently in clinical trials (Sarkissian et al. 2021). ML is being utilised in designing antibodies and enzymes for targeted cancer therapies and gene therapy. In the biofuel and biorenewable chemicals sector, engineered enzymes play a crucial role in breaking down plant biomass into fermentable sugars. Companies such as Arzeda and Ginkgo Bioworks use ML-driven platforms to design efficient cellulases and ligninases for industrial biorefineries.

AI is also being utilised to develop novel metabolic pathways and enzymes for producing biofuels and bio-based chemicals from renewable feedstocks, thereby promoting sustainable alternatives to petroleum-based processes (Choi et al. 2019). AI-designed enzymes play a crucial role in agriculture, enhancing digestibility, reducing phosphate pollution, and protecting crops. They can degrade pesticides or act as biopesticides. Machine learning (ML) is used to optimise gene editing systems, such as CRISPR-Cas, which offers higher specificity, efficiency, and novel targeting capabilities. This is due to the use of large datasets of gRNA efficiency and off-target effects, leading to safer and more precise tools in agriculture (Dang et al. 2023).

The food and beverage industry is utilising AI-designed enzymes to enhance processes and produce several products. These include proteases and peptidases for reducing milk allergens, transaminases for synthesising artificial sweeteners, and enzymes for enhancing flavour or modifying texture. Lactases engineered for refrigeration temperatures enable the production of lactose-free milk, reducing energy consumption and costs. Additionally, AI models are being used to engineer enzymes, such as PETases and laccases, for enhanced activity, stability, and improved production in industrial wastewater (Dang et al. 2023). This is used to develop "super-enzymes" that can break down plastic waste or toxic chemicals, offering a biocatalytic solution to pollution. The integration of automated synthesis is expected to accelerate enzyme engineering and biocatalyst production, paving the way for a sustainable, bio-based economy.

9.4 AI/ML in Protease Engineering

Random mutagenesis and high-throughput screening have been used for protease engineering, which hydrolyses peptide bonds. AI and ML processes are rational, predictive, and faster. AI/ML models identify complex, non-linear relationships using massive datasets of protease sequences, structures, and biochemical properties. They anticipate the effects of mutations, develop novel protease variations, and plan experiments in regions of the potential sequence-function landscape. This paradigm shift enables proteases to be precisely used for optimal catalytic efficiency, stability, and specificity, thereby facilitating new biological, industrial, and technological applications (Yang et al. 2019; Bedbrook et al. 2023a, b). Supervised learning techniques, such as Random Forests, gradient boosting machines, and deep neural networks, are trained on protease variant datasets using sequence or structural information, along with observed properties (Wu et al. 2019). Variety autoencoders (VAEs) are unsupervised learning methods that compress and represent protease sequences to capture functional protein features and produce novel, diverse sequences that differ from natural homologs (Alley et al. 2019). Reinforcement learning can teach protein engineering rules through computer trial and error (Angermueller et al. 2020). The advent of deep learning has dramatically accelerated progress in structural biology. As illustrated in Fig. 9.1, complex neural networks, such as convolutional and transformer models, can now accurately predict a protein's three-dimensional structure from its amino acid sequence and are increasingly being used to design novel protein folds and functions.

Protein language models (pLMs), such as ESM-2 and ProtTrans, have evolutionary-scale protein sequence representations in AI/ML models. These models integrate each amino acid in a sequence with a high-dimensional vector that specifies its structure and function, thereby enabling the identification of

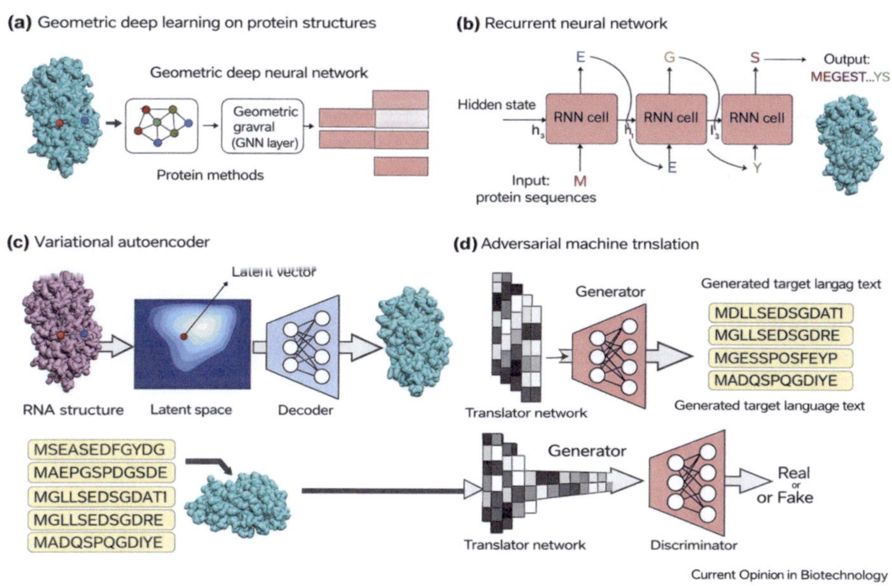

Fig. 9.1 Deep learning for protein structure prediction and design

statistical patterns of evolution and biophysical constraints (Rives et al. 2021; Hie et al. 2022a, 2022b). 3D convolutional neural networks (3D-CNNs) or graph neural networks (GNNs) can represent protein structure as a graph with nodes and edges and predict the effects of mutations on stability and binding using data-implied physical principles, such as hydrophobic packing or hydrogen bonding network energetic (Shi et al. 2022). Complex models, such as deep neural networks, are generally considered "black box" models; hence, explainable AI (XAI) approaches are being developed to uncover significant residues or qualities that influence predictions, thereby bridging the gap between forecasts (Luo et al. 2021). AI and ML in protease engineering enable the synthesis of enzymes for diverse applications by enhancing prediction accuracy and facilitating optimized design.

9.4.1 Protease Catalytic Efficiency

The key parameter of proteases, the specificity constant (kcat/KM), can be modified to enhance catalytic efficiency, facilitate faster reaction rates, reduce enzyme doses, and yield better diagnostic results. AI and ML widely use these models to describe complex molecular efficiency factors and predict changes that maximise substrate binding, transition state stability, and product release. Protein structural trade-offs and complex epistasis make designing for catalytic efficiency a challenging task. In complex fitness contexts, AI/ML models learn the sequence-activity relationship by evaluating the activity of thousands of single and multiple mutants using mutational scanning datasets. A machine learning algorithm trained on a high-throughput activity screen found mutation combinations that boosted target substrate activity in subtilisin protease research. ML can predict highly active variants with non-obvious synergistic alterations far from the active site, making it more efficient than blindly screening large libraries. AI can anticipate substrate specificity and develop proteases that work differently on diverse peptide sequences (Bastian et al. 2021).

Mutations that enhance surface complementarity and electrostatic interactions have created broad-spectrum antivirals from the hepatitis C virus NS3/4A protease, which efficiently cleaves a viral substrate from a different genotype (Oshima and Honig 2022). Molecular dynamics (MD) simulations and ML reveal catalytic efficiency factors at the atomic level. These models describe the catalytic cycle, but analysing terabyte-scale data is challenging. ML approaches, such as state-free reversible VAMP nets (SRVs), can identify the slowest collective variables and reveal significant conformational changes that affect catalytic rates (Wang and Chan 2022). AI helps engineers identify bottlenecks in the product release cycle and suggest improvements to reduce hydrophobic contact or enhance electrostatic repulsion. These models can predict the effects of point mutations and add protein domains or loops to create synthetic, multifunctional proteases. This could lead to hyper-efficient proteases for the digestion of protein-based waste or the ultra-specific processing of therapeutic proteins in biomanufacturing.

9.4.2 Protease Stability

Industrial and medicinal applications require thermodynamic, kinetic, and operational stability of proteases. AI/ML predicts mutations that enhance protease design against destabilising influences without altering catalytic activity, enhancing protease engineering. Stability designs using disulfide bridges, salt bridges, and prolines can limit conformational flexibility and activity. AI/ML models can learn stability determinants from protease family or protein-specific data. Supervised learning models can be trained using thermal shift assay or differential scanning calorimetry datasets, employing variant sequences as input and observed unfolding temperature as output. A Gaussian process regression model trained on hundreds of mutations predicted new variant thermostability in a fungal protease study (Benedix et al. 2020). Data-driven methods infer evolutionary constraints from unsupervised machine learning to discover non-intuitive stabilising processes. The evolutionary model of sequence fitness predicts the stability and function-critical residues, as well as their placements. Deep learning methods, such as Potts models or variational autoencoders, may learn complex co-evolutionary patterns. Sequences with higher fitness scores are more "native-like" and stable (Figliuzzi et al. 2018).

Phylogenetic analysis and consensus design use evolutionary optimisation, whereas in silico screening of millions of mutations enhances model fitness. Autoproteolysis, a key concern for protease stability during recombinant expression, is predicted and mitigated using machine learning AI models that anticipate cleavage locations for flexible linker portions, thereby inactivating proteases. Engineers can rationally modify vulnerable peptide bonds or stabilise exposed loops to decrease cleavage using substrate profiling data (Varadi et al. 2022). AI-driven stability engineering develops proteases for detergents, biocatalysis, and therapeutics. Stability assay records can teach ML models a 40% dimethylformamide stability sequence. They can directly build "extremophilic" proteases without a template from extremophilic species. GANs can generate stable, functional protease sequences under user-defined constraints, thereby improving synthetic enzyme design (Repecka et al. 2021). AI provides a comprehensive computational framework allowing researchers to build a reliable protease for real-world use.

9.4.3 Characterisation of Evolved Proteases

9.4.3.1 Substrate Profiling

Substrate profiling is the systematic determination of a protease's specificity, which is crucial for its evolution and adaptation. Modern, machine-learning-powered substrate profiling involves a comprehensive, proteome-wide analysis, moving beyond simple peptide libraries. Phage display and synthetic peptide libraries are used for specificity profiling, as they expose a vast library of peptides to the protease and identify cleaved sequences through sequencing. The machine learning models, such as position-weight matrices (PWMs), support vector machines (SVMs), or deep neural networks. These models learn the protease's recognition motif and output a "cleavage logo" visually representing the preference for each amino acid at each position (Bastian et al. 2021). Profiling and comparing the evolved variant to the wild-type protease provides a quantitative assessment of how mutations altered recognition (Götz et al. 2022). Proteomic analysis techniques, such as proteomic identification of cleavage specificity (PICS), use proteome-derived peptide libraries or native proteomes for analysis. PICS involves incubating a proteome digest with a protease and using mass spectrometry to identify the generated neo-N-termini. AI models can develop highly accurate specificity models and predict the protease's degradome, which is crucial for assessing potential off-target effects of therapeutic proteases or validating the precision of engineered proteases for diagnostic applications (Schilling and Overall 2022).

9.4.3.2 Non-specific Cleavage Activity

Industrial and medicinal applications require thermodynamic, kinetic, and operational stability of proteases. AI/ML predicts mutations that enhance protease design against destabilising influences without compromising catalytic activity, thereby improving protease engineering. Stability designs using disulfide bridges, salt bridges, and prolines can limit their flexibility and activity. AI/ML models can learn stability from the protease family or protein-specific data. Supervised learning models can be trained using thermal shift assay or differential scanning calorimetry datasets, employing variant sequences as input and observed unfolding temperature or $\Delta\Delta G$ as output. A Gaussian process regression model trained on hundreds of mutations predicted new variant thermostability in a fungal protease study (Benedix et al. 2020). Data-driven methods infer evolutionary constraints from unsupervised machine learning to discover non-intuitive stabilising processes.

The evolutionary model of sequence fitness predicts the stability and function-critical residues, as well as their placements. Deep learning methods, such as Potts models or variational autoencoders, may learn complex co-evolutionary patterns. Sequences with higher fitness are more "native-like" and stable (Figliuzzi et al. 2018). Phylogenetic analysis and consensus design use evolutionary optimisation, whereas in silico screening of millions of mutations enhances model fitness. Autoproteolysis is a key concern for protease stability during recombinant expression, which is predicted and mitigated using machine learning AI models that anticipate cleavage locations for flexible linker portions that inactivate proteases. Engineers can rationally modify vulnerable peptide bonds or stabilise them to decrease cleavage using substrate profiling data (Varadi et al. 2022). AI-driven stability engineering develops proteases for detergents, biocatalysis, and therapeutics. Stability assay records can teach ML models a 40% dimethylformamide stability sequence. GANs can generate stable, functional protease sequences under user-defined constraints, thereby improving synthetic enzyme design (Repecka et al.

2021). AI provides a computational framework for a reliable protease for real-world use.

9.4.4 *Emerging Applications of Engineered Proteases*

AI/ML-enabled precision engineering of proteases is transforming them from simple degradative enzymes into sophisticated tools for synthetic biology and next-generation analytics, expanding their applications beyond traditional detergents, food processing, and leather treatment, and transforming them into programmable molecular devices.

9.4.4.1 Proteases in Gene Circuits

Synthetic biology utilises genetic circuits to reprogram cells for new functions. Slow, restricted, and cross-talking transcription factors are traditional. Engineered proteases enable rapid and robust post-translational regulation of gene circuits. The key idea is to create a protease that cleaves explicitly a circuit protein node, such as a transcriptional activator or repressor. External cues can regulate protease activity, resulting in complex, dynamic, and orthogonal control systems. AI-driven specificity engineering is necessary because a protease must detect a unique, non-native degron tag attached to the target protein and avoid interacting with host proteins. Proteases with restricted specificity are designed using machine learning models based on vast degradome profiling data. ML may improve the specificity of a naturally orthogonal protease (Grimm et al. 2023). Engineering allosteric control sites can induce protease activity using machine learning algorithms to predict the impacts of mutations on the allostery process. This permits fine circuit temporal control. Complex synthetic programs wired into cascades that delay or pulse gene expression can be created using protease-based regulators. The quick degradation of repressor proteins can cause quick and intense gene expression cycles, making them useful for synthetic feedback loops or oscillators (Brockman and Deans 2021). A protease-based "kill switch" could be constructed for therapeutic cells, such as CAR-T cells, to activate a small molecule medication that cleaves and inactivates a key survival protein, thereby prompting programmed cell death if the therapy becomes harmful. The next generation of complicated genomic circuits in therapeutic and bioproduction needs AI-optimised proteases.

9.4.4.2 Proteases for Next-Gen Protein Sequencing

NGS technology has revolutionised DNA sequencing, but protein sequencing remains a challenge. This developing area will rely on engineered proteases that sequentially release amino acids from a single peptide chain using a highly processive exopeptidase. Fluorescent labelling or nanopore detection must identify each released amino acid immediately. A protease with such extreme engineering requirements cannot be built without the aid of AI/ML. To sequence whole proteins, the enzyme must be extremely processive, homogenous, able to handle all 20 amino acids without halting, and stable for long periods. Designing this "ideal" enzyme from scratch requires AI. Supervised learning algorithms trained on single-molecule enzymology data can predict and select candidates based on their expected kinetic parameters. In contrast, generative models can generate thousands of candidate sequences for a processive exopeptidase fold (Alley et al. 2019).

Machine learning–based molecular dynamics simulations can examine the ability of enzymes to tolerate amino acid side chains without compromising their function. Bottom-up single-molecule protein sequencing, such as shotgun DNA sequencing, requires engineered proteases. These methods separate and sequence peptides by cleaving a protein molecule at specified residues. Computationally reassembling the sequence involves a suite of proteases with predicted specificities. AI-driven design creates ultra-specific proteases that function well on surface-immobilised or nanopore-threaded proteins. These sequencing systems enable the accurate detection of post-translational modifications using advanced machine learning methods (Swaminathan et al. 2018). AI for enzyme engineering and data analysis will transform next-generation protein sequencing, enabling researchers to read a single cell's protein complement easily. This will allow applications in proteomics, diagnostics, and medicine.

9.5 Current Limitations of ML-Aided Protease Engineering

Protease engineering has grown predictive and knowledge-driven due to ML, substituting directed evolution. This computational method reveals complex sequence-structure-function interactions that impact protease activity, specificity, and stability. The field faces challenges with data availability, quality, sequence-conformation modelling gaps, protease fitness landscape complexity, interpretability, and integration gaps between in silico predictions and wet-lab validation (Yang et al. 2019). Training ML models without high-quality, well-curated, and standardised data are complex. ImageNet is used for computer vision and natural language processing, whereas pricey lab experiments develop functional protease data. Fragmented data, multiple experimental settings, and diverse substrates make aggregation and meta-analysis problematic (Wittmann et al. 2021a, 2021b). A model trained on a few hundred or thousands of mutant sequences memorises the noise and quirks of the restricted training set, resulting in poor generalisation on novel sequences (Bedbrook et al. 2023a, b). These protease models can hinder their ability to predict unbalanced input. They utilise transfer learning to fine-tune models from large protein sequence databases; however, this approach may misrepresent active site and dynamic constraints. AlphaFold2 and RoseTTA Fold's ab initio predictions for single-chain proteins highlight the challenges in predicting protease structure and function (Jumper et al. 2021). The Protein Data Bank uses static ground-state structures for structure prediction models, ignoring critical dynamics, conformational ensembles, and free energy landscapes. Accurate models for protease design, including those with cofactors and binding inhibitors, are crucial for predicting the effects of mutations (Langan et al. 2022).

Molecular dynamics (MD) simulations offer temporal precision but are computationally expensive and limited in testing choices compared to ML. Physics-based simulations offer deep mechanistic insight but are computationally intensive for large-scale screenings. ML models offer high-throughput sequence-to-function mapping. Protease fitness is challenging due to its complexity and intricate interactions (Weinreich et al. 2006). The limited training areas of machine learning models make it difficult to extend multi-mutant variant design to unknown sequence spaces or predict the effects of mutation combinations. Advanced models and intense neural networks hinder scientific discovery, rendering highly predictive models for their sequence properties or interactions (Ribeiro et al. 2016). Engineers often make costly mistakes, but models help scientists analyse issues and learn mechanical principles for future design cycles through various mechanisms (Lundberg and Lee 2017). High-fitness variations are challenging to achieve due to the integration gap between machine learning and traditional engineering methods. The absence of automated mechanisms for evaluating ML predictions and communicating model advances delays iterative learning and hinders discoveries until robotic high-throughput testing becomes more accessible (Shivange et al. 2022). The study reveals that machine learning-aided protease engineering has potential but faces challenges such as data availability, protein folding and dynamics physics, biological fitness landscape complexity, model interpretability, and experimental validation throughput. The integration of artificial intelligence (AI) has revolutionised the field of enzyme engineering. As summarised in Table 9.1, machine learning (ML) models are now being deployed to predict protease function from sequence, guide directed evolution campaigns, and optimise production conditions, significantly accelerating the design of proteases with enhanced stability, activity, and specificity.

9.6 Future Prospects in AI-Driven Enzyme Design

Machine learning–aided protease engineering faces growth limitations, but AI-designed enzymes could revolutionise predictive biology. Deep generative models, exponential growth in biological data, AI architecture, automated robotic laboratories, and physics-based modelling will enhance precision in enzyme engineering (Lopez & Mukhopadhyay, 2023). Protein science utilises generative AI, specifically protein language models and generative adversarial networks, to create novel, thermodynamically stable, and foldable protein sequences that respect biological constraints and are unrelated to real proteins, rather than predictive models (Madani et al. 2023; Faure and Kaczmarski 2023). Cloud-based robotic laboratories

Table 9.1 Artificial intelligence and machine learning in protease engineering and optimisation

S. no	Applications	Methods	Advantages	Industrial relevance
1	Protein sequence analysis	AI-based sequence classifiers, deep learning models	Predicts protease activity and function directly from amino acid sequences	Enables rapid classification of novel proteases for pharmaceutical and industrial use
2	Structure prediction	Alpha fold 2, molecular modelling tools	Accurate 3D structures help visualise active sites	Improves the rational design of proteases for drug discovery and the food industry
3	Mutant design and optimisation	Machine learning mutation prediction, directed evolution with AI guidance	Suggests beneficial mutations for stability, specificity, and tolerance	Enhances the performance of proteases in detergents, textiles, and therapeutics
4	Enzyme–substrate interaction prediction	Docking simulations, ML-based affinity models	Identifies substrates, inhibitors, and binding affinities	Crucial for developing targeted drugs and efficient biocatalysts
5	Stability and activity prediction	ML regression models, thermodynamic simulations	Predicts performance under pH, temperature, and solvents	Facilitates the use of proteases in harsh industrial environments
6	High-throughput screening	AI-driven virtual screening, automation tools	Accelerates screening of large protease libraries	Reduces cost and time in enzyme discovery for biotech
7	Industrial applications	AI/ML-integrated enzyme optimisation platforms	Designs task-specific proteases with improved efficiency	Applied in detergents, pharmaceuticals, food processing, and waste management

enable large-scale experiments, continuously cycling DBTL through a self-learning and optimisation loop, where AI models propose new versions, robots test them, and findings are used to improve models (Stein et al. 2022).

Closed-loop automation will enhance the visualisation of complex protease higher-order epistatic effects. Data-driven AI models will be developed using sequences, structures, molecular dynamics, spectroscopy data, and quantum mechanical reaction mechanisms (Bennett and Coventry 2024). They also predict protein conformational ensemble and catalytic kinetics, predicting loop flexibility, collective motions for substrate binding and product release, and catalytic efficiency-affecting electric field and pKa fluctuations, based on coarse-grained and quantum mechanical/molecular mechanical simulations (Noé and Husic 2022). Next-generation AI focuses on multi-objective optimisation and interpretability, aiming for active, thermostable, high-yield, and non-toxic proteases (Lundberg et al. 2020). AI-driven enzyme design improves proteomics and their catalytic proteases for sustainable chemistry, medicine, and biotechnology. The practical implementation of artificial intelligence in protease development follows a structured, iterative cycle. The Fig. 9.2 outlines the initial protease data, machine learning models, predictive

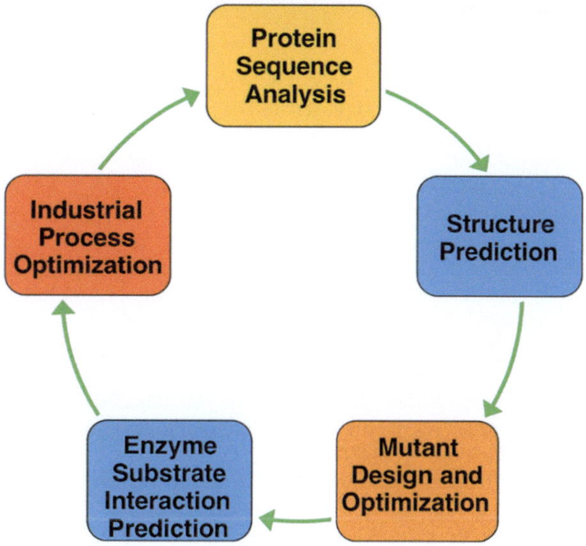

Fig. 9.2 Application of AL/ML in protease engineering

designs that are experimentally validated, with the results in a continuous loop of improvement.

9.7 Conclusion

Artificial intelligence (AI) and machine learning (ML) are revolutionising protease engineering, transforming the field from traditional labour-intensive methods to a data-driven, predictive, and efficient discipline. These technologies, including supervised learning, generative models, and advanced computational tools, have enabled the optimisation of proteases for enhanced catalytic efficiency, stability, specificity, and novel functionalities. They leverage large-scale datasets from evolutionary, structural, and biochemical sources, enabling accurate mutation effects prediction, de novo design of enzyme variants, and intelligent navigation of sequence spaces. AI-driven approaches have accelerated the engineering process and reduced reliance on costly experimental screening, facilitating the development of proteases for various industrial, therapeutic, and diagnostic applications. The emergence of AI and integrative models of protease mechanics and epistatic interactions enhances design precision. Despite challenges such as data scarcity and model interpretability, the continued advancement of AI and ML with automated robotic laboratories and closed-loop Design-Build-Test-Learn cycles holds immense potential for overcoming these limitations. The synergy between AI and protease engineering is poised to drive innovations in sustainable chemistry, precision medicine, synthetic biology, and next-generation protein sequencing, contributing to a bio-based economy and biocatalysis.

References

Alley EC, Khimulya G, Biswas S, AlQuraishi M, Church GM (2019) Unified rational protein engineering with sequence-based deep representation learning. Nat Methods 16(12): 1315–1322. https://doi.org/10.1038/s41592-019-0598-1

Angermueller C, Dohan D, Belanger D, Deshpande R, Murphy K, Colwell L (2020) Model-based reinforcement learning for biological sequence design. Intl Conf Learn Represent (ICLR):1–12

Arnold FH (2018) Directed evolution: bringing new chemistry to life. Angew Chem Int Ed 57(16): 4143–4148. https://doi.org/10.1002/anie.201708408

Arnold FH, Giver L, Wells JA (2018) Directed evolution of a subtilisin that functions in dimethylformamide. ACS Synth Biol 7(4):1120–1125. https://doi.org/10.1021/acssynbio.8b00023

Ashkenazy H, Abadi S, Martz E, Chay O, Mayrose I, Pupko T, Ben-Tal N (2016) ConSurf 2016: an improved methodology to estimate and visualise evolutionary conservation in macromolecules. Nucleic Acids Res 44(W1):W344–W350. https://doi.org/10.1093/nar/gkw408

Baker M (2015) Reproducibility crisis: blame it on the antibodies. Nature 521(7552):274–276. https://doi.org/10.1038/521274a

Bastian F, Schaefer H, Roth Z (2021) Machine learning for protease specificity prediction. J Chem Inform Model 61(5):2177–2186. https://doi.org/10.1021/acs.jcim.1c00169

Bedbrook CN, Yang KK, Robinson JE, Mackey ED, Lee KK, Arnold FH (2023a) Machine learning-guided channelrhodopsin engineering enables minimally invasive optogenetics. Nat Methods 20(1):123–130. https://doi.org/10.1038/s41592-022-01688-9

Bedbrook CN, Yang KK, Robinson JE, Mackey ED, Lee KK, Arnold FH (2023b) Machine learning-guided protein engineering. ACS Synth Biol 12(1):16–34. https://doi.org/10.1021/acssynbio.2c00455

Benedix A, Becker CM, de Groot BL, Caflisch A (2020) Predicting the effect of mutations on protein stability using machine learning. J Chem Theory Comput 16(4):2724–2732. https://doi.org/10.1021/acs.jctc.9b01250

Bennett NR, Coventry B (2024) Multi-modal foundation models for protein engineering: challenges and opportunities. Curr Opin Struct Biol 84:102764. https://doi.org/10.1016/j.sbi.2023.102764

Bloom JD, Arnold FH (2009) In the light of directed evolution: pathways of adaptive protein evolution. Proc Natl Acad Sci USA 106(Suppl 1):9995–10000. https://doi.org/10.1073/pnas.0901522106

Bolognesi M, Lebioda L (2018) Rational protein design. In: Lutz S, Bornscheuer UT (eds) Protein engineering. Springer, Cham, pp 1–21

Bornscheuer UT, Huisman GW, Kazlauskas RJ, Lutz S, Moore JC, Robins K (2019) Engineering the third wave of biocatalysis. Nature 485(7397):185–194. https://doi.org/10.1038/nature11117

Brandes N, Ofer D, Peleg Y, Rappoport N, Linial M (2022) ProteinBERT: a universal deep-learning model of protein sequence and function. Bioinformatics 38(8):2102–2110. https://doi.org/10.1093/bioinformatics/btac020

Brännström K, österlund N, Gharibyan AL, Olofsson A, Lundgren J (2018) The role of protein engineering in biomedical and industrial biotechnology. Q Rev Biophys 51:e10. https://doi.org/10.1017/S0033583518000056

Brockman IM, Deans TL (2021) Engineering transcriptional regulation for cell-based therapies. Curr Opin Biotechnol 70:88–96. https://doi.org/10.1016/j.copbio.2020.12.010

Cadwell RC, Joyce GF (1992) Randomisation of genes by PCR mutagenesis. PCR Methods Appl 2(1):28–33. https://doi.org/10.1101/gr.2.1.28

Chen K, Arnold FH (2020) Engineering new catalytic activities in enzymes. Nature Catalysis 3(3): 203–213. https://doi.org/10.1038/s41929-019-0385-5

Choi KR, Shin JH, Cho JS, Yang D, Lee SY (2019) Systems metabolic engineering of microorganisms for food and cosmetics production. Nat Rev Microbiol 17(11):671–689. https://doi.org/10.1038/s41579-019-0256-9

Costello Z, Martin HG (2023) A machine learning approach to predict metabolic pathway dynamics from time-series multiomics data. NPJ Syst Biol Appl 9(1):1–12. https://doi.org/10.1038/s41540-023-00272-4

Crawshaw R, Crossley AE, Johannissen L, Burke AJ, Hay S, Scrutton NS (2022) Engineering an efficient and enantioselective enzyme for the Morita–Baylis–Hillman reaction. Nat Chem 14(3):313–320. https://doi.org/10.1038/s41557-021-00833-9

Dang L, Jiang T, Zhang Y (2023) Machine learning for CRISPR guide RNA design and optimisation. ACS Synth Biol 12(1):1–17. https://doi.org/10.1021/acssynbio.2c00439

Dill KA, MacCallum JL (2012) The protein-folding problem, 50 years on. Science 338(6110):1042–1046. https://doi.org/10.1126/science.1219021

Faure AJ, Kaczmarski JA (2023) Deep mutational scanning and machine learning reveal structural constraints and evolutionary landscapes of proteins. Annu Rev Biophys 52:123–145. https://doi.org/10.1146/annurev-biophys-092722-085852

Figliuzzi M, Jacquier H, Schug A, Tenaillon O, Weigt M (2018) Coevolutionary landscape inference and the context-dependence of mutations in codon usage bias. Mol Biol Evol 35(2):433–446. https://doi.org/10.1093/molbev/msx298

Götz M, Wortmann L, Schmitt A, Herten DP (2022) Engineering the specificity of TEV protease using machine learning. ACS Synth Biol 11(2):816–827. https://doi.org/10.1021/acssynbio.1c00577

Grey VE, Hause RJ, Luebeck J, Shendure J, Fowler DM (2018) Quantitative missense variant effect prediction using large-scale mutagenesis data. Cell Syst 6(1):116–124. https://doi.org/10.1016/j.cels.2017.11.003

Grimm AR, Barth ZK, Silva DA (2023) De novo design of protein specificity by a combination of deep learning and energy-based optimisation. Protein Sci 32(2):e4562. https://doi.org/10.1002/pro.4562

Gupta A, Zou J (2019) Feedback GAN for DNA optimises protein functions. Nat Mach Intellig 1(2):105–111. https://doi.org/10.1038/s42256-019-0017-4

Hie BL, Shanker VR, Xu D, Bruun TUJ, Weidenbacher PA, Tang S (2022a) Efficient evolution of human antibodies from general protein language models. Nat Biotechnol 40(11):1710–1717. https://doi.org/10.1038/s41587-022-01231-3

Hie BL, Yang KK, Kim PS (2022b) Evolutionary velocity with protein language models predicts evolutionary dynamics of diverse proteins. Cell Syst 13(4):274–285. https://doi.org/10.1016/j.cels.2021.09.003

Jumper J, Evans R, Pritzel A, Green T, Figurnov M, Ronneberger O et al (2021) Highly accurate protein structure prediction with AlphaFold2. Nature 596(7873):583–589. https://doi.org/10.1038/s41586-021-03819-2

Langan RA, Boyken SE, Ng AH, Samson JA, Dods G, Westbrook AM et al (2022) De novo design of protein structure and function with RFdiffusion. Science 379(6635):1053–1062. https://doi.org/10.1126/science.adf2061

López CE, Mukhopadhyay A (2023) The self-driving laboratory: the convergence of robotics, artificial intelligence, and data science for accelerated discovery. Trends in Chemistry 5(4):259–270. https://doi.org/10.1016/j.trechm.2022.12.003

Lundberg SM, Lee SI (2017) A unified approach to interpreting model predictions. Adv Neural Inf Proces Syst 30:1–10

Lundberg SM, Erion G, Chen H, DeGrave A, Prutkin JM, Nair B et al (2020) From local explanations to global understanding with explainable AI for trees. Nat Mach Intell 2(1):56–67. https://doi.org/10.1038/s42256-019-0138-9

Luo Y, Jiang G, Yu T, Liu Y, Vo L, Ding Y, Tang H (2021) ECNet is an evolutionary context-integrated deep learning framework for protein engineering. Nat Commun 12(1):5743. https://doi.org/10.1038/s41467-021-25976-8

Lutz S, Bornscheuer UT (2009) Protein Engineering Handbook, *Volume 1*. Wiley-VCH, Weinheim

Madani A, Krause B, Greene ER, Subramanian S, Mohr BP, Holton JM et al (2023) Large language models generate functional protein sequences across diverse families. Nat Biotechnol 41(8): 1099–1106. https://doi.org/10.1038/s41587-022-01618-2

Nelson DL, Cox MM (2021) Lehninger principles of biochemistry, 8th edn. W.H. Freeman, New York

Noé F, Husic BE (2022) Machine learning for molecular dynamics simulations: foundations and best practices. arXiv preprint arXiv:2202.07069:1–32

Oshima H, Honig B (2022) Molecular dynamics simulations and machine learning techniques for the prediction of protease substrate specificity. Curr Opin Struct Biol 73:102338. https://doi.org/10.1016/j.sbi.2022.102338

Packer MS, Liu DR (2015) Methods for the directed evolution of proteins. Nat Rev Genet 16(7): 379–394. https://doi.org/10.1038/nrg3927

Rao R, Liu J, Verkuil R, Meier J, Canny JF, Abbeel P, Sercu T (2021) MSA transformer. In: Proceedings of the 38th international conference on machine learning (ICML), vol 139, pp 8844–8856

Reetz MT, Carballeira JD (2007) Iterative saturation mutagenesis (ISM) for rapid directed evolution of functional enzymes. Nat Protoc 2(4):891–903. https://doi.org/10.1038/nprot.2007.72

Repecka D, Jauniskis V, Karpus L, Rembeza E, Zokrataite I, Zrimec J (2021) Expanding the functional protein space with generative adversarial networks. Nature Mach Intellig 3(4): 324–333. https://doi.org/10.1038/s42256-021-00310-5

Ribeiro MT, Singh S, Guestrin C (2016) Explaining the predictions of any classifier. In: Proceedings of the 22nd ACM SIGKDD international conference on knowledge discovery and data mining, pp 1135–1144

Rives A, Meier J, Sercu T, Goyal S, Lin Z, Liu J et al (2021) Biological structure and function emerge from scaling unsupervised learning to 250 million protein sequences. Proc Natl Acad Sci USA 118(15):e2016239118. https://doi.org/10.1073/pnas.2016239118

Romero PA, Krause A, Arnold FH (2013) Navigating the protein fitness landscape with Gaussian processes. Proc Natl Acad Sci USA 110(3):E193–E201. https://doi.org/10.1073/pnas.1215251110

Ryu JY, Kim HU, Lee SY (2019) Deep learning enables high-quality and high-throughput prediction of enzyme commission numbers. Proc Natl Acad Sci USA 116(28):13996–14001. https://doi.org/10.1073/pnas.1821905116

Sarkissian CN, Gamez A, Scriver CR (2021) What we know that would make sense to design into a therapeutic enzyme for PKU: a perspective from the bench. Mol Genet Metab 133(1):1–7. https://doi.org/10.1016/j.ymgme.2021.02.002

Schilling O, Overall CM (2022) Proteome-derived, database-searchable peptide libraries for identifying protease cleavage sites. Nat Biotechnol 40(1):152–161. https://doi.org/10.1038/s41587-021-01010-6

Shi Y, Li Z, Zhang Y (2022) Active learning for protein engineering. Curr Opin Struct Biol 72:242–249. https://doi.org/10.1016/j.sbi.2021.11.013

Shivange AV, Marienhagen J, Mundhada H, Schenk A, Schwaneberg U (2019) Advances in generating functional diversity for directed protein evolution. Curr Opin Chem Biol 53:90–96. https://doi.org/10.1016/j.cbpa.2019.08.004

Shivange AV, Marienhagen J, Mundhada H, Schenk A, Schwaneberg U (2022) Advances in ultrahigh-throughput screening technologies for directed enzyme evolution. Curr Opin Struct Biol 73:102338. https://doi.org/10.1016/j.sbi.2022.102338

Sinai S, Jain N, Church GM, Kelsic ED (2020) Generative AAV designs enabled by deep learning. bioRxiv 2020.07.21.214163:1–15

Stein HS, Gregoire JM, Kusne AG (2022) Progress and prospects for accelerating materials science with automated and autonomous workflows. Chem Rev 122(11):10015–10051. https://doi.org/10.1021/acs.chemrev.1c00720

Swaminathan J, Boulgakov AA, Marcotte EM (2018) A theoretical justification for single-molecule peptide sequencing. PLoS Comput Biol 14(10):e1006520. https://doi.org/10.1371/journal.pcbi.1006520

Tucker JB, Zilinskas RA (2006) The promise and perils of synthetic biology. The New Atlantis 12:25–45

Turner NJ (2009) Directed evolution drives the next generation of biocatalysts. Nat Chem Biol 5(8):567–573. https://doi.org/10.1038/nchembio.203

Varadi M, Anyango S, Deshpande M, Nair S, Natassia C, Yordanova G et al (2022) AlphaFold protein structure database: massively expanding the structural coverage of protein-sequence space with high-accuracy models. Nucleic Acids Res 50(D1):D439–D444. https://doi.org/10.1093/nar/gkab1061

Vig J, Madani A, Varshney LR, Xiong C, Socher R, Rajani NF (2021) BERTology meets biology: interpreting attention in protein language models. In: Proceedings of the 9th international conference on learning representations (ICLR), pp 1–15

Virnekäs B, Ge L, Plöckthun A, Schneider KC, Wellnhofer G, Moroney SE (1994) Trinucleotide phosphoramidites: ideal reagents for the synthesis of mixed oligonucleotides for random mutagenesis. Nucleic Acids Res 22(25):5600–5607. https://doi.org/10.1093/nar/22.25.5600

Wang J, Chan C (2022) Discovering reaction pathways and slow collective variables from molecular dynamics simulations with state-free reversible VAMP nets. J Chem Phys 156(3):034104. https://doi.org/10.1063/5.0078149

Weinreich DM, Delaney NF, Depristo MA, Hartl DL (2006) Darwinian evolution in mutational paths to fitter proteins. Science 312(5770):111–114. https://doi.org/10.1126/science.1123539

Wijma HJ, Floor RJ, Janssen DB (2014) Structure- and sequence-analysis inspired engineering of proteins for enhanced thermostability. Curr Opin Struct Biol 23:588–594. https://doi.org/10.1016/j.sbi.2013.06.015

Wittmann BJ, Johnston KE, Wu Z, Arnold FH (2021a) Advances in machine learning for directed evolution. Curr Opin Struct Biol 69:11–18. https://doi.org/10.1016/j.sbi.2021.01.008

Wittmann BJ, Yue Y, Arnold FH (2021b) Informed training set design enables efficient machine learning-assisted directed protein evolution. Cell Syst 12(11):1026–1045. https://doi.org/10.1016/j.cels.2021.07.008

Wu Z, Kan SBJ, Lewis RD, Wittmann BJ, Arnold FH (2019) Machine learning-assisted directed protein evolution for improved catalytic efficiency. Nature Catalysis 2(2):95–103. https://doi.org/10.1038/s41929-018-0204-4

Yang KK, Wu Z, Arnold FH (2019) Machine-learning-guided directed evolution for protein engineering. Nat Methods 16(8):687–694. https://doi.org/10.1038/s41592-019-0496-6

Zimmerman JF, Parameswaran R, Murray G (2016) Microfluidic-based directed evolution of biocatalysts. Curr Opin Biotechnol 42:159–164. https://doi.org/10.1016/j.copbio.2016.05.005

Part II
Emerging Trends in Microbial Protease Applications and Future Prospects

Chapter 10
Exploring the Multifaceted Applications of Microbial Proteases in Dermatology and Cosmetics

M. Swathi, Sai Sankeerthana Bedadakota, Richa Sikligar, B. U. Ananya, Pranav Rao, Andhere Animish, and M. A. Jayasri

Abstract The fields of dermatology and cosmetics seek microbial proteases as alternative treatments because these enzymes provide flexible and sustainable solutions for skin and hair care. Microorganisms, including bacteria, fungi, and yeasts, produce proteolytic enzymes through which they split proteins, which drive fundamental biological processes, including tissue reshaping, protein recycling, and cell regulation. The biochemical features of these agents, which include substrate specificity alongside thermostability, together with adaptability to multiple biotechnological methods, make them essential for fulfilling diverse requirements in cosmetic and therapeutic applications. The application of enzyme exfoliation through microbial proteases leads to dead cell elimination, which produces a better skin look and starts the regenerative process. The skin healing process, along with scar reduction and anti-ageing treatments, benefits from enzymes that include collagenases and keratinases since these enzymes boost collagen production and enhance skin elasticity. Proteases control melanin formation, which helps address skin problems, including hyperpigmentation and skin tone unevenness, and keratinases restore damaged hair structure while strengthening it. These proteases demonstrate medical potential for treating dermatological conditions, including acne, psoriasis, and xerosis, through specific protein targeting of disease-related proteins. Nanotechnology, along with enzyme engineering and delivery technologies, has enhanced protease stability and specificity and overall functionality, which allows their wider use. However, certain challenges persist. To manage enzyme instability and skin irritation potential, together with allergenic reactions, thorough evaluation and management are required. *Staphylococcus aureus* and *Streptococcus pyogenes*–derived proteases from pathogens increase skin inflammation, thus requiring controlled use with strict safety assessments. Despite challenges, microbial proteases

M. Swathi · S. S. Bedadakota · R. Sikligar · B. U. Ananya · P. Rao · A. Animish · M. A. Jayasri (✉)
Marine Biotechnology and Bioproducts Laboratory, School of Biosciences and Technology, Vellore Institute of Technology, Vellore, Tamil Nadu, India
e-mail: jayasri.ma@vit.ac.in

V. Zambare et al. (eds.), *Biotechnological Advances and Applications in Microbial Proteases*, Interdisciplinary Biotechnological Advances,
https://doi.org/10.1007/978-981-95-9011-7_10

represent a transformative advancement that brings new possibilities for skincare and cosmetics. The chapter examines their natural and sustainable features and current research and technological development to create innovative validated solutions for contemporary dermatological and cosmetic problems.

Keywords Microbial proteases · Cosmetics · Dermatology · Skin care applications

10.1 Introduction

Microbial proteases are protease enzymes secreted by microbes, such as bacteria and fungi. They are increasingly valued in dermatology and cosmetic sciences for their functions in therapeutic and aesthetic applications (Mahnashi et al. 2022). Traditionally used in medicine for debridement and tissue remodelling, these enzymes are now being studied for applications in skin renewal, anti-ageing treatments, scar reduction, and scalp care (He et al. 2024). In dermatology, microbial proteases such as collagenase aid in healing wounds by breaking down the necrotic tissue, reducing inflammation, and promoting tissue regeneration. They are valuable assets in managing chronic wounds and minimizing scarring, due to their ability to regulate extracellular matrix components and modulate immune responses (Amadeh et al. 2025). In cosmetics, enzymes such as keratinase and elastase aid in exfoliating dead skin cells, remodelling collagen and elastin fibre, to help enhance skin permeability, which will further enhance delivery and effectiveness of active ingredients. These activities lead to smoother skin texture, reduced wrinkles, and improved skin firmness (Trevi sol et al. 2022). Proteases also improve hair and scalp care by eliminating the protein build-up, increasing shine, and restoring scalp health and balance in cases such as dandruff (Takami et al. 1992). They also align with green cosmetic trends by being eco-friendly, produced sustainably, and biodegradable (Singh and Bajaj 2017). Overall, microbial proteases offer non-invasive, non-toxic, and sustainable solutions in both the fields of dermatology and cosmetics, displaying increased potential in next-generation skincare and other related therapeutic products.

10.2 Microbial Proteases in Skin Exfoliation and Renewal

Exfoliating skin can help get rid of dry, dead corneocytes on the stratum corneum, which contributes to cell turnover, a quality texture, and better absorption of products (Lee et al. 2023). Widely used procedures such as scrubs and acid peels are mostly effective but can trigger irritation and skin barrier disruption. Selective destruction of proteins that augment cell adhesion can be achieved by microbial proteases, in particular subtilisin and keratinases, which are easier to apply and cause gentler desquamation without damaging neighbouring cells (Li 2021; Gupta et al. 2019a, b).

Synthesized by microorganisms, including *Bacillus*, *Chryseobacterium*, and *Stenotrophomonas*, these enzymes break the peptide bond and disulfide bonds in keratin, which weakens the structure of corneocytes (Li 2021). Such exfoliation has been effectively achieved with keratinases produced by *Chryseobacterium camellia* Dolsongi-HT1 in model skins at 30–60 °C and pH 8 with no toxic effects to lower layers (Kim et al. 2025). Subtilisin, a proteolytic enzyme, produced by *Bacillus subtilis* is also the enzyme used in enzyme peels due to its non-irritating effects, which are beneficial to sensitive skin (Lee et al. 2023).

When compared to plant-derived proteases such as papain and bromelain, microbial proteases have improved stability and may be genetically engineered using recombinant technology (Măgeruşan et al. 2023). They resemble natural desquamation and stimulate keratinocyte turnover and skin beauty (Lee et al. 2015). They are also a green alternative to rough exfoliants since they are biodegradable and occur via environmentally friendly processes (Măgeruşan et al. 2023; Gupta et al. 2013). Table 10.1 displays the comparative properties of microbial proteases used in exfoliation.

10.3 Functions of Microbial Proteases in Skin Cell Turnover

The elimination of dead corneocytes is important to smooth textural properties and protective effects. The exfoliants, physical and chemical, tend to provoke side effects, such as irritation or barrier disruption (Trevisol et al. 2022; Măgeruşan et al. 2023). Yet, microbial proteases provide a less stressful, well-controlled alternative because they mimic natural desquamation pathways (Li 2021).

Bacillus, *Chryseobacterium*, *Stenotrophomonas*, and fungi-produced proteases, such as keratinases and subtilisin, hydrolyze keratin structure as well as other structural proteins (involucrin and corneodesmosin), separating dead skin cells without damage to the cytoplasm of the live epidermis (Gupta et al. 2013; Singh et al. 2023).

Proteases, both microbial and plant, have the capability of acting under skin-compatible conditions. In peels and their cleanser products, they increase smoothness and the penetration of products (Măgeruşan et al. 2023). The example of keratinases produced by *Chryseobacterium* is that it is more effective to penetrate the topmost layer of skin without affecting deeper parts (Gupta et al. 2013). Natural exfoliation is facilitated by the proteases secreted by the commensal skin microbe, including *Staphylococcus* and *Corynebacterium* (Byrd et al. 2018). Such enzymes control inflammation and prevent the formation of bacterial biofilms, promoting skin health (Capozza et al. 2022). Keratinases are very versatile, and they operate at pH 8–12 and up to 100 ° C, typically after two stages, disulfide bond reduction and proteolysis (Li 2021).

Table 10.1 Comparative properties of microbial proteases used in exfoliation

Enzyme	Microbial source	Target substrate	Optimal pH and temperature	Action on skin	Commercial use	References
Keratinase	*B. subtilis*	Keratin	8–10, 40–50 °C	Break down dead skin	Enzyme creams	Song et al. (2004)
Subtilisin	*B. licheniformis*	Keratin, casein	7–8, 37 °C	Gentle exfoliation	Peels, masks	Lee et al. (2023)
Actinidin	*Actinidia* sp.	Protein bonds	5–7, 30–40 °C	Mild exfoliation	Plant-based peels	Măgerușan et al. (2023)
Papain	*Carica papaya*	Peptides	5–6.5, 37 °C	Brightening action	Enzyme scrubs	Trevisol et al. (2022)

Table 10.2 Mechanism of action of keratin-degrading enzymes in skin desquamation

Protease	Action on SC	Breakdown product	Cosmetic form	Effect	Citations
Keratinase	Degrades bridges	Peptides, amino acids	Creams, serums	Smooth texture	Li (2021)
Proteinase K	Breaks corneodesmosomes	Peptides	Peels	Cell turnover	Tsai et al. (2023)
Ficin	Cleaves peptide bonds	Amino acids	Cleanser	Surface polishing	Măgeruşan et al. (2023)

Such methods as nanoencapsulation and blending enzymes using LPMO and disulfide reductases enhance formulation efficacy and stability of enzymes (Tóth et al. 2023). These enzymes also break down keratin to oligopeptides and maintain a healthy skin turnover (Gupta et al. 2013; Măgeruşan et al. 2023). Compared to the advantages, dosage reading, stability, and the parameters of allergy have yet to reach optimal conditions (Trevisol et al. 2022). Table 10.2 displays the mechanism of action of keratin-degrading enzymes in skin desquamation.

10.4 Microbial Proteases in Exfoliation Products

Exfoliation cosmetics have started using microbial protease to exfoliate gently and effectively. *Bacillus subtilis* DPI keratinase is active at pH 8–12 and temperature 20 to 50 ° C and can be mixed with such ingredients as sodium stearate, which helps exfoliating creams work effectively without irritating (Sanghvi et al. 2016).

B. subtilis AMR produces crude keratinase, which was proven to degrade human hair to create peptides in skincare (Mazotto et al. 2010). These enzymes do not just break down hard keratins but also promote keratolytic therapies (Mazotto et al. 2010). *B. subtilis* S14 expresses a keratinase that operates at the pH of 5.5–10.5 and is resistant to surfactants such as SDS and Tween, and can thus be used as a cleanser. Keratoclean® Hydra PB contains subtilisin *B. licheniformis*, which exfoliates but does not disrupt the skin barrier by causing skin-damaging inflammation (e Silva et al. 2014). Daily exfoliants also include pumpkin proteases fortified by microbe activity, which is commonly supplemented by vitamins A and E to allow enhanced radiance (Dermascope. 2025). Thickened skin (on the elbow and heels) is obtained by adding microbial keratin hydrolysates to the products for thickened skin, providing regeneration resources with amino acid-rich peptides (Gupta et al. 2013). These enzymes are also used in eco-friendly leather processing, showcasing their utility under harsh conditions, attributes that translate well into durable skincare products (Nounou et al. 2017). Table 10.3 displays the microbial proteases used in exfoliation products.

Table 10.3 Microbial proteases used in exfoliation products

Microbial Source	Enzyme	Product/ application	Formulation type	Key feature	Citations
Bacillus subtilis DP1	Keratinase	Enzyme-based depilatory cream	Cream formulation	Alkaline pH stability, effective keratin degradation, mild skin action	Gupta et al. (2013)
Bacillus subtilis AMR	Keratinase	Exfoliating peptides from human hair	Hydrolysate ingredient	Generates small peptides (800–2600 Da), active at pH 8–9	Mazotto et al. (2010)
Bacillus subtilis S14	Keratinase	Enzyme-based cleansers and gels	Liquid surfactant-based	Active over a wide pH range (5.5–10.5), surfactant-stable	Lee et al. (2023)
Bacillus licheniformis	Subtilisin (cross-linked)	Gentle enzyme peels	Gel/serum enzyme peel	Cross-linked for prolonged stability, suitable for sensitive skin	Lee et al. (2023)
Bacillus licheniformis	Keratinase	Keratoclean® hydra PB-acne/dandruff care	Lotion/solution for skin/ scalp	Corneocyte softening, comedolytic action, anti-flaking	Gupta et al. (2013)
Fungal or bacterial sources	Papain, bromelain, actinidin	Enzyme peels for sensitive skin	Cream/gel peel formulations	Mild exfoliation, good for daily/weekly use	Trevisol et al. (2022)
Microbial-enhanced pumpkin enzyme	Pumpkin protease	Daily exfoliating cleansers	Cleanser	Vitamin-rich, gentle enzymatic action	Dermascope. (2025)
Various bacillus spp.	Keratinase	Creams for hyperkeratotic skin (heels/ elbows)	Ointment/ smoothing cream	Produces keratin hydrolysates to soften and regenerate skin	Gupta et al. (2013)
Bacillus subtilis SCK6	Keratinase	Depilatory/ keratolytic candidate	Industrial origin, cosmetic potential	Dehairing at 33–35 °C in 6 h, no sulfide needed	Mazotto et al. (2010)

10.5 Microbial Proteases in the Treatment of Skin Disorders

Microbial proteases have demonstrated high potential in the treatment of various skin disorders through the modulation of host immunity, degrading pathogenic structures, and promoting tissue remodelling (Bond 2019). Proteases released on the skin surface by residing bacteria such as *Staphylococcus epidermidis* and *Cutibacterium acnes* affect skin homeostasis and immune signalling (Chua et al. 2022). *C. acnes* relies on proteolysis to metabolise arginine, releasing peptides that upregulate sebocyte activity and contribute to acne pathophysiology (Lee et al. 2015). *S. epidermidis* expresses the cysteine protease EcpA, which destabilises desmoglein-1 in the epidermis, compromising barrier integrity and aggravating inflammatory conditions such as eczema (Dayel et al. 2024). Fungal pathogens such as *Candida albicans* and *Malassezia* spp. produce secretory aspartyl proteases which break down keratin, collagen, and host defense proteins, aiding in colonization and infection (Krysan et al. 2005). These enzymes play a key role in the pathogenesis of skin mycoses and are recently being studied as an antifungal target due to increasing resistance to fluconazole among more than 50 percent of clinical isolates (Silva et al. 2025).

Proteases secreted by *Serratia marcescens* and *Pseudomonas* spp. disrupt the extracellular matrix, causing tumour and fibrotic tissue degradation, a mechanism currently used in wound healing (Lindsay et al. 2017). For example, carboxypeptidase G1 from *Pseudomonas* interferes with the folate metabolism in abnormal tissue (Lindsay et al. 2017). In the treatment of conditions such as psoriasis, ichthyosis, and xerosis, proteases such as alkaline aspartic proteases and bacterial keratinases from *Bacillus licheniformis*, *Thermoanaerobacter*, and *Nesterenkonia* are used (Gupta et al. 2019a,b).

Postbiotics, the non-viable microbial products, improve skin condition by regulating immunity, returning skin pH, promoting tight junction proteins, and reducing inflammation (Prajapati et al. 2025). They are advantageous to sensitive skin and reduce the risk of infection associated with probiotics (Prajapati et al. 2025). Lipoteichoic acid and LactoSporin are among the postbiotics that have been effective in reducing acne and oiliness (Prajapati et al. 2025). Microbial keratinases are already present in commercial enzymatic peels and regeneration products and facilitate exfoliation and healing of stretch marks, scars, and photodamage (Gupta et al. 2019a, b). These proteases, therefore, address the therapeutic gaps that are in the conventional dermatological treatments (Fornbacke and Clarsund 2013).

10.6 Microbial Proteases in Acne Management

Newer approaches towards the treatment of acne are available in the form of microbial proteases, specifically breaking down the keratin plug and microbial biofilms that clog pores (Gupta et al. 2013). A frequently occurring pathogen in acne, *Staphylococcus epidermis* also contains proteases that help it to develop biofilms, which puts the bacteria in both the crusader and target categories when it comes to treatment with enzymes (Martínez-García et al. 2018).

Cutibacterium acnes releases proteases that bind to PAR-2 receptors on the keratinocytes, therefore causing the production of inflammation and MMP, which causes scarring and tissue damage (Kim et al. 2025). Keratin plugs can be broken

Table 10.4 Dual role of microbial proteases in acne pathogenesis and treatment

Role	Protease source	Pathogenic/ beneficial	Effect	Citations
Biofilm formation	*C. acnes*, *S. epidermidis*	Pathogenic	Follicular blockage	Almoughrabie et al. (2023)
PAR-2 activation	*C. acnes* protease	Pathogenic	Inflammation, MMP release	Kim et al. (2025)
Keratin plug degradation	Bacterial keratinase	Beneficial	Comedolysis	Gupta et al. (2013)
Biofilm disruption	Proteinase K (exogenous)	Beneficial	Enhances treatment delivery	Capozza et al. (2022)

down using externally administered microbial enzymes, whose main interest is keratinase. The prediction is that the keratinases of *Meiothermus taiwanensis* and *Pseudomonas aeruginosa* can work synergistically with current drugs used to treat acne, such as adapalene and tretinoin (Gupta et al. 2013). These enzymes are thermostable, which means that they can withstand the temperature commonly seen in acne environments that are alkaline in nature, which are also seen in acne, and are already commercially available in products such as Keratoclean® Hydra PB (Gupta et al. 2013). Anti-acne liposomal delivery systems have been shown to exhibit better efficacy by disrupting biofilms and otitis due to co-delivery with proteases and retinoic acid (Tsai et al. 2023).

Another protective mechanism is represented by native sequences such as Esp serine protease in *S. epidermidis*, which inhibits *S. aureus* and inflammation (Byrd et al. 2018). In the meantime, *C. acnes* propionic acid regulates the barrier repair and autophagy (Megyeri et al. 2018: Almoughrabie et al. 2023). Though promising, further clinical validation, formulation stability, and safety studies are needed to optimize these enzymes for routine acne therapy (Adamiak et al. 2024). Table 10.4 displays the role of microbial proteases in acne pathogenesis and treatment. Figure 10.1 depicts the overview of microbial proteases in acne treatment.

10.7 Role of Microbial Proteases in Wound Healing and Scar Reduction

Any disruption in the skin or tissues that results from trauma (cuts, scrapes, or punctures), surgery, or even some pathological reasons, such as ulcers and infections, can be defined as a wound. It's mainly classified into two: One can be open/acute (e.g., cuts and abrasions), and the other closed/chronic (e.g., bruises and contusions) (Shi et al. 2023). An acute wound is healed within a timely manner and undergoes what is called a healing process with four phases: haemostasis, inflammation, proliferation, and remodelling. On the other hand, if the wound exhibits a prolonged inflammatory phase, then it can be known as a chronic wound, which usually takes much longer. This condition mainly occurs due to the

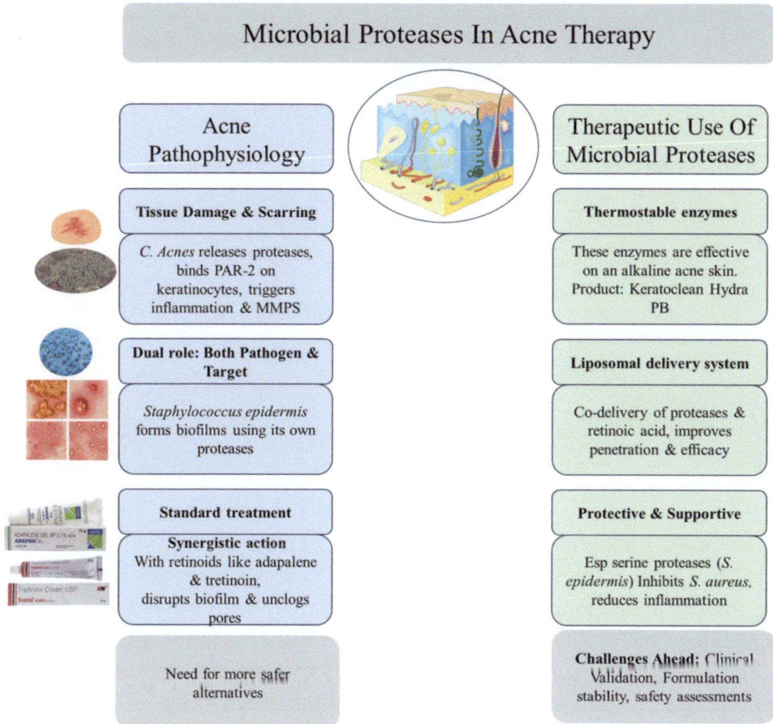

Fig. 10.1 Overview of microbial proteases in acne treatment: mechanisms, delivery, and benefits

dysbiosis of the extracellular matrix at the infection site (Mathew-Steiner et al. 2021).

10.7.1 Mechanism

A primary class of enzymes that play a crucial role in the wound healing process is known as matrix metalloproteinases (McCarty and Percival 2013). This works on the ability of collagenase, mainly produced by *Clostridium histolyticum* and various other microbial species, to generate enzymatic debridement that breaks down the collagen present in the wounded tissues; thus, by clearing the area, it slowly promotes healing in that specific region (S. Shi et al. 2023). This action clears the wound bed of devitalized tissue, allowing for better oxygenation and reducing bacterial load (Shi et al. 2010). Apart from this, collagenase is responsible for the migration of keratinocytes and fibroblast cells into the wounded tissues, which induces re-epithelialization and promotes granulation for tissue formation. And it accelerates the healing process by modulating the inflammatory response and creating an environment favourable for tissue remodelling and ultimately causes minimal

Table 10.5 Different types of microbial proteases, along with their source, which play a role in the wound healing process

Proteases	Microbial source	Role in wound healing	References
Collagenase	*Clostridium histolyticum, Vibrio* spp.	Debridement of wounded tissues cleaves collagen and promotes re-epithelialization	Shi et al. (2023)
Serratiopeptidase	*Serratia marcescens*	Reduces fibrin and necrotic debris, enhances tissue remodelling	Bhagat et al. (2013)
Fibrinolysin	*Streptococcus pyogenes*	Dissolves fibrin clots and necrotic eschar	Kotb (2013)
Keratinase	*Bacillus licheniformis, Bacillus subtilis*	Aids chronic wound healing	Gupta et al. (2019a,b)
Elastase	*Pseudomonas aeruginosa*	Selectively degrades elastin	McCarty and Percival (2013)

damage to surrounding healthy tissue (Shi et al. 2023). The application of collagenase injections has been shown to accelerate wound healing in several in vitro and in vivo experiments. Collagenases can more successfully cure wounds, including those from third-degree burns, because of their unique ability to break down collagen that is difficult to break down. This ability is due to the property of inflammatory cells like macrophages, which help collagenases to break down older collagen fibrils (Chung et al. 2004). For this reason, *C. collagenase*-containing ointments are applied to wounded areas for the removal of necrotic tissues. As the presence of such dead tissue on the wound site increases the inflammatory phase, applying the collagenase can help in a proactive way (Cioce et al. 2024). Hence, this approach enhances a healthier healing environment by reducing inflammation along with wound debridement. Consequently, the application of collagenase can be used in the healing of chronic wounds (Alipour et al. 2016). Table 10.5 displays the different types of microbial proteases, along with their source, that play a role in the wound healing process, and Fig. 10.2 depicts the application of collagenases in the wound healing process.

10.8 Protease Role in Addressing Hyperpigmentation

Hyperpigmentation is caused by the fact that UV light, inflammation, or hormonal fluctuations stimulate the production of excessive melanin (Markiewicz et al. 2022). Traditional agents such as hydroquinone are generally irritating, cause photosensitivity, and do not have a long-term application as they produce pigment rebound (Ni et al. 2025). Biologically more compatible products such as microbial peptides and proteases will produce safer results because they can regulate melanin synthesis and facilitate epidermal turnover (Prajapati et al. 2025). Microbial fermentation or

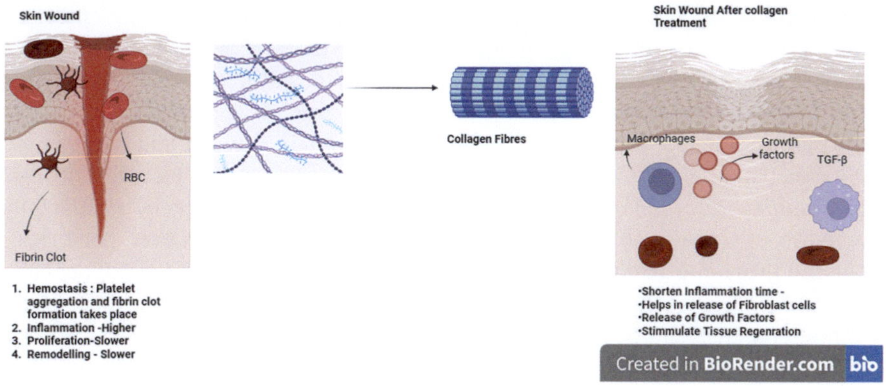

Fig. 10.2 The application of collagenases in the wound healing process

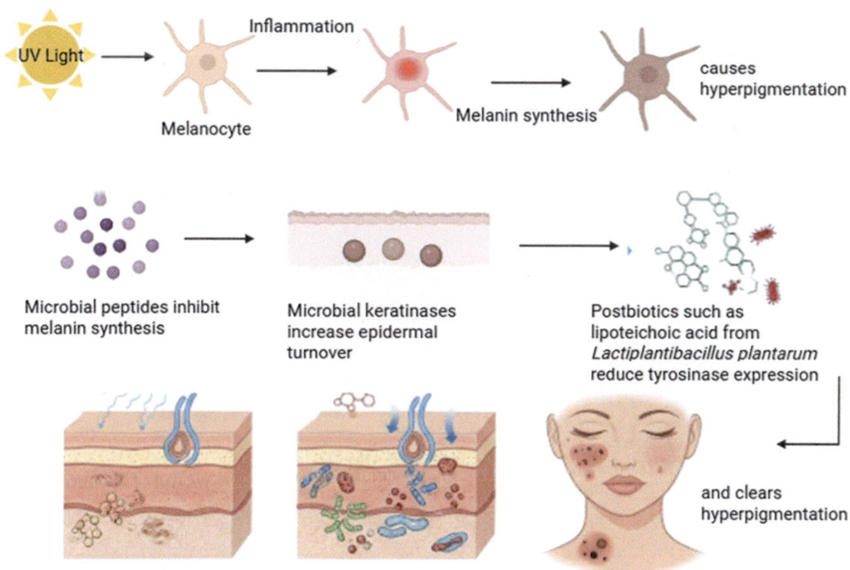

Fig. 10.3 UV light causes inflammation in the skin, which leads to overproduction of melanin. Microbial proteases inhibit tyrosinase activity and reduce hyperpigmentation on the skin surface

enzymatically generated peptides that are tyrosinase inhibitors and MC1R inhibitors, such as decapeptide-12, oligopeptide-68, and nonapeptide-1 (melanostatine-5), decrease melasma, post-inflammatory hyperpigmentation (PIH), and age spots with fewer side effects (Sekino et al. 2024). A cyclic peptide, cyclo(L-Pro-L-Tyr), a metabolite of *Corynebacterium tuberculostearicum*, is another compound that fits into the active site of tyrosinase, directly inhibiting it (Sekino et al. 2024). Figure 10.3 displays the role of proteases in addressing hyperpigmentation.

In addition to tyrosinase inhibition, microbial keratinases accelerate epidermal turnover by exfoliating melanin-loaded keratinocytes and are less abrasive than chemical exfoliants like glycolic acid (Gupta et al. 2019a,b). In comparison to their chemical counterparts, these enzyme-based exfoliants are suitable for chronic hyperpigmentation of sensitive skin, since they maintain skin pH and barrier function (Prajapati et al. 2025). Furthermore, postbiotics such as lipoteichoic acid from *Lactiplantibacillus plantarum* have shown anti-melanogenic effects by reducing the tyrosinase expression in melanocytes (Prajapati et al. 2025). Other postbiotics, including propionic acid from *C. acnes*, reduce pigmentation while simultaneously maintaining the microbiome balance (Prajapati et al. 2025). Combined, these bioactive microbial agents combine melanin inhibition and enhanced turnover with low toxicity, making them ideal next-generation treatments for pigmentation disorders (Tang et al. 2025).

10.9 Enzymatic Pathways for Melanin Regulation

Melanins are heterogeneous pigments formed by the oxidative polymerization of phenolic and indolic compounds. These pigments are found in a diverse range of organisms, including some fungi and bacteria. Melanins are divided into distinct categories, namely eumelanin or DOPA-melanin (dihydroxyphenylalanine-melanin), pheomelanin (found in animal hair, feathers), neuromelanin (found exclusively in human neurons), and allomelanin, based on their chemical properties, synthesis routes, and enzymes involved (Pavan et al. 2020).

10.9.1 Why Is Melanin Produced by Microorganisms?

Microorganisms that can synthesize and produce melanin confer great benefits to the host, such as stress resistance, UV protection, and virulence. The pigment has been recognized to serve as: (a) an antioxidant and radical scavenger, (b) a photoprotector that efficiently absorbs and dissipates solar radiation in the form of heat, and (c) an absorber that chelates metals and binds organic compounds (Tran-Ly et al. 2020a,b). Melanin's ability to scavenge free radicals can reduce the oxidative burst of host cells, and it was seen for the plant pathogen *R. solanacearum*. It has been proposed that protection against oxidative stress could assist the bacterium in coping with plant defense mechanisms (Ahmad et al. 2016). In melanin-producing *V. cholerae*, the synthesis of the pigment enhances cholera toxin and pilus expression, which in turn aids in host colonization (Valeru et al. 2009). Melanin also plays a key role in the interactions of bacterial biofilms and other species. The biofilms of the marine bacteria *Pseudoalteromonas lipolytica* generate melanin that inhibits the larval settling and metamorphosis of mussels, resulting in reduced biofouling activity (Zeng et al. 2017). One of the key features attributed to melanin produced by microorganisms is their ability to protect cells from UV irradiation and damage. A study conducted by Han and colleagues proved this feature by observing that the melanin-producing mutants of strains of *Bacillus anthracis* were more resistant to UV radiation as compared to the normal cells that were unable to produce this pigment (Han et al. 2015).

10.9.2 Pathways of Melanin Production

Bacteria primarily produce allomelanin, a category that includes several non-nitrogenous subgroups of melanin generated from various catecholic and dihydroxynaphtalene precursors, such as 1,8-dihydroxynaphthalene (DHN)-melanin and pyomelanin, while some microorganisms can also produce DOPA-melanin, which is normally seen in mammalian melanin synthesis. In the DOPA-melanin pathway, tyrosinase and laccase (in the case of *Azospirillum lipoferum*) convert tyrosine to L-DOPA, which is then transformed into dopaquinone. Dopaquinones are highly active compounds that spontaneously oxidize and polymerize to form melanin. The DHN pathway, seen in *Streptomyces griseus,* involves malonyl-coenzyme A, a precursor produced within the organism. The polyketide synthase enzyme initiates successive decarboxylation of five molecules of malonyl-CoA to form 1,3,6,8-tetrahydroxynaphthalene (THN). THN is then reduced and dehydrated to form 1,8-dihydroxynaphthalene (DHN). The polymerization of DHN produces DHN-melanin as the product.

In terms of high-yield melanin production, microorganisms using the DHN pathway are not preferred since the pigment is generated endogenously and is tightly connected to the inner side of the cell wall. This makes melanin extraction extremely challenging and can result in artifacts caused by severe extraction chemicals (Toledo et al. 2017). As a result, many bacteria rely on exogenous tyrosine or tyrosine derivative substrates for melanin production. This is significant to scientists studying microbial melanization since melanin is synthesized extracellularly, allowing for less severe extraction. Various microorganisms producing several types of melanin are summarized in Table 10.6

10.9.3 Factors Influencing the Production of Melanin

There are many different pathways for melanin synthesis, but all involve the oxidation of hydroxylated aromatic compounds. Some of these compounds are synthesized by tyrosinases, laccases, or polyketide synthases in dedicated biosynthetic pathways, subject to specific regulation. Pigment formation in bacteria that produce melanin through the oxidation of L-DOPA is linked to the regulation of enzyme synthesis driven by nutrient factors such as the presence of Cu, nitrogen sources, and oxygen availability, as well as by general regulatory networks that

Table 10.6 Melanin-producing microorganisms

Microorganism	Melanin class	Tyrosine as a precursor	Substrate(s)	Source
Cryptococcus neoformans	DOPA-melanin (eumelanin)	Yes	L-DOPA, catecholamines	Wang et al. (2019)
Aspergillus nidulans	DHN-melanin	No	Acetyl-CoA→polyketide pathway	Sato et al. (2009)
Bacillus thuringiensis	Pyomelanin	Yes	Tyrosine → HGA (homogentisic acid)	Yang et al. (2018)
Shewanella colwelliana	Pyomelanin	Yes	Tyrosine → HGA	Turick et al. (2003)
Pseudomonas aeruginosa	Pyomelanin	Yes	Tyrosine → HGA	Rodríguez-Rojas et al. (2009)
Rhizobium spp.	Pyomelanin	Yes	Tyrosine	Chatfield and Cianciotto (2007)
Wangiella dermatitidis	DHN-melanin	No	Acetyl-CoA → polyketide	Paolo et al. (2006)

control cellular processes such as nitrogen fixation, stress responses, or morphogenesis.

A study of melanin-deficient *Sinorhizobium meliloti* transposon mutants discovered a link between thioredoxin and melanin synthesis. The gene coding for thioredoxin (trxL) is increased in melanin-producing circumstances, and trxL mutants are unable to create pigment due to a drastically decreased tyrosinase activity (Castro-Sowinski et al. 2007). Peptone contributes towards the increase of the tyrosinase activity but not with tyrosine, and the presence of Fe and Ni greatly enhances the expression of tyrosinase and melanin formation, according to the most recent study on dietary factors influencing pigmentation in a *Streptomyces* isolate (Wang et al. 2019). Temperature also plays an important role in regulating melanin biosynthesis enzyme activity. According to another study, high temperatures (42 °C) stimulate melanin synthesis in *Bacillus thuringiensis*, enabling the bacteria to resist UV rays and maintain their bioinsecticide properties. These detrimental effects can be lessened by high temperatures, which might suggest a competing nature between melanin formation and other physiological processes. Heavy metals (cadmium, chromium, nickel, copper, etc.) could stimulate melanin production in microorganisms, as a defensive mechanism by binding and immobilizing these metals, thereby reducing their toxicity and facilitating bioremediation (Muñoz-Torres et al. 2024).

A recent study found that a *P. aeruginosa* transcription factor produced in response to exposure to surface acoustic waves boosted melanin production via an unknown mechanism, demonstrating the numerous and diverse factors that influence melanin synthesis (Ben-David et al. 2018).

10.10 Role of Microbial Proteases in Anti-ageing

Skin, which is the most important part of the body, acts as a barrier as well as protects against various environmental factors and various changes as we age. Skin ageing mainly occurs due to both intrinsic and extrinsic factors (Guo et al. 2022). Hence, it's a multifactorial process. Intrinsic factors include genetics, which mainly determine the skin type and its thickness, hormonal changes, and cell death, which reduces cellular turnover. Skin microbes also highly impact the health of the skin. Apart from these, various extrinsic factors include environmental influences such as sun exposure, pollution, excessive cleansing, and harsh skincare. These also involve diet and smoking habits (Sgarbieri and Pacheco 2017). Both intrinsic and extrinsic factors can contribute to skin ageing by influencing the decrease in collagen and sebum production, which makes skin more fragile and leads to the appearance of wrinkles over time.

Microbial proteases aid in the breakdown of peptides and proteins, mostly produced by bacteria, fungi, or other microorganisms (Shirzad et al. 2018). Through this process, they aid in the elimination of damaged protein that causes wrinkles to form and lessen the characteristics of wrinkled skin (Gupta et al. 2019a,b). Major functions include enzymatic skin exfoliation, which often happens due to the enzymatic action of proteases that break down, promote skin elasticity, reduce roughness, and promote regeneration. This reduces wrinkles and regulates scar formation on the skin. When paired with hyaluronic acid, microbial collagenases, especially those derived from *Vibrio alginolyticus*, have a strong foundation for collagen remodelling, which is selective (Gupta et al. 2019a,b). When used as an anti-ageing treatment, it focuses primarily on the texture, elasticity, and wrinkles of the skin tissue. Another type of microbial protease, which also has an anti-ageing property, is elastase. These proteolytic enzymes mainly include serine proteases and a few metalloproteases, which are responsible for the breakdown of elastin, which is primarily responsible for providing skin elasticity. These enzymes can remove damaged or aged elastin fibres and thus give rise to the renewal of new elastin proteins that lead to more elastic skin and prevent skin shagging. The most used Bacterial elastases are produced from *Pseudomonas aeruginosa*, *Bacillus licheniformis,* and *Bacillus coagulans.* It aids skin remodelling, provides firmness, reduces wrinkles, and acts as an anti-ageing agent. Remodelling of the aged collagen is a major function, as these proteases are responsible for sending signals to the fibroblasts, which helps in the synthesis of new collagen and thus revitalizes the skin. It makes the skin more permeable and aids in improved and quicker absorption of hyaluronic acid and other anti-ageing lotions. Consequently, spots are lessened, and the skin becomes even softer. In addition, a combination of proteases and probiotics may be crucial for reducing pro-inflammatory cytokines and ultimately lowering the quantity of reactive oxygen species (ROS) to lessen oxidative stress (Jiratchayamaethasakul et al. 2020).

10.11 Commercially Used Microbial Proteases

The most common source of proteases that is being used is alkaline aspartic proteases, which are extracted from various alkaliphilic bacteria. Bacteria can treat various skin diseases like dry skin, scaly skin, and even skin inflammation (Shirzad et al. 2018). Another type of protease which is used is keratinase, mainly known for the treatment of scar tissue and stretch marks; it hastens the process of healing by accelerating the regeneration of epithelial cells. One of the most used microbial strains for the commercial production of keratinase is *Bacillus licheniformis* (Brandelli et al. 2010). Keratinase can also be produced by a few thermophilic species (Gupta et al. 2019a,b). Apart from this, it can also be produced by various Gram-positive bacteria, such as *Lysobacter*, *Nesterenkonia*, *Kocuria*, and *Microbacterium*, as well as Gram-negative bacteria such as *Vibrio*, *Xanthomonas*, *Stenotrophomonas*, *Chryseobacterium*, *Fervidobacterium*, *Thermoanaerobacter*, *and Nesterenkonia*. The well-known keratinophilic fungi *Microsporum*, *Aspergillus niger*, *Epidermophyton*, *Trichophyton*, and *Chrysosporium* may also be able to break down keratin fibres. In addition to enzymes, a few peptides, which are responsible for the breakdown of protein, have also been employed extensively in cosmetic treatments. Gels, emulsions, powders, and lotions contain soluble peptides, whereas face masks contain insoluble peptides. These peptides are produced for commercial use by the regulated action of proteases, which are mainly produced by several *Bacillus* species (Shirzad et al. 2018). Additionally, pentapeptides are frequently utilized to lessen wrinkles and roughness of the face.

10.12 Collagen Remodelling and Elasticity Improvement

One of the essential proteins that results in better and healthier skin is collagen. This makes up most of the structure of the skin. This gives it various properties, such as support, enhances elasticity, and strength (Mathew-Steiner et al. 2021). Apart from this, it also gives various benefits to skin health.

There has been an extensive use of non-specific proteolytic enzymes in clinical practices. As a result, several diseases are brought on by the unidentified build-up of collagen, resulting in fibrous skin and scars, or by the development of necrotic connective tissues, which can happen in the case of burns or chronic ulcers (Suleman 2016). Using certain specific proteolytic enzymes will be considered a much better approach. The major types of collagens I, III, IV, andV are responsible for collagen remodelling and improving elasticity (Song et al. 2023). Collagenases have a unique ability to remodel collagen-rich extracellular matrix (ECM) and improve the elasticity of skin tissue. Microbial proteases, specifically collagenase and elastases produced mainly from bacterial sources such as *Clostridium histolyticum* and *Vibrio alginolyticus*, have shown these abilities through extensive research on them (Han et al. 2015). This microbial collagenase particularly targets type I and III collagen, attacks on the native triple helical structure, which is present on the wound tissue, and breaks them down into smaller peptides and proteins, leading to degradation of aged collagen when administered topically, injected, or utilized in tissue engineering techniques (Suleman 2016). Hence, fibres that are excessive or defective and obstruct normal tissue mechanics are removed by the enzymatic cleavage of these proteolytic enzymes (Varani et al. 2006).

These collagen fibres are cleared out by the process of phagocytosis, which eventually breaks open the ECM and reveals bioactive regions. These regions produce fibroblasts. Once these fibroblasts are stimulated, they release various growth factors, such as TGF-β, that act on collagen and help to clear out, resulting in the synthesis of new collagen fibres. The skin develops healthy, stable collagen with the regained ability of proper elasticity and structurally integrated fibres, which occurs due to this whole balanced cycle of degradation and regeneration of collagen (Varani et al. 2006).

Another class of microbial enzymes is some elastases that act on elastin and break down and synthesize the elastin at the wound site (Han et al. 2015). Thus, with growing uses in wound care, regenerative medicine, and aesthetic dermatology, microbial proteases provide a potential, bio-specific method to alter the extracellular matrix, improve collagen remodelling, and restore tissue elasticity (Varani et al. 2006).

10.13 Hair Care Applications

10.13.1 Applications in Hair Texture and Shine Enhancement

Microbial proteases also play a crucial role in hair quality enhancement by enabling specific protein changes and executing mild cleansing activities. Keratinases are a unique protease with exceptional efficiency in enhancing hair quality by demolishing only damaged keratin structures with accuracy while preserving the integrity of healthy hair shafts (Sharma and Devi 2018). The mode of action of such enzymes is through a multi-step process that disrupts deranged disulfide links and weak protein structures characteristic of chemically treated or damaged hair (Navone and Speight 2018).

Studies have shown microbial protease enzymatic activity helps to remove excessive keratin build-up that may cause hair to look dead and dull, while at the same time encouraging the development of a smoother cuticle surface that will allow greater reflection of light and contributing to easily manageable and lustrous hair (Presti, 2013; Bhogal et al. 2014).

Clinical trials have indicated that hair treated with keratinase has improved mechanical properties, including increased tensile strength and flexibility (Navone and Speight 2018). Controlled enzymatic breakdown process is carefully administered, attacking only the broken protein structures and preserving intact healthy keratin. The outcome is hair that is more lustrous, softer, and manageable (Bhogal et al. 2014; Liu et al. 2023). In addition to this, microbial proteases allow conditioners to penetrate the hair shaft by opening minute channels within damaged cuticle scales (Navone and Speight 2018). The improved absorption enables the effectiveness of follow-up conditioning treatments to result in more long-lasting alterations in the texture and appearance of the hair (Liu et al. 2023).

10.13.2 Function in Minimizing Dandruff and Scalp Health

The application of microbial proteases for the treatment of dandruff represents a sophisticated means of scalp health maintenance that attacks the source of biochemical abnormalities responsible for flaking and irritation. Kallikrein proteases (KPs), particularly hK5 and hK7, are critical in the normal process of desquamation of the scalp (Caubet et al. 2004). However, when these proteases become overactive, they are responsible for the excessive flaking characteristic of dandruff diseases (Caubet et al. 2004).

Recent studies have shown that dandruff scalps are characterized by an imbalance of proteases and anti-proteases that result in overproduction and abnormal scaling of the stratum corneum. This biochemical disorder causes the visible white scales and inflammatory association, which causes the redness, burning, and overall irritation typical of dandruff (Caubet et al. 2004; Turner et al. 2012).

Therapeutic interventions with protease inhibitors have proved very effective in controlling the symptoms of dandruff. Studies have shown that selectively inhibiting proteases can potentially restore normal balance in the breakdown of corneodesmosomes to enhance the function of the scalp barrier with limiting shedding (Caubet et al. 2004). It involves inhibiting excessive protein degradation underlying cell-to-cell adhesion in the outer scalp (Caubet et al. 2004).

Zinc-based products have been reported to be particularly beneficial in protease inhibition for dandruff management (Bhogal et al. 2014). Research indicates that zinc salts can reduce protease activity within hair follicles and surrounding tissues significantly, therefore leading to improved hair retention and reduced scalp irritation (Turner et al. 2012). The combined effect of protease inhibition and antimicrobial activity makes zinc-based products effective in overall scalp care. Application of protease inhibitors with traditional antifungal drugs concurrently has proved more effective than the use of one drug alone (Bhogal et al. 2014). Trichogen® (zinc gluconate) combined with climbazole has better protease inhibition activity and hair retention strength, as shown by in vitro and ex vivo tests (Bhogal et al. 2014).

10.14 Advancement in Green Cosmetic Formulations

10.14.1 Green and Sustainable Production Practices

Microbial protease-derived biodegradable cosmetics are a significant breakthrough in green beauty science. Microbial enzyme production has several important environmental advantages over conventional chemical manufacture, including reduced use of energy, elimination of harmful solvents, and production of biodegradable by-products (Dahiru et al. 2024).

Biocatalytic production of cosmetic ingredients is done under mild conditions, which substantially reduces the carbon footprint of production processes. Enzymatic reactions can be performed under physiological pH and temperature as opposed to high temperatures, high-concentration acids, and organic solvents, which are common in most traditional chemical synthesis, saving up to 60% of energy compared to traditional processes (Boaz and Clendennen 2009). Microbial protease production from renewable raw materials enhances the sustainability attributes of the raw materials. Crop waste products such as feather meals, wheat bran, and food-processing by-products are available as cheap substrates for protease production (Feyissa 2025). This circular economy model not only reduces the cost of production but also addresses the waste management issue of different industries.

Solid-state fermentation (SSF) methods have become especially eco-friendly methods of protease production (Feyissa 2025). These methods need little water, produce less liquid waste, and can successfully transform agricultural waste into useful, enzymatic products. Research has shown that SSF can produce protease yields equivalent to or even better than those of conventional submerged fermentation, while concurrently significantly lessening the environmental impact (Feyissa 2025). The biodegradability of protease-derived cosmetics is a natural advantage towards the solution of environmental problems associated with personal care products. Microbial proteases, as opposed to petrochemical-derived surfactants and chemicals that may persist in the aquatic environments, are completely biodegradable, avoiding the possibility of bioaccumulation and ecological harm (Mouro et al. 2024). The requirements of green chemistry mandate the production of protease-derived cosmetic materials from renewable feedstocks, avoiding harmful chemicals, and are designed for biodegradability. Enzymatic cosmetic ingredient production classification by the Presidential Green Chemistry Award confirms the environmental advantage of such processes and promotes their broader use in the industry (Boaz and Clendennen 2009).

Recent developments in enzyme immobilization and stabilization have expanded the commercial potential of protease-containing cosmetics. Sophisticated formulation techniques allow the manufacturing of stable, long-lasting products with retained enzymatic activity during shelf life and full biodegradation upon use (Ojo-Omoniyi et al. 2024). Application of life cycle assessment (LCA) techniques in protease-based cosmetic development ensures careful evaluation of environmental performance from raw material procurement to end-of-life disposal. Such studies repeatedly confirm improved environmental attributes of enzymatic methods compared to conventional chemical synthesis, further supporting the movement towards sustainable cosmetic production (Ojo-Omoniyi et al. 2024).

10.15 Microbial Proteases in Peptide-Based Cosmetics

The bioactive peptides in the cosmeceuticals involved in rejuvenation, hydration, and pigmentation are synthesized with microbial proteases (Tang et al. 2025). These peptides penetrate and moisturize the stratum corneum, stimulating the fibroblast to build collagen, elastin, and glycosaminoglycans to improve skin elasticity and smooth skin texture (Badilli and Inal 2025). A pentapeptide, KTTKS, found in 1993, and its derivatives, such as palmitoyl-KTTKS (also known as Matrixyl) and palmitoyl tripeptide-5 (also known as Syn(R)-Coll), resemble natural growth factors, such as TGF-β, that promote ECM production (Badilli and Inal 2025). Peptides that can be used as carriers, such as Cu-GHK, aid in wound healing and antioxidant defense, along with their function in supporting collagen and elastin production (Badilli and Inal 2025). Recombinant microbial systems such as *E. coli* and *Pichia pastoris* are preferred for the biosynthesis of peptides due to their high yield, precision, and ease of modification (Tang et al. 2025).

Proteases contribute to "smart delivery" systems for cosmeceuticals, as they activate pro-peptides upon skin contact, minimize degradation, and improve activity (Tang et al. 2025). Microneedles and liposomes act as delivery vehicles, which, when combined with protease-sensitive peptides, increase dermal absorption (Badilli and Inal 2025). As well as antioxidant peptides such as L-carnosine that shield the skin against photoaging and oxidative stress, are synthesized by these microbial proteases (Badilli and Inal 2025). Enzymatic synthesis with microbial proteases is more environmentally friendly as compared to chemical methods and is safer, with greater compatibility (Gupta et al. 2019a,b). With the increased demand for natural and sustainable cosmetics, microbial proteases play the key role in the future of innovative, secure, and efficient skin care products (Gupta et al. 2019a, b).

10.16 Conclusion

Microbial proteases emerge as valuable tools in dermatology and cosmetic science, offering effective, sustainable solutions for skin health and beauty. In skincare, they enable gentle exfoliation by breaking protein bonds between dead skin cells, avoiding the harshness of physical or chemical methods. Enzymes such as agarase, derived from microbes, are commonly used to improve skin texture and barrier function. Beyond exfoliation, microbial proteases contribute to acne treatment by disrupting *Cutibacterium* acnes biofilms and reducing inflammation. They also aid wound healing by removing necrotic tissue and promoting collagen remodelling, helping minimize scars. In hyperpigmentation, proteases regulate pathways such as PAR2, which control melanin transfer, offering an alternative to chemical brightening agents and enhancing skin tone. For anti-ageing, proteases stimulate collagen production and influence matrix metalloproteinases (MMPs), improving skin elasticity and reducing fine lines. *Bacillus subtilis* enzymes are particularly effective in supporting skin firmness and renewal. In hair care, keratinases help break down excess keratin on the scalp, reducing dandruff and improving scalp health.Their production through environmentally friendly fermentation aligns with the rising demand for sustainable beauty products. Proteases also help synthesize bioactive peptides with better skin adsorption and fewer synthetic additives. Recent enzyme-based products target skin brightening by reducing melanin formation and clearing pigmented dead cells, offering a safer alternative to harsh treatments. Despite their potential, challenges remain in stability, allergenicity, and formulation compatibility. Ongoing research in protein engineering and delivery systems is key to overcoming these issues. With continued innovation, microbial proteases are set to shape the future of safe, effective, and eco-friendly skincare and cosmetics.

References

Adamiak K, Tadrowski T, Kurzawa M, Sionkowska A (2024) Novel methods in acne treatment and skin care: a review. J Drugs Dermatol: JDD 23(12):e181–e185. https://doi.org/10.36849/jdd.7757

Ahmad S, Lee SY, Kong HG, Jo EJ, Choi HK, Khan R, Lee W (2016) Genetic determinants for Pyomelanin production and its protective effect against oxidative stress in Ralstonia solanacearum. PLoS One 11(8):e0160845. https://doi.org/10.1371/journal.pone.0160845

Alipour H, Raz A, Zakeri S, Djadid ND (2016) Therapeutic applications of collagenase (metalloproteases): a review. Asian Pac J Trop Biomed 6(11):975–981. https://doi.org/10.1016/j.apjtb.2016.07.017

Almoughrabie S, Cau L, Cavagnero K, O'Neill AM, Li F, Roso-Mares A et al (2023) Commensal Cutibacterium acnes induce epidermal lipid synthesis important for skin barrier function. Sci Adv 9(33):eadg6262. https://doi.org/10.1126/sciadv.adg6262

Amadeh A, Mohebbi N, Amadeh Z, Jamshidbeigi A (2025) Comparative efficacy of autolytic and collagenase-based enzymatic debridement in chronic wound healing: a comprehensive systematic review. Int Wound J 22(4):e70177. https://doi.org/10.1111/iwj.70177

Badilli U, Inal O (2025) Current approaches in cosmeceuticals: peptides, biotics and marine biopolymers. Polymers 17(6):798. https://doi.org/10.3390/polym17060798

Ben-David Y, Zlotnik E, Zander I, Yerushalmi G, Shoshani S, Banin E (2018) SawR a new regulator controlling pyomelanin synthesis in *Pseudomonas aeruginosa*. Microbiol Res 206: 91–98. https://doi.org/10.1016/j.micres.2017.10.004

Bhagat S, Agarwal M, Roy V (2013) Serratiopeptidase: a systematic review of the existing evidence. Int J Surg 11(3):209–217. https://doi.org/10.1016/j.ijsu.2013.01.010

Bhogal RK, Mouser PE, Higgins CA, Turner GA (2014) Protease activity, localization and inhibition in the human hair follicle. Int J Cosmet Sci 36(1):46–53. https://doi.org/10.1111/ics.12091

Boaz NW, Clendennen SK (2009) A green, solvent-free biocatalytic method to produce cosmetic esters. Cosmet Toiletries 124(7)

Bond JS (2019) Proteases: history, discovery, and roles in health and disease. J Biol Chem 294(5): 1643–1651. https://doi.org/10.1074/jbc.TM118.004156

Brandelli A, Daroit DJ, Riffel A (2010) Biochemical features of microbial keratinases and their production and applications. Appl Microbiol Biotechnol 85:1735–1750. https://doi.org/10.1007/s00253-009-2398-5

Byrd AL, Belkaid Y, Segre JA (2018) The human skin microbiome. Nat Rev Microbiol 16(3): 143–155. https://doi.org/10.1038/nrmicro.2017.157

Capozza M, Stefania R, Dinatale V, Bitonto V, Conti L, Grange C, Skovronova R, Terreno E (2022) A novel PSMA-targeted probe for NIRF-guided surgery and photodynamic therapy: synthesis and preclinical validation. Int J Mol Sci 23(21):12878. https://doi.org/10.3390/ijms232112878

Castro-Sowinski S, Matan O, Bonafede P, Okon Y (2007) A thioredoxin of Sinorhizobium meliloti CE52G is required for melanin production and symbiotic nitrogen fixation. Mol Plant-Microbe Interact 20(8):986–993. https://doi.org/10.1094/MPMI-20-8-0986

Caubet C, Jonca N, Brattsand M, Guerrin M, Bernard D, Schmidt R et al (2004) Degradation of corneodesmosome proteins by two serine proteases of the kallikrein family, SCTE/KLK5/hK5 and SCCE/KLK7/hK7. J Invest Dermatol 122(5):1235–1244. https://doi.org/10.1111/j.0022-202X.2004.22512.x

Chatfield CH, Cianciotto NP (2007) The secreted pyomelanin pigment of legionella pneumophila confers ferric reductase activity. Infect Immun 75(8):4062–4070. https://doi.org/10.1128/iai.00489-07

Chua W, Poh SE, Li H (2022) Secretory proteases of the human skin microbiome. Infect Immun 90(1):e00397–e00321. https://doi.org/10.1128/IAI.00397-21

Chung L, Dinakarpandian D, Yoshida N, Lauer-Fields JL, Fields GB, Visse R, Nagase H (2004) Collagenase unwinds triple-helical collagen prior to peptide bond hydrolysis. EMBO J 23(15): 3020–3030. https://doi.org/10.1038/sj.emboj.7600318

Cioce A, Cavani A, Cattani C, Scopelliti F (2024) Role of the skin immune system in wound healing. Cells 13(7):624. https://doi.org/10.3390/cells13070624

Dahiru MM, Abdulhamid AA, Abaka AM (2024) Current perspectives on enzyme applications in medicine, agriculture, and industries. Asian J Trop Biotechnol 21(1). https://doi.org/10.13057/biotek/c210102

Dayel SB, Hussein RS, Gafar HH (2024) The role of botulinum neurotoxin BoNT-A in the management of oily skin and acne vulgaris: a comprehensive review. Medicine 103(8): e37208. https://doi.org/10.1097/MD.0000000000037208

Dermascope (2025). Pumpkin enzyme exfoliation for mature skin: clinical benefits. Dermascope. https://www.dermascope.com/boldijarre-koronczay-and-brian-goodwin.

e Silva LAD, Macedo AJ, Termignoni C (2014) Production of keratinase by Bacillus subtilis S14. Ann Microbiol 64:1725–1733. https://doi.org/10.1007/s13213-014-0816-0

Feyissa BH (2025) Screening, identification, and optimization of protease producing Bacillus pumilus strain DT-15 from tannery waste disposal site in Addis Ababa, Ethiopia. Int J Microbiol 2025(1):7176092. https://doi.org/10.1155/ijm/7176092

Fornbacke M, Clarsund M (2013) Cold-adapted proteases as an emerging class of therapeutics. Infect Dis Ther 2(1):15–26. https://doi.org/10.1007/s40121-013-0002-x

Guo J, Huang X, Dou L, Yan M, Shen T, Tang W, Li J (2022) Aging and aging-related diseases: from molecular mechanisms to interventions and treatments. Signal Transduct Target Ther 7(1): 391. https://doi.org/10.1038/s41392-022-01251-0

Gupta R, Sharma R, Beg QK (2013) Revisiting microbial keratinases: next generation proteases for sustainable biotechnology. Crit Rev Biotechnol 33(2):216–228. https://doi.org/10.3109/07388551.2012.685051

Gupta PL, Rajput M, Oza T, Trivedi U, Sanghvi G (2019a) Eminence of microbial products in cosmetic industry. Nat Prod Bioprospect 9:267–278. https://doi.org/10.1007/s13659-019-0215-0

Gupta M, Rani S, Kumar V (2019b) Advances in microbial proteases for cosmetic applications. Appl Microbiol Biotechnol 103(12):4777–4794. https://doi.org/10.3389/fbioe.2019.00110

Han Y, Li X, Zhou Q, Jie H, Lao X, Han J et al (2015) FTY720 abrogates collagen-induced arthritis by hindering dendritic cell migration to local lymph nodes. J Immunol 195(9):4126–4135. https://doi.org/10.4049/jimmunol.1401842

He X, Gao X, Guo Y, Xie W (2024) Research progress on bioactive factors against skin aging. Int J Mol Sci 25(7):3797. https://doi.org/10.3390/ijms25073797

Jiratchayamaethasakul C, Ding Y, Hwang O, Im ST, Jang Y, Myung SW et al (2020) In vitro screening of elastase, collagenase, hyaluronidase, and tyrosinase inhibitory and antioxidant activities of 22 halophyte plant extracts for novel cosmeceuticals. Fish Aquat Sci 23:1–9. https://doi.org/10.1186/s41240-020-00149-8

Kim EM, Oh S, Choi H, Park WS (2025) Characterization of keratinase from *Chryseobacterium camelliae* Dolsongi-HT1 and efficacy on skin exfoliation. Enzym Microb Technol 186:110605. https://doi.org/10.1016/j.enzmictec.2025.110605

Kotb E (2013) Activity assessment of microbial fibrinolytic enzymes. Appl Microbiol Biotechnol 97:6647–6665. https://doi.org/10.1007/s00253-013-5052-1

Krysan DJ, Ting EL, Abeijon C, Kroos L, Fuller RS (2005) Yapsins are a family of aspartyl proteases required for cell wall integrity in Saccharomyces cerevisiae. Eukaryot Cell 4:1364–1374. https://doi.org/10.1128/ec.4.8.1364-1374.2005

Lee SE, Kim JM, Jeong SK, Choi EH, Zouboulis CC, Lee SH (2015) Expression of protease-activated receptor-2 in SZ95 sebocytes and its role in sebaceous lipogenesis, inflammation, and innate immunity. J Invest Dermatol 135(9):2219–2227. https://doi.org/10.1038/jid.2015.151

Lee CH, Lee KWA, Chan LKW, Lee KFV (2023) Review of chemical peeling in photoaging skin. J Cosmet Med 7(1):1–5. https://doi.org/10.25056/JCM.2023.7.1.1

Li Q (2021) Structure, application, and biochemistry of microbial keratinases. Front Microbiol 12: 674345. https://doi.org/10.3389/fmicb.2021.674345

Lindsay S, Oates A, Bourdillon K (2017) The detrimental impact of extracellular bacterial proteases on wound healing. Int Wound J 14(6):1237–1247. https://doi.org/10.1111/iwj.12790

Liu Y, Liu J, Xiao J (2023) Enzymatic crosslinking of amino acids improves the repair effect of keratin on hair fibre. Polymers 15(9):2210. https://doi.org/10.3390/polym15092210

Măgeruşan ŞE, Hancu G, Rusu A (2023) A comprehensive bibliographic review concerning the efficacy of organic acids for chemical peels treating acne vulgaris. Molecules 28(20):7219. https://doi.org/10.3390/molecules28207219

Mahnashi MH, Muddapur UM, Turakani B, Shaikh IA, Al Awadh AA, Alshahrani MM et al (2022) A review on versatile eco-friendly applications of microbial proteases in biomedical and industrial applications. Sci Adv Mater 14(4):622–637. https://doi.org/10.1166/sam.2022.4264

Markiewicz E, Karaman-Jurukovska N, Mammone T, Idowu OC (2022) Post inflammatory hyperpigmentation in dark skin: molecular mechanism and skincare implications. Clin Cosmet Investig Dermatol 15:2555–2565. https://doi.org/10.2147/CCID.S385162

Martínez-García S, Rodríguez-Martínez S, Cancino-Diaz ME, Cancino-Diaz JC (2018) Extracellular proteases of Staphylococcus epidermidis: roles as virulence factors and their participation in biofilm. APMIS 126(3):177–185. https://doi.org/10.1111/apm.12805

Mathew-Steiner SS, Roy S, Sen CK (2021) Collagen in wound healing. Bioengineering 8(5):63. https://doi.org/10.3390/bioengineering8050063

Mazotto AM, Lage Cedrola SM, Lins U, Rosado AS, Silva KT, Chaves JQ et al (2010) Keratinolytic activity of Bacillus subtilis AMR using human hair. Lett Appl Microbiol 50(1): 89–96. https://doi.org/10.1111/j.1472-765X.2009.02760.x

McCarty SM, Percival SL (2013) Proteases and delayed wound healing. Adv Wound Care 2(8): 438–447. https://doi.org/10.1089/wound.2012.0370

Megyeri K, Orosz L, Bolla S, Erdei L, Rázga Z, Seprényi G et al (2018) Propionibacterium acnes induces autophagy in keratinocytes: involvement of multiple mechanisms. J Invest Dermatol 138(4):750–759. https://doi.org/10.1016/j.jid.2017.11.018

Mouro C, Gomes AP, Gouveia IC (2024) Microbial exopolysaccharides: structure, diversity, applications, and future frontiers in sustainable functional materials. Polysaccharides 5(3): 241–287. https://doi.org/10.3390/polysaccharides5030018

Muñoz-Torres P, Cárdenas-Ninasivincha S, Aguilar Y (2024) Exploring the agricultural applications of microbial melanin. Microorganisms 12(7):1352. https://doi.org/10.3390/microorganisms12071352

Navone L, Speight R (2018) Understanding the dynamics of keratin weakening and hydrolysis by proteases. PLoS One 13(8):e0202608. https://doi.org/10.1371/journal.pone.0202608

Ni X, Luo X, Jiang X, Chen W, Bai R (2025) Small-molecule Tyrosinase inhibitors for treatment of hyperpigmentation. Molecules 30(4):788. https://doi.org/10.3390/molecules30040788

Nounou MI, Zaghloul TI, Ahmed NA, Eid AA, El-Khordagui LK (2017) Skin permeability enhancement by Bacillus subtilis alkaline protease: application to transdermal drug delivery. Int J Pharm 529(1–2):423–432. https://doi.org/10.1016/j.ijpharm.2017.06.057

Ojo-Omoniyi OA, Moro DD, Afolab OB (2024) Microbial proteases: sources, significance and industrial applications. Int J Curr Microbiol App Sci 13:1–23. https://doi.org/10.20546/ijcmas.2024.1306.001

Paolo WF, Dadachova E, Mandal P, Casadevall A, Szaniszlo PJ, Nosanchuk JD (2006) Effects of disrupting the polyketide synthase gene WdPKS1 in Wangiella [Exophiala] dermatitidis on melanin production and resistance to killing by antifungal compounds, enzymatic degradation, and extremes in temperature. BMC Microbiol 6:1–16. https://doi.org/10.1186/1471-2180-6-55

Pavan ME, López NI, Pettinari MJ (2020) Melanin biosynthesis in bacteria, regulation and production perspectives. Appl Microbiol Biotechnol 104(4):1357–1370. https://doi.org/10.1007/s00253-019-10245-y

Prajapati SK, Lekkala L, Yadav D, Jain S, Yadav H (2025) Microbiome and postbiotics in skin health. Biomedicine 13(4):791

Presti RA (2013) U.S. Patent No. 8,413,666. Washington, DC: U.S. Patent and Trademark Office

Rodríguez-Rojas A, Mena A, Martín S, Borrell N, Oliver A, Blazquez J (2009) Inactivation of the hmgA gene of Pseudomonas aeruginosa leads to pyomelanin hyperproduction, stress resistance and increased persistence in chronic lung infection. Microbiology 155(4):1050–1057. https://doi.org/10.1099/mic.0.024745-0

Sanghvi G, Patel H, Vaishnav D, Oza T, Dave G, Kunjadia P, Sheth N (2016) A novel alkaline keratinase from Bacillus subtilis DP1 with potential utility in cosmetic formulation. Int J Biol Macromol 87:256–262. https://doi.org/10.1016/j.ijbiomac.2016.02.067

Sato K, Kido N, Murakami Y, Hoover CI, Nakayama K, Yoshimura F (2009) Lipopolysaccharide biosynthesis-related genes are required for colony pigmentation of Porphyromonas gingivalis. Microbiology 155(4):1282–1293

Sekino Y, Yamamoto I, Watanabe M, Kuramochi K, Furuyama Y (2024) Cyclo(L-pro-L-Tyr) isolated from the human skin commensal Corynebacterium tuberculostearic inhibits tyrosinase. Int J Mol Sci 25(13):7365. https://doi.org/10.3390/ijms25137365

Sgarbieri VC, Pacheco MTB (2017) Healthy human aging: intrinsic and environmental factors. Braz J Food Technol 20:e2017007. https://doi.org/10.1590/1981-6723.00717

Sharma R, Devi S (2018) Versatility and commercial status of microbial keratinases: a review. Rev Environ Sci Biotechnol 17(1):19–45. https://doi.org/10.1007/s11157-017-9454-x

Shi L, Ermis R, Garcia A, Telgenhoff D, Aust D (2010) Degradation of human collagen isoforms by clostridium collagenase and the effects of degradation products on cell migration. Int Wound J 7(2):87–95. https://doi.org/10.1111/j.1742-481X.2010.00659.x

Shi S, Wang L, Song C, Yao L, Xiao J (2023) Recent progresses of collagen dressings for chronic skin wound healing. Collagen and Leather 5(1):31. https://doi.org/10.1186/s42825-023-00136-4

Shirzad M, Hamedi J, Motevaseli E, Modarressi MH (2018) Anti-elastase and anti-collagenase potential of lactobacilli exopolysaccharides on human fibroblast. Artif Cells Nanomed Biotechnol 46(sup1):1051–1061. https://doi.org/10.1080/21691401.2018.1443274

Silva LBB d, Castilho IG, Souza Silva FA d, Ghannoum M, Garcia MT, Carmo PHF d (2025) Antimicrobial photodynamic therapy for superficial, skin, and mucosal fungal infections: an update. Microorganisms 13(6):1406. https://doi.org/10.3390/microorganisms13061406

Singh S, Bajaj BK (2017) Potential application spectrum of microbial proteases for clean and green industrial production. Energy Ecol Environ 2(6):370–386. https://doi.org/10.1007/s40974-017-0076-5

Singh D, Rai V, Agrawal DK (2023) Regulation of collagen I and collagen III in tissue injury and regeneration. Cardiol Cardiovasc Med 7(1):5. https://doi.org/10.26502/fccm.92920302

Song JY, Kang HA, Kim MY, Park YM, Kim HO (2004) Damage and recovery of skin barrier function after glycolic acid chemical peeling and crystal microdermabrasion. Dermatologic Surg 30(3):390–394

Song P, Zhang X, Wang S, Xu W, Wang F, Fu R, Wei F (2023) Microbial proteases and their applications. Front Microbiol 14:1236368. https://doi.org/10.3389/fmicb.2023.1236368

Suleman L (2016) Extracellular bacterial proteases in chronic wounds: a potential therapeutic target? Adv Wound Care 5(10):455–463. https://doi.org/10.1089/wound.2015.0673

Takami H, Nakamura S, Aono R, Horikoshi K (1992) Degradation of human hair by a thermostable alkaline protease from alkaliphilic Bacillus sp. no. AH-101. Biosci Biotechnol Biochem 56(10): 1667–1669. https://doi.org/10.1271/bbb.56.1667

Tang Y, Nie T, Zhang L, Liu X, Deng H (2025) Peptides in cosmetics: from pharmaceutical breakthroughs to skincare innovations. Cosmetics 12(3):107. https://doi.org/10.3390/cosmetics12030107

Toledo AV, Franco MEE, Lopez SMY, Troncozo MI, Saparrat MCN, Balatti PA (2017) Melanins in fungi: types, localization and putative biological roles. Physiol Mol Plant Pathol 99:2–6. https://doi.org/10.1016/j.pmpp.2017.04.004

Tóth GD, Koplányi G, Kenéz B, Balogh-Weiser D (2023) Nanoformulation of therapeutic enzymes: a short review. Period Polytech Chem Eng 67(4):624–635. https://doi.org/10.3311/PPch.22826

Tran-Ly AN, Reyes C, Schwarze FW, Ribera J (2020a) Microbial production of melanin and its various applications. World J Microbiol Biotechnol 36:1–9. https://doi.org/10.1007/s11274-020-02941-z

Tran-Ly AN, Ribera J, Schwarze FW, Brunelli M, Fortunato G (2020b) Fungal melanin-based electrospun membranes for heavy metal detoxification of water. Sustain Mater Technol 23: e00146. https://doi.org/10.1016/j.susmat.2019.e00146

Trevisol TC, Henriques RO, Souza AJA, Furigo A Jr (2022) An overview of the use of proteolytic enzymes as exfoliating agents. J Cosmet Dermatol 21(8):3300–3307. https://doi.org/10.1111/jocd.14673

Tsai MJ, Lin CY, Trousil J, Sung CT, Loo MH, Fang JY, Yang SC (2023) Proteinase K/retinoic acid-loaded cationic liposomes as multifunctional anti-acne therapy to disorganize biofilm and regulate keratinocyte proliferation. Int J Nanomedicine:3879–3896. https://doi.org/10.2147/IJN.S416966

Turick CE, Caccavo F Jr, Tisa LS (2003) Electron transfer from Shewanella algae BrY to hydrous ferric oxide is mediated by cell-associated melanin. FEMS Microbiol Lett 220(1):99–104. https://doi.org/10.1016/S0378-1097(03)00096-X

Turner GA, Hoptroff M, Harding CR (2012) Stratum corneum dysfunction in dandruff. Int J Cosmet Sci 34(4):298–306. https://doi.org/10.1111/j.1468-2494.2012.00723.x

Valeru SP, Rompikuntal PK, Ishikawa T, Vaitkevicius K, Sjöling A, Dolganov N et al (2009) Role of melanin pigment in expression of vibrio cholerae virulence factors. Infect Immun 77(3): 935–942. https://doi.org/10.1128/iai.00929-08

Varani J, Dame MK, Rittie L, Fligiel SE, Kang S, Fisher GJ, Voorhees JJ (2006) Decreased collagen production in chronologically aged skin: roles of age-dependent alteration in fibroblast function and defective mechanical stimulation. Am J Pathol 168(6):1861–1868. https://doi.org/10.2353/ajpath.2006.051302

Wang L, Li Y, Li Y (2019) Metal ions driven production, characterization and bioactivity of extracellular melanin from Streptomyces sp. ZL-24. Int J Biol Macromol 123:521–530. https://doi.org/10.1016/j.ijbiomac.2018.11.061

Yang W, Ruan L, Tao J, Peng D, Zheng J, Sun M (2018) Single amino acid substitution in homogentisate dioxygenase affects melanin production in bacillus thuringiensis. Front Microbiol 9:2242. https://doi.org/10.3389/fmicb.2018.02242

Zeng Z, Guo XP, Cai X, Wang P, Li B, Yang JL, Wang X (2017) Pyomelanin from Pseudoalteromonas lipolytica reduces biofouling. Microb Biotechnol 10(6):1718–1731. https://doi.org/10.1111/1751-7915.12773

Chapter 11
Harnessing the Potential of Microbial Proteases for Sustainable Biorefinery Applications

Rajat Kani Ghosal, Sonali Mandal, Archita Vyas, Anne Jana, Vasundhara, Goverdhini, Andhere Animish, and K. Suthindhiran

Abstract The global transition to sustainable industrial processes has accelerated the investigation of microbial proteases as adaptable biocatalysts in biorefinery applications. Microbial proteases provide economical and environmentally beneficial substitutes for traditional chemical processes, aiding biofuel generation, waste reuse, and biomass conversion. They are crucial to the development of biotechnology because of their wide range of catalytic capabilities, stability in various environments, and potential for genetic optimization. The origins, categorization, and processes of microbial proteases are examined in this study, emphasizing their critical functions in protein hydrolysis, biopolymer breakdown, and the creation of value-added products. We also review current biotechnological developments, the difficulties of large-scale use, and potential future developments for improving enzyme efficiency and industrial scalability. The worldwide bioeconomy may be greatly enhanced by using microbial proteases for sustainable biorefinery applications, which minimize environmental effects while optimizing resource use.

Keywords Hydrolytic enzymes · Biocatalyst · Sustainability · Microbial proteases · Biorefinery

---

R. K. Ghosal · S. Mandal · A. Vyas · A. Jana · Vasundhara · Goverdhini · A. Animish · K. Suthindhiran (✉)
Marine Biotechnology and Bioproducts Laboratory, Department of Bio Medical Sciences, School of Biosciences and Technology, Vellore Institute of Technology, Vellore, Tamil Nadu, India
e-mail: ksuthindhiran@vit.ac.in

V. Zambare et al. (eds.), *Biotechnological Advances and Applications in Microbial Proteases*, Interdisciplinary Biotechnological Advances,
https://doi.org/10.1007/978-981-95-9011-7_11

11.1 Introduction

11.1.1 Overview of Biorefineries and Their Importance

A cellulose refinery is an enterprise that utilizes equipment and procedures to convert feedstock into valuable chemical compounds, fuels, and energy. A biorefinery is like a modern petroleum refinery, transforming petroleum into a range of fuels and products (Gupta et al. 2024). A biorefinery maximizes the value gained from the raw material, biomass, by using all the elements of bioenergy and their precursors to create a variety of products. A biorefinery may create a low-value, high-volume liquid transportation fuel, such as biodiesel or bioethanol, and one or more low-volume, high-value chemical compounds. It can generate energy and process heat for its use, and maybe enough to sell to local utilities, simultaneously using CHP technology. High-value items increase profitability, high-volume fuel meets energy demands, and power production lowers greenhouse gas (GHG) emissions and energy prices in conventional power plant facilities (Gupta et al. 2024).

Biorefineries can tackle waste management and greenhouse gas emission issues while also helping maximize organic waste's energy potential (Leong et al. 2021). With the right enzymatic or chemical processing, waste from biomass can be transformed into liquid or gaseous fuels (Kataya et al. 2023). The preliminary treatment techniques employed in biorefining yield paper pulp, HFCS, cleaners, alcohol, resin-based laminates, binding agents, flavor chemicals, carbon, activated carbon, fuel supplements, unprocessed sugars, and various other items that are usually disregarded in traditional processes. This technique may use a range of resources derived from biomass, whether generated from plants or animals, further increasing its appropriateness (Banerjee 2023). Figure 11.1 displays a sustainable approach to producing fuels and chemicals from renewable biomass resources, including agricultural residues, forestry products, and municipal solid waste.

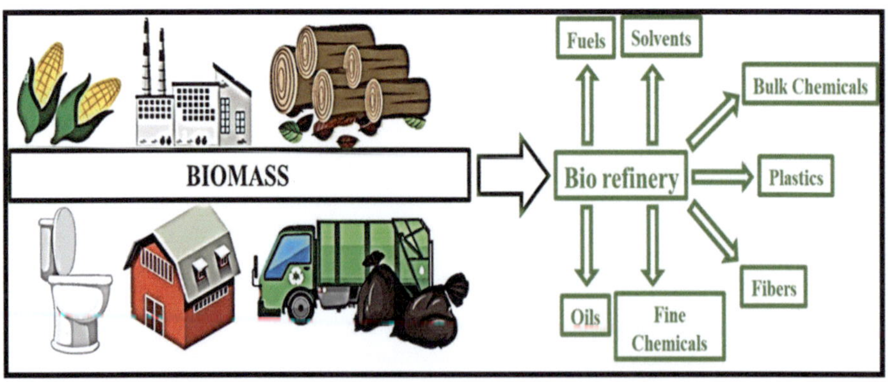

Fig. 11.1 The biorefinery: turning biomass waste into wealth through sustainable innovation

11.2 Role of Enzymes in Sustainable Biorefineries

Enzymes are natural proteins that speed up chemical reactions in living organisms. These biological catalysts are widely utilized in food processing, textiles, pharmaceuticals, and biofuels due to their high efficiency, specificity, and ability to operate in mild conditions. Industrial enzymes provide for more cost-effective and ecologically friendly manufacturing processes by removing the need for toxic chemicals, high temperatures, and high-pressure settings.

Enzymes are essential in biorefinery because they break down biomass into fermentable sugars, which are then converted into ethanol or biodiesel (Sohail 2024). Utilizing enzymes helps to optimize the production process, boost energy efficiency, and reduce environmental waste (Khambhaty 2020). The food industry uses enzymes to improve product quality and extend shelf life, and the pharmaceutical industry produces drugs more accurately and with fewer by-products (Kumar et al. 2024).

Biocatalysis, using microorganisms in chemical synthesis, is becoming increasingly popular because it may improve reaction effectiveness and specificity while using less hazardous chemicals (O'Connell et al. 2024). This is consistent with the principles of environmental synthetic chemistry, which aims to reduce the harm that chemical production does to the natural world and to the well-being of people.

11.3 Significance of Microbial Proteases in Industrial Applications

Microbial proteases have multiple applications in various industries (Fig. 11.2).

Since the late 1990s, protease enzymes, which make up about 20% of all enzyme sales worldwide, have been incorporated into surfactants to raise the caliber and effectiveness of detergents. By breaking down food, they shorten the time needed for agitation and soaking. Enzymes called Serine alkaline protease are utilized for trash management and housekeeping tasks. However, handling them requires caution due to their high cost and probable respiratory sensitivity (Solanki et al. 2021). The leather industry uses protease enzymes to lower pollutants and enhance leather quality. Alkaline proteases are becoming more popular because of their capacity to lessen chemical reduction during the leather-making process, as well as their keratinolytic and elastolytic qualities. Microbial alkaline proteases support environmentally friendly operations and safer enzymatic processes (Khambhaty 2020).

Waste materials of all kinds, including toxic, liquid, and solid, are broken down by protease enzymes. They help with metabolic functions like clearing clogged pipes of hair. Alkaline protease enzymes lower cleaning expenses and environmental hazards when employed in the food industry and at home to handle waste (Rao et al. 1998). Myofibrillar and connective tissue proteins are impacted by proteases in meat, which partially break down muscle proteins and covalent connections.

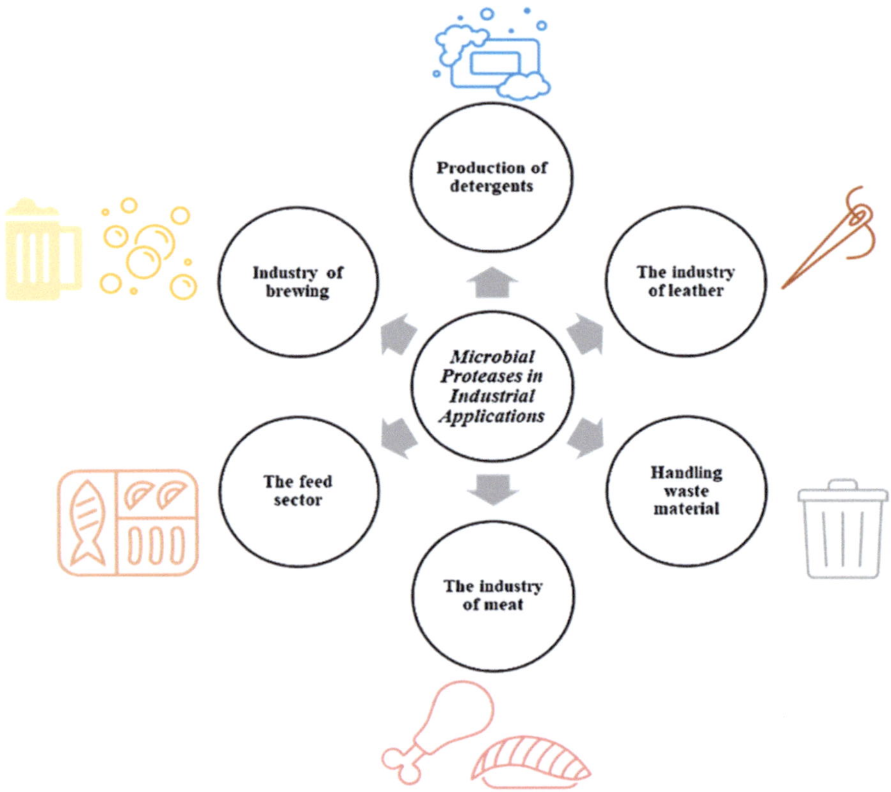

Fig. 11.2 Microbial proteases for various industrial applications

Proteolytic enzymes become more accessible when meat is softer, hydrated, and protein-digestible, improving its flavor, texture, and color. Protease-supplemented meat may help stop lipid oxidation (Tipton et al. 2018).

This type of polypeptide and proteolytic enzymes may naturally remove fibrous animal-based protein from the nails, hair, horns, and feathers to make them suitable for animal feed. Enzymes that digest proteins improve animal feed's digestibility, according to studies on amino acids in livestock. Lactic acid bacteria (LAB) are widely used in industrial, medicinal, and agricultural uses because of their potent antibacterial activity and lytic qualities (Raman et al. 2022).

In the brewing industry, amino acids are used to break down into proteins, prevent wine from hazing, and enhance clarity and nourishment. Malted grains require natural proteases, but uniform protein dispersal can make the brew taste bland and ineffective, which makes filtering challenging (Erdmann-Pham et al. 2020).

11.4 An Overview of Microbial Proteases

11.4.1 Types and Classification of Proteases

According to the groups of amino acids in their sites of action, proteases are divided into six categories: metallo, glucose, aspartic, cysteine, serine, and threonine (López-Otín and Bond 2008). Proteases are divided into four distinct groups of Group 3 (hydrolases) by the International Union of Biochemistry and Molecular Biology's Nomenclature Committee. Protease is further divided into two groups: exopeptidase and endopeptidase, based on where the peptide bonds are broken (Li 2021). They are further divided into serine proteases, amino acid proteases, disulfide protein proteases, and metalloproteases based on their mechanism of action (Jisha et al. 2013). The various protease types and their modes of action are shown in Table 11.1 (Rani et al. 2012).

11.4.2 Sources of Microbial Protease

Microbial peptidases can be obtained and extracted from various sources, and their nomenclature often varies depending on their origin. The primary substrates of proteases are discussed in the following section (Jisha et al. 2013).

11.4.2.1 Microbial Protease

Since microorganisms can grow rapidly and are easier to cultivate than plants or animals, they are often used for high-scale production of proteases. Proteases are an important product in many industries; these include detergents, medicines, leather, food, and agriculture (Song et al. 2023). They constitute two-thirds of the global enzyme market. The industrial enzymes account for over 60% of the over $3 billion in global sales. Microorganisms release proteinases and peptidases that are used to break down proteins (Song et al. 2023). Proteases exist in different forms depending on the producer species. Both intracellular and extracellular proteases are crucial for cellular functions and protein utilization, and alkaline proteases are helpful in environmentally friendly applications (Ward 2011).

It can be distinguished into two subclasses: (1) fungal protease, (2) bacterial protease.

1. Fungal protease

These enzymes have drawn the attention of researchers because of their great variety, wide substrate specificity, and stability in harsh environments. They also have the benefit of separating mycelium via simple filtering. The solid-state

Table 11.1 Different types of proteases and their mechanism of action

Sl. no.	Protease	Mechanism of action
1	Exopeptidase	Cleaves near the terminus of the peptide chain Based on their site at the N and C terminus, they are classified as aminopeptidase and carboxypeptidase, respectively
2	Endopeptidase	According to their action on the inner region of the peptide bond (away from the N and C terminal) They are divided into four subgroups: (1) aspartic protease, (2) serine protease, (3), cysteine proteases, (4) metalloprotease
3	Dipeptidase	Type of exopeptidase specific for dipeptides
4	Aspartic protease	This aspartic acid protease, a type of endopeptidase, depends on aspartic acid residues for its catalytic activity Works on low pH (3–4) Mostly found in viruses rather than bacteria
5	Serine protease	The presence of a serine molecule in the active area is a characteristic of serine proteases Serine peptidase is widely found in exopeptidase, endopeptidase, oligopeptidase, and omega peptidase groups Bacteria, molds, fungi, and yeasts widely produce it Serine peptidase activity is increased in neutral and alkaline pH conditions (7–11) It can be irreversibly inhibited by phenylmethylsulfonyl (PMSF), diisopropyl fluorophosphate (DFP), and tosyl-L-lysine chloromethyl ketone (TLCK)
6	Cysteine protease	It occurs in both prokaryotes and eukaryotes Depending on the cysteine and histidine residue, this is classified into 20 subgroups Based on their side chain specificity, they are broadly divided into three groups: (1) papain-like, (2) trypsin with specificity for arginine residue, and (3) glutamic acid specificity They are mainly susceptible to sulfhydryl agents such as PCMB but are unaffected by DFP and metal-chelating agents such as EDTA and have neutral pH optima
7	Metalloprotease	These subclasses use divalent metal ions for their catalytic mechanisms This group of proteases can be found in higher organisms

fermentation technique makes it easy to produce fungal proteases. Food proteins can also be modified by fungal proteases (Ward 2011).

2. Bacterial protease

Due to their strong catalytic activity and manufacturing capability, they are used in the silk and silk industries (Nguyen et al. 2019). Its defining characteristics are the high activity of bacterial alkaline proteases at alkaline pH (8–12) and an ideal temperature range of 50–70 °C (Gautam 2024). Based on their characteristics, bacterial alkaline proteases are suitable for use in the detergent industry.

- *Bacillus thuringiensis* proteases

Many ecological environments are inhabited by the gram-positive bacterium *Bacillus*. Because they produce proteins, these bacteria are important for various industrial uses and are considered safe by food and medicine regulatory agencies (Raddadi et al. 2012). Few *Bacillus* species are toxic, but most can produce enzymes, antibodies, and insecticides. Especially notable for its insecticidal proteins, *B. thuringiensis* is a common biopesticide and an important source of proteases.

- Alkaline proteases from *Bacillus* sp.

Alkaline proteases, mostly secreted by *Bacillus* bacteria under alkaline conditions, are significant to many industries, mainly as detergents, because they cleave proteins (Song et al. 2023). Two *Bacillus* species are significantly efficient in producing these enzymes, namely *RGR-14* and *B. firmus MTCC7728* (Ortiz and Sansinenea 2022). One of the intracellular proteases characterized among the proteases of *Bacillus thuringiensis* has been observed to have high activity at 60 °C (Mohan and Gujar 2003). The generation of alkaline proteases depends on specific growth conditions and nutrient sources, especially during the stationary growth phase. Typically, enzyme synthesis is related to the growth phase of the organism and varies among different *Bacillus* species (Sun et al. 2023).

11.4.3 Unique Properties of Microbial Peptidase

Many studies have been carried out on alkaline and alkalophilic microbial proteases. *Bacillus RV. B2.90* has been developed for different bio-industries producing foodstuffs, drugs, tanneries, and detergents. Alkaline proteases are most active in alkaline conditions (Nigam 2013). Details of microbial peptidase strains, their sources, optimal temperatures, and applications are presented in Table 11.2.

11.5 Mechanisms of Protease Action

11.5.1 Hydrolytic Mechanism of Protein Breakdown

Protease Ti is a novel energy-dependent protease found in *Escherichia coli* that breaks down ATP and peptides simultaneously. Its selectivity and ATP function are catalyzed by distinct components. Fluorogenic peptides such as N-succinyl (Suc)-Leu-Tyr-ami-domethylcoumarin (AMC), Suc-Ala-Ala-Phe-AMC, and Suc-Leu-Leu-Val-Tyr-AMC are hydrolyzed by protease Ti. Nevertheless, it has minimal efficacy against hydrophobic or basic peptides, particularly those broken down by protease (Mechesso et al. 2025). Protease is the most well-characterized of the two different ATP-mediated protein proteases that have been discovered in *E. coli* cells (Rotanova et al. 2006).

Table 11.2 Geographical distribution, strain origin, temperature preference, and application spectrum of microbial peptidases isolated from various regions

Country	State	Strains	Temperature	Application
India	Tamil Nadu, West Bengal	*Bacillus licheniformis*, *Pseudomonas aeruginosa*	25–45 °C	Bioremediation, food processing
United States	California, Wisconsin	*Bacillus subtilis*, *Aspergillus oryzae*	20–50 °C	Detergents, pharmaceuticals
Japan	Kyoto, Hokkaido	*Aspergillus sojae*, *Bacillus amyloliquefaciens*	15–40 °C	Soy sauce production, Enzyme supplements
Germany	North Rhine-Westphalia, Saxony	*Bacillus clausii*, *Thermoactinomyces vulgaris*	10–50 °C	Medical enzymes, textile processing
Brazil	São Paulo, Bahia	*Aspergillus niger*, *Penicillium* spp.	25–45 °C	Bioethanol, dairy industry
South Africa	Gauteng, Mpumalanga	*Bacillus cereus*, Thermophilic bacteria	20–60 °C	Waste management, biotechnology
Australia	Victoria, Queensland	Halophilic bacteria, Fungal proteases	15–50 °C	Salt-tolerant industries, biotechnology
China	Shandong, Yunnan	*Bacillus subtilis*, *Actinobacteria*	10–40 °C	Detergents, food industry
Russia	Krasnoyarsk, Altai Republic	Psychrophilic bacteria, *Bacillus* sp.	−5 °C to 20 °C	Cold-active enzymes, biotechnology
Canada	Yukon, Alberta	Cold-adapted fungi, *Bacillus thuringiensis*	−10 °C to 15 °C	Agriculture, biofuels

Depending on how many amino acids act as disintegration sites, proteases are categorized as certain enzymes. Certain enzymes hydrolyze peptide bonds that sit alongside one or two types of amino acids, yet their hydrolysates frequently contain a large number of particular peptides. Certain structural characteristics or certain amino acids can occasionally cause impulsive cleavages in intestines, peptide hormones, and vasoactive gastrointestinal peptides (Lopes et al. 2022). When proteins are hydrolyzed by enzymes, spontaneous cleavages may happen even at extremely low reaction rates (Butré et al. 2015).

The ubiquitin-induced process of ATP-mediated protein degradation and the potential importance of ubiquitin–histone conjugates in chromatin are covered in detail by the lysosomal system considering the evidence for a variety of protein mechanisms of degradation in vivo (Ciechanover et al. 1984). The reader is directed to previous reviews (Butré et al. 2015; Ciechanover et al. 1984; Woo et al. 1989; Fang et al. 2019; Pogson et al. 2009) and articles from recent symposia for in-depth considerations of the numerous features and physiological implications of intracellular protein breakdown. An earlier review presented information on the proteolytic system before 1981 (Alexander et al. 2015).

Enzyme Ti, an ATP-dependent protease found in *Escherichia coli*, concurrently breaks down proteins and ATP (Hwang et al. 1987). With a variety of fluorogenic peptides as substrates, the enzyme's capacity for selectivity and the role of ATP were investigated. Protease Ti hydrolyzed N-succinyl (Suc)-Leu-Tyr-aminomethyl coumarin (AMC), Suc-Ala-Ala-Phe-AMC, and Suc-Leu-Leu-Val-Tyr-AMC, but it degraded relatively few basic or polar peptides (Tallorin et al. 2018). The same peptides as the total enzyme are quickly broken down by component P's serine-active region. Peptide hydrolysis was not considerably impacted by the inclusion of element A, which houses the ATP-hydrolyzing site. N-ethylmaleimide, which deactivates the ATPase, did not raise component A's ATPase activity or prevent peptide hydrolysis. Suc-Leu-Tyr-AMC and diisopropyl fluorophosphate, respectively, inhibited the dehydration of both protein and polymer substrates at elevated levels (Woo et al. 1989).

11.5.2 Factors Influencing Protease Activity and Stability

Chemical reactions can be accelerated by enzymes, which are partially biological catalysts. Enzymes are necessary for the majority of metabolic activities to proceed rapidly enough. Enzymes in the cell control the metabolic pathway to sustain life. Biological substances called enzymes quicken biological and metabolic processes. They function as chemical catalysts, accelerating these internal and external reactions within the organism's cells.

Furthermore, enzymes carry out several tasks. For organisms to survive, they are primarily in charge of inter-chemical conversion. Nature has several kinds of enzymes. Every enzyme functions differently on its own. Enzymes can break down polymeric molecules into their constituent monomers or vice versa. Furthermore, certain enzymes change the molecule's conformational structure. The type of function the enzyme performs determines the names assigned to it.

Large protein molecules are broken down by proteases and separated into shorter chains. The peptide in the bridges that connect the amino acids in the peptide chain that makes up the protein is hydrolyzed by the enzyme, which breaks down the lengthy protein chain. One of the enzymes that aids in digestion is protease, which breaks down proteins in the stomach and intestines into their constituent amino acids. Proteases are of two types: internal and external.

The various factors influencing the milk sorting medium's ability to produce the protease enzyme from the *B. cereus* isolate are as follows:

(a) *pH*

The ideal pH for growth and enzyme production was found by preparing the skim milk agar medium and using a pH meter to adjust the pH to 4, 6, 7, 8, and 9. The bacterial isolate *B. cereus* was then cultivated in Petri dishes using sterile media and, in terms of nutrients, broth medium, and then incubated for 24 h at 37 °C. A hole formed in the milk sorting dish with a diameter of 1 cm was then filled with 40 μL. The ideal pH was ascertained by measuring the diameter of the transparent zone during a 24-h incubation period at 37 °C (Woo et al. 1989).

(b) *Temperature*

In a 1 cm diameter hole in the middle of the dish, 40 μL of bacterial culture were added to the skim milk agar medium. Next, for 24 h, the dishes were incubated at various temperatures: 30, 37, 45, 50, 55, and 60 °C. This was done to ascertain the ideal temperature for the diameter of the transparent zone, which was measured after the *B. cereus* bacteria produced the protease enzyme to calculate the ideal temperature (Woo et al. 1989).

(c) *Incubation*

By measuring the diameter of the transparent zone that formed surrounding the bacterial growth, the *B. cereus* isolate was tested on skim milk medium at 37 °C for 24–48 h, which is the ideal incubation period for enzyme synthesis (Alexander et al. 2015).

(d) *Optimal Ventilation*

To find the ideal ventilation conditions for enzyme production, 40 μL of bacterial inoculum were added to the middle of the skim milk agar medium. The medium was then incubated in a standard incubator for 24 to 48 to 72 h at 37 °C. The diameter of the transparent zone was determined (Alexander et al. 2015).

The ongoing search for new strains with better qualities and yield is progressing to satisfy the increasing demand for these enzymes. Microorganisms must produce enzymes under the right circumstances and with optimization. The most popular way for optimizing the output parameters to raise the enzyme yield is the traditional "one-at-a-time approach." Statistical optimization techniques such as Plackett and Burman or Response Surface Method (RSM) are now commonly utilized because they are time-consuming. This method includes a summary of regression and factorial design, aids in assessing the important variables, creates models to comprehend the interaction, and chooses the best variables for the intended outcome (Andrade 2024).

11.5.3 Engineering Proteases for Enhanced Efficiency

Keratinolytic proteases are important in waste management, leather processing, feed additives, and prion hydrolysis, as they can hydrolyze proteins with many disulfide bonds (Vidmar and Vodovnik 2018). Wild-type microorganisms used for traditional production lack the desired cell density and enzyme production. Optimizations of the medium and strategies for protein engineering have been established but are associated with several disadvantages, such as low extracellular yield, specificity, and thermostability (Listov et al. 2024). Protein engineering and heterologous expression have been utilized to increase the stability of keratinolytic proteases in complex environments, but this is limited. The N-terminal property for *E. coli* has increased the secretion and maturation of keratinolytic protease (Zhao et al. 2024). This work will use protein engineering to design an engineered mutant with high specific activity and to secrete it extracellularly (Fang et al. 2019).

Proteases are a vital element of human genes necessary for cell survival and functioning. They could recognize a restricted collection of peptide sequences, making it helpful for industrial applications and biological research. Their selectivity and specificity are unaffected, but they might also be employed as therapeutic agents, such as tissue plasminogen activators. Engineering exactly the protease substrate specificity should lead to using designed proteases as medications, hence decreasing dosage and treatment costs (Meister et al. 2023). This has led to firms such as Catalyst Biosciences and DIREVO Biotech (Pogson et al. 2009).

Soy protein is hydrolyzed in phases by extracellular proteases of *Aspergillus oryzae*. Additionally, as the salt-free soy sauce is a result of *A. oryzae* fermenting in an ethanol medium, its flavor impact is rather comparable to that of the commercial soy sauce fermentation stage since the ideal pH of alkaline protease and acid protease is 4.0 or 9.0, respectively (Hu et al. 2020).

One zinc atom per molecule makes up the class of metallic endo proteinases known as the three types of proteases released by *A. oryzae*, which are classified as acid, alkaline, and neutral proteases based on the pH optimum. In an acidic environment, alkaline protease will eventually lose its action. Neutral proteases (NPs) that exhibit the best activity (at pH 7.0) should take the lead, particularly in the second.

Three NPs were identified by the *A. oryzae RIB40* genome sequencing result: NPI, NPII, and NPIII. In applied research, NPI has received increased attention (Zhong et al. 2018). According to gel filtration, NPII's molecular weight of 19.8 kDa is significantly lower than NPI's. NPI has significant activities on common laboratory substrates such as milk casein, hemoglobin, albumin, and gelatine, but NPII has poor activities on them (Cupp-Enyard 2008). The transcription factor controls the expression of NPIII, a secreted zinc protease that belongs to the M36 protein family (Li 2021). The NPI gene encodes 634 amino acid residues, including single peptides (1–18 aa), properties (19–245 aa), and mature polypeptides (246–634 aa), with an open reading frame of 1905 bp. *Pichia pastoris* showed significant levels of expression, and glycosylation analysis of *P. pastoris* has been reported (Li 2021). Although NPI's ideal pH is 7.5, acidic conditions rendered it inactive, and at pH <6.0, its enzyme activity rapidly declined (Salelles et al. 2021). Therefore, an acidic pH will significantly limit the NPs' enzymatic activity and lower the raw materials' utilization efficiency throughout the fermentation process. Therefore, it is essential to use the right breeding method to increase the activities of NPs at acidic pH to produce high-quality, low-salt, and salt-free soy sauce (Ngoune Tandzi et al. 2018).

Direct evolution is one of the most important technologies for examining the structure–function connections of enzymes (Banerjee and Ray 2017). Neutral enzymes' characteristics or activity have been altered using this technique to produce more beneficial qualities (Solanki et al. 2021). Several researchers adopted in vivo laboratory evolution to explore nonnatural features, including guided evolution and rational design (Andraos and Matlack 2022). Using the method of rational design, *Candida rugosa lipases* (CRL) can have the most thermostable mutant,

Asp457Phe, which exhibits a 5.5-fold extended half-life at 50 °C, an increase in the ideal temperature range (González-Siso 2019). In their structure-based rational design of a thermostable phytase (PhyAn), Zhou et al. obtained the mutant Y65H enzyme, which improved stability in more acidic settings and changed the pH from 5.0 to 3.0 (Zhou et al. 2019).

The error-prone PCR to create a random mutated library of NPI genes with 2104 variants to adapt to the severe industrial environment, and 50% of the variants still exhibited no protease activity (McCullum et al. 2010). An enzyme strain with a mutation whose pH optimum was 5.5 was eliminated (Hu et al. 2020). DNA sequencing revealed the mutation of three amino acids (Singh 2020). Site-specific mutation was carried out at three different sites to confirm which site offered the biggest influence on the ideal reaction pH (Ikebe et al. 2021). Error-prone PCR was used to initially alter the recombinant NPI gene to increase catalytic effectiveness at acidic pH, which would be more useful in the fermentation field (Hu et al. 2020).

11.6 Detailed Overview of Proteases' Industrial Applications

11.6.1 Protease in Industrial Sludge Disposal Management

One worldwide industry-related concern is the disposal of industrial sludge (Shete and Deshpande 2022). Most industries today use a lot of land for sludge disposal, one of the four main methods (landfill, incineration, sea discharge, and land application). According to a recent US EPA estimate, 54% of the 6500 municipal landfill sites would be closed during the next 5 years, and half of the landfills were closed by the end of the preceding century. The ultimate disposal pathway for surplus sludge produced by the treatment of wastewater also evolved into a significant problem, mostly because of the increase in population, the buildup of sludge in major cities, and the complexity and volume of associated industrial activity. Consequently, it is imperative to decrease the quantity of sludge produced by enhancing the effectiveness of the treatment processes or introducing novel technologies that can utilize sludge as a raw material for producing biomolecules, enzymes, compost, and energy (Ahmad et al. 2017). Protease can be found in sludge, a concentrated, semiliquid mass that remains after treating wastewater or industrial water (Karn and Kumar 2015).

Using industrial sludge for protease recovery is a cost-efficient and ecologically acceptable method to bio-convert sludge into value-added products, and is increasingly important, since sludge management poses a considerable environmental challenge owing to stringent ecological constraints on sludge. Proteases may now be effectively removed from sludge due to the development of biotechnological techniques, which may also expand their many applications to provide long-term environmental protection (Karn and Kumar 2015). Because they can break apart bonds in organic compounds and promote their conversion into less harmful and more biodegradable forms, proteases are also an excellent tool for improving bioremediation (Razzaq et al. 2019). Proteases and proteolytic enzymes, which belong to a broad family of catalysts that accelerate the breakdown of amino acids and are essential for both physiological and commercial reasons, account for over 60% of all enzyme sales globally (Sharma et al. 2017).

11.6.2 Protease in the Food Industry

Numerous food-related processes employ proteases. The main applications of proteases in the food industry include the preparation of baby food, the debittering of protein hydrolysates and protein supplements, the development of various fermented foods and drinks, such as cheese and beer, the removal of turbidity, and so on (Banerjee and Ray 2017). Cold-adapted proteases are produced by psychrophilic and psychrotolerant bacteria, which thrive at temperatures between 20 and 35 °C. By hydrolyzing proteins, softening cold meat and improving flavor, accelerating the maturing of cheese, and digesting haze-producing proteins in beer, these enzymes are used to create functional food components (Kuddus and Ramteke 2012).

The dairy sector mostly uses proteases to produce cheese. Microbial milk coagulant proteases comprise an acid aspartic protease family, with molecular weights ranging from 30,000 to 40,000 (Rao et al. 1998). The primary enzyme that has historically been employed to cause milk to coagulate while manufacturing cheese is chymosin, also known as rennin (Nicosia et al. 2022). The primary function of acid proteases in cheese manufacturing is to cleave peptide bonds to generate para-K-casein and macro peptides. Because of its high casein specificity, chymosin is used to make cheese. Aspartic proteases produced by GRAS (seen as genetically safe) bacteria such as *Endothia parasitica*, *B. subtilis*, and *Mucor miehei* are quickly replacing chymosin in the manufacturing of cheese (Mamo and Assefa 2018).

Numerous microbial proteolytic enzymes break down the connective tissue proteins and myofibrils present in muscle (Rao et al. 1998). Aspartic and neutral proteases from *Aspergillus oryzae* and *Bacillus subtilis* are used to tenderize meat. The most well-known bacterial enzymes that soften meat are subtilisin and the neutral proteases produced by *B. subtilis*, *B. alcalophilus*, *B. licheniformis*, and *B. lentus* (Ao et al. 2018). *Chryseobacterium* soli-derived freeze-activated peptidases were utilized for myofibril breakdown or beef tenderization. The proteolytic enzyme's rate of myofibril disintegration rose quickly to 221%. A *Pseudoalteromonas* sp. protease improves the quality of cold meat by releasing more essential and tasty amino acids than mesophilic proteases (Mageswari et al. 2017). When soybean proteins are treated with alkaline (Novozymes), the hydrolysates produced have a high protein yield, low bitterness, and excellent solubility (Rao et al. 1998). Serine proteases from *Bacillus licheniformis*, *Thermus mediterraneanus*, and *Thermoactinomyces vulgaris* are assessed in bread goods.

11.6.3 Proteases in the Leather Industry

One of the significant businesses directly tied to daily life is leather processing. Even though the leather business significantly boosts the economy, it also causes substantial environmental contamination by using chemicals and discharging numerous harmful substances. Hair removal from hides and skins is the first stage of leather processing in the beam house process (Khambhaty 2020). Hair removal from hides is known as unhairing. When using the lime sulfide-unhairing technique, lime's high alkalinity and suspended particles cause wastewater issues. However, this sulfide also releases toxic hydrogen sulfide, posing a major risk to sewer men and tannery workers (Malone Rubright et al. 2017).

In the leather business, various wastes are created during the hide-making process. Over 90% of this debris is made up of keratin, which is extremely inert, insoluble in water, and not biodegradable by the majority of proteolytic enzymes, including papain, pepsin, and trypsin (Akhter et al. 2020; Onifade et al. 1998). The costly, non-sustainable, and energy-intensive traditional methods of disposing of keratin, such as incineration, landfilling, and chemical treatments, lead to pollution of the environment and the depletion of natural resources (Onifade et al. 1998; Zhou et al. 2018). Protease, an enzyme that carries out synthetic and biodegradative tasks, is ubiquitously present in microbes, flora, and fauna.

In the biodegradation of collagen, microbial keratin-digesting proteases are considered a promising alternative to chemical techniques due to their accessibility, rapid development, high yield, low cost, minimal space need, and long-lasting nature (Lange et al. 2016; Vidmar and Vodovnik 2018). Bacteria may produce huge amounts of keratinolytic proteases, mostly from outside sources (Lange et al. 2016). Gram-positive *Actinomycetes*, *Streptomyces*, *Arthrobacter*, *Bacillus*, and *Clostridium* are the most powerful and dynamic makers of keratino-lytic proteases; they convert refractory keratin into the simplest molecules with outstanding catalytic capacity and specificity for the substrate (Turk 2006).

11.6.4 Proteases in Biofuel

Biofuels are considered the preferred fuels of the future and renewable resources since they are derived from biomass. Biofuels are produced by anaerobic digestion using biological components from bacteria, plants, animals, and agriculture (Love 2022). Biomass is an alternate energy source with three categories of organic materials: solid, gaseous, and liquid. Due to their considerable success, affordability, and massive energy production, solid and liquid biofuels are widely utilized for energy generation (Shah and Deka 2024). Carbon-neutral and sustainable biofuels are made from biomass, including biohydrogen, bioethanol, biodiesel, biomethane, and bioelectricity. Biofuels come in four varieties: the initial, subsequent, the third, and the final one (Shah and Deka 2024).

Third-and fourth-generation biofuels are more desirable than their predecessors because of their inherent properties, which include greater lipid content, biomass yield, high manufacturing rates, enhanced capacity to collect CO_2, less space and time requirements, and so on.

Various techniques, including thermochemical, mechanical, physical, enzymatic, and microemulsion, could be utilized for biofuel production. The enzymatic approach is regarded as the most environmentally friendly, economical, simple, safe, and clean of these procedures. Enzymatic techniques are becoming more widely accepted and are intended to overcome the drawbacks of traditional chemical catalysts, enhance productivity and efficacy, lessen adverse ecological effects, and enhance the quality of the generated biofuels; from an economic perspective, repeated use of the enzyme is crucial. Enzymatic hydrolysis, for example, has a far lower utility cost than other acidic hydrolysis techniques because it is performed under moderate circumstances and does not require further treatment procedures.

The enzyme known as protease uses proteolysis to break down the amino acids in the feedstock that is utilized to make biofuel. They can be either exopeptidases, which extract amino acid sequences from both the nitrogen and carbon terminals, or endopeptidases, which disrupt the bonds that are within chains of polypeptide (López-Otín and Bond 2008).

Proteases are produced by many bacteria, yeasts, and fungi, including *Bacillus subtilis*, *Aspergillus oryzae*, and *Penicillium roqueforti*, to name a few. By giving microbes nitrogen and higher alcohols produced from longer-chain 2-keto acids for biofuel conversion, proteins improve their efficiency in the production of biofuels (Lee et al. 2016).

11.7 Sustainability Implications

11.7.1 Environmental Benefits of Using Microbial Proteases

Microbial proteases have emerged as promising solutions when integrated with biorefinery applications, offering significant environmental benefits compared to traditional chemical methods (Song et al. 2023). Their application reduces the reliance on harsh chemical treatments and optimizes resource use in various bio-based production systems (Kurniawati et al. 2023). Since they solve waste management issues and minimize ecological footprints, microbial proteases positively affect the environment and make substantial contributions to the development of sustainable approaches in various sectors.

By efficiently processing high-protein waste, microbial proteases provide environmentally friendly and sustainable waste management options (Uekert et al. 2025). The tannery microbiota-derived enzyme PersiProtease1 has been used to process whey protein, chicken feathers, and X-ray film waste, as well as to break down the proteins in tannery wastewater and dehair sheepskins (Song et al. 2023). Additionally, it has been shown that proteases can effectively deproteinize shrimp waste to extract chitin. While neutral proteases increase the effectiveness of sludge dewatering, alkaline proteases encourage the formation of short-chain fatty acids from waste-activated sludge, lowering carbon emissions (Song et al. 2023). Moreover, microbial proteases efficiently control the odor-causing skatole in pig manure, while keratinases convert poultry feathers into organic fertilizers, reducing pollution. Additionally, proteases ensure waste management and environmental safety by eliminating protein residues from food processing facilities (Razzaq et al. 2019).

Microbial proteases are essential in sustainable waste management, particularly in degrading protein-rich waste from food processing and household activities (Razzaq et al. 2019). These enzymes are highly effective at breaking down tough materials, such as feathers, which are challenging to degrade naturally (Barclay and Acharya 2023). The process begins with the enzyme binding to the substrate, breaking down the loosely bound proteins. The more tightly bound core of the waste is then solubilized more slowly, depending on the enzyme's diffusion rate and the structure of the substrate (Nomura et al. 2007). *Bacillus* species are known for producing keratinases that can efficiently degrade keratin, making them ideal for feather waste management (Li 2021). This process addresses waste disposal and transforms it into valuable resources, such as protein-rich material for animal feed. The use of microbial proteases can reduce environmental pollution, offering a more sustainable and eco-friendlier alternative to traditional waste management methods (Razzaq et al. 2019).

11.7.2 Reduction of Energy and Chemical Input in Biorefineries

Microbial proteases establish a foundation for a future defined by enhanced sustainability and renewable energy sources. Microbial proteases have shown great potential in improving the sustainability of biorefineries by significantly reducing energy and chemical usage (Razzaq et al. 2019). A primary benefit of microbial proteases is their ability to function effectively at lower temperatures, thus cutting down on energy consumption. For instance, cold-adapted proteases perform well at ambient temperatures, eliminating the need for the high energy costs associated with heating during traditional enzyme processes. This is particularly useful in biorefineries where temperature control can be a major energy expenditure (Shahzad et al. 2024).

Moreover, proteases can break down complex protein-rich wastes into simpler components without harsh chemicals. This makes them an ideal tool for waste management in industries like food processing, where enzymes help solubilize proteinaceous materials, which can then be converted into valuable products like livestock feed or biofuels (Sharma et al. 2024). Using proteases in such contexts reduces reliance on chemical treatments that might otherwise require significant energy inputs or produce toxic by-products (Song et al. 2023).

Beyond waste processing, microbial proteases also enhance the overall efficiency of bioconversion processes in biorefineries. For example, proteases are instrumental in degrading difficult-to-process substrates, such as keratin in feathers or shrimp shells, into usable compounds, making the recycling process more sustainable (N. Jisha et al. 2013). Proteases are enzymes that are utilized by biorefineries to lessen their environmental impact and support a circular economy, which minimizes waste and efficiently reuses resources (Sakarika et al. 2024). Because of these benefits, microbial proteins are a great tool for biorefinery operations that want to increase production while lowering their footprint on the environment.

11.7.3 *Life Cycle Assessment of Protease-Driven Processes*

All proteins have a life cycle driven by proteases, which recycle used or unnecessary proteins while transporting and activating freshly created, would-be proteins into their functional state. Proteases are essential for fundamental physiology and control across several systems biology levels, far from their reputation as enzymes that break down proteins. Since proteases are closely linked to illness, effective treatments are thought to target the control of proteolytic activity (Verhamme et al. 2019).

Since enzymatic procedures are targeted, quick to respond, and frequently save raw materials, energy, chemicals, and/or water compared to traditional processes, they have been used in many sectors in recent decades (Hevilla et al. 2021). In the last 15 years, several comparative environmental assessment studies have been undertaken to assess if these enzymatic process features enhance the environment and facilitate the transition to cleaner industrial production (Jisha et al. 2013).

11.8 Advances in Microbial Protease Technology

11.8.1 *The Substrate Used for Protease Production*

The following substrates were utilized to produce the protease enzyme by fermentation: the bran from wheat, the bran of rice, bagasse from sugar cane, maize bran, corn food, and the husk of corn.

11.8.2 *Microorganisms Used for Protease Production*

The capacity of various microbes to create proteases varies. A variety of microbes can thus produce proteases that humans require. The food, pharmaceutical industry, and health sectors now significantly depend on protease fermentation technology. Many microorganisms, including fungi, bacteria, and yeasts, are employed to manufacture proteases. Submerged fermentation is the primary method for producing proteases. The *Bacillus* sp. is mainly utilized in the industrial synthesis of proteases, e.g., *Bacillus subtilis*. The primary benefits of employing this microbe include that it is quicker to acquire and simpler to manage, with simplicity of handling. It provides optimum output and may digest a range of cheap basic ingredients like sugar and starch (Song et al. 2023).

11.8.3 Sources for Protease Production

Proteases are from bacteria, such as *Bacillus subtilis* and *Bacillus licheniformis* strains.

11.8.3.1 Fungal Proteases

Industrial production of fungal proteases is by *Aspergillus flavus* species, *Aspergillus wentii*, *Aspergillus oryzae*, *Mucor miehei*, *Mucor delmarxianus*, and *Amylomyces rouxii*.

11.8.3.2 Animal Protease

The most prevalent animal proteases are Trypsin, Chymotrypsin, Renin, and Pancreatic Pepsin. Pepsin is found in the bellies of vertebrates as well. From the gastrointestinal digestion path, this enzyme is Chymotrypsin from an animal's pancreas extract.

11.8.3.3 Plant Proteases

Papain is a compound that is a renowned plant protease. Papain is from papaya fruit, aspartic protease is from potato leaves, and bromelain is from stems and pulp of pineapples, which are thiol-containing proteases from the Crown Leaf (Sharma et al. 2017).

11.8.4 Techniques of Protease Production

11.8.4.1 Solid-State Fermentation

Microorganisms cultivated on solid substrates chosen for this purpose produce most compounds from the commercial process known as solid-state fermentation (SSF).

This method of cultivating microorganisms is a substitute for submerged or liquid fermentation, and it is mostly used in industry (e.g., cheese maturation, bread).

11.8.4.2 Submerged Fermentation

An industrial method known as "submerged fermentation" involves immersing catalysts and other volatile compounds in a liquid, such as alcohol or nourishing broth. The technique is employed in the commercial production of protease enzymes, among other uses.

11.8.5 Stages of Protease Production

Upstream, real fermentation, and downstream processes are the three most crucial steps in industrial protease fermentation, which are necessary to complete product fermentation.

11.8.5.1 Upstream Process of Fermentation

The first stages of the fermentation process are included in the upstream process.

To increase productivity and yield, the upstream process of protease fermenting includes preliminary steps to screen and identify appropriate microorganisms, improve them, choose substrates, and prepare culture media. It also includes maintaining strain purity, preparing inoculum, and sterilizing media and the bioreactor to increase the process's economic efficiency.

11.8.5.2 Actual Fermentation Process

Fermentation uses microbes to break down complex chemicals into simpler ones. All the ingredients are combined in a bioreactor to create the intended product. Preserving the ideal circumstances needed to produce citric acid, such as temperature, pH, oxygen concentration, and nutrients, is provided for the development and proliferation of microorganisms within a bioreactor.

11.8.5.3 Downstream Process

The downstream stage of fermentation is the retrieval of by-products from the barrel by various procedures, including the extraction process, division, purification process, and product packing.

11.8.5.4 Factors Affecting Industrial Protease Production

(a) *Temperature*

There are several ways that temperature affects the synthesis of proteases. Protease activity was most active at 40 °C, with significant activity occurring between 30 °C and 50 °C. The activity dropped to 97% of the total and 85% at 50 °C and 60 °C, respectively. The development and activity of microorganisms can be influenced by temperature. The progress of the fermentation process increases as the temperature rises within the ideal temperature range, as does the microorganisms' growth and metabolism. The microbe's activity is rapidly inactivated, and the pace of fermentation is decreased as the temperature rises over the ideal temperature range. Therefore, maintaining the ideal temperature guarantees regular fermentation (Sharma et al. 2017).

(b) *pH*

The pH varies depending on the kind of substrate, media, and microorganisms. pH can impact the activity of bacteria, digestive enzymes, and production rate. The ideal pH range for gastrointestinal proteases was around 9 and 11, while the peak activity of the crude proteolytic extract was discovered at 7.0. Additionally, the medium's nutritional breakdown will be impacted by pH. As a result, the fermentation broth's pH needs to be managed (Sharma et al. 2017).

(c) *Dissolved oxygen concentration*

Since *Aspergillus niger* constitutes an atmospheric microbe, oxygen is necessary for aerobic fermentation, which depends on the availability of oxygen. The oxygen requirement of the citric acid fermentation process is satisfied by agitation and aeration. The agitation process keeps the uniform culture environment by continuously mixing; thus, an important quantity of oxygen needs to be constantly supplied to the fermentation process of the broth, and mixing can increase the dissolution of oxygen in the fermentation broth. Agitation is crucial for heat transfer, sufficient mixing, and mass exchange. As the agitation speed rose, so did the dissolved oxygen content (Sharma et al. 2017).

(d) *Concentration of nutrients*

The proliferation of microorganisms and the accumulation of metabolites are directly impacted by the concentration of different minerals in the fermentation broth, particularly the proportion of carbon dioxide to nitrogen-based fertilizers, inorganic salts, and vitamins. These nutrients include carbon sources such as glucose, sucrose, fructose, maltose, mannose, and starch. Sources of nitrogen are nitrate ions or ammonium. Additionally, in trace metals, including the elements iron, manganese, zinc, and copper, as well as additional minerals such as phosphate, low levels of potassium, magnesium, and sulfate, for instance, the metabolic pathway will be impacted by variations in the NH_4^+ concentration through the process of fermentation of glutamate. Thus, adequate nutrients are provided for fermentation during the process (Sharma et al. 2017).

11.9 Challenges and Limitations

Microbial proteases have emerged as important biocatalysts in the search for sustainable biorefinery applications. Historically, proteases were regarded as facilitators of protein digestion. As differentiation techniques have advanced over time, our understanding of how they operate has expanded, leading to numerous practical applications (Vidmar and Vodovnik 2018; Turk 2006; Lange et al. 2016). In various sectors, including food, detergent, leather, pharmaceuticals, beauty products, silk dehydrating, silver recovery, chemical sector, and wastewater treatment, these enzymes can function as biocatalysts in the breakdown of proteins and amino acids. Microbial proteases have a lot of benefits, but some challenges and limitations need to be addressed before they are used in industry. This chapter will explore three key areas of concern: stability and scalability in industrial applications, cost efficiency and economic ability, and regulatory and safety considerations (Akhter et al. 2020).

11.9.1 Stability and Scalability in Industrial Conditions

One of the primary challenges in using microbial proteases in biorefineries is keeping them stable and working well under industrial conditions (Song et al. 2023). Proteases are sensitive to temperature, pH, and ionic strength, greatly affecting their efficiency (Yao et al. 2023). For instance, the pharmaceutical business needs highly purified fractions of proteases that have increased selectivity. In contrast, the leather and detergent companies prioritize affordability and stability in their proteases. Meanwhile, the food industry seeks proteases that are not only stable and specific but also free from toxic contaminants (Niyonzima and More 2015). Ultimately, the selection of commercial proteases depends on their specificity, stability, and cost-effectiveness for a particular industrial application.

Microbial proteases are more stable, but temperature changes to too hot or too cold, and pH changes can cause proteolytic enzymes to break down and lose their activity, making them less efficient in biorefining applications. Studies have found that many industrial processes require enzymes that can withstand high temperatures, and an appropriate pH is required due to the nature of the materials being processed. As a result, there is a growing need for proteases resistant to heat and pH (Zhou et al. 2018). Certain substances, such as ions and metabolic by-products, can slow or stop protease activity. To work effectively, proteases must resist these inhibitors in the environment (Song et al. 2023).

While scientists often look for new microbial proteases, turning lab discoveries into large-scale industrial processes can be difficult. This is because making large amounts of proteases can be expensive, and it is hard to get high amounts of the enzyme. Usually, the raw materials used to produce microbial proteases are expensive; hence, alternative cost-effective materials such as waste biomass help reduce cost (Song et al. 2023).

Another limitation is that to get a high yield of microbial proteases in biorefineries, researchers cannot get a high yield of microbial proteases in biorefineries because of variations in environmental conditions and microbial performance. To overcome this, they maintain proper microbial strains, growth conditions, and extraction methods to enhance yield and activity (Song et al. 2023).

Fermentation processes that work on a small scale may not work well in a large-scale industrial setting. Because of large-scale production, a high amount of heat and nutrient distribution occurs. To solve this problem, researchers are working to develop special microbes and improve fermentation processes. They are also using genetic engineering to make proteases more stable and active, which could help them work better in industrial conditions (Song et al. 2023).

11.9.2 Cost Efficiency and Economic Viability

Another challenge or limitation of the microbial proteases in biorefinery was widespread adoption because of their high-cost efficiency and economic viability. The initial investment in enzyme production may seem justified based on the benefits offered. The production of microbial proteases often involves advanced fermentation technologies, and important raw material increases the cost. Factors such as yield, enzyme purity, and recovery during the purification process contribute to the financial feasibility of microbial proteases. The manufacturing and processing costs of the enzyme's proteases exceed those of traditional chemical processes.

The demand for eco-friendly products is increasing, but people get them only if they are affordable. The most important challenge and limitation is to adopt cost-effective strategies to enhance the profitability of microbial proteases in biorefineries.

Several approaches can help reduce costs and enhance the economic viability of microbial protease production: (1) utilization of agricultural and industrial residues as fermentation substrates; (2) advancements in protein engineering to improve enzyme stability and specificity, thereby reducing the need for extensive downstream processing; (3) optimization of fermentation techniques to increase yield and lower operational costs; (4) co-production of microbial proteases with other value-added products to improve overall process economics; and (5) supportive government policies and funding that contribute to the economic feasibility of microbial protease production.

11.9.3 Regulatory and Safety Considerations

The effectiveness and eco-friendly characteristics make microbial proteases an important biocatalyst in biorefinery (Razzaq et al. 2019). However, this requires careful attention to regulation and safety measures. Regulatory agencies often require extensive safety evaluation and risk assessments before approving new commercial bioproducts. The genetic modifications of microorganisms, frequently utilized to produce more efficient proteases, can bring additional regulatory procedures due to potential environmental and health impacts (Razzaq et al. 2019).

Incorporating microbial digestive enzymes into manufacturing processes is also impacted by safety and legal issues. While regulatory agencies such as the US Environmental Protection Agency (EPA) and the European Food Safety Authority (EFSA) are crucial for testing and approving new biological products, they can also pose serious obstacles to the quick dissemination of bacteria-derived enzymes (Razzaq et al. 2019).

Regulatory and safety considerations also influence the integration of microbial proteases into industrial applications. Microbial proteases from genetically modified organisms (GMOs) should have proper governance using the regulatory bodies. Public perception and ethical concerns regarding the use of GMOs may interfere with acceptance and market growth (Razzaq et al. 2019).

11.10 Conclusion

Microbial protease plays an important role as a biocatalyst for making an efficient biorefinery. The utilization of microbial proteases possesses great potential for sustainable biorefinery applications. They have a versatile use in making biomass and for making various value-added products in the industry. These enzymes can play a crucial role in efficiently converting lignocellulosic biomass into usable by-products that contribute to developing a greener and eco-friendly chemical industry. The economic viability of biorefinery processes can be improved by the cost-effective production of naturally occurring enzymes from various microbes. They not only reduce the utilization of natural resources but also reduce the emission of carbon footprint. Additionally, the genetically engineered microbes strengthen the sustainability of biorefinery applications. The advancements in protein engineering, fermentation technologies, and fluorescent staining technology are stable challenges due to cost, scalability, and susceptibility. In addition to the variable economic and regulatory conditions, safety guidelines can pose some problems in their worldwide adoption. Since the demand for sustainable alternatives to fossil-based resources continues to increase, the strategic harnessing of the potential of microbial proteases in biorefinery operations will be important in bringing about the change toward eco-friendly industries. Their use in industrial biorefineries production makes a big step toward a clean, green, and sustainable future.

Acknowledgments The authors wish to extend heartfelt thanks to the management of Vellore Institute of Technology for providing the necessary facilities and support.

Competing Interests The authors have no conflicts of interest to declare that are relevant to the content of this chapter.

References

Ahmad T, Ahmad K, Alam M (2017) Sludge quantification at the water treatment plant and its management scenario. Environ Monit Assess 189(9):453. https://doi.org/10.1007/s10661-017-6166-1

Akhter M, Wal Marzan L, Akter Y, Shimizu K (2020) Microbial bioremediation of feather waste for keratinase production: an outstanding solution for leather dehairing in tanneries. Microbiol Insights 13. https://doi.org/10.1177/1178636120913280

Alexander SP, Davenport AP, Kelly E, Marrion N, Peters JA, Benson HE, Faccenda E, Pawson AJ, Sharman JL, Southan C, Davies JA (2015) The Concise Guide to Pharmacology 2015/16: G protein-coupled receptors. Br J Pharmacol 172(24):5744–5869. https://doi.org/10.1111/bph.13348

Andrade C (2024) Understanding factorial designs, main effects, and interaction effects: simply explained with a worked example. Indian J Psychol Med 46(2):175–177. https://doi.org/10.1177/02537176241237066

Andraos J, Matlack AS (2022) Introduction to Green Chemistry. CRC Press. https://doi.org/10.1201/9781003033615

Ao X, Yu X, Wu D, Li C, Zhang T, Liu S, Chen S, He L, Zhou K, Zou L (2018) Purification and characterization of neutral protease from Aspergillus oryzae Y1 isolated from naturally fermented broad beans. AMB Express 8(1):96. https://doi.org/10.1186/s13568-018-0611-6

Banerjee N (2023) Biomass to energy — an analysis of current technologies, prospects, and challenges. Bioenergy Res 16(2):683–716. https://doi.org/10.1007/s12155-022-10500-7

Banerjee G, Ray AK (2017) Impact of microbial proteases on biotechnological industries. Biotechnol Genet Eng Rev 33(2):119–143. https://doi.org/10.1080/02648725.2017.1408256

Barclay A, Acharya KR (2023) Engineering plastic eating enzymes using structural biology. Biomolecules 13(9):1407. https://doi.org/10.3390/biom13091407

Butré CI, Buhler S, Sforza S, Gruppen H, Wierenga PA (2015) Spontaneous, non-enzymatic breakdown of peptides during enzymatic protein hydrolysis. Biochimica et Biophysica Acta (BBA) Prot Proteom 1854(8):987–994. https://doi.org/10.1016/j.bbapap.2015.03.004

Ciechanover A, Finley D, Varshavsky A (1984) The ubiquitin-mediated proteolytic pathway and mechanisms of energy-dependent intracellular protein degradation. J Cell Biochem 24(1):27–53. https://doi.org/10.1002/jcb.240240104

Cupp-Enyard C (2008) Sigma's non-specific protease activity assay – casein as a substrate. J Vis Exp 19. https://doi.org/10.3791/899

Erdmann-Pham DD, Dao Duc K, Song YS (2020) The key parameters that govern translation efficiency. Cell Syst 10(2):183–192.e6. https://doi.org/10.1016/j.cels.2019.12.003

Fang Z, Sha C, Peng Z, Zhang J, Du G (2019) Protein engineering to enhance keratinolytic protease activity and excretion in Escherichia coli and its scale-up fermentation for high extracellular yield. Enzym Microb Technol 121:37–44. https://doi.org/10.1016/j.enzmictec.2018.11.003

Gautam S (2024) A review of bacillus species alkaline protease production and industrial applications. Int J Ther Innov 2(6):266–273. https://doi.org/10.55522/ijti.v2i6.0091

González-Siso M-I (2019) Editorial for the special issue: thermophiles and thermozymes. Microorganisms 7(3):62. https://doi.org/10.3390/microorganisms7030062

Gupta, A., Sahni, T., & Kumar, S. (2024). Biorefineries: an analogue to petroleum refineries (pp. 1–29). https://doi.org/10.1007/978-981-99-8224-0_1

Hevilla V, Sonseca A, Echeverría C, Muñoz-Bonilla A, Fernández-García M (2021) Enzymatic synthesis of polyesters and their bioapplications: recent advances and perspectives. Macromol Biosci 21(10). https://doi.org/10.1002/mabi.202100156

Hu Y, Li T, Tu Z, He Q, Li Y, Fu J (2020) Engineering a recombination neutral protease I from Aspergillus oryzae to improve enzyme activity at acidic pH. RSC Adv 10(51):30692–30699. https://doi.org/10.1039/D0RA05462C

Hwang BJ, Park WJ, Chung CH, Goldberg AL (1987) Escherichia coli contains a soluble ATP-dependent protease (Ti) distinct from protease La. Proc Natl Acad Sci 84(16): 5550–5554. https://doi.org/10.1073/pnas.84.16.5550

Ikebe J, Suzuki M, Komori A, Kobayashi K, Kameda T (2021) Enzyme modification using a mutation site prediction method for enhancing the regioselectivity of substrate reaction sites. Sci Rep 11(1):19004. https://doi.org/10.1038/s41598-021-98433-7

Jisha VN, Smitha RB, Pradeep S, Sreedevi S, Unni KN, Sajith S, Priji P, Josh MS, Benjamin S (2013) Versatility of microbial proteases. Adv Enzyme Res 01(03):39–51. https://doi.org/10.4236/aer.2013.13005

Karn SK, Kumar A (2015) Hydrolytic enzyme protease in sludge: recovery and its application. Biotechnol Bioprocess Eng 20(4):652–661. https://doi.org/10.1007/s12257-015-0161-6

Kataya G, Cornu D, Bechelany M, Hijazi A, Issa M (2023) Biomass waste conversion technologies and its application for sustainable environmental development—A review. Agronomy 13(11): 2833. https://doi.org/10.3390/agronomy13112833

Khambhaty Y (2020) Applications of enzymes in leather processing. Environ Chem Lett 18(3): 747–769. https://doi.org/10.1007/s10311-020-00971-5

Kuddus M, Ramteke PW (2012) Recent developments in production and biotechnological applications of cold active microbial proteases. Crit Rev Microbiol 38(4):330–338. https://doi.org/10.3109/1040841X.2012.678477

Kumar A, Dhiman S, Krishan B, Samtiya M, Kumari A, Pathak N, Kumari A, Aluko RE, Dhewa T (2024) Microbial enzymes and major applications in the food industry: a concise review. Food Prod Process Nutr 6(1):85. https://doi.org/10.1186/s43014-024-00261-5

Kurniawati A, Toth G, Ylivainio K, Toth Z (2023) Opportunities and challenges of bio-based fertilizers utilization for improving soil health. Org Agric 13(3):335–350. https://doi.org/10.1007/s13165-023-00432-7

Lange L, Huang Y, Busk PK (2016) Microbial decomposition of keratin in nature—a new hypothesis of industrial relevance. Appl Microbiol Biotechnol 100(5):2083–2096. https://doi.org/10.1007/s00253-015-7262-1

Lee D, Lee S, Jang E, Shin H, Moon B, Lee C (2016) Metabolomic profiles of Aspergillus oryzae and Bacillus amyloliquefaciens during rice koji fermentation. Molecules 21(6):773. https://doi.org/10.3390/molecules21060773

Leong HY, Chang C-K, Khoo KS, Chew KW, Chia SR, Lim JW, Chang J-S, Show PL (2021) Waste biorefinery towards a sustainable circular bioeconomy: a solution to global issues. Biotechnol Biofuels 14(1):87. https://doi.org/10.1186/s13068-021-01939-5

Li Q (2021) Structure, application, and biochemistry of microbial keratinases. Front Microbiol 12. https://doi.org/10.3389/fmicb.2021.674345

Listov D, Goverde CA, Correia BE, Fleishman SJ (2024) Opportunities and challenges in design and optimization of protein function. Nat Rev Mol Cell Biol 25(8):639–653. https://doi.org/10.1038/s41580-024-00718-y

Lopes S, Alves M, Rodrigues P (2022) Vasoactive intestinal peptide tumor as the cause of persistent diarrhea: a diagnostic challenge. Cureus. https://doi.org/10.7759/cureus.29130

López-Otín C, Bond JS (2008) Proteases: Multifunctional Enzymes in Life and Disease. J Biol Chem 283(45):30433–30437. https://doi.org/10.1074/jbc.R800035200

Love J (2022) Microbial pathways for advanced biofuel production. Biochem Soc Trans 50(2): 987–1001. https://doi.org/10.1042/BST20210764

Mageswari A, Subramanian P, Chandrasekaran S, Karthikeyan S, Gothandam KM (2017) Systematic functional analysis and application of a cold-active serine protease from a novel Chryseobacterium sp. Food Chem 217:18–27. https://doi.org/10.1016/j.foodchem.2016.08.064

Malone Rubright SL, Pearce LL, Peterson J (2017) Environmental toxicology of hydrogen sulfide. Nitric Oxide 71:1–13. https://doi.org/10.1016/j.niox.2017.09.011

Mamo J, Assefa F (2018) The role of microbial aspartic protease enzyme in food and beverage industries. J Food Qual 2018:1–15. https://doi.org/10.1155/2018/7957269

McCullum, E. O., Williams, B. A. R., Zhang, J., & Chaput, J. C. (2010). Random mutagenesis by error-prone PCR (pp. 103–109). https://doi.org/10.1007/978-1-60761-652-8_7

Mechesso AF, Zhang W, Su Y, Xie J, Wang G (2025) Segment-based peptide design reveals the importance of n-terminal high cationicity for antimicrobial activity against gram-negative pathogens. Probiotics Antimicrob Proteins 17(1):15–34. https://doi.org/10.1007/s12602-024-10376-3

Meister SW, Parks L, Kolmar L, Borras AM, Ståhl S, Löfblom J (2023) Engineering of TEV protease variants with redesigned substrate specificity. Biotechnol J 18(11). https://doi.org/10.1002/biot.202200625

Mohan M, Gujar GT (2003) Characterization and comparison of midgut proteases of Bacillus thuringiensis susceptible and resistant diamondback moth (Plutellidae: Lepidoptera). J Invertebr Pathol 82(1):1–11. https://doi.org/10.1016/S0022-2011(02)00194-5

Ngoune Tandzi L, Mutengwa CS, Ngonkeu ELM, Gracen V (2018) Breeding maize for tolerance to acidic soils: a review. Agronomy 8(6):84. https://doi.org/10.3390/agronomy8060084

Nguyen TP, Nguyen QV, Nguyen V-H, Le T-H, Huynh VQN, Vo D-VN, Trinh QT, Kim SY, Van Le Q (2019) Silk fibroin-based biomaterials for biomedical applications: a review. Polymers 11(12):1933. https://doi.org/10.3390/polym11121933

Nicosia FD, Puglisi I, Pino A, Caggia C, Randazzo CL (2022) Plant milk-clotting enzymes for cheesemaking. Foods 11(6):871. https://doi.org/10.3390/foods11060871

Nigam P (2013) Microbial enzymes with special characteristics for biotechnological applications. Biomolecules 3(3):597–611. https://doi.org/10.3390/biom3030597

Niyonzima FN, More S (2015) Detergent-compatible proteases: microbial production, properties, and stain removal analysis. Prep Biochem Biotechnol 45(3):233–258. https://doi.org/10.1080/10826068.2014.907183

Nomura T, Araki S, Nagao T, Konishi Y (2007) Resource recovery treatment of waste sludge using a solubilizing reagent. J Mat Cycles Waste Manag 9(1):34–39. https://doi.org/10.1007/s10163-006-0169-2

O'Connell A, Barry A, Burke AJ, Hutton AE, Bell EL, Green AP, O'Reilly E (2024) Biocatalysis: landmark discoveries and applications in chemical synthesis. Chem Soc Rev 53(6):2828–2850. https://doi.org/10.1039/D3CS00689A

Onifade AA, Al-Sane NA, Al-Musallam AA, Al-Zarban S (1998) A review: potentials for biotechnological applications of keratin-degrading microorganisms and their enzymes for nutritional improvement of feathers and other keratins as livestock feed resources. Bioresour Technol 66(1):1–11. https://doi.org/10.1016/S0960-8524(98)00033-9

Ortiz A, Sansinenea E (2022) The industrially important enzymes from Bacillus species, pp 89–99. https://doi.org/10.1007/978-3-030-85465-2_4

Pogson M, Georgiou G, Iverson BL (2009) Engineering next generation proteases. Curr Opin Biotechnol 20(4):390–397. https://doi.org/10.1016/j.copbio.2009.07.003

Raddadi N, Crotti E, Rolli E, Marasco R, Fava F, Daffonchio D (2012) The most important bacillus species in biotechnology. In: Bacillus thuringiensis biotechnology. Springer Netherlands, pp 329–345. https://doi.org/10.1007/978-94-007-3021-2_17

Raman J, Kim J-S, Choi KR, Eun H, Yang D, Ko Y-J, Kim S-J (2022) Application of lactic acid bacteria (lab) in sustainable agriculture: advantages and limitations. Int J Mol Sci 23(14):7784. https://doi.org/10.3390/ijms23147784

Rani K, Rana R, Datt S (2012) Review on latest overview of proteases. https://www.researchgate.net/publication/265865007.

Rao MB, Tanksale AM, Ghatge MS, Deshpande VV (1998) Molecular and biotechnological aspects of microbial proteases. Microbiol Mol Biol Rev 62(3):597–635. https://doi.org/10.1128/MMBR.62.3.597-635.1998

Razzaq A, Shamsi S, Ali A, Ali Q, Sajjad M, Malik A, Ashraf M (2019) Microbial proteases applications. Front Bioeng Biotechnol 7. https://doi.org/10.3389/fbioe.2019.00110

Rotanova TV, Botos I, Melnikov EE, Rasulova F, Gustchina A, Maurizi MR, Wlodawer A (2006) Slicing a protease: structural features of the ATP-dependent Lon proteases gleaned from investigations of isolated domains. Protein Sci 15(8):1815–1828. https://doi.org/10.1110/ps.052069306

Sakarika M, Matassa S, Carvajal-Arroyo JM, Ganigué R (2024) Editorial: microbial biorefineries for a more sustainable, circular economy. Front Bioeng Biotechnol 12. https://doi.org/10.3389/fbioe.2024.1512756

Salelles L, Floury J, Le Feunteun S (2021) Pepsin activity as a function of pH and digestion time on caseins and egg white proteins under static in vitro conditions. Food Funct 12(24):12468–12478. https://doi.org/10.1039/D1FO02453A

Shah M, Deka D (2024) Emerging sustainable technologies for biofuel production. Springer Nature Switzerland. https://doi.org/10.1007/978-3-031-52167-6

Shahzad HMA, Asim Z, Khan SJ, Almomani F, Mahmoud KA, Mustafa MRU, Rasool K (2024) Thermochemical and biochemical conversion of agricultural waste for bioenergy production: an updated review. Discov Environ 2(1):134. https://doi.org/10.1007/s44274-024-00171-w

Sharma KM, Kumar R, Panwar S, Kumar A (2017) Microbial alkaline proteases: optimization of production parameters and their properties. J Genet Eng Biotechnol 15(1):115–126. https://doi.org/10.1016/j.jgeb.2017.02.001

Sharma V, Singh A, Grenier M, Singh V, Thakur M (2024) Waste Valorization in Food Industries: A Review of Sustainable Approaches, pp 161–183. https://doi.org/10.1007/978-3-031-46046-3_9

Shete RS, Deshpande SK (2022) Characteristic study of industrial sludge, treatment and their disposal mechanism. EPRA Int J Multidiscip Res (IJMR):1–11. https://doi.org/10.36713/epra11329

Singh R (2020) A report on DNA sequence determinants in gene expression. Bioinformation 16(5):422. https://doi.org/10.6026/97320630016422

Sohail M (2024) Enzymes in biorefinery fermentation. 10(10):521. https://doi.org/10.3390/fermentation10100521

Solanki P, Putatunda C, Kumar A, Bhatia R, Walia A (2021) Microbial proteases: ubiquitous enzymes with innumerable uses. 3 Biotech 11(10):428. https://doi.org/10.1007/s13205-021-02928-z

Song P, Zhang X, Wang S, Xu W, Wang F, Fu R, Wei F (2023) Microbial proteases and their applications. Front Microbiol 14. https://doi.org/10.3389/fmicb.2023.1236368

Sun B, Zou K, Zhao Y, Tang Y, Zhang F, Chen W, Tang X, Chang C, Zheng Y (2023) The fermentation optimization for alkaline protease production by Bacillus subtilis BS-QR-052. Front Microbiol 14. https://doi.org/10.3389/fmicb.2023.1301065

Tallorin L, Wang J, Kim WE, Sahu S, Kosa NM, Yang P, Thompson M, Gilson MK, Frazier PI, Burkart MD, Gianneschi NC (2018) Discovering de novo peptide substrates for enzymes using machine learning. Nat Commun 9(1):5253. https://doi.org/10.1038/s41467-018-07717-6

Tipton KD, Hamilton DL, Gallagher IJ (2018) Assessing the role of muscle protein breakdown in response to nutrition and exercise in humans. Sports Med 48(S1):53–64. https://doi.org/10.1007/s40279-017-0845-5

Turk B (2006) Targeting proteases: successes, failures and future prospects. Nat Rev Drug Discov 5(9):785–799. https://doi.org/10.1038/nrd2092

Uekert T, Bleem AC, Holmes EC, Pal A, Johnson CW, Beckham GT (2025) Coupling waste feedstocks to microbial protein production in a circular food system. ACS Sustain Chem Eng 13(2):709–724. https://doi.org/10.1021/acssuschemeng.4c06516

Verhamme IM, Leonard SE, Perkins RC (2019) Proteases: pivot points in functional proteomics, pp 313–392. https://doi.org/10.1007/978-1-4939-8814-3_20

Vidmar B, Vodovnik M (2018) Microbial keratinases: enzymes with promising biotechnological 4 applications. Food Technol Biotechnol 56(3). https://doi.org/10.17113/ftb.56.03.18.5658

Ward OP (2011) Proteases. In: Comprehensive biotechnology. Elsevier, pp 604–615. https://doi.org/10.1016/B978-0-444-64046-8.00187-7

Woo KM, Chung WJ, Ha DB, Goldberg AL, Chung CH (1989) Protease Ti from Escherichia coli requires ATP hydrolysis for protein breakdown but not for hydrolysis of small peptides. J Biol Chem 264(4):2088–2091. https://doi.org/10.1016/S0021-9258(18)94145-1

Yao H, Liu S, Liu T, Ren D, Zhou Z, Yang Q, Mao J (2023) Microbial-derived salt-tolerant proteases and their applications in high-salt traditional soybean fermented foods: a review. Bioresour Bioprocess 10(1):82. https://doi.org/10.1186/s40643-023-00704-w

Zhao M, Ma J, Zhang L, Qi H (2024) Engineering strategies for enhanced heterologous protein production by Saccharomyces cerevisiae. Microb Cell Factories 23(1):32. https://doi.org/10.1186/s12934-024-02299-z

Zhong Y, Lu X, Xing L, Ho SWA, Kwan HS (2018) Genomic and transcriptomic comparison of Aspergillus oryzae strains: a case study in soy sauce koji fermentation. J Ind Microbiol Biotechnol 45(9):839–853. https://doi.org/10.1007/s10295-018-2059-8

Zhou C, Qin H, Chen X, Zhang Y, Xue Y, Ma Y (2018) A novel alkaline protease from alkaliphilic Idiomarina sp. C9-1 with potential application for eco-friendly enzymatic dehairing in the leather industry. Sci Rep 8(1):16467. https://doi.org/10.1038/s41598-018-34416-5

Zhou S, Liu Z, Xie W, Yu Y, Ning C, Yuan M, Mou H (2019) Improving catalytic efficiency and maximum activity at low pH of *Aspergillus neoniger* phytase using rational design. Int J Biol Macromol 131:1117–1124. https://doi.org/10.1016/j.ijbiomac.2019.03.140

Chapter 12
Role of Microbial Protease in Mitigating Microbial Biofilm Development and Enhancing Effective Wound Repair

Asharfi Pathela, Andhere Animish, Neha Sinha, Sejal Mali, Shruti Rangari, Sofia Mohanty, and K. Suthindhiran

Abstract The breakdown of biofilm-associated barriers and necrotic tissue, along with proteases, aids in restoring the natural evolution of the wound healing stages. Moreover, some microbial-derived proteases exhibit immunomodulatory properties, stimulating cytokine production and angiogenesis, which are important for healing wounds and growing new tissue. This chapter investigates the contribution of microbial proteases in the breaking up of biofilms for the enhancement of healing wounds. Microbial proteases play a vital role in the breakdown of biofilms to facilitate healing of wounds, as observed among the molecular docking studies, illustrating the effectiveness of a metalloprotease in *Paenibacillus polymyxa* against the biofilm constituents of *Staphylococcus aureus* and *Pseudomonas aeruginosa*. Nevertheless, there should be an equilibrium between proteases and their inhibitors to prevent complications resulting from either excessive degradation or excessive deposition of extracellular matrix. Other developments, such as protease-modulating dressings, matrix metalloproteinase (MMP) modulators, and signaling molecules, are promising in regulating protease action. Also, proteases are used as bifunctional agents in wound healing and immunological regulation by stimulating immune mediators to act on proteolytic activation of neural cells. The use of proteases combined with anti-biofilm peptides and antimicrobial agents is a new way of treating chronic wounds and improving the results of patients. We further talk about microbial proteases in therapeutic use, sophisticated delivery platforms, use of nanoformulations and hydrogels, combination therapy with antibiotics, and plant threats to improve disruption of biofilms and wound healing. Moreover, the chapter lists a thorough and effective description of microbial proteases in biofilm disruption and the contribution toward wound repair, giving the dual roles of microbial proteases in biofilm disruption and immunomodulation. It further discusses new

A. Pathela · A. Animish · N. Sinha · S. Mali · S. Rangari · S. Mohanty · K. Suthindhiran (✉)
Department of Bio-Medical Sciences, School of Biosciences and Technology, Vellore Institute of Technology, Vellore, Tamil Nadu, India
e-mail: ksuthindhiran@vit.ac.in

V. Zambare et al. (eds.), *Biotechnological Advances and Applications in Microbial Proteases*, Interdisciplinary Biotechnological Advances,
https://doi.org/10.1007/978-981-95-9011-7_12

delivery mechanisms and combination therapies that are revolutionizing the area of wound healing toward achieving better clinical results.

Keywords Microbial proteases · Anti-biofilm activity · Wound repair · Microbial biofilm · Biofilm disintegration

12.1 Introduction

Proteases (also called proteinases or peptidases) are protein-hydrolyzing enzymes that cleave proteins by hydrolyzing peptide bonds. Proteolytic enzymes or proteases are categorized in relation to the location where they act on the protein. The peptide linkages that hold the protein molecule together are broken by endopeptidases, resulting in shorter peptide fragments. Examples include pepsin and trypsin. Exopeptidases instead work at protein chain termini, hydrolyzing amino acids at the N-terminal or the C-terminus (Chang and Yang 2000). These include enzymes such as aminopeptidases, which act against the N-terminal, and carboxypeptidases, which act on the C-terminal (Chang and Yang 2000). Of the proteases, microbial proteases are especially prized for their efficacy in different environments and their wide range of applications in the industrial sector, including the food, pharmaceutical, detergent, and textile industries (Chang and Yang 2000). Chronic wounds, such as bedsores, diabetic foot ulcers, and ulcers on the legs that are caused due to poor circulation, are a serious medical issue because they heal exceptionally slowly and are often vulnerable to microbial infection. This slow healing owes much to ongoing inflammation as well as the infection of microbial populations (Cavallo et al. 2024).

One of the key issues in these injuries is biofilm formation, organized colonies of bacteria covered in a protective matrix, which complicates treatment and further prolongs healing. The causative bacteria may be aerobic and anaerobic and include species such as *Staphylococcus aureus*, *Pseudomonas aeruginosa*, *Enterococcus faecalis*, and other *Streptococcus* species (Cavallo et al. 2024; Suleman 2016). These microbes continue to stimulate the wound environment, causing the release of inflammatory molecules and degradative enzymes by the body. Sadly, such substances harm essential healing elements such as the growth factor and extracellular matrix, eventually providing a hostile environment for tissue repair and regeneration (Suleman 2016). In chronic wounds, there is high enzyme activity, oxidative stress, and inadequate oxygen supply, which interfere with the necessary healing mechanisms, including new skin formation, angiogenesis (the growth of new blood vessels), and regeneration of tissue. Such a multifactorial and long-standing state necessitates novel treatment approaches (Ghoreishi et al. 2022).

One of the more promising methods is the application of microbial proteases, enzymes derived from nonpathogenic or genetically modified microorganisms. These enzymes break up biofilms, and this action eliminates large groups of bacteria housed in the biofilm matrix. The microbial proteases also contribute to debridement of the wound by hydrolyzing necrotic tissue, thus cleansing the wound, and preparing it for healing. When used correctly, these enzymes are able to assist in reactivating the healing process in chronic wounds, restoring them from stagnant, nonhealing lesions to wounds that are able to regenerate healthy tissue (Chang and Yang 2000; Tseng et al. 2018).

Existing treatment methods, including debridement and systemic antibiotics, are short-lived and generally do not cure biofilms but rather temporarily alleviate the symptoms. Surgical procedures may eradicate the obvious biomass but tend to leave residual microorganisms that can easily re-form the biofilm. The misuse of antibiotics also gives rise to multidrug-resistant strains, further compounding the issue and reducing the effectiveness of the conventional pharmacological strategies. These constraints have compelled the exploration for biofilm-specific drug targets, particularly those able to modulate or break down the biofilm matrix directly (Borges et al. 2020). Microbial protease selectivity also helps in preventing collateral injury to host tissue. Proteases are not likely to cause injury to intact cells through indiscriminate bonding compared to some chemicals or broad-spectrum antibiotics.

Proteases can be designed or selected for selective activity against biofilm architecture. Protein engineering and synthetic biology advances have made the design of custom proteases with increased stability, activity, and substrate specificity possible under conditions of wound microenvironment, including low pH and high levels of protease inhibitors (Borges et al. 2020; Lahiri et al. 2022a). Enhancement of novel therapeutic means using microbial proteases thus holds a bright future in surmounting the limitations of the existing interventions of wound care. With their ability to destroy biofilms, control immunities, and induce tissue healing, they are a multidimensional agent of chronic wound curing (Bi et al. 2021). As the research in the sector continues, there is an opportunity that protease-based therapy can become a fundamental pillar of the future generation of wound care protocols that yield better patient outcomes for different patients with chronic wounds caused by biofilms.

12.2 Microbial Biofilms in Wound Environment

Microbial biofilms refer to a community of organized microorganisms that adhere to surfaces, and they are covered by the film, which is composed of extracellular polymeric substances (EPS) that the biofilm secretes. These polysaccharides, proteins, lipids, and extracellular DNA-rich matrix form a protective environment, which promotes the survival of microbes in unfavorable conditions (Flemming et al. 2016).

Biofilms can be deposited on both dead and alive surfaces, this can include the presence of human tissues, medical implants, and industrial equipment. Biofilm formation is a sequential process. It initiates with forming reversible conjugation of free-floating cell microbes with a surface, then irreversible conjugation, production of a matrix of EPS, maturation to a complex mixture, and lastly spread of cells to cover new regions (Flemming et al. 2016; Hall-Stoodley et al. 2004). During the maturation process, quorum sensing, a population-based chemical signaling between cells, allows synchronized gene expression. This organization secures the stability, together with the functional arrangement of the biofilm. The microbial cells in the biofilm also exist in varying physiological states, which makes the population heterogeneous and thus better adapted to survive the change of the environment (Donlan 2001).

Biofilms persist longer after the formation and after exposure to antibiotics than individual planktonic (free-floating) cells due to increased antibiotic resistance and reduced response to the immune system (through biofilm matrices and mechanisms). This resistance is said to be increased because of the EPS barrier, and the existence of slow-growing cells or persister cells in itself is less susceptible to antimicrobials (Flemming et al. 2016; Hall-Stoodley et al. 2004). Well-known biofilm infections concern chronic wound infections, urinary tract infections, and infections of healthcare devices whose presence requires durable installation, such as catheters or prosthetic joints. The disparity in biofilms is valuable to comprehend so as to draft specific measures to prevent their occurrence or take advantage of their benefits (Donlan 2001). In practice, there are new methods under investigation, including cleaving the microbial proteases, enzymes that break down the proteins of the EPS, which are used in clinical practice. These enzymes have the potential to enhance the efficacy of antibiotics and promote healing of tissues, since weakening the biostructure is a way to enhance the entry of antibiotics.

12.2.1 Stages of Biofilm Formation

A biofilm is a highly structured community of attached microorganisms that are bound together in an organized coherent structure enclosed by a self-produced matrix, EPS (O'Toole et al. 2000). This matrix primarily comprises extracellular DNA, proteins, lipids, and polysaccharides collectively forming a shield and adhesive microbial habitat. Biofilms may form on a very diverse surface, such as biotic surfaces, which refer to biological surfaces (living tissue, e.g., skin or mucosal lining) and abiotic surfaces (e.g., medical implants, catheters, wound dressing) (Flemming et al. 2016; O'Toole et al. 2000).

In contrast to planktonic bacteria, microorganisms in biofilm show phenotypic changes such as modification of gene expression, growth rate, and resistance to environmental stressors. Besides holding the cells together and to the surface, the EPS matrix acts as a physical barrier that guards the microbial population against antimicrobial agents, immune system attack, desiccation, and shear (O'Toole et al. 2000). Biofilms are especially significant in those conditions when infections that are chronic and wound healing are involved, causing delays in tissue regeneration, chronic inflammation, and antibiotic tolerance. In chronic and acute wounds, such as in pressure sores or diabetic ulcers, biofilm-forming pathogens such as *Staphylococcus aureus* and *Pseudomonas aeruginosa* can produce an unfavorable microenvironment that prevents normal wound healing. These biofilms prevent the immigration of immune cells and promote the persistence of low-grade inflammation, causing chronicity of the wound. The critical aspect of biofilm biology is to

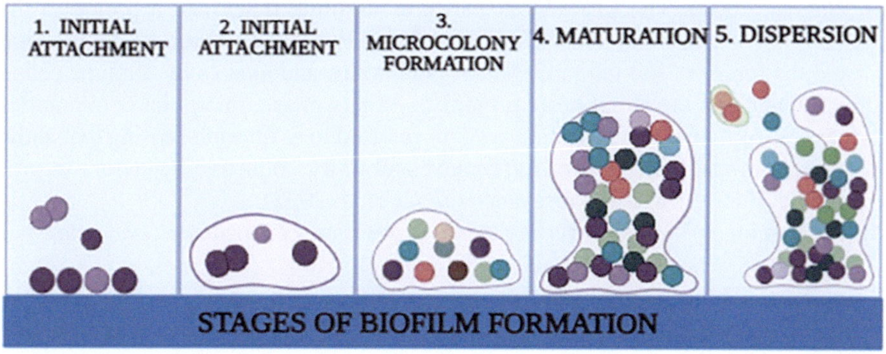

Fig. 12.1 Phases of biofilm development

understand optimal methods of preventing the formation of biofilms or disrupting them before they occur, especially under clinical conditions where biofilms pose a significant cause of failure to treatments and lengthy recovery, as shown in Fig. 12.1 (Shree et al. 2023; Suleman 2016). The following are the different stages involved in the formation of biofilm.

(a) *Primary interface response and transient adhesion*

The formation of biofilms starts with the ability of independent microbial cells that are in a free state and bind to a surface in a nonpermanent fashion. This first attachment is usually weak and reversible and is often propelled by a wide variety of physicochemical forces, such as hydrophobic effects, van der Waals forces, hydrogen bonds, and electrostatic forces (Karatan and Watnick 2009; O'Toole et al. 2000).

(b) *Irreversible adhesion:- development of secure, stable surface bonding*

Following the consideration of environmental factors, microorganisms enter another stage of irreversible adhesion, and they attach themselves to the surface fully. The stage is associated with the expression of certain adhesive agents and biofilm-enabling factors that intensify their hold and activate the subsequent moves in biofilm development. The microbial cells begin producing EPS, a complex sticky grid consisting of polysaccharides, lipids, proteins, and extracellular DNA. Besides binding the cells firmly to each other, this matrix also fixes the surface and constitutes the framework of a growing biofilm (Donlan and Costerton 2002; Karatan and Watnick 2009).

(c) *Priming biofilm and maturation: development of microcolonies and matrix enlargement*

Once a microbe is firmly attached to a surface, it begins to grow and develops into tiny colonies, referred to as microcolonies. When such groups are large, the EPS matrix secreting activity also increases. This matrix is an elaborate blend that consists of polysaccharides, extracellular DNA, enzymes, and structural proteins, which operate together with each other to guard the cells and yield support to the growing biofilm (Donlan and Costerton 2002; Karatan and

Watnick 2009). The EPS matrix performs multiple functions: it promotes the retention of water and nutrients, shields the cells against environmental pressure and the forces of the immune system (antibiotics and attacks by immune cells), and maintains the structure of a biofilm. At this phase, microbial communities begin to initiate communication systems referred to as quorum sensing that allow the synchronization of gene expression and unify collective behaviors to promote biofilm growth and resistance (Sauer et al. 2022).

(d) *Maturation of biofilm development: sophisticated structure and maximal resistance*

As the biofilm grows, it transforms into a well-structured and complex microbial community. During this stage, quorum sensing becomes increasingly active, deliberating processes such as matrix maintenance, virulence expression, and preparation for cell release. The biofilm reaches its highest level of defense against antibiotics and immune responses. This heightened resistance is the shielding effect of the EPS matrix and the presence of slow-dividing or dormant cells that are naturally more tolerant to treatment (Donlan and Costerton 2002; Karatan and Watnick 2009; Sauer et al. 2022).

(e) *Dispersal of biofilm: release with colonization at new regions*

During late-stage biofilm growth, some microbial cells regain movement and differ in form compared with the mature biofilm. This is referred to as dispersal, and it aids the organisms to evade unfavorable conditions such as a lack of nutrition or crowding and to begin colonizing new grounds (Karatan and Watnick 2009). Scattering may be either by passive agents, e.g., a shearing flow of fluid, abrasion, or nutrient gradients, or by an active process where the bacteria become actively involved in breaking down the biofilm matrix with an enzyme such as proteases or nucleases. These enzymes aid in destroying the EPS, and this causes the cell to detach and wander. After being released, these free-floating cells revert to a planktonic habitat and can start the development of new biofilms in some other place (Donlan and Costerton 2002; Sauer et al. 2022).

12.2.2 Composition of Extracellular Polymeric Substances (EPS)

EPS is a gelatinous, sticky compound produced by microorganisms of a biofilm. It accelerates the stickiness of surfaces, does not get worn off, and can tolerate a challenging environment. The big component of this matrix is polysaccharides, which are a form of sugar molecules that bind this biofilm together and frequently bind to metals to enhance their strength. In another instance, *Pseudomonas aeruginosa* synthesizes alginate that helps to maintain its biofilm and render it more difficult to be eliminated by the immune system (Ryder et al. 2016). PNAG (poly-N-acetylglucosamine) is another molecule that enables the microbes to bind to a surface, evade the immune system, and withstand antibiotics. Other components of the matrix are proteins, including adhesins, fibers, and enzymes (including proteases), and extracellular DNA, which binds everything together and provides a shield to the cells (Lahiri et al. 2022b; Vu et al. 2009). The whole structure may change depending on the environment, as illustrated in Table 12.1, although the small components, such as lipids and surfactants, are significant in the movement and fragmentation of the biofilm. This poses a challenge when it comes to its treatment, and also opens up new avenues in treatment using microbial proteases to break up the matrix and aid healing.

12.2.3 Clinical Relevance: Biofilms in Chronic Wounds

Microbial biofilms easily colonize chronic wounds including diabetic foot ulcers and pressure sores and delay healing by establishing an extracellular matrix providing shelter against immunological cells and antimicrobial therapies. These biofilms trigger a prolonged inflammatory response by attracting immunological cells, including macrophages and neutrophils; however, the matrix prevents effective microbial clearance, causing the sustained secretion of cytokines that promote inflammation such as IL-6 and TNF-α. This chronic inflammation damages tissue, impairs the function of key repair cell types, including keratinocytes and fibroblasts, and inhibits the genesis of new skin and blood vessels. Moreover, enzymes secreted by biofilms, including bacterial proteases and lipases, disintegrate the extracellular matrix constituents such as collagen, further hindering wound repair. As a result, these wounds persist in a nonhealing, inflamed state, often requiring prolonged treatment and posing risks of infection or systemic complications. Microbial protease-based therapies offer a promising approach by degrading the biofilm matrix, reducing the microbial load, and facilitating the removal of necrotic tissue, thereby creating a more favorable environment for healing and improving outcomes in chronic wound care.

12.3 Microbial Proteases

Proteases are degradative enzymes that facilitate the hydrolysis of peptide linkages across multiple proteins, including those within proteins (Gao et al. 2015). They are fundamental to the selective modification of proteins, such as the activation of zymogen enzyme forms by proteolysis, blood clotting, lysis of fibrin clots, tumor growth, protein catabolism, inflammation, and the processing and transport of secretory proteins across membranes. These enzymes hold a pivotal position due to their applications in industries such as food, feed, leather, textile, and pharmaceutical. The primary advantage of microbial protease preference lies in its economic feasibility, substrate specificity, and ease of genetic manipulation for functionality

Table 12.1 Composition and functional roles of EPS components in microbial biofilms

Component	Subtypes	Structural role	Functional role	Genetic regulation	References
Polysaccharides	Homopolymers (e.g., cellulose, mutan), heteropolymers (e.g., alginate, Pel, Psl, PNAG, colanic acid, glucans)	Scaffold formation, moisture retention, cross-linking with divalent ions	Surface adhesion, immune evasion, antimicrobial resistance, nutrient retention, quorum sensing support	Synthesized under control of microbial operons (e.g., *alg*, *bcs*, *icaADBC*); regulated by environmental and internal signals	Flemming and Wingender (2010)
Alginate	β-D-mannuronic acid + α-L-guluronic acid	Gel-like consistency, chelates Ca^{2+} for cross-linking	Adhesion to epithelial surfaces, immune evasion (binds complement proteins, antibodies), resistance in chronic infections (e.g., cystic fibrosis lungs)	Controlled by the *alg* operon; epimerization and acetylation influence viscosity	Ryder et al. (2016)
Cellulose	β-1,4-Glucan	Tensile strength, crystalline structure	Adherence to surfaces, resistance to enzymatic degradation, hydrodynamic and antimicrobial resistance	Synthesized via the *bcs* operon; regulated by c-di-GMP	Ghafoor et al. (2011)
Poly-N-acetylglucosamine (PNAG)	Also called PIA	Electrostatic network formation	Enhances cell–cell cohesion, immune evasion, limits antibiotic penetration	Synthesized by *icaADBC* operon; used by *S. aureus*, *E. coli*, etc.	Vu et al. (2009)
Other polysaccharides	Pel (cationic, glucose-rich), Psl (mannose, rhamnose, glucose), colanic acid, glucans, and mutans	Pel: forms a fibrous scaffold with eDNA Psl: promotes surface attachment Colanic acid: capsule integration under stress	Microcolony formation, acid resistance (e.g., in dental plaque), matrix cohesion in anaerobic environments	Individually regulated by species-specific genetic systems	Lahiri et al. (2022b)
Proteins	Adhesins, Curli fibers (*E. coli, Salmonella*), Fap amyloids (*Pseudomonas*),	Structural fibers, scaffold crosslinkers, dynamic matrix remodeling	Attachment and cohesion, EPS degradation and dispersal, signal transduction,	Expression is environmentally regulated and crucial for maintaining and dispersing biofilms	Wozniak and Ethan (2015)

	enzymes (proteases, nucleases), lectins (LecA, LecB)		quorum sensing, and stress response		
Extracellular DNA (eDNA)	Released via lysis or secretion	Electrostatic linker, hydrogen bond formation	Matrix reinforcement, binds and neutralizes antimicrobials, and genetic exchange (horizontal gene transfer)	Targeted for biofilm disruption using extracellular DNases	Whitchurch et al. (2002)
Lipids/ biosurfactants/ small molecules	Lipopolysaccharides (LPS), outer membrane vesicles (OMVs), rhamnolipids, siderophores, toxins, and signaling molecules	Hydrophobic domains, modulate surface tension	Surface colonization, biofilm detachment, intercellular signaling, nutrient acquisition	Secreted in small quantities, modulate biofilm fluidity and dispersal	Hall-Stoodley et al. (2004)
Dynamic regulation of EPS	Early biofilms: simpler proteins/polysaccharides Mature biofilms: complex polysaccharides/proteins	Adapts to the biofilm lifecycle stage	Enhances resistance under stress, adjusts architecture and composition in hostile environments	EPS components are regulated via quorum sensing, nutrient status	Leid et al. (2005)

(Chang and Yang 2000). The global sector value of industrial enzymes is presently estimated to be approximately $1 billion, out of which hydrolytic enzymes account for nearly 75% of sales. Proteases constitute one of the three commercially significant enzymes, accounting for 60% of the total enzyme market (Al-maqtari and Mahdi 2019).

12.3.1 Classification of Microbial Proteases

Proteases are classified in subgroup 4 of group 3 according to the Nomenclature Committee of the International Union of Biochemistry and Molecular Biology. Microbial proteases may be categorized according to their source, catalytic mechanism, specificity, or the characteristics of the reactive group inside the site of catalysis, as seen in Table 12.2. Proteases are categorized into two groups based on the site of action on polypeptide chains: exopeptidases and endopeptidases (Chang and Yang 2000). Exopeptidases act at the termini of polypeptide chains and are categorized as carboxypeptidases and aminopeptidases, according to their activity at the carboxyl or amino end, respectively. Endopeptidases hydrolyze internal peptide linkages of the polypeptide chains, which are away from both end termini. This can be further distinguished into six (6) categories, namely serine, aspartic, cysteine, metallo, glutamic acid, and threonine proteases, according to the catalytic residue present in the active site (Hartley 1960).

12.3.1.1 Exopeptidases

Exopeptidases only act at or near the ends of polypeptide chains, and this is further classified into two categories according to the action on polypeptide chains, namely aminopeptidases and carboxypeptidases (Chang and Yang 2000).

12.3.1.1.1 Aminopeptidases

Aminopeptidases are enzymes that facilitate the elimination of amino acids or short peptides from the N-terminus of polypeptide chains. These possess the ability to cleave the N-terminal methionine in exogenously expressed proteins. Aminopeptidases are predominantly intracellular and are widely distributed across microbial taxa, including bacteria and fungi. A rare phenomenon of extracellular production occurs in *Aspergillus oryzae* and *Escherichia coli*, which require Mg^{2+} or Mn^{2+} for optimal activity of aminopeptidase (Chang and Yang 2000; Labbé et al. 1974).

Table 12.2 Classification and functional comparison of proteases based on catalytic residues and site of action

Protease type	Catalytic residue/ requirement	Site of action	Optimum pH	Inhibitors	Features and examples	References
Aminopeptidases	–	N-terminal (exopeptidase)	Variable	–	Remove amino acids from N-terminus; often intracellular; *Aspergillus oryzae* produces extracellularly, e.g., *E. coli* aminopeptidase I (Mg^{2+}/Mn^{2+}-dependent)	Chang and Yang (2000) and Labbé et al. (1974)
Carboxypeptidases	Serine, metallo, or cysteine	C-terminal (exopeptidase)	Variable	Varies by type	Remove amino acids from the C-terminus; species include *Penicillium*, *Saccharomyces*, and *Aspergillus*	Félix and Brouillet (1966)
Serine proteases	Serine (with His and Asp catalytic triad)	Endo-/ exopeptidase	7–11	PMSF, DFP, TLCK, 3,4-DCI, E64	Broad specificity; found in all domains; examples: chymotrypsin, subtilisin, carboxypeptidase C	Chang and Yang (2000)
Aspartic proteases	Two aspartic acids	Endopeptidase	3–4	Pepstatin, DAN, EPNP + Cu^{2+}	Active at acidic pH; families: pepsin (A1), retropepsin (A2); microbial forms include *Aspergillus*, *Rhizopus*, *Mucor*	Chang and Yang (2000)
Cysteine proteases	Cysteine–histidine dyad	Endopeptidase	Neutral (some acidic)	PCMB, iodoacetic acid	Require reducing agents (HCN, DTT); examples: papain, clostripain (Ca-dependent), streptopain	Chang and Yang (2000)
Metalloproteases	Divalent metal ion (Zn^{2+}, Co^{2+}, Mn^{2+})	Endopeptidase	Variable (neutral to alkaline)	EDTA	30 families; metal-binding motifs (e.g., HEXXH-E); examples: thermolysin, collagenase, elastase, MMPs	Chang and Yang (2000), Song et al. (2023), and Weaver et al. (1977)

12.3.1.1.2 Carboxypeptidases

Exopeptidases known as carboxypeptidases catalyze the cleavage of dipeptides or amino acids (singly) from polypeptide chains at the carboxy terminus. The three categories of these enzymes, serine, metallo, and cysteine carboxypeptidases, are distinguished by the characteristics of their catalytic residues. Similar substrate specificities are shared by serine carboxypeptidases, which are found in *Aspergillus*, *Saccharomyces*, and *Penicillium*. However, they differ in terms of molecular weight, stability, pH, inhibitor sensitivity, and optima (Félix and Brouillet 1966).

12.3.1.2 Endopeptidases

Endopeptidases are distinguished by their ability to cleave polypeptide chains inside their internal regions rather than at the N or C terminus. This enzyme's activity is negatively impacted by the existence of a free carboxyl or amino group. Endopeptidases are classified into four classes according to their catalytic mechanisms, which include metalloproteases, cysteine-, aspartic-, and serine proteases. According to Rawlings and Barrett, each family of peptidases has been given a code letter that indicates the catalytic type: S stands for serine, C for cysteine, A for aspartic, and M for metallo (Rawlings and Barrett 1993).

12.3.1.2.1 Serine Proteases

Serine residue on their active site identifies proteases. These are very dispersed in prokaryotes, viruses, and eukaryotes. They are divided into many subgroups, such as omega peptidases, oligopeptidases, endopeptidases, and exopeptidases. Carboxypeptidase C, subtilisin, and chymotrypsin all share the catalytic trio of serine, histidine, and aspartate residues, though serine proteases have four different evolutionary origins. The protein folding differs, which serves to demonstrate convergent evolution. The pH range in which these proteases exhibit best action is 7–11 (Chang and Yang 2000). They have a broad substrate specificity, both amidase and esterolytic activities. These are normally inhibited by compounds such as diisopropyl fluorophosphate, 3,4-dichloroisocoumarin (3,4-DCI), L-3-carboxytrans-2,3-epoxypropyl-leucylamido-(4-guanidine) butane, tosyl-L-lysine chloromethyl ketone, and phenylmethylsulfonyl fluoride (Chang and Yang 2000).

12.3.1.2.2 Aspartic Proteases

Aspartic protease uses two aspartic acid residues for catalytic action. These are also called acidic proteases as they are active in acidic conditions (pH 3–4). They are classified into three families: pepsin, pararetroviral enzymes, and retropepsin. They have a bilobed structure in which the active site is among the lobes. The catalytic aspartic residue is encompassed in the conserved motif Asp-Xaa-Gly (where Xaa is Ser or Thr). Aspartic proteases are somewhat less specific than pepsin, typically demonstrating substrate selectivity for peptide bonds flanked by nonpolar or massive amino acid residues. Pepstatin and chemicals such as DAN and EPNP are amalgamated with copper ions, suppressing aspartic proteases. Microbial aspartic proteases are further classified into two types: pepsin-like enzymes (made by genera such as *Aspergillus*, *Penicillium*, *Rhizopus*, and *Neurospora*) and rennin-like enzymes (found in species such as *Endothia* and *Mucor*) (Chang and Yang 2000).

12.3.1.2.3 Cysteine/Thiol Proteases

Cysteine proteases (thiol proteases) are proteolytic enzymes that are found in prokaryotes and eukaryotes. Their catalytic activity is dependent on a cysteine–histidine dyad and is activated by reducing agents such as HCN, cysteine, or DTT. These enzymes are blocked by sulfhydryl-reactive compounds such as PCMB and iodoacetic acid. These proteases are further grouped based on substrate specificity into papain-like, trypsin-like, glutamic acid-specific, and other types. The best recognized member of the cysteine protease family is papain, often used in the food business. These digestive enzymes function optimally at a neutral pH, whereas some enzymes, such as those found in lysosomes, require an acidic environment. They significantly contribute to the breakdown of misfolded proteins, such as prions, and are resilient to serine protease inhibitors and metal chelators like EDTA. The mandatory requirement for calcium is observed in Clostripain, from *Clostridium histolyticum*, which targets arginyl residues, whereas Streptopain from *Streptococcus* spp. exhibits a wider specificity (Chang and Yang 2000).

12.3.1.2.4 Metalloproteases

Metalloproteases exhibit the greatest and diverse class of proteolytic enzymes, requiring divalent metal ions, such as Zn^{2+}, Co^{2+}, or Mn^{2+}, due to their specific catalytic activity. Just like cysteine proteases, even metalloproteases tend to be inhibited by chelating agents like EDTA and demonstrate extensive substrate specificity. These enzymes have a significant industrial relevance, particularly in the pharmaceutical sector. Approximately 30 families of metalloproteases are identified. These are organized into clans according to metal-binding motifs, such as the HEXXH-E motif characteristic of clan MA. They can even be distinguished by unique substrate preferences and optimal pH range into neutral, alkaline, and myxobacterial types (Chang and Yang 2000). Some of the prominent members of metalloproteases include thermolysin, a thermostable protease from *Bacillus stearothermophilus*; collagenase, which specifically targets collagen and gelatin; and elastase and alkaline proteases derived from *Pseudomonas* species (Song et al. 2023; Weaver et al. 1977). The matrix metalloproteases (MMPs) carry out applications such as tissue development, wound healing, disease advancement such as cancer and arthritis, and so on (Song et al. 2023).

12.3.2 Mechanism of Proteolytic Action

Proteases are diverse enzymes that utilize various strategies to catalyze the hydrolysis of peptide bonds and depend on the characteristics of structural complexities and active-site residues. Purification of enzymes to homogeneity has played a crucial part in elucidating their processes, including methods such as affinity-, ion-exchange chromatography, electrophoretic methods, and gel filtration. Protease is mainly categorized into the following groups because of their numerous structural compositions: serine proteases, aspartic proteases, metalloproteases, and cysteine proteases.

The activation of serine proteases is a two-step process, which implicates deacylation and acetylation. When the nucleophilic serine breaks down the bond of the peptide in a normal Ser-His-Asp catalytic triad, a covalent enzyme–peptide intermediate is created. Water undergoes a nucleophilic reaction, resulting in the hydrolysis of the peptide linkage and the restoration of the active enzyme. Serine proteases are classified based on their substrate specificity at the P1 site into three subclasses: trypsin-like proteases, which cleave basic amino acids; elastase-like proteases, which prefer small hydrophobic residues; and chymotrypsin-like proteases, which target and cleave after bulky hydrophobic residues (Masaki et al. 1978). Substrate specificity can be governed by interactions at the substrate-binding sites flanking the enzyme's catalytic core.

However, it has been observed that variations exist in the catalytic machinery within this class. Subtilisin's have a similar triplet with an alternative residue sequence (Asp-His-Ser), and some entail thiol cofactors for efficiency. Furthermore, instead of using the whole triplet, certain serine proteases use a Ser-base dyad. Serine-dependent carboxypeptidases with acidic pH optima include a Glu residue before the catalytic Ser, contributing to their distinct activity profile (Proteinase et al. 1986).

The bilobal framework of aspartic proteases, particularly pepsin and endothiapepsin, features two aspartic acid residues that facilitate general acid–base catalysis. The catalytic Asp residues each lobe offers are usually found inside an Asp-Thr-Gly pattern. Enzymes like these cleave hydrophobic substrates, generally requiring peptides of at least six residues to bind well. Catalysis occurs when the Asp dyad activates a water molecule, making it easier for a nucleophile to attack the scissile bond. Retropepsins, on the other hand, have only one catalytic Asp and work via dimorization (Li et al. 2015).

The activity of metalloproteases relies on divalent metal ions, mainly zinc. By coordinating the metal ion, the conserved HEXXH motif polarizes the carbonyl moiety of the target peptide link, allowing an activated, energized water molecule to initiate a nucleophile attack. Glu residues in the proximity aid in this process, as demonstrated by thermolysin. Dialysis can inactivate metalloproteases, which are sensitive to metal chelators (Chang and Yang 2000).

As demonstrated by papain, cysteine proteases function through a catalytic dyad or trio, primarily involving a Cys-His or Cys-His-Asn motif. A covalent acyl–thiol intermediate emerges as part of the action, and then it is hydrolyzed by a water molecule. This chemical pathway like that of serine proteases, except the nucleophilic attack, uses a thiol instead of a hydroxyl group. Papain is the prototype of cysteine proteases that has been used as a structural and mechanistic model to demonstrate the fundamental principles of substrate binding and catalysis in this enzyme class (Ozhelvaci and Steczkiewicz 2023).

12.3.3 Sources of Microbial Proteases

12.3.3.1 Bacteria

Bacteria are one of the most essential and industrially exploited sources of microbial proteases. Some of the characteristics that contribute to its significance are rapid growth, ease of large-scale cultivation, and susceptibility to genetic engineering. *Bacillus* species, specifically *B. subtilis*, *B. licheniformis*, and *B. amyloliquefaciens*, are often used in the industrial synthesis of neutral and alkaline proteases due to their well-established genetic systems, effective mechanisms for protein secretion, and ability to synthesize enzymes on a large scale. Neutral proteases are commonly utilized in the food business since they produce protein hydrolysates with reduced bitterness, and they are effective at a pH of 5–8. Alkaline proteases are widely applicable in the detergent industry due to their high stability and broad substrate specificity, and they exhibit optimal activity in a pH range of approximately 10 (Chang and Yang 2000).

12.3.3.2 Fungi

Fungi are the chief sources of microbial proteases due to their diverse metabolism, e.g., *Aspergillus*, *Penicillium*, *Rhizopus*, *Mucor*, and *Thermomyces*. These can be acidic, alkaline, or neutral and function in a wide range of pH, but they possess lower thermal stability. *Aspergillus oryzae* is a well-known protease producer that can be used in the food and fermentation industry (Chang and Yang 2000; Ojo Omoniyi et al. 2024).

12.3.3.3 Actinomycetes

Streptomycetes species have a diverse group of substrate specificities and applications. *Actinomycetes* produce both serine and metalloproteases. They have broad pH tolerance and high thermostability as they have harsh environment tolerance and protein degradation ability (Ojo Omoniyi et al. 2024).

12.4 Protease-Mediated Biofilm Disruption

Biofilms are sticky, protective layers formed by microbial colonies. The extracellular matrices constitute those layers that offer support and protection to the microbial community against environmental hazards and the immune reaction of the host. Proteins, among other components of the matrix, are needed to provide the stability of the biofilm. This protein structure can be disrupted by exposing it to protease enzymes, and this will break down the biofilm that makes the microbes resistant. In simple terms, the proteases disrupt the protein scaffold that biofilms are held together by making the bacteria susceptible.

12.4.1 Proteolysis of EPS Matrix Proteins

Biofilm assists in structural support, and it also protects the microorganisms against external stress such as antibiotics, desiccation, mechanical pressure, and the host immune system. Each species has a particular composition of EPS and characterization based on environmental conditions, contributing to an overall stability of the biofilm (Flemming and Wingender 2010; Ramírez-Larrota and Eckhard 2022). These proteins anchor the bacteria to the EPS, mediate adhesion to biotic and abiotic surfaces, and confer resilience to mechanical shear and chemical activity (Zhang et al. 2015).

These proteins can effectively be degraded by the proteinase-catalyzed process of proteolysis, which is the breakdown of proteins. These matrix proteins are both cleaved and weakened in their attachment to the polysaccharides by endogenous bacterial proteinases such as LasB elastase of *P. aeruginosa* and exogenous proteinases (Saggu et al. 2019). The degradation of matrix proteins renders the biofilm structure weak, leading to dispersal of bacterial cells and enhanced access of antibiotics (Alsuwat et al. 2025). Besides, proteolysis is able to increase access to the immune response, thereby promoting the elimination of biofilm-related infections (Ramírez-Larrota and Eckhard 2022). This renders the protease-mediated degradation a possible approach in clinical and industrial biofilm control.

As an example, in *Pseudomonas aeruginosa*, LasB protease degrades the matrix adhesin CdrA, resulting in dysfunctional biofilms and greater antibiotic sensitivity (Cooley et al. 2016). This has also been proved in metalloproteases of other bacterial genera, which break down matrix proteins, with far-reaching capabilities to interfere with biofilms because of the activity of proteases (Saggu et al. 2019). Overall, proteolytic degradation of the matrix proteins by EPS has a potent biofilm-destabilizing effect and a fundamental approach to improving antimicrobial therapies.

12.4.2 Synergy with Antibiotics and Antimicrobial Agents

Biofilms present a formidable obstacle in infection management because bacteria embedded within these communities are remarkably tolerant to antibiotics. Unlike free-floating (planktonic) cells, biofilm-resident bacteria are encased in an EPS matrix that limits the access of antibiotics and evades immune defenses. Recently, combining biofilm-disrupting agents, especially proteases, with antibiotics has shown promise in overcoming this challenge. Such combinations often exhibit synergy, where proteases enhance antibiotic penetration and improve bacterial eradication more effectively than antibiotics alone (Ramírez-Larrota and Eckhard 2022).

12.4.2.1 Biofilm Resistance to Antibiotics

Biofilm resistance is multifaceted. The EPS matrix forms a dense barrier that restricts diffusion of antimicrobial agents, leading to sublethal concentrations inside the biofilm. Furthermore, bacteria within biofilms grow more slowly or enter dormant states, reducing the effectiveness of antibiotics that focus on proliferating cells. Importantly, biofilms harbor "persister cells," a small fraction of highly tolerant bacteria capable of surviving high antibiotic doses and recolonizing once treatment ends (Flemming and Wingender 2010; Zhang et al. 2015).

12.4.2.2 Role of Proteases in Biofilm Disruption

Proteolytic enzymes can break down matrix proteins, thereby weakening the structural integrity of the EPS. For example, *Pseudomonas aeruginosa* secretes the protease LasB, which cleaves matrix-associated proteins such as CdrA, which is crucial for maintaining biofilm cohesion. By disrupting these key proteins, proteases facilitate deeper antibiotic penetration and lower bacterial tolerance within biofilms (Ramírez-Larrota and Eckhard 2022; Zhang et al. 2015).

12.4.2.3 Evidence of Synergy Between Proteases and Antibiotics

A growing number of studies report that pairing proteases with antibiotics leads to significantly improved biofilm control compared to either treatment alone. These synergistic effects have been demonstrated across multiple bacterial species and combinations of proteases and antibiotics. The summary of notable findings is in

Table 12.3 Synergistic effects of protease–antibiotic combinations against bacterial biofilms

Bacterial species	Protease type	Antibiotic(s)	Reported synergistic effect	References
Pseudomonas aeruginosa	Metalloprotease (from *Microbacterium* sp. SKS10)	Ciprofloxacin	Enhanced biofilm degradation and significantly increased bacterial killing (via CFU reduction)	Saggu et al. (2019)
Staphylococcus aureus	Proteinase K (bacterial source)	Vancomycin	Marked reduction in biofilm biomass and improved antibiotic efficacy	Rao (2015)
Staphylococcus aureus, *Escherichia coli*, *Pseudomonas aeruginosa*	Recombinant human trypsin	Ceftriaxone, ciprofloxacin	Two- to eight-fold reduction in MIC; enhanced biofilm clearance observed	Banar et al. (2016)
Pseudomonas aeruginosa	Endogenous LasB (elastase)	Tobramycin	Improved antibiotic penetration resulted in a significantly reduced bacterial burden in a murine lung model	Suleman (2016)
Pseudomonas aeruginosa	Neutrophil elastase (host-derived)	Meropenem and others	Synergistic matrix degradation; increased susceptibility of embedded cells	Fleming and Rumbaugh (2017)

CFU colony forming units, *MIC* minimum inhibitory concentration

Table 12.3. These studies highlight how protease treatment degrades the proteinaceous biofilm matrix, loosens cell adhesion, and allows antibiotics to reach and kill embedded bacteria more effectively.

12.4.2.4 Mechanistic Insights into Synergy

The synergy arises from both physical disruption and biochemical modulation of the biofilm matrix. Proteases lower the amount of matrix and degrade the adhesive proteins, physically exposing the bacteria to greater concentration of antibiotics. In addition, certain proteases can disrupt the bacterial signaling pathways, including quorum sensing, which induces a process of dispersal and sensitization to therapeutic antibiotics (Ramírez-Larrota and Eckhard 2022; Zhang et al. 2015).

12.4.2.5 Clinical and Industrial Implications

The use of protease antimicrobial synergy shows prospects to address drug-resistant biofilm infections, such as those that lead to chronic wounds, lung infections linked to cystic fibrosis, and device-related infections. This can also be used beyond medical purposes as a biofouling controller on water systems, pipelines, and medical instruments, which could result in reduced costs of maintenance and improved cleanliness (Banar et al. 2016; Saggu et al. 2019).

The combination of antibiotics and proteolytic enzymes represents a strong measure that can be taken to overcome the stubbornness of biofilms. The scale-up of these findings in the clinics and industry will also require additional research to find the best combinations of enzymes and antibiotics, and the ways of their delivery.

12.4.3 In Vitro and In Vivo Evidence of Biofilm Degradation

Diverse model systems, including microtiter plate experiments, flow-cell assays, and organoid cultures to examine channels of biofilm degradation in vitro and mouse wound infection models, catheter infection models, or respiratory infection in vivo models have been described (Stepanović et al. 2007). Although in vitro investigations are essential sources of information that can be studied in tightly controlled conditions, it is important to confirm them in living organisms because biofilms, unlike in vitro settings, are characterized by different behaviors in host conditions. In vivo models incorporate complex biological factors such as the immune response, blood flow, and tissue architecture, all of which impact biofilm development and treatment outcomes. These models are crucial for confirming the efficacy of biofilm-disrupting agents, such as proteases, in real infections, thereby supporting the development of effective therapies.

Integrating evidence from both in vitro and in vivo studies enhances our understanding of biofilm dynamics and therapeutic potential, facilitating the translation of laboratory research into clinical and industrial applications.

12.5 Role of Microbial Proteases in Wound Healing

Wound healing involves a complex, multilayered physiological process that involves hemostasis, inflammation, proliferation, and remodeling (maturation). There is a range of cellular messages and enzymatic activity in each stage. Among all the obligatory enzymes, proteases are a key player in making the degradation of the broken-down extracellular matrix (ECM) possible (Karsdal et al. 2013). While matrix metalloproteases are indeed known to positively contribute to wound healing, proteases produced by some bacteria have in the past been viewed as detrimental due to their ability to degrade tissue and their involvement in persistent, nonhealing ulcers. Understanding their operation at different periods of wound healing and comparing them with host or synthetic enzymes allows for the maximum comprehension of their potential.

12.5.1 *Phases of Wound Healing and Enzymatic Requirements*

Wound healing is a dynamic, finely coordinated process involving cellular and molecular responses that aim to reestablish tissue integrity. It progresses through four primary phases, overlapping with each other: hemostasis, inflammation, proliferation, and remodeling. Both steps are essential for successful tissue healing. Hemostasis involves the stop of the bleeding due to clot formation, followed by the immune inflammatory phase, an event during which the immune cells eliminate any foreign materials and bacteria. During the proliferative stage, the tissue is regenerated, angiogenesis and reepithelialization occur, and during the remodeling stage, the extracellular matrix is reinforced and reorganized with an attempt to rebuild the functionality and structure (Coma et al. 2021; Karsdal et al. 2013). These stages also depend on enzymes such as proteases that serve as regulatory elements governing clot stability, immune response, tissue growth, and turnover of the matrix.

These proteases, of which matrix metalloproteinases (MMPs) are among many, are capable of proteolytically breaking down the ECM. MMPs play a considerable role at all stages of wound recovery by remodeling the wound matrix, making migration of the cells, and remodeling of the tissue (McCarty and Percival 2013). Proteases such as neutrophil elastase and cathepsins help in the excision of the necrotic tissue during the inflammatory phase. Such enzymes also clear a path for the entry of immune cells by loosening the ECM. As healing progresses to the proliferative stage, the MMPs facilitate the migration of the keratinocytes and fibroblasts that break down the physical barriers in the ECM. Migration is needed to produce new tissue and blood vessels, which is mediated by growth factors such as vascular endothelial growth factor (VEGF) and transforming growth factor-beta (TGF-β) that are activated or secreted by proteolytic activity (Coma et al. 2021; Karsdal et al. 2013).

The last stage is maturation. This stage can be termed remodeling phase where proteases help in smoothing extracellular matrix by breaking down temporary components and helping the collagen fibers to get rearranged to end with strengthened healed tissue. Tissue inhibitors of metalloproteinases (TIMPs), which is responsible in regulating the activity of protease, are stringently regulated in this process.

The lack of balance resulting in either excess or deficiency in activity of protease might lead to adverse consequences on the healing process and ultimately lead to the formation of chronic wounds (Karsdal et al. 2013; McCarty and Percival 2013). Proteases play a crucial role in the repair of efficiently damaged tissues, but this process should be limited with respect to eliciting detrimental effects. Increased knowledge on how these enzymes play their roles during every step will help the researchers and clinicians to come up with better solutions of treating both chronic type and acute type wounds. The destabilization of this equilibrium leads to the increase in chronicity of wounds due to excessive amounts of proteases and reduced amounts of protease inhibitors (McCarty and Percival 2013).

12.5.1.1 Hemostasis

The early-stage wound causes vascular damage and then bleeding, and the body must respond to such a stress immediately, in the form of vasoconstriction, the formation of platelet plugs, and coagulation. However, upon the occurrence of an injury, this layer is exposed to the circulating blood components. The platelets, in turn, promptly attach to the bare region and start aggregating, creating an initial plug of hemostasis. This temporary seal assists in ceasing blood loss in the area of injury (Lindsay et al. 2017).

Platelets, in order to stick to the injury site, need special proteins on their surface called integrin receptors. These integrins mediate the sticking of platelets to the injured vascular wall, together with the aggregation of the platelets to each other, which is vital in the formation of clots. By entering the clot, platelets stop and degranulate, which releases a variety of different substances, such as inflammatory cells chemoattractants, local fibroblasts, endothelial cell activation factors, and vasoconstrictors (Lindsay et al. 2017). Examples of such liberated compounds are chemokine (C-C motif) ligand 5 (CCL5), thrombin, transforming growth factor-beta (TGF-β), as well as a platelet-derived growth factor (PDGF), and vascular endothelial growth factor (VEGF) (Paterlini 2002).

The most important substance promoting blood clotting is called fibrin, and it not only allows the access of immunological cells to the area and the activation of the complement cascade but also interacts with the receptors CD11b/CD18 on the cells. Also, it binds to FGF-2 and VEGF, which facilitates vascularization of the tissue in the healing process. Fibrin helps in tissue repair by binding to IGF-1, which stimulates stromal cells (supportive tissue cells) to multiply and grow (Lindsay et al. 2017).

12.5.1.2 Inflammation

Following hemostasis, the inflammatory response is initiated through vasodilation and activation of the coagulation and complement pathways. Bradykinin, along with complement proteins C3a and C5a, increases vascular permeability, facilitating the efficient influx of immunological cells to the locus of injury. These complement components also activate mast cells, promoting the release of inflammatory mediators, such as histamine and leukotrienes. Collectively, these events initiate the early inflammatory phase, which is critical for effective wound healing (Ng 2010).

The endothelial cells of blood vessels become more permeable, allowing the migration of neutrophils, the primary immune cells, from the circulatory system into the damaged tissue. This function is promoted by thrombin and interleukin-8, which relax endothelial junctions and promote the movement of immunological cells. At the wound site, neutrophils kill pathogens by releasing proteases, reactive oxygen species, and antimicrobial peptides. When inflammation ceases, neutrophils also undergo apoptosis, a process mediated by cathepsin D-D, which activates the caspase enzymes that cause DNA breakdown and cell death. Besides removing pathogens, neutrophils also secrete markers that direct macrophages to organize the next process of healing. Macrophages are essential for wound healing; their depletion delays tissue repair by impairing debris clearance and reducing fibroblast activity (Ng 2010; Paterlini 2002).

12.5.1.3 Proliferative Phase

This phase begins roughly 2 days after injury and coincides with the inflammatory phase. It consists of a breakdown of the fibrin-platelet matrix and infiltration by endothelial cells and fibroblasts, facilitated by proteases. Fibroblasts, which are activated by PDGF and TGF-β, are activated from mast cells and macrophages, and become major by days 3–5. Fibroblasts secrete fibronectin, hyaluronic acid, proteoglycans, and collagen, creating a new ECM for keratinocyte migration (Ng 2010; Paterlini 2002). Granulation tissue develops on day 4 and includes fibroblasts, macrophages, and new vessels in loose ECM. Angiogenesis, which is directed by VEGF, bFGF, angiopoietin-1, and CXCL8, is integrin αvβ3-dependent, produced by fibroblasts, keratinocytes, and endothelial cells. Myofibroblasts appear around day 4 under TGF-β, CXCL8, and fibronectin-EDA to induce contraction of the wound through focal adhesions. Epithelialization is stimulated by EGF, TGF-β, and other cytokines (PDGF, FGF, TNF-α, KGF, CXCL8), which stimulate proliferation and migration of keratinocytes (Ladwig et al. 2002).

12.5.1.4 Remodeling Phase

The remodeling phase is the last and longest process of wound healing, which may last up to 1 year after injury. It is characterized by massive extracellular matrix (ECM) turnover, a rapid reduction in cellularity through the apoptosis of remaining inflammatory cells and myofibroblasts, and the involution of recently established blood vessels. Important processes occurring at this stage are wound contraction, which is mainly facilitated by myofibroblasts, and collagen remodeling governed by matrix metalloproteinases (MMPs). In the majority of adult and children's wounds, this period ends in the development of a scar. These include using TGF-β3-based treatments, antibodies that neutralize TGF-β, and inhibitors of connexin 43, a key player in TGF-β signaling (Lindsay et al. 2017). Molecules such as fibromodulin and decorin help regulate TGF-β activity, supporting regenerative healing and while decreasing the risk of fibrosis. Additional antifibrotic approaches involve the use of recombinant growth factors, cytokines, protease inhibitors, and agents that block profibrotic signals or limit excess collagen production. In total, the remodeling stage not only completes structural rehabilitation but also establishes functional and aesthetic properties of the repaired tissue (Ng 2010; Paterlini 2002).

12.5.2 Enzymatic Debridement: Microbial Proteases as Wound Cleaners

Debridement refers to the excision of dead tissue and nonviable cells, foreign material, and aberrant healing tissue from the wound. Debridement can be performed in several ways. Autolytic debridement is a wound cleansing technique in which the enzymes of the body break down and destroy dead tissue. Special bandages are used to create an optimal environment for this process by either adding or absorbing moisture, depending on the wound's state (Tarnuzzer and Schultz 1996).

In larval debridement, disinfected green bottle fly larvae (*Lucilia sericata*) are employed. These digestive enzymes are released by the larvae, which act selectively to degrade necrotic and dead tissue without damaging healthy tissue. In mechanical debridement, a damp gauze is applied to the wound and allowed to dry, allowing it to adhere to the dead tissue. When the gauze is stripped off, dead tissue comes off with it (Sinclair and Ryan 1994). The technique, however, can be extremely painful and will also damage healthy tissue. It gently removes dead tissue and wound debris, making the patient more comfortable and less painful. Dead tissue is cut away in sharp or episodic debridement by a scalpel, forceps, or scissors. Enzymatic debridement is the process of using proteolytic enzymes topically on the wound to dissolve dead tissue (Malik et al. 2017).

Enzymatic debridement with microbial protease aids the inflammatory process by debriding necrotic tissue and reducing the microbial load. This not only manages infection and inflammation but also conditions the wound bed during the proliferative phase, enabling granulation and tissue regeneration.

They act on wound debris, such as fibrin clots, damaged collagen, and eschar, all of which slow down the growth of healthy tissue. The *Bacillus subtilis* D10 strain produces a highly selective serine protease. When used in conjunction with silver sulfadiazine (SSD), a common burn treatment, it accelerates healing by stimulating cell growth, reducing inflammation, promoting the formation of new blood vessels, and facilitating tissue regeneration, as demonstrated in mice (Al-Dhuayan et al. 2021). By clearing away damaged proteins and improving blood flow, the enzyme facilitates the formation of new vessels and attracts skin cells, resulting in faster wound closure. In another study with rats, keratinase worked as well as bromelain in removing dead tissue but was safer and more effective at clearing eschar (Xu et al. 2023).

Despite their promise, using microbial proteases in clinical settings presents challenges. If not controlled, the enzymes can damage healthy tissue. It is also challenging to maintain the optimal pH and enzyme levels in wounds. Careful control of enzyme activity is necessary to prevent irritation, particularly in sensitive or inflamed wounds. This means precise formulation and delivery are key for safe and effective treatment.

12.5.3 Microbial Proteases in Immune Modulation and Tissue Regeneration

Although proteases are primarily associated with wound healing through the breakdown of dead tissue during debridement, recent studies suggest they do more than that. Their presence also influences how cells behave at the wound site, playing a significant role in reducing inflammation and supporting tissue repair as part of the body's intrinsic healing mechanisms. Some even contribute directly to tissue regeneration by encouraging angiogenesis, fibroblasts, and skin cells. This broader biological activity highlights their dual potential, not only as debriding agents but also as active participants in the healing process (Al-Dhuayan et al. 2021; Xu et al. 2023).

12.5.3.1 Modulation of Inflammatory and Immune Responses

This is a pivotal initial part with regard to wound healing, involving the recruitment of immunological cells to eliminate pathogens, dead tissue, and debris. Although necessary, this phase must be tightly regulated; if it becomes prolonged, it has the potential to degrade healthy tissue and hinder the healing process. Chronic wounds, most notably, are typically locked into a state of inflammation due to high concentrations of pro-inflammatory cytokines, which inhibit fibroblast function and tissue repair.

More recent studies indicate that microbial proteases could be modulating this immune response. Certain enzymes affect macrophage function, promoting a transition from the inflammatory M1 cell type to the anti-inflammatory tissue-repairing M2 form. Other cytokines can directly degrade excess cytokines, aiding in the reduction of inflammation and promoting the healing proliferative phase (Malik et al. 2017).

Microbial proteases are not only debriding agents for necrotic tissue but also play an active role in modulating the immune environment of wounds. A new study identified that in a rat model of burn injury, application of keratinase isolated in *Fusarium* species increased wound healing through recruitment of macrophages and polarization toward the M2 phenotype. By day 7 after injuries, the M2:M1 ratio rose in the keratinase-treated group by 2.47 compared to 1.33 in the control group, signifying a clear bias toward the immune patterns of anti-inflammation and pro-healing. There was also a considerable reduction in pro-inflammatory cytokines, including IL-1 and TNF-α, which contributed to the attenuation of the inflammatory status or pathway and the prompting of the proliferative phase (Xu et al. 2023).

To back up this immunomodulatory activity, an earlier in vitro study showed that dentilisin, an extracellular protease of *Treponema denticola*, had a direct cleaving action on the inflammatory cytokines IL-1b, IL-6, and TNF-α in a dose-dependent manner (Xu et al. 2023). Dentilisin contains a pathogenicity factor despite the rather general possibility of microbial proteases to alter immunity via cytokine cleavage. In aggregate, these findings indicate that microbial enzymes, with appropriate design and control, could be applied to not only debride wounds but also achieve immune homeostasis and tissue repair (Xu et al. 2023).

Consider dentilisin, a protein produced by the bacterium *Treponema denticola*, which has been linked to gum disease. In experimental lab tests, this enzyme was determined to break down key pro-inflammatory cytokines, including IL-1β, TNF-α, and IL-6, in a dose-dependent fashion (Xu et al. 2023). Now, while its impact on disease is unfavorable, allowing for the avoidance of immune detection by the pathogen, what is interesting is how specifically it targets inflammatory molecules. What this implies is that, with targeted adjustments, the same proteases can be engineered into treatments someday to limit excessive inflammation in chronic wounds. It is a wake-up call that even pathogenic microbes may hold the key to improved healing technologies, if properly utilized.

These two microbial proteases, *Fusarium* sp. keratinase and *Treponema denticola* dentilisin, exemplify the remarkable adaptability of microbial enzymes in wound healing. Keratinase has been demonstrated to have definite therapeutic potential in increasing tissue regeneration and altering immune responses toward repair (Xu et al. 2023). Dentilisin, although secreted by a pathogen, illustrates the potential of microbial enzymes to target and modulate specific inflammatory cytokines. Together, they propose that proteases from microbes provide more than just simple enzymatic debridement; they may also modulate immune behavior, alleviate chronic inflammation, and create a microenvironment that promotes quicker and healthier healing (Xu et al. 2023).

12.5.3.2 Promotion of Granulation Tissue and Reepithelialization

After the wound bed is adequately debrided, the following critical step in healing is tissue regeneration, i.e., the generation of granulation tissue and reepithelialization. Proteases of microbes, which were first utilized to remove necrotic debris and slough, remain a potent force even after cleaning. Their enzymatic activity pretreats the wound environment both physically and biologically, facilitating regeneration. Granulation tissue consists of newly developed connective tissue, including fibroblasts, collagen, and budding capillaries, all of which play an important role in supporting new tissue genesis. Keratinase-treated wounds have also demonstrated enhanced vascularization, suggesting that microbial enzymes may indirectly promote angiogenesis by reducing inflammatory barriers and improving oxygen availability. This revascularization is a critical aspect of tissue regeneration, as it ensures adequate nutrient delivery and supports the metabolic demands of the healing process (Xu et al. 2023).

In addition to granulation, microbial proteases also influence the stage of reepithelialization, during which keratinocytes migrate from the wound border to rebuild the epidermal barrier. Proper debridement with microbial enzymes enables the exposure of a healthy, hydrated bed, which is essential for the attachment and migration of keratinocytes. In the same keratinase experiment, not only was epithelial coverage more rapid in the treated group but histological examination also reflected more mature and stratified epidermal layers than the control group. This indicates that microbial enzymes are not only helping to clean but also aiding in cellular communication and tissue patterning necessary for skin regeneration. Fibroblasts induced into the wound secrete signaling molecules that include growth factors and cytokines, which bring keratinocytes into the site of injury and enable wound closure (Sinclair and Ryan 1994; Tarnuzzer and Schultz 1996).

However, most existing data target keratinase; comparable regenerative results have been reported with other microbial proteases, such as subtilisin, known to enhance epithelial migration and closure of wounds in previous experimental conditions. Such proteases, when delivered in suitable formulations such as gels or hydrogels, provide sustained enzymatic activity and controlled degradation, and are hence viable contenders for inclusion in chronic wound care protocols. Together, microbial proteases serve not just as enzymatic agents for wound cleansing but also as biological promoters that drive the wound ahead into the reparative phase by facilitating the establishment of granulation tissue, vascularization, and new epithelial layers.

12.6 Therapeutic Applications and Delivery Systems

During the healing process, necrotic (dead) tissue can accumulate in the wound, which often hinders repair by harboring bacteria and containing high concentrations of inflammatory mediators. These factors promote chronic inflammation and impede cellular migration, which is necessary for healing (Ruseva et al. 2019).

To promote faster and more effective recovery, it is essential to remove necrotic tissue through debridement. Autolytic and enzymatic debridement are the safest and most commonly used methods in wound care because they selectively and efficiently remove dead tissue. Polymers such as poly(sulfobetaine methacrylate) (pSBMA) and poly(carboxybetaine methacrylate) (pCBMA) are zwitterionic polymers (ZPs) that uniquely swell more in salty solutions such as wound exudate. Unlike traditional polyelectrolyte dressings that shrink in salt solutions, ZPs expand, providing better absorption and patient comfort by avoiding pain and wound disruption (Ruseva et al. 2019).

Thermal burns are skin injuries caused by exposure to hot objects such as liquids, fire, or heated equipment, affecting people of all ages worldwide. They have a high incidence, especially in underdeveloped countries, with an estimated 6.6 million burn cases annually in India alone. The initial injury involves protein denaturation, inflammation, and ischemia, leading to burns of varying depths. Secondary complications such as edema, infection, and cell death contribute to wound progression, known as burn wound conversion. The healing process involves hemostasis, inflammation, proliferation, and remodeling phases, largely dependent on the dermis, which contains cells that regulate epidermal regeneration (Sahu et al. 2017).

Chronic wounds affect over 6.5 million patients worldwide, driven largely by diseases such as diabetes, cardiovascular issues, and immunosuppression. Chronic wounds pose a significant economic and social burden because of high prevalence, amputation, and recurrence rates. Despite the progress in clinical remedies, there has been an insignificant development in the field of effective wound management during the previous century. Knowledge of the anatomy of the skin and the contrast between the processes of acute and chronic wound healing is necessary to achieve better results during treatment (Zhang et al. 2020).

Hydrogels are omnipresent classes of materials in the form of polymer networks that can retain a lot of water and hence be used as a delivery system in many applications, especially in wound healing (Jacob et al. 2021). Fibroblast growth factor (FGF) family proteins have been under active development as drugs since the late twentieth century, and some of their members have already achieved approval to be used clinically in wound treatment. To illustrate, basic-FGF (bFGF) loading into bioinspired hydrogels exhibited the extended presence of growth factor, augmented fibroblast proliferation in vitro, accelerated wound healing in scratch assays, and promoted healing in full-thickness skin incisions in mice (Zhang et al. 2020). Even greater innovations have come up using hydrogels in combination with other delivery methods, such as liposomes and microspheres, to form two-layered systems to extend the release of encapsulated growth factors such as bFGF. The result of such sustained delivery is better wound closure, a more orderly collagen deposition, enhanced vascularization, granulation tissue formation, myofibroblast activity, and lower inflammation (Jacob et al. 2021; Zhang et al. 2020).

Another promising method of hydrogels concerns liquid-phase hydrogels, which can offer better wound coverage. As an example, heparin-poloxamer hydrogels can be applied topically, due to their liquid nature at 4 °C and rigid consistency at body temperature (Zöller et al. 2025). These thermo-sensitive hydrogels emphasized the significance of the drug–scaffold interaction, where acidic FGF was more effective in tissue regeneration than basic FGF, although both scaffolds used heparin-binding domains. Sensors sensitive to temperature have also been produced in microneedle hydrogel technologies that target the delivery of VEGF to inflamed wound areas, resulting in granulation, collagen deposition, developing angiogenesis, and minimized inflammation (Jacob et al. 2021).

As wound healing is seen to depend on an intricate interaction of numerous molecular signals, and not only growth factors, recent developments have been in delivering growth factor mixtures or combining growth factor molecules with other bioactive molecules (Nurkesh et al. 2020; Zöller et al. 2025). In another study, the bilateral administration of substance P and TGF-β-1 microparticles in injectable hydrogels achieved faster skin repair and layering in irradiated mice through the infiltration of mesenchymal stem cells (Nurkesh et al. 2020).

12.6.1 Nanoformulations and Encapsulation Technologies

The wound infection may incorporate one or many microbial species, and has the characteristic of generating biofilms. These biofilms minimize the efficiency of therapies and enhance resistance to treatments. Biofilms promote different activities between bacteria as well as between bacteria and hosts, such as competition and cooperation, since they act as self-sustaining ecosystems. Alternative therapies are therefore urgently necessitated by the rising trend of chronic wound cases and biofilm infections related to chronic wounds. Nanotechnology has potential solutions because of the inbuilt antimicrobial properties of nanoparticles or even as drug carriers. They have a wide range of antiviral, antibacterial, and antifungal activity, and their low doses can deal with an extremely broad range of infectious agents with limited risk of resistance-causing pathogenic mutations, which is made possible by nanoparticles and nanostructured coating (Mihai et al. 2018).

The skin represents the largest visible organ of the human body, which has a covering area of around 1.5–2 m^2. It is the central boundary between the body and the outside environment that fulfils essential duties of stabilizing body temperature, preventing microbes, maintaining electrolyte equilibrium, and protecting against injuries, chemicals, and UV light. Moreover, skin is an important factor in the uptake of the drug that affects effectiveness. The physical nature of the skin is the epidermis, dermis, and subcutaneous tissue. It has added features of hair follicles, sebaceous glands, and sweat glands, and has blood vessels, lymphatic vessels, and nerves (Mihai et al. 2018).

There are two main ways through which drugs can enter the skin, namely through the skin appendages and the epidermis. There exist two primary pathways of penetration: transcellular, substances move through the keratinocytes and within the lipids and must diffuse through both hydrophilic and hydrophobic interfaces, and the intercellular path in which drugs pass a tortuous route around the keratinocytes through the lipid matrix. Most medications are thought to follow the intercellular route in the majority of instances since it is somewhat accessible (Nurkesh et al. 2020). Transdermal drug delivery systems (TDDS) offer a mode of drug administration through the skin in order to deliver locally or systemically. Among the potential positive features of TDDS as a promising sphere of pharmaceutical studies, one can distinguish convenience, noninvasiveness, and enhanced patient adherence (Mihai et al. 2018).

Irrespective of all these advantages, the natural barrier of the skin usually limits the capability of drug delivery via TDDS and can restrict the delivery of drug concentrations to achieve therapeutic effects. Such preparations are pain-free, produce little skin damage without affecting skin barrier properties, and add permeability of macromolecular medicines. Nanoformulations encompass various vesicles, including transfersomes, liposomes, niosomes, ethosomes, and invasomes, as well as nanoparticles, such as polymeric nanoparticles, lipid nanoparticles, and nanoemulsions.

Polymeric nanoparticles (PNPs) have gained considerable attention in the target drug delivery because they increase both the water-soluble and insoluble drugs and biologically active substances. PNPs improve the bioavailability and safety of drugs, biocompatibility, decrease toxicity, and retain therapeutic effect (Cheng et al. 2018; Mihai et al. 2018). The proper construction of such systems depends on the organ to be targeted, the type of drug, and its biocompatibility with biodegradable polymers. The chances of target tissue penetration by polymeric nanomedicines depend on the size, surface charge, and hydrophobic or hydrophilic nature. The size of the particle is very important in the bioavailability in overcoming biological barriers to the site of action. The surface charge determines whether the nanoparticles will stick to systemic circulation or they will respond to the cell membranes (Zöller et al. 2025). To deliver targeted delivery, PNPs need to circulate over long periods, and therefore are often altered to prevent clearance by the mononuclear phagocyte system (MPS) in a short time (Swami et al. 2019).

12.6.2 Controlled Release and Targeted Delivery

There is a significant healthcare problem of nonhealing wounds, which is exacerbated by the increased prevalence of obesity as well as diabetes. The roles of growth factors, enzymes, and cytokines (therapeutic proteins) during the wound healing process, hemostasis, to tissue regeneration are important. However, their effective release is commonly restricted by the unwelcoming, protease-existing environment of the wound, as well as barriers to physiology, e.g., the skin (Duarte et al. 2025). It is acknowledged that therapeutic proteins will play a promising role in wound management, but ongoing barriers of availability, affordability, and appropriateness need to be surmounted. The significance of the interdisciplinary cooperation is underlined in creating innovative, practical approaches to improving wound healing treatments (Hofmanová et al. 2019).

As an antibiotic, clindamycin hydrochloride was chosen and entrapped into Carbopol nanogel. Being a representative of the lincosamide antibiotics group, clindamycin is structurally similar to lincomycin and demonstrates a high efficiency concerning Gram-negative bacteria, especially *Staphylococcus aureus* (Duarte et al. 2025; Hofmanová et al. 2019). Povidone-iodine (PVP-I) refers to a pharmaceutical that consists of polyvinylpyrrolidone carrying free iodine, which is used in wound cleaning and disinfection of hands and preventing infections on the skin (before and after surgery) (Duarte et al. 2025; Weldrick et al. 2019). Other widely used biocides that have been applied against harmful microorganisms include benzalkonium chloride (BKC). Its use has, however, been limited by the rising number of allergic reactions and cases of bacterial resistance. Side effects related to BKC are cytotoxic interference with neutrophil functionalities and leukocyte responses at sites of inflammation. Due to some studies of toxicity to mammalian cells, it is argued that BKC is not dangerous when used in the correct manner (Weldrick et al. 2019).

Chloroxylenol is used as a disinfectant for skin wounds, sterilization of surgical instruments, and domestic cleaning products. It can interfere with the membranes of *S. aureus* cells because of its ionizable qualities and consequently raise the fluidity and permeability. It also disrupts business proteins and enzymes, improving the susceptibility of bacteria to biocides (Weldrick et al. 2019). One more alternative is the use of cetrimide as an antiseptic against wound infection; this option is a cationic surfactant synthesized by quaternary ammonium salts. It is more resistant to bacteria, as compared to PVP-I, and even triggers epithelial cell toxicity. It was, however, proved effective against Gram-negative and Gram-positive bacteria, as it maintains its antimicrobial effect over time (Weldrick et al. 2019).

The wound healing is a multiphased, complex and coordinated process, consisting of hemostasis, inflammation, proliferation, maturation, and remodeling (Sahu et al. 2017). This is a synchronized effort in healing all layers of skin through cellular proliferation and extracellular matrix (ECM) remodeling (Jacob et al. 2021; Sahu et al. 2017; Zöller et al. 2025). The exudate, a fluid with a high content of water, nutrients, electrolytes, proteins, and cells, is formed in the body during the healing process and provides favorable conditions of tissue repair (Sahu et al. 2017). In chronic wounds though, delayed healing may be associated with the resulting formation of a scar and this may negatively impact on the appearance of the skin and the psychological well-being of the patient, resulting in a subsequent compromise in the quality of life. Because scars inhibit complete skin functionality restoration and ensure its recovery, it is important to work out new materials that will ensure faster healing and deter scarring (Jacob et al. 2021; Sahu et al. 2017). Experiments on biocompatible drug carriers are required to enhance skin regeneration and wound healing.

Silk proteins represent one of the natural biopolymeric drug carriers that has attracted interest in wound healing applications on the basis of their tunable physical and chemical characteristics such as biodegradability, biocompatibility, and hydrophilicity (Bernardes et al. 2024; Jacob et al. 2021; Sahu et al. 2017). Silk, a protein polymer produced by various insects, including spiders, silkworms, and bees, has broad biomedical applications, especially in drug delivery (Bernardes et al. 2024; Weldrick et al. 2019). *Bombyx mori* silkworm silk is the most widely used commercially because of its scalability (Bernardes et al. 2024).

12.6.3 Combination Therapies: Protease + Antibiotics or Plant Extracts

Nonhealing chronic wounds are a significant complication, and amputations of the lower extremities occur every year in the United States due to diabetes, with over 70,000 people affected. The underlying reasons for impaired healing in diabetic wounds are not fully understood, and effective therapies to promote repair remain limited. In previous studies using db/db mice, two active forms of matrix metalloproteinases (MMPs), MMP-8 and MMP-9, were identified in wounds. It was proposed that MMP-8 may contribute positively to the wound healing process, while MMP-9 is associated with pathological damage. In this study, we demonstrate that ND-336, a novel, highly selective inhibitor targeting gelatinases (MMP-2 and MMP-9) and MMP-14, accelerates diabetic wound healing by reducing inflammation and promoting angiogenesis and reepithelialization, effectively reversing the pathological state (Gao et al. 2015).

The increase in bacterial strains resistant to antibiotic treatments has challenged the initial optimism surrounding antimicrobial discovery. Today, many clinically significant bacteria exhibit resistance not only to single drugs but also to multiple antibiotics, a consequence of decades of antimicrobial overuse and misuse. Plants have long been recognized for their significant medicinal value, containing chemical compounds that elicit physiological effects beneficial to human health. Importantly, some plants produce compounds that inhibit multidrug resistance (MDR), enhancing the efficacy of antibiotics. For example, *Catha edulis* extracts have been shown to lower the minimum inhibitory concentrations (MICs) using penicillin G and tetracycline to combat bacteria that have developed resistance in the mouth. Various plant-derived compounds, including polyphenols, diterpenes, and flavones, have been shown to modulate antibiotic resistance, particularly against resistant *Staphylococcus aureus* strains (Aiyegoro et al. 2009).

Numerous examples of plant species have been extensively studied in regards to their therapeutic potential in traditional medicine. *Achillea millefolium* L., or Yarrow, has been used among Europeans, Asia, North Africa, and North America to treat injuries, stomach problems, cold, and swelling. Known as *Achillea millefolium* or Achilles, the plant was named after the allegedly named Greek hero Achilles, who would use it to stop bleeding and heal wounds. The flavonoids contained in yarrow are known to provide it with its potential antioxidant and anti-inflammatory properties. These plant extracts have demonstrated positive tyrosinase-inhibiting, antibacterial, and skin-protective activity, enhancing the skin thickness and expression levels of major epidermal health proteins (Aiyegoro et al. 2009; Albahri et al. 2023).

Aloe vera (*Aloe barbadensis*), as a plant in the Liliaceae, has a history of more than 2000 years and is known to have numerous medicinal as well as cosmetic uses. Ancient Egyptians regarded Aloe vera as a symbol of immortality because of its ability to heal wounds, which is the reflection of its antibacterial, anti-inflammatory, antitumor, and wound-healing effects, and that is why it is present in so many products, including skincare creams or laxatives (Albahri et al. 2023).

12.7 Challenges and Limitations

12.7.1 *Protease Overactivity and Risk of Tissue Damage*

Proteases are required to aid the process of debridement and reconstruction of the extracellular matrix (ECM) in wound healing. Their hyperactivity, however, poses a major challenge to wound healing because of the likelihood of causing innocuous tissue damage. Besides the necrotic tissue, other biologically active, viable ECM such as collagen, fibronectin, and growth factors, which are fundamental to granulation and epithelialization processes, can also be degraded by redundant proteases (Trengove et al. 1999). Elevated levels of serine proteases and MMPs are commonly seen in chronic wounds. These enzymes can continue a harmful cycle in which the products of tissue breakdown increase the release of proteases by stimulating inflammatory mediators (Liu et al. 2009).

Because diabetic ulcers and pressure injuries disrupt the regulation of protease inhibitors such as tissue inhibitors of metalloproteinases (TIMPs), elevated protease concentrations are especially problematic in these conditions (Lobmann et al. 2002). This imbalance hinders proper closure by degrading freshly produced tissue and delaying recovery. To prevent enhancing pathogenic proteolytic conditions, the dosage and duration of exogenous microbial proteases used therapeutically must be carefully regulated.

Creating protease formulations with time-controlled activity, utilizing delivery matrices that release enzymes gradually, and coadministering specific inhibitors to maintain proteolytic balance are some methods to reduce overactivity (McCarty and Percival 2013). Additionally, tracking wound biomarkers, such as neutrophil elastase levels and MMP-9, can assist in customizing protease therapy to the unique wound biochemistry of each patient. Even with encouraging developments, a significant obstacle in protease-based wound repair treatments is preserving the delicate balance between preventing collateral tissue injury and achieving successful debridement.

12.7.2 *Enzyme Inhibition by Wound Exudates*

A complex combination of plasma proteins, inflammatory mediators, protease inhibitors, and cellular debris makes up wound exudate. Although exudate aids in immunological protection and moist wound healing, it also significantly reduces the effectiveness of exogenous proteases employed in debridement. Wound fluids are rich in protease inhibitors, such as tissue inhibitors of metalloproteinases (TIMPs) and alpha-2-macroglobulin, which can deactivate applied proteases and reduce their therapeutic effect (McCarty and Percival 2013).

Protease concentrations and inhibitor levels are both raised in chronic wounds due to the ongoing inflammatory condition (Trengove et al. 1999). Therefore, the debridement potential of microbial proteases may be diminished as exudate components quickly neutralize them. Additionally, the viscosity of exudate can prevent proteases from diffusing into necrotic tissue, requiring larger dosages or more frequent applications (Eming et al. 2007).

Encasing proteases in delivery methods, such as hydrogels or nanoparticles, which preserve enzyme activity until they reach the wound surface, is one way to overcome this inhibition (Kolahreez et al. 2024). As an alternative, experimental research has indicated promise in creating inhibitor-decoy compounds or engineering protease variations with inhibitor resistance. However, it remains technically challenging to achieve adequate proteolytic activity in the environment of chronic wounds, which is rich in proteases but also contains numerous inhibitors.

12.7.3 Host Immune Responses and Allergenicity

Although proteases have several benefits for infection prevention and wound debridement, their microbial origin presents immunological difficulties. The host immune system can identify exogenous proteases as foreign antigens, triggering both innate and adaptive immunological responses. These reactions may show up as erythema, pruritus, local inflammation, or even systemic hypersensitivity reactions (Anderson et al. 2008).

By directly breaking down skin barrier proteins and promoting antigen penetration, proteases can function as allergens and raise the risk of IgE-mediated responses. For example, following repeated exposure, several serine proteases from *Bacillus* or fungal sources have been shown to sensitize patients. Additionally, by cleaving protease-activated receptors, proteases can indirectly activate mast cells and basophils, hence intensifying inflammation (Steinhoff et al. 2005).

Manufacturers frequently thoroughly purify proteases to eliminate endotoxins and other microbial residues to lower immunogenicity. Furthermore, PEGylation and site-directed mutagenesis have been investigated to conceal immunogenic epitopes, increase circulation duration, and lessen the likelihood of allergic reactions (Veronese and Pasut 2005). Protease-based wound remedies require careful patient selection, patch testing, and close observation for side effects. Regulatory criteria require preclinical and clinical testing to assess immunotoxicity. Although protein engineering has advanced technologically, host immunological responses and allergenicity are still substantial obstacles to the broad use of exogenous proteases in wound care.

12.8 Future Prospects

Microbial proteases have considerable potential as next-generation treatments for chronic wound management. To address the complex wound microenvironment, characterized by variable pH, abundant protease inhibitors, and ongoing inflammation, future research is likely to focus on engineering proteases with higher specificity and stability. The creation of designer enzymes that can withstand inactivation by wound exudates while retaining regulated activity to avoid harming healthy tissues may be made possible by advancements in protein engineering and synthetic biology. Proteases can also be used in conjunction with smart delivery technologies, such as thermoresponsive carriers and zwitterionic polymer hydrogels, to improve patient comfort, adjust dosage, and sustain enzyme release. Since conventional enzyme preparations are easily removed from heavily oozing wounds, such systems are well-positioned to address current issues.

To develop synergistic formulations that target multiple barriers to healing, microbial proteases can be integrated with complementary therapies, such as growth factors, antimicrobial peptides, and biofilm-disrupting compounds. To confirm the long-term effectiveness, immunogenicity, and safety of these therapies, regulatory science studies are required. Future research can optimize protease formulations for specific wound types, such as pressure injuries or diabetic foot ulcers, with the aid of more advanced preclinical models and enhanced characterization of wound biochemistry. Protease-based therapies, customized to the distinctive features of every patient's healing profile, may be made possible by integrating diagnostic technologies to track wound biomarkers in real-time. When considered as a whole, these developments suggest that microbial proteases will be incorporated into comprehensive, flexible wound care plans that address both biofilm persistence and compromised tissue regeneration.

12.9 Conclusion

Microbial proteases are showing promise as a novel treatment for biofilm-complicated chronic wounds. These enzymes have two functions: breaking the biofilm barrier for antimicrobial therapy effectiveness and removal of dead tissue for faster recovery. Moreover, it increases blood flow, reduces inflammation, and induces blocked healing. However, the uses of protease have several challenges, such as manufacturing, reducing immunological reaction, maintaining stability, and ensuring enzyme activity to prevent healthy tissue damage. Some advanced methods, such as liposomal carriers and hydrogels containing enzymes, are introduced to improve stability and precision. Some studies suggest that dealing with both chronic inflammation and biofilm is crucial for healing. Microbial protease is a potential in wound care; however, further studies are required for advanced formulations. Their capacity to facilitate tissue regeneration will enhance recovery and help in chronic wound crises.

Acknowledgments The authors wish to extend heartfelt thanks to the management of Vellore Institute of Technology for providing the necessary facilities and support.

Competing Interests The authors have no conflicts of interest to declare that are relevant to the content of this chapter.

References

Aiyegoro OA, Afolayan AJ, Okoh AI (2009) Synergistic interaction of Helichrysum pedunculatum leaf extracts with antibiotics against wound infection associated bacteria. Biol Res 42(3): 327–338. https://doi.org/10.4067/S0716-97602009000300007

Albahri G, Badran A, Hijazi A, Daou A, Baydoun E, Nasser M, Merah O (2023) The therapeutic wound healing bioactivities of various medicinal plants. Life 13(2). https://doi.org/10.3390/life13020317

Al-Dhuayan I, Kotb E, Alqosaibi A, Mahmoud A (2021) Histological studies on a newly isolated bacillus subtilis d10 protease in the debridement of burn wound eschars using mouse model. Pharmaceutics 13(7). https://doi.org/10.3390/pharmaceutics13070923

Al-maqtari QA, Mahdi AA (2019) Microbial enzymes produced by fermentation and their applications in the food industry – a review. August

Alsuwat MA, Shah AA, Ullah S, Khan RU, Alissa M, Khan MS (2025) Microbial biofilm formation to mitigate foodborne pathogens strategies and control measures. Indian J Microbiol:0123456789. https://doi.org/10.1007/s12088-025-01461-4

Anderson JM, Rodriguez A, Chang DT (2008) Foreign body reaction to biomaterials. Semin Immunol 20(2):86–100. https://doi.org/10.1016/j.smim.2007.11.004

Banar M, Emaneini M, Satarzadeh M, Abdellahi N, Beigverdi R, Van Leeuwen WB, Jabalameli F (2016) Evaluation of mannosidase and trypsin enzymes effects on biofilm production of Pseudomonas aeruginosa isolated from burn wound infections. PLoS One 11(10):1–13. https://doi.org/10.1371/journal.pone.0164622

Bernardes BG, Veiga A, Barros J, García-González CA, Oliveira AL (2024) Sustainable silk-based particulate systems for the controlled release of pharmaceuticals and bioactive agents in wound healing and skin regeneration. Int J Mol Sci 25(6). https://doi.org/10.3390/ijms25063133

Bi Y, Xia G, Shi C, Wan J, Liu L, Chen Y, Wu Y, Zhang W, Zhou M, He H, Liu R (2021) Therapeutic strategies against bacterial biofilms. Fundam Res 1(2):193–212. https://doi.org/10.1016/j.fmre.2021.02.003

Borges A, Meireles A, Mergulhão F, Melo L, Simões M (2020) Biofilm control with enzymes. In: Recent Trends Biofilm Sci Technol, pp 249–271. https://doi.org/10.1016/B978-0-12-819497-3.00011-8

Cavallo I, Sivori F, Mastrofrancesco A, Abril E, Pontone M, Gino E, Domenico D, Pimpinelli F (2024) Approaches, pp 1–19

Chang HY, Yang X (2000) Proteases for cell suicide: functions and regulation of caspases. Microbiol Mol Biol Rev 64(4):821–846. https://doi.org/10.1128/mmbr.64.4.821-846.2000

Cheng H, Fu Y, Chang Q, Zhang N, Bu M, Niu Y, Wu Q, Chen C, Verpoort F (2018) Synthesis, biochemical evaluation and computational simulations of new cytochrome bc1 complex inhibitors based on N-(4-aryloxyphenyl) phthalimides. Chin Chem Lett 29(12):1897–1900. https://doi.org/10.1016/j.cclet.2018.10.008

Coma M, Fröhlichová L, Urban L, Zajícek R, Urban T, Szabo P, Novák S, Fetissov V, Dvoránková B, Smetana K, Gál P (2021) Molecular changes underlying hypertrophic scarring following burns involve specific deregulations at all wound healing stages (inflammation, proliferation and maturation). Int J Mol Sci 22(2). https://doi.org/10.3390/ijms22020897

Cooley RB, Smith TJ, Leung W, Tierney V, Borlee BR, O'Toole GA, Sondermann H (2016) Cyclic di-GMP-regulated periplasmic proteolysis of a Pseudomonas aeruginosa type Vb secretion system substrate. J Bacteriol 198(1):66–76. https://doi.org/10.1128/JB.00369-15

Donlan RM (2001) Biofilms and device-associated infections. Emerg Infect Dis 7(2):277–281. https://doi.org/10.3201/eid0702.010226

Donlan RM, Costerton JW (2002) Biofilms: survival mechanisms of clinically relevant microorganisms. Clin Microbiol Rev 15(2):167–193. https://doi.org/10.1128/CMR.15.2.167-193.2002

Duarte, S. O. D., Lima, A. R., Brandão, P., & Fonte, P. (2025). Chapter 8 - Delivery of therapeutic proteins for wound healing (P. Fonte & for W. H. Brandão (Eds.); Drug delivery systems for wound healing pp. 219–250). Academic Press. https://doi.org/10.1016/B978-0-323-85840-3.00008-X

Eming SA, Krieg T, Davidson JM (2007) Inflammation in wound repair: molecular and cellular mechanisms. J Invest Dermatol 127(3):514–525. https://doi.org/10.1038/sj.jid.5700701

Félix F, Brouillet N (1966) Purification et proprietes de deux peptidases de levure de brasserie. Biochim Biophys Acta (BBA) – Enzymol Biol Oxid 122(1):127–144. https://doi.org/10.1016/0926-6593(66)90096-8

Fleming D, Rumbaugh KP (2017) Approaches to dispersing medical biofilms. Microorganisms 5(2). https://doi.org/10.3390/microorganisms5020015

Flemming HC, Wingender J (2010) The biofilm matrix. Nat Rev Microbiol 8(9):623–633. https://doi.org/10.1038/nrmicro2415

Flemming HC, Wingender J, Szewzyk U, Steinberg P, Rice SA, Kjelleberg S (2016) Biofilms: an emergent form of bacterial life. Nat Rev Microbiol 14(9):563–575. https://doi.org/10.1038/nrmicro.2016.94

Gao M, Nguyen TT, Suckow MA, Wolter WR, Gooyit M, Mobashery S, Chang M (2015) Acceleration of diabetic wound healing using a novel protease anti-protease combination therapy. Proc Natl Acad Sci USA 112(49):15226–15231. https://doi.org/10.1073/pnas.1517847112

Ghafoor A, Hay ID, Rehm BHA (2011) Role of exopolysaccharides in pseudomonas aeruginosa biofilm formation and architecture. Appl Environ Microbiol 77(15):5238–5246. https://doi.org/10.1128/AEM.00637-11

Ghoreishi FS, Roghanian R, Emtiazi G (2022) Novel chronic wound healing by anti-biofilm peptides and protease. Adv Pharm Bull 12(3):424–436. https://doi.org/10.34172/apb.2022.047

Hall-Stoodley L, Costerton JW, Stoodley P (2004) Bacterial biofilms: from the natural environment to infectious diseases. Nat Rev Microbiol 2(2):95–108. https://doi.org/10.1038/nrmicro821

Hartley BS (1960) Proteolytic enzymes. Annu Rev Biochem 29:45–72. https://doi.org/10.1146/annurev.bi.29.070160.000401

Hofmanová JK, Rajabi-Siahboomi A, Haque S, Mason J, Teckoe J, To D, Batchelor HK (2019) Developing methodology to evaluate the oral sensory features of pharmaceutical tablet coatings. Int J Pharm 562:212–217. https://doi.org/10.1016/j.ijpharm.2019.03.046

Jacob S, Nair AB, Shah J, Sreeharsha N, Gupta S, Shinu P (2021) Emerging role of hydrogels in drug delivery systems, tissue engineering and wound management. Pharmaceutics 13(3). https://doi.org/10.3390/pharmaceutics13030357

Karatan E, Watnick P (2009) Signals, regulatory networks, and materials that build and break bacterial biofilms. Microbiol Mol Biol Rev 73(2):310–347. https://doi.org/10.1128/mmbr.00041-08

Karsdal MA, Nielsen MJ, Sand JM, Henriksen K, Genovese F, Bay-Jensen AC, Smith V, Adamkewicz JI, Christiansen C, Leeming DJ (2013) Extracellular matrix remodeling: the common denominator in connective tissue diseases possibilities for evaluation and current

understanding of the matrix as more than a passive architecture, but a key player in tissue failure. Assay Drug Dev Technol 11(2):70–92. https://doi.org/10.1089/adt.2012.474
Kolahreez D, Ghasemi-Mobarakeh L, Liebner F, Alihosseini F, Quartinello F, Guebitz GM, Ribitsch D (2024) Approaches to control and monitor protease levels in chronic wounds. Adv Ther 7(Issue 5). https://doi.org/10.1002/adtp.202300396
Labbé JP, Rebeyrotte P, Morel E, Turpin M (1974) Extracellular leucine aminopeptidases (EC 3.4.1.1) of Aspergillus oryzae (IP 410). Glycoprotein nature of the fraction leucine aminopeptidase 1. C R Hebd Seances Acad Sci Ser D 279(26):2139–2142
Ladwig GP, Robson MC, Liu R, Kuhn MA, Muir DF, Schultz GS (2002) Ratios of activated matrix metalloproteinase-9 to tissue inhibitor of matrix metalloproteinase-1 in wound fluids are inversely correlated with healing of pressure ulcers. Wound Repair Regen 10(1):26–37. https://doi.org/10.1046/j.1524-475X.2002.10903.x
Lahiri D, Nag M, Dey A, Sarkar T, Ray RR, Rebezov M, Shariati MA, Thiruvengadam M, Simal-Gandara J (2022a) Immobilized enzymes as potent antibiofilm agent. Biotechnol Prog 38(5): 1–10. https://doi.org/10.1002/btpr.3281
Lahiri D, Nag M, Dutta B, Dey A, Ray RR (2022b) Bacterial extracellular polysaccharides in biofilm formation and function. Appl Biofilms Appl Microbiol 3(3):1–23. https://doi.org/10.1016/B978-0-323-90513-8.00003-0
Leid JG, Willson CJ, Shirtliff ME, Hassett DJ, Parsek MR, Jeffers AK (2005) The exopolysaccharide alginate protects pseudomonas aeruginosa biofilm bacteria from IFN-γ-mediated macrophage killing. J Immunol 175(11):7512–7518. https://doi.org/10.4049/jimmunol.175.11.7512
Li M, Gustchina A, Cruz R, Simões M, Curto P, Martinez J, Faro C, Simões I, Wlodawer A (2015) Structure of RC1339/APRc from rickettsia conorii, a retropepsin-like aspartic protease. Acta Crystallogr D Biol Crystallogr 71:2109–2118. https://doi.org/10.1107/S1399004715013905
Lindsay S, Oates A, Bourdillon K (2017) The detrimental impact of extracellular bacterial proteases on wound healing. Int Wound J 14(6):1237–1247. https://doi.org/10.1111/iwj.12790
Liu Y, Min D, Bolton T, Nubé V, Twigg SM, Yue DK, Mclennan SV (2009) Increased matrix metalloproteinase-9 predicts poor wound healing in diabetic foot ulcers. Diabetes Care 32(1): 117–119. https://doi.org/10.2337/dc08-0763
Lobmann R, Ambrosch A, Schultz G, Waldmann K, Schiweck S, Lehnert H (2002) Expression of matrix-metalloproteinases and their inhibitors in the wounds of diabetic and non-diabetic patients. Diabetologia 45(7):1011–1016. https://doi.org/10.1007/s00125-002-0868-8
Malik K, Ahlawat S, Mehta S (2017) Microbial proteases: current perspectives and potential applications. Ann Biol 33(2):207–210
Masaki T, Soejima M, Nakamura K, Isono M (1978) A new proteolytic enzyme from achromobacter lyticus m497-1. Agric Biol Chem 42(7):1443–1445. https://doi.org/10.1271/bbb1961.42.1443
McCarty SM, Percival SL (2013) Proteases and delayed wound healing. Adv Wound Care 2(8): 438–447. https://doi.org/10.1089/wound.2012.0370
Mihai MM, Preda M, Lungu I, Gestal MC, Popa MI, Holban AM (2018) Nanocoatings for chronic wound repair—modulation of microbial colonization and biofilm formation. Int J Mol Sci 19(4). https://doi.org/10.3390/ijms19041179
Ng MFY (2010) The role of mast cells in wound healing. Int Wound J 7(1):55–61. https://doi.org/10.1111/j.1742-481X.2009.00651.x
Nurkesh A, Jaguparov A, Jimi S, Saparov A (2020) Recent advances in the controlled release of growth factors and cytokines for improving cutaneous wound healing. Front Cell Dev Biol 8 (July):1–7. https://doi.org/10.3389/fcell.2020.00638
O'Toole G, Kaplan HB, Kolter R (2000) Biofilm formation as microbial development. Ann Rev Microbiol 54:49–79. https://doi.org/10.1146/annurev.micro.54.1.49
Ojo Omoniyi OA, Dighitoghi Moro D, Bridget Afolabi O (2024) Microbial proteases: sources, significance and industrial applications. Int J Curr Microbiol App Sci 13(6):1–23. https://doi.org/10.20546/ijcmas.2024.1306.001

Ozhelvaci F, Steczkiewicz K (2023) Identification and classification of papain-like cysteine proteinases. J Biol Chem 299(6):1–23. https://doi.org/10.1016/j.jbc.2023.104801
Paterlini MG (2002) Structure modeling of the chemokine receptor CCR5: implications for ligand binding and selectivity. Biophys J 83(6):3012–3031. https://doi.org/10.1016/S0006-3495(02)75307-1
Proteinase K, Jany K, Lederer G, Mayer B (1986) 1-S2.0-0014579386804677-Main.Pdf. 199(2): 139–144
Ramírez-Larrota JS, Eckhard U (2022) An introduction to bacterial biofilms and their proteases, and their roles in host infection and immune evasion. Biomolecules 12(2). https://doi.org/10.3390/biom12020306
Rao SKS, TS (2015) Staphylococcus aureus biofilm removal by targeting biofilm-associated extracellular proteins. 146(March):517–520. https://doi.org/10.4103/ijmr.IJMR_410_15
Rawlings ND, Barrett AJ (1993) Evolutionary families of peptidases. Biochem J 290(1):205–218. https://doi.org/10.1042/bj2900205
Ruseva K, Nedkov P, Alexandrova R, Dinev D, Shestakova P, Hristov P, Vassileva E (2019) Polyzwitterionic hydrogels as wound dressings with enzymatic debridement functionality for highly exuding wounds. Polym Int 68(9):1626–1635. https://doi.org/10.1002/pi.5867
Ryder C, Byrd M, Wozniak DJ (2016) Development 10(6):644–648
Saggu SK, Jha G, Mishra PC (2019) Enzymatic degradation of biofilm by metalloprotease from microbacterium sp. Sks10. Front Bioeng Biotechnol 7(AUG). https://doi.org/10.3389/fbioe.2019.00192
Sahu K, Kaurav M, Pandey RS (2017) Protease loaded permeation enhancer liposomes for treatment of skin fibrosis arisen from second degree burn. Biomed Pharmacother 94:747–757. https://doi.org/10.1016/j.biopha.2017.07.141
Sauer K, Stoodley P, Goeres DM, Hall-Stoodley L, Burmølle M, Stewart PS, Bjarnsholt T (2022) The biofilm life cycle: expanding the conceptual model of biofilm formation. Nat Rev Microbiol 20(10):608–620. https://doi.org/10.1038/s41579-022-00767-0
Shree P, Singh CK, Sodhi KK, Surya JN, Singh DK (2023) Biofilms: understanding the structure and contribution towards bacterial resistance in antibiotics. Med Microecol 16(June). https://doi.org/10.1016/j.medmic.2023.100084
Sinclair RD, Ryan TJ (1994) Proteolytic enzymes in wound healing: the role of enzymatic debridement. Australas J Dermatol 35(1):35–41. https://doi.org/10.1111/j.1440-0960.1994.tb01799.x
Song P, Zhang X, Wang S, Xu W, Wang F, Fu R, Wei F (2023) Microbial proteases and their applications. Front Microbiol 14(September):1–24. https://doi.org/10.3389/fmicb.2023.1236368
Steinhoff M, Buddenkotte J, Shpacovitch V, Rattenholl A, Moormann C, Vergnolle N, Luger TA, Hollenberg MD (2005) Proteinase-activated receptors: transducers of proteinase-mediated signaling in inflammation and immune response. Endocr Rev 26(1):1–43. https://doi.org/10.1210/er.2003-0025
Stepanović S, Vuković D, Hola V, Di Bonaventura G, Djukić S, Ćirković I, Ruzicka F (2007) Quantification of biofilm in microtiter plates: overview of testing conditions and practical recommendations for assessment of biofilm production by staphylococci. APMIS 115(8): 891–899. https://doi.org/10.1111/j.1600-0463.2007.apm_630.x
Suleman L (2016) Extracellular bacterial proteases in chronic wounds: a potential therapeutic target? Adv Wound Care 5(10):455–463. https://doi.org/10.1089/wound.2015.0673
Swami A, Aggarwal A, Pathak A, Patnaik S, Kumar P, Singh Y, Gupta KC (2019) Retraction notice to "Imidazolyl-PEI modified nanoparticles for enhanced gene delivery". Int J Pharm. (2007) 335(1–2) (180–192), (S0378517306009732) 560:416. https://doi.org/10.1016/j.ijpharm.2019.03.023
Tarnuzzer RW, Schultz GS (1996) Biochemical analysis of acute and chronic wound environments. Wound Repair Regen 4(3):321–325. https://doi.org/10.1046/j.1524-475X.1996.40307.x

Trengove NJ, Stacey MC, Macauley S, Bennett N, Gibson J, Burslem F, Murphy G, Schultz G (1999) Analysis of the acute and chronic wound environments: the role of proteases and their inhibitors. Wound Repair Regen 7(6):442–452. https://doi.org/10.1046/j.1524-475X.1999.00442.x

Tseng BS, Reichhardt C, Merrihew GE, Araujo-hernandez SA, Harrison JJ, Maccoss MJ, Parsek R (2018) A biofilm matrix-associated protease inhibitor protects *Pseudomonas aeruginosa* from proteolytic attack. mBio 9:e00543-18

Veronese FM, Pasut G (2005) PEGylation, successful approach to drug delivery. Drug Discov Today 10(21):1451–1458. https://doi.org/10.1016/S1359-6446(05)03575-0

Vu B, Chen M, Crawford RJ, Ivanova EP (2009) Bacterial extracellular polysaccharides involved in biofilm formation. Molecules 14(7):2535–2554. https://doi.org/10.3390/molecules14072535

Weaver LH, Kester WR, Matthews BW (1977) A crystallographic study of the complex of phosphoramidon with thermolysin. A model for the presumed catalytic transition state and for the binding of extended substances. J Mol Biol 114(1):119–132. https://doi.org/10.1016/0022-2836(77)90286-8

Weldrick PJ, Hardman MJ, Paunov VN (2019) Enhanced clearing of wound-related pathogenic bacterial biofilms using protease-functionalized antibiotic nanocarriers. ACS Appl Mater Interfaces 11(47):43902–43919. https://doi.org/10.1021/acsami.9b16119

Whitchurch CB, Tolker-Nielsen T, Ragas PC, Mattick JS (2002) Extracellular DNA required for bacterial biofilm formation. Science 295(5559):1487. https://doi.org/10.1126/science.295.5559.1487

Wozniak DJ, Ethan EM (2015) Pseudomonas biofilm matrix composition and niche biology. Pseudomonas 33(4):395–401. https://doi.org/10.1111/j.1574-6976.2011.00322.x

Xu Y, Hu K, Liu C, Du P, Zhou F, Lu Y, Fu Q, Xu J, Lyu G (2023) Eschar dissolution and the immunoregulator effect of keratinase on burn wounds. Sci Rep 13(1):1–10. https://doi.org/10.1038/s41598-023-39765-4

Zhang W, Sun J, Ding W, Lin J, Tian R, Lu L, Liu X, Shen X, Qian PY (2015) Extracellular matrix associated proteins form an integral and dynamic system during pseudomonas aeruginosa biofilm development. Front Cell Infect Microbiol 5(MAY):1–10. https://doi.org/10.3389/fcimb.2015.00040

Zhang X, Shu W, Yu Q, Qu W, Wang Y, Li R (2020) Functional biomaterials for treatment of chronic wound. Front Bioeng Biotechnol 8(June):1–15. https://doi.org/10.3389/fbioe.2020.00516

Zöller K, To D, Bernkop-Schnürch A (2025) Biomedical applications of functional hydrogels: innovative developments, relevant clinical trials and advanced products. Biomaterials 312(July 2024). https://doi.org/10.1016/j.biomaterials.2024.122718

Chapter 13
Novel Approaches in Microbial Proteases for Biotransformation: Production, Commercial Aspect, and Applications

Slimane Mokrani and El-hafid Nabti

Abstract In a global context of sustainability and food security, various scientific researchers are being carried out all over the world. The use of enzymes in bioprocessing is both to meet the need to increase production and to solve various technological problems. Microorganisms and their enzymes represent many intersecting characteristics such as elevating growth speed, capacity of cultivation at high amounts, safety, and sustainability. In particular, proteases are important enzymes in the industrial field that can have different origins. Bioprocessing using microbial proteases opens up promising prospects for remedying these various problems. Proteases can be used in waste recycling/environmental protection and many other applications such as bioremediation, industrial bioconversions, livestock feed, pharmaceutical, and biotechnological significance. This chapter attempts to explain and deals with various features allied to the use of microbial proteases in biotransformation, as well as the future prospects and recommendations in this field of research.

Keywords Applications · Biotransformation · Emerging trends · Microbial proteases

S. Mokrani
Department of Agronomy, Laboratory of Research on Biological Systems and Geomantic (L.R.S.B.G.), University of Mustapha Stambouli, Mascara, Algeria

Laboratoire de Maitrise des Energies Renouvelables, Faculté des Sciences de la Nature et de la Vie, Université de Bejaia, Bejaia, Algeria

E.-h. Nabti (✉)
Laboratoire de Maitrise des Energies Renouvelables, Faculté des Sciences de la Nature et de la Vie, Université de Bejaia, Bejaia, Algeria

V. Zambare et al. (eds.), *Biotechnological Advances and Applications in Microbial Proteases*, Interdisciplinary Biotechnological Advances,
https://doi.org/10.1007/978-981-95-9011-7_13

13.1 Introduction

The first instance of microbial biotransformation of steroids was performed in 1937 (Okafor 2007). Biotransformation is the process by which organic materials change their shape so as to decrease their persistence and toxicity. Numerous microorganisms and their products, including fungus, bacteria, and enzymes, contribute to this process. If there are difficulties with synthetic methods, biotransformation can also be utilized to generate chemicals or materials (Smitha et al. 2017). Microbial biotransformation is frequently employed for modifying a variety of pollutants or substances, including metals, hydrocarbons, and medications (Karigar and Rao 2011). It has long been known that microbial biotransformation is an essential method for lowering the production of a number of chemicals used in the food, agrochemical, and other fields (Smitha et al. 2017).

Among the three main categories of industrial enzymes, proteases make about 60% of the world's overall enzyme sales (Sandhya et al. 2005). Proteases belong to class 3 enzymes, which are hydrolases, and subclass 3.4, which is known as peptidases or peptide hydrolases, according to the International Union of Biochemistry and Molecular Biology's Nomenclature Committee. The committee recommends that the term "peptidase" be used interchangeably beside "peptide hydrolase" for any enzyme capable of breaking down peptide bonds (Sandhya et al. 2005; Yegin and Dekker 2013). They are effectively regarded as an eco-friendly indicator of the environment and a substitute for chemicals. Based on their protein sequences, the developmental connections among neutral, acidic, and alkaline proteases have been examined; however, there is little information available on the factors influencing the variation in their specificity. Because microbial proteases can withstand hard environments, have substrate selectivity, stability at ideal pH, and strategies to inhibit autoproteolytic activity, researchers are actively looking for them (Razzaq et al. 2019).

Despite the fact that microorganisms' proteolytic enzymes are the favored source for commercial enzyme applications because of their technologies and benefits to the economy (Saran et al. 2007). Microbial proteases have been extensively studied and are among the most important hydrolytic enzymes; they make up one of the most important classes of commercial enzymes (Zambare et al. 2011; Zambare and Nilegaonkar 2016). Researchers are able to change the bacterial genome using a variety of tools for biotechnology to increase the protease output with the intended properties, as well as employ various substrates to produce microbial proteases that are both cost-effective and high-quality (Razzaq et al. 2019). This chapter's major goal is to address different aspects related to the microbial proteases in biotransformation.

13.2 Nomenclature and Classification

Proteases, also known as peptidases or proteinases, are a type of hydrolases that break down chains of proteins (Song et al. 2023). Growing understanding of the catalytic mechanism and structure has led to the development of a comprehensive categorization system. Proteases are divided into many clans and families based on their evolutionary ties and amino acid sequences (Rawlings et al. 2017).

13.2.1 Protease Sources

Proteases are a ubiquitous component, which may be recovered from bacteria, animals, and plants. Proteases act like catalysts for breaking the peptide bond in the amino acid polypeptide chain (Barrett and McDonald 1986). However, manufacturing of enzymes from plant and animal sources has been restricted because of low-efficiency production methods, ethical concerns, and environmental issues (Jisha et al. 2013). Researchers have successfully synthesized proteases from several microbiological sources. Around the world, two-thirds of proteases sold commercially come from microbes (Beg and Gupta 2003). Microbial proteases are readily accessible in marketplace because to their elevate yield, low space and time requirements, significant levels of genetic modification, and affordability, making them appropriate for biotechnological applications (Ali et al. 2016).

The inclusion of all the qualities needed for industrial applications makes these microbial proteases advantageous to plant and animal proteases (Palsaniya et al. 2012; Sathishkumar et al. 2015). In comparison to proteases from plants and animals, microbial proteases have several benefits: significant biochemical variety, including a wide range of pH, temperature, and substrate specificity; potential for genetic modification; and viability for large-scale manufacturing (Raval et al. 2012). Microbial enzymes are widely used commercially because of their many biochemical diversities, as well as their scientific and commercial advantages (Jisha et al. 2013).

13.2.1.1 Bacteria

Gram-negative bacteria of the genus *Pseudomonas* mostly generate refined proteases and alkaline proteolytic enzymes (Jellouli et al. 2008). Bacteria have been shown to include three of the four endoprotease classes: metalloproteases, cysteine proteases, and serine proteases. Nevertheless, prokaryotes have not yet been shown to include aspartate protease (Bond and Butler 1987). The kind of protease that *P. fluorescens* produces has not yet been identified as belonging to any of the four aforementioned types (Liao and McCallus 1998). A variety of phenotypic traits have led to the division of the highly diverse and broad group *P. fluorescens* into five biotypes (Palleroni 1984).

The suggested use of alkaline proteases in the identification of *Pseudomonas aeruginosa* in ELISA testing is an original use for them. According to current reports, the majority of alkaline proteases are pH tolerant, thermostable, and compatible with conventional laundry detergents (Anwar and Saleemnudden 2000). A thermostable extracellular protease, AprX, can be produced by *Pseudomonas*. Finding efficient methods to reduce, identify, and deactivate AprX can be aided by a thorough understanding of the protein's biological characteristics, regulation, proteolytic potential, and effects on UHT milk. AprX is noteworthy as a representative of all the extracellular metalloproteases that are generated in milk by psychrotrophic bacteria (Zhang et al. 2019).

Proteases in bacteria most likely originate from the genus *Bacillus*. It has the capacity to produce considerable quantities of alkaline and neutral proteolytic enzymes with exceptional qualities, including high stability against oxidizing agents, organic solvents, detergents, and extreme temperatures (Contesini et al. 2018). Among microorganisms, many *Bacillus* species are extensively researched for their widespread protease synthesis and are utilized in a variety of fields, including detergent, textile, leather, and pharmaceuticals (Singhal et al. 2012; Rehman et al. 2017). Numerous commercial proteases have also been identified and isolated from various *Bacillus* species (Contesini et al. 2018). Some *Bacillus* strains are among the most significant commercial enzyme producers due to their ability to manufacture and release high amounts (20–25 g/L) of extracellular enzymes. To meet a range of specific requirements, many new commercial enzyme products with the necessary temperature, pH, activity, and stability characteristics have been devised due to the genus's thermophiles and the capacity of various species to ferment in the pH ranges of alkaline, neutral, and acid (Schallmey et al. 2004).

Furthermore, the gram-negative bacteria *Serratia* releases extracellular protease into the media. A variety of *Serratia* strains, such as E-15, generate extracellular metalloprotease, an anti-inflammatory substance. Also, three zinc ligands and an active site were discovered when the matching gene was revealed in *E. coli* and *S. marcescens*. Another study demonstrated that *S. marcescens*' extracellular serine protease was eliminated by way of *E. coli*'s outer membrane. The nucleotide sequence indicated that it generated a 112 kDa preproenzyme made up of both N-terminal and C-terminal sequences (Dumas et al. 2013).

13.2.1.2 Fungi

Cells located close to the hyphae's apex release fungal enzymes. They can be used to make useful products in industrial preparations (Lange 2017). Due to their great diversity, fungal proteases have been thoroughly investigated (Damare et al. 2006). There are several fungi from which proteases have been extracted, including *Myceliophthora thermophile*, *Schizophyllum commune*, *Thermomyces lanuginosus*, *Pleurotus ostreatus*, *Thermomyces ibadanensis*, *Sporotrichum thermophiles*,

Aspergillus flavus, *Saccharomyces pombe*, *Phanerochaete chrysosporium*, *Candida mogii*, and *Neurospora crassa* (Hoffmeister and Keller 2007; Kumar 2020).

The cost-effective synthesis of fungal-derived proteases has several advantages, including faster production, simplicity of enzyme modification, and the ease of filtering out mycelium. Although submerged fermentation has been used to produce proteases, solid-state fermentation conditions provide a number of potential benefits for the synthesis of fungal enzymes (Souza et al. 2015). Since mycelium is easily removed by filtering, fungal-derived proteases have an advantage over bacterial proteases. Additionally, because fungi are usually known as GRAS (generally regarded as safe), utilizing them to manufacture enzymes is less hazardous than using bacteria (Germano et al. 2003). Emerging data showed that CRISPR/Cas9 technology has been developed successfully for editing particular genes in a variety of filamentous fungus and higher fungi. In addition to their application in medicine, the food sector uses fungal proteases extensively to prepare butter, fruits, beverages, and cheese and to prolong their period of storage (Naeem et al. 2022).

13.2.1.3 Actinomycetes

Despite the fact that actinomycetes are thought to be one of the most significant manufacturers of antibiotics (Avitha et al. 2010). The current understanding of actinomycetes' proteases indicates that they are the most significant and extensively used class of secondary metabolites (Limkhada et al. 2010). One important class of microorganisms found in soil, plant tissues, and marine settings are actinomycetes. They generate a large number of useful extracellular enzymes that cleave a wide range of organic molecules. Novel enzymes with significant industrial potential are known to be produced by actinomycetes found in harsh settings (Suthindhiran et al. 2014). Actinomycetes, such as *Streptomyces*, *Nocardia*, and *Nocardiopsis*, have been shown in several investigations to produce proteases (Wietzorrek and Bibb 1997).

The proteolytic enzymes of mesophilic actinomycetes have been the subject of several investigations (Collins et al. 1995). Utilizing an enrichment approach, six strains of thermophilic *Streptomyces* were recovered from soil utilizing feathers as the only source of carbon and nitrogen. One strain, provisionally identified as *Streptomyces thermonitrificans*, had the highest level of protease activity. Recently, *Nocardiopsis* spp. NCIM 5124 alkaline protease has been identified and purified. *Streptomyces nogalator* strain Ac 80, an actinomycete that produces extracellular proteases, was isolated from soil. Fulzele (2011) identified *Streptomyces* sp. from Brazilian Cerrado soil that produces thermophilic proteases.

13.2.1.4 Other Sources

The majority of protease enzymes come from microbial fermentation process; however, protease is naturally found in all living things, including prokaryotes, eukaryotes, plants, animals, and humans (Gurumallesh et al. 2019). However, low-efficiency manufacturing methods, ethical concerns, and environmental issues have limited the production of plant and animal-based enzymes. Commercially, the wide range of biochemical diversity and scientific and commercial benefits of microbial enzymes make them attractive (Jisha et al. 2013).

The food and pharmaceutical industries make substantial use of plant proteases derived from a variety of plants. The most studied enzymes include papain, ficin, and bromelain, which are isolated from various fruits such as *Carica papaya*, *Ficus caric*, and *Ananas comosus*. Protease, which is abundant in these fruits, is utilized extensively in brewing, meat softening, digestion, milk coagulation, cancer treatment, tissue damage, and virus treatment. Compared to mature papaya, unripe papaya fruit has higher grade papain proteases (Gonzalez-Rabade et al. 2011).

Animal proteases, including chymotrypsin, pepsin, rennin, and pancreatic trypsin, are produced from animals. Since trypsin is an enzyme that comes from animal intestines, it functions as an intestinal digestive enzyme. It has a variety of uses, including some medicinal ones and the biocontrol of insects and other pests. The applications for pancreatic chymotrypsin include analysis and diagnosis. The laundry industry uses pepsin, an enzyme that is isolated from vertebrate intestines and has an acidic nature, to convert rennet into active rennin. In the dairy industry, rennin enzymes are employed to prepare curd with a good flavor (Verma 2014).

13.3 Production

Established fermentation techniques allow for the relatively quick cultivation of huge amounts of proteases, which also yield an abundant and consistent supply of the needed output (Gupta et al. 2002). Unlike proteases derived from plants and animals, the proteases produced by microbes are exogenous in nature, and the producer releases them straight into the fermentation broth, making easier subsequent processing of the enzyme (Savitha et al. 2011). They have been made using both solid-state fermentations (SSF) and submerged fermentations (SmF), each of which offers unique benefits that the other cannot achieve (Sandhya et al. 2005; Sun and Xu 2009). SSF features a number of benefits over traditional SmF, including lower production costs, reduced energy and space requirements, and the utilization of raw materials as substrates, fewer issues with downstream processing, product stability because of less medium dilution, and more productive manufacturing (Das and Mukherjee 2007; Sun and Xu 2009). Process management and the straightforward recovery of mycelia, spores, or extracellular enzymes are two benefits of SmF. Enzymatic extracts, however, may be less stable than those from SSF, and the products are diluted (Souza et al. 2015).

In particular, variables like the change in the C/N ratio are crucial for the progression of the fermentation process (Souza et al. 2015). In addition, the presence of a few readily metabolized carbohydrates, such glucose, and metal ions, in addition to a number of other physical parameters, including temperature, incubation duration, pH, aeration, and inoculum density. Rapidly metabolizable nitrogen sources, including amino acids in the medium, can have an impact on protease production (Gupta et al. 2002). Enzyme manufacturing is followed by the extremely difficult process of purification. Several techniques are planned for their purification. Enzymes with added value are recovered using a variety of methods. Whether the enzyme comes from an external or intracellular source determines the strategy to use (Mienda and Yahya 2011).

13.4 Commercial Applications

Enzymes' economic and environmental benefits are driving a sharp rise in demand in industrial industries (Singh et al. 2016). Approximately US$ 2.5–3 billion is the anticipated annual size of the industrial enzyme industry, which is composed of enzymes for nonfood/feed enzyme employments (US$ 1.4–1.7 billion), manufacturing of food (US$ 800 million), and animal feed (US$ 400 million) (Ward 2011). Of the 60% of enzymes sold globally, 20% are proteases (Singhal et al. 2012). Especially, metalloprotease (EC. 3.4.24), aspartic proteases (EC. 3.4.23), cysteine (thiol) proteases (EC. 3.4.22), and serine protease (EC. 3.4.21) are among the most significant classes of industrial enzymes (Nascimento and Martins 2004; Beg and Gupta 2003; Ellaiah et al. 2003). In this regard, microbial proteases have been reported by Beg and Gupta (2003) that they are more valuable commercially than enzymes derived from plants and mammals. He mentioned that websites and others offer useful information on the main uses of proteases, along with connections to technical papers and studies and new research and technology. Other companies comprise 27–30% of the total, with the majority located in North America, Europe, and Asia. An estimated US$ 1.5–1.8 billion is spent annually on proteases. Proteases are most frequently used in detergents, where alkaline protease sales total around US$ 1 billion annually (Ward 2011).

13.4.1 Proteases in Biotransformation

Waste generation as a result of human activity is one of the most important issues confronting our society and the natural systems throughout the world. Waste production is rising quickly, and this is causing societal changes as nearly every country transition from rural, agrarian civilizations to urban, modern ones (Lu et al. 2024). However, agricultural product preharvesting, harvesting, and processing processes generate agricultural waste, even in agro-industrial wastewater. Wastewater and agro/food waste are produced by industrial food processing operations and agricultural processing activities. A lot of waste effluent from livestock and the processing of agricultural goods is among the many agro-industrial wastewaters that have a high content of organic contaminants (Lin and Ying 2009). In addition, the collection and processing of garbage produced by natural people' residential, commercial, and construction activities by municipalities is known as municipal solid waste (MSW) (Karak et al. 2012).

Serious environmental problems including soil, water, and air pollution might arise in the absence of an integrated waste management program (Zhu et al. 2014). Additionally, both short-term and long-term exposure to industrial waste's toxic substances can have a negative impact on human health (Islam et al. 2016). Due to filthy living conditions and polluted surroundings, poor waste management in shanty communities in developing nations has had a major negative influence on public health and contributed to widespread outbreaks of illnesses such as typhoid, malaria, and cholera (Kitole et al. 2024). Furthermore, waste production has a significant effect on society. People who live close to these locations, who are frequently in a vulnerable socioeconomic condition, also have to deal with the stigma attached to being close to degraded regions, which can lower property values and people' self-esteem (de Lima et al. 2021). In addition, the existence of waste pickers at landfills, who labor in hazardous and unhealthy conditions, emphasizes the absence of successful governmental policies for recycling-based revenue generation and social inclusion (da Silva and Bandeira 2025).

In order to transform these diverse waste streams into useful chemicals and fuels, valorization techniques, both biological and chemical, are needed (Lad et al. 2022). Microbial proteases have significant implications for environmental, economic, and social sustainability in the field of dairy waste management. Protease-producing bacteria offer an economically viable way to convert waste into useful products. Protease-producing bacteria can be used to drastically lower the chemical and biochemical oxygen demands (BOD and COD) of dairy waste, thus decreasing its negative environmental effects (Ghatage et al. 2024). For instance, tannery wastewater microbiota was tested for metagenome-derived PersiProtease1. The new PersiProtease1 was isolated from the microbiota and used for biodegrade waste X-ray films, whey protein, chicken feathers, dehairing sheepskins, and tannery wastewater protein (Ariaeenejad et al. 2022). Proteases have been shown in several studies to have exceptional deproteinization abilities for converting shrimp waste into chitin (Doan et al. 2019). Alkaline protease has been utilized for waste management in a variety of food manufacturing establishments as well as for residential activities as protein-based residues are a major source of pollutants in food bioprocessing industries (such as those that process meat and milk) (Majumder et al. 2015).

Table 13.1 summarizes the applications of microbial proteases in biotransformation including biotechnological and industrial bioconversions (detergent industry, food industry like meat tenderization; dairy industry; baking, juice production, and brewing); livestock feed (livestock feed, poultry feed, and aquafeed); medical and pharmaceutical applications; bioremediation, waste recycling, and environmental protection.

Table 13.1 Examples of microbial protease utilizations

Producing microorganisms	Utilizations	References
Bacillus thuringiensis strain-MA8	Dairy industry (cheese manufacturing)	Abdella et al. (2023)
Bacillus sp. FW2	Food waste degradation	Pham et al. (2021)
Pseudomonas fluorescens BJ-10	Dairy industry (degradation on milk proteins)	Zhang et al. (2015)
Serratia marcescens FS-3	Deproteinization of crab shell waste for chitin extraction	Jo et al. (2008)
Aspergillus niveus	Animal feed	De Oliveira Simas et al. (2024)
Chrysosporium keratinophilum	Bioremediation of keratinic wastes	Singh (2003)
Rhodococcus ruber C208	Effect of biofilm formation of the plastic-degrading actinomycete	Gilan and Sivan (2013)
Bacillus pumilus 3–19	Effect of growth parameters and gut microbiome of broiler chickens	Pudova et al. (2020)
Bacillus subtilis E20	Aquafeed (improvement in the growth performance of white shrimp, *Litopenaeus vannamei*)	Liu et al. (2009)
Escherichia coli	Pharmaceutical medicine (produce recombinant human growth hormone)	Wojtowicz-Krawiec et al. (2014)

13.4.2 Biotechnological and Industrial Bioconversions

The majority of enzymes on the market are alkaline proteases, which are also a highly valuable class of enzymes utilized mostly as additives for detergents (Sharma et al. 2017). Laundry detergents can be enhanced with alkaline proteases to aid in the removal of proteinaceous material from stains (Tanwar et al. 2022). With the exception of pyridine, which inhibits protease activity, halophilic enzymes are receiving increased interest in biotechnological uses because of their resilience to heat and capacity to maintain function withstands severe organic solvent stress. Even after at least 30 min at 50, 55, and 60 °C, the enzyme activity was constant at up to 80% (Xue et al. 2012). Protease's primary functions in detergent are their isoelectric points (PI). The best-performing proteases should have a detergent PI that coincides; this backs the usage of enzymes in detergent. *Bacillus* species that are alkalophilic and produce proteases can tolerate larger pH ranges because they have very high PI. It is anticipated that combining lipases, amylase, and cellulase will improve the effectiveness of laundry detergent proteases (Rao et al. 1998). Every detergent protease now on the market is a serine protease made by a *Bacillus* strain. Fungal alkaline proteases have an advantage over their bacterial counterparts in that they are easier to process downstream to create an enzyme free of microbes. Alkaline protease produced by *Conidiobolus coronatus* is in harmony with commercial detergents made in India (Phadatare et al. 1993) and maintained 43% of its capacity after 50 min at 50 °C with 25 mM Ca^{2+} and 1 M glycine present (Bhosale et al. 1995).

Enzymes are used by the food industry in several processes, such as dairy, brewing, baking, and juice production (Singh et al. 2016). In the food industry, enzymes can be utilized to solve problems of a high-quality food source. Enhancing food production and its constituents—taste, fragrance, color, texture, appearance, and nutritional value—is effectively accomplished by these biomolecules (Neidleman 1984). The main function of acid proteases in food manufacturing is to help milk clot while making cheese. Curds are formed when the milk proteins coagulate, which are solid masses. Then, the whey is extracted to produce cheese (Theron and Divol 2014). Likewise, proteases are crucial for degrading down the structural proteins in connective tissues, which lessens meat's toughness. Tenderizing meat is another application for bacterial proteases. Bacterial proteases effectively decompose collagen and elastin by proteolysis, but they barely break down myofibrillar proteins (Arshad et al. 2017). Animal circulatory systems are frequently treated with solutions of proteases 10–30 min prior to slaughter (Kalisz 1988), comprising thermophilic alkaline protease (Wilson et al. 1992) and alkaline elastase (Chen and He 2006).

13.4.3 Animal Feed

The rise of feed enzymes happened steadily in order to satisfy the growing demand for milk and meat consumers worldwide. The 1980s saw the introduction of enzymes into animal diets, and their utilization skyrocketed in the 1990s. Feed enzymes are becoming more and more important because they help improve animal feed utilization and nutrient digestibility (Choct 2006). In animal feeds, proteases decompose proteins into their component amino acids, thereby overcoming anti-nutritional effects. The impact of these feed enzymes in lowering feed costs and enhancing meat quality is becoming more significant, in addition to enhancing the nutritional value of feed for improved animal conversion of feed (Lei and Stahl 2001; Adrio and Demain 2014). For instance, the addition of *Lactobacillus licheniformis* in low-protein infant food may greatly increase the intestinal morphology, performance in terms of growth, and nutrient digestibility of weaned pigs (Park et al. 2020).

Given that the pancreatic protease activity of the chicken is low at hatching and rises until age 21, proteases may help poultry grow more efficiently (Jin et al. 1998). Digestive protease may be restricting the digestion of proteins in young birds because of the immaturity of their digestive tracts. Exogenous proteases may speed up the breakdown of intestinal proteins and supplement pancreatic enzymes (Mahmood et al. 2017). Besides, supplemental proteases can increase protein digestion and decrease the need for amino acids (Isaksen et al. 2010). For examples, Dalev (1994) has reported that in the poultry industry, *B. subtilis*'s alkaline keratinase may be helpful for feather processing. In a similar study, Verma et al. (2014) have also reported that *Thermoactinomyces* sp. RS1's feather-degrading alkaline keratinase is useful for preparing feed for chicken farms.

Two well-known nutritional techniques used to address issues with the sustainability and scarcity of protein supplies in aquaculture are low-fishmeal and protein-saving diets. However, these diets have been linked to negative outcomes for aquatic animals' illness resistance, feed utilization, and growth performance. Meals with lower protein content or fishmeal alternatives have benefited from the application of exogenous protease in order to lessen these difficulties. This has improved the bioavailability of nutritious components (Chen et al. 2024).

13.4.4 Medical and Pharmaceutical Applications

The roles of enzymes in the pharmaceutical and diagnostic industries are numerous and vital. They are extensively utilized as therapeutic medications for conditions related to digestive disorders and enzymatic deficiencies, in addition to diabetes testing kits and diagnostic procedures like ELISA (Mane and Tale 2015). The two main benefits of using microbial proteases to synthesize pharmaceutical peptides are cost-effectiveness and practicality (Raval et al. 2012). They are a promising target for the development of therapeutic medicines because of their involvement in the life cycle of many pathogenic organisms and combat serious illnesses such as AIDS, cancer, malaria, and bacterial infections (Chanalia et al. 2011). Additionally, protease is an essential anti-inflammatory agent. Serratiopepetidase, a unique protease produced by the bacterium *Serratia* sp., has the ability to quickly alleviate inflammation (Banerjee and Ray 2017).

In the medical sector, these microbially derived proteolytic enzymes have enormous potential, particularly for development of therapies, as biomarkers for diagnosis, and as medications (Nair et al. 2025). They are used as cancer treatment, wound debridement, prion degradation, contact lens cleaning, biofilm removal, and fibrinolytic agents. Additionally, proteases are necessary for the treatment of cardiovascular disease, which emphasizes the need for safe, affordable, and effective fibrinolytic drugs. Protease biosensors and proteolytic enzymes are increasingly being used in diagnostic and therapeutic contexts. Sensors based on nanomaterials are among the revolutionary technologies being researched to increase the sensitivity, specificity, and adaptability of protease biosensors (Jamal et al. 2025).

13.4.5 Bioremediation, Waste Recycling, and Environmental Protection

A biotechnological approach called bioremediation uses biological processes to manage contaminants through mineralization and detoxification while also restoring the recalcined environment. The application of microbial enzymes in prevention of persistent organic pollutants (POPs) is thought to be novel, economical, environmentally benign, and promising (Tahri et al. 2013). Enzymes are widely used in waste management, and several enzymes contribute to the breakdown of harmful contaminants. Numerous chemical commodities found in household garbage and industrial effluents are harmful or dangerous to living things and the ecological system. Whether used either individually or in combination, microbial enzymes are employed to decompose or bioconvert harmful chemical compounds into harmless chemicals in order to handle industrial effluents containing phenols, nitriles, aromatic amines, and other substances (Raj et al. 2006; Rubilar et al. 2008; Pandey et al. 2011).

Keratinase (E.C. 3.4.21/24/99.11) can degrade keratin proteins and be applied to poultry waste bioremediation by degrading and using keratinous wastes to make valuable by-products. Significant activity has been demonstrated by the protease enzyme keratinase, which is generated from *Stenotrophomonas maltophilia* KB13, in the biodegradation of chicken feathers (Bhange et al. 2016). Keratinases from microbial sources are highly beneficial for waste management because of their wide substrate specificity. The keratinases often turn feathers or additional waste materials (such as wool, hair, scale, etc.) into keratin, which is highly sought after as a feed element in the poultry and industries (Deivasigamani and Alagappan 2008; Motyan et al. 2013). Keratinases have been frequently employed as nematocidal agents to produce biofuel and to make fertilizer from poultry manure because of their unique properties (Verma et al. 2017). Protease enzymes are also used in the bioremediation of marine crustacean wastes that have undergone deproteinization in order to remove chitin. Shrimp wastes are 75% deproteinized by the alkaline protease produced by *Bacillus licheniformis* MP1 (Jellouli et al. 2011).

13.5 Challenges and Future Prospects

Enzymatic modulation, crystallography, and molecular characterization by genetic engineering, bioinformatics tools, and algorithms should continue to be employed in "omics" and biological technology to expand basic understanding of microbial ecology (e.g., enzymes, their development, and significance in several industrial fields) (Song et al. 2023).

To generate better strains of protease, the researchers are using innovative methods including protein/genetic engineering, molecular biology, and computational biology. In vitro evolutionary alterations in the main structure of the protein will be used to develop bacterial strains with desired traits. The development of bacterial proteases with properties such as increased yield, changed substrate specificity, improved thermal stability, altered optimal pH, and avoidance of autoproteolytic inactivation is one of the main objectives of scientists (Razzaq et al. 2019). Particularly, finding new proteases through genetic engineering and immobilization techniques, generating highly efficient and effective enzyme systems, optimizing the activities of already-existent proteases, and using less energy and resources while still producing maximum product yields are all priorities (Song et al. 2023).

Recently, rapid screening of new enzymes from unculturable microbes, particularly those found in hostile habitats such as desert, saline, thermophilic, and arctic locations, has been made possible by culture-independent techniques including metagenomics, metatranscriptomics, and metaproteomics (Kennedy et al. 2010; Mukhtar et al. 2017). These technologies offer novel perspectives on the functional metagenomic (metaproteomics) analysis of a specific environment and assist in identifying the functional features of a microenvironment using cutting-edge techniques such as substrate-induced gene expression screening (SIGEX) and preamplification inverse-PCR (PAIPCR) (Olsen et al. 2000). Even though microbial proteases are now widely used in industry, scientists are still searching for a more effective novel wild-type biocatalyst that can work with industrial applications and process settings. These enzymes have considerably more potential and will probably be used in more processes in the future due to their economic success (Mienda et al. 2014).

13.6 Conclusion

Enzymes are undoubtedly very important tools in biotransformation. In particular, microbial enzymes are of considerable commercial interest worldwide. They offer a number of advantages, such as being eco-friendly and less costly, and featuring a wide range of microbial biodiversity. These enzymes have numerous applications in biotechnological and industrial bioconversions; animal feed; medical and pharmaceutical applications; and bioremediation, waste recycling, and environmental protection. Current research on microbial proteases, combining different techniques and aspects, holds great promise for improving our knowledge of these biomolecules, enhancing their use in a wide range of sectors, and addressing the various challenges involved in their exploitation.

References

Abdella MA, Ahmed SA, Ibrahim OA (2023) Statistical improvement of protease production from a new isolate *Bacillus thuringiensis* strain-MA8 and its application in the production of enzyme-modified cheese. Int J Biol Macromol 225:361–375. https://doi.org/10.1016/j.ijbiomac.2022.11.073

Adrio JL, Demain AL (2014) Microbial enzymes: tools for biotechnological processes. Biomolecules 4(1):117–139. https://doi.org/10.3390/biom4010117

Ali N, Ullah N, Qasim M, Rahman H, Khan SN, Sadiq A, Adnan M (2016) Molecular characterization and growth optimization of halo-tolerant protease producing *Bacillus subtilis* strain

BLK-1.5 isolated from salt mines of Karak. Pak Extremophiles 20:395–402. https://doi.org/10.1007/s00792-016-0830-1

Anwar A, Saleemnudden M (2000) Alkaline protease from Spilosoma oblique: potential application in bioformulation. Biotechnol Appl Biochem 31(2):85–89. https://doi.org/10.1042/ba19990078

Ariaeenejad S, Kavousi K, Mamaghani ASA, Ghasemitabesh R, Hosseini Salekdeh G (2022) Simultaneous hydrolysis of various protein-rich industrial wastes by a naturally evolved protease from tannery wastewater microbiota. Sci Total Environ 815:152796. https://doi.org/10.1016/j.scitotenv.2021.152796

Arshad MS, Kwon JH, Imran M, Sohaib M, Aslam A, Nawaz I, Amjad Z, Khan U, Javed M (2017) Plant and bacterial proteases: a key towards improving meat tenderization, a mini review. Cogent Food Agric 2(1):1–10. https://doi.org/10.1080/23311932.2016.1261780

Avitha A, Prabhakar P, Vijayalakshmi M, Venkateswarlu Y (2010) Purification and biological evaluation of the metabolites produced by *Streptomyces* sp. TK-VL_333. Res Microbiol 161(5):335–345. https://doi.org/10.1016/j.resmic.2010.03.011

Banerjee G, Ray AK (2017) Impact of microbial proteases on biotechnological industries. Biotechnol Genet Eng Rev 33(2):119–143. https://doi.org/10.1080/02648725.2017.1408256

Barrett AJ, McDonald JK (1986) Nomenclature: protease, proteinase and peptidase. Biochem J 237:935. https://doi.org/10.1042/bj2370935

Beg QK, Gupta R (2003) Purification and characterization of an oxidation stable, thiol-dependent serine alkaline protease from *Bacillus mojavensis*. Enzym Microb Technol 32:294–304. https://doi.org/10.1016/S0141-0229(02)00293-4

Bhange K, Chaturvedi V, Bhatt R (2016) Feather degradation potential of *Stenotrophomonas maltophilia* KB13 and feather protein hydrolysate (FPH) mediated reduction of hexavalent chromium. 3 Biotech 6(1):1–9. https://doi.org/10.1007/s13205-016-0370-5

Bhosale SH, Rao MB, Deshpande VV (1995) Thermostability of high-activity alkaline protease from *Conidiobolus coronatus* (NCL 86.8.20). Enz Microbial Technol 17:136–139. https://doi.org/10.1016/0141-0229(94)00045-s

Bond JS, Butler PE (1987) Intracellular proteases. Annu Rev Biochem 56:333–364. https://doi.org/10.1146/annurev.bi.56.070187.002001

Chanalia P, Gandhi D, Jodha D, Singh J (2011) Applications of microbial proteases in pharmaceutical industry: an overview. Rev Med Microbiol 22(4):96–101. https://doi.org/10.1097/mrm.0b013e3283494749

Chen S, Maulu S, Wang J, Xie X, Liang X, Wang H, Wang J, Xue M (2024) The application of protease in aquaculture: prospects for enhancing the aquafeed industry. Anim Nutr 16:105–121. https://doi.org/10.1016/j.aninu.2023.11.001

Chen Q, He G (2006) Optimization of elastolysis conditions and elastolytic kinetic analysis with elastase from Bacillus licheniformis ZJUEL31410. J Zhejiang Univ SCIENCE B 7(6):482–490. https://doi.org/10.1631/jzus.2006.B0482

Choct M (2006) Enzymes for the feed industry: past, present and future. Worlds Poult Sci J 62:5–15. https://doi.org/10.1079/wps200480

Collins CH, Lyne PM, Granje JM (1995) Microbiological method. Butterworth and Heinemann Publishers, London

Contesini FJ, Melo RRD, Sato HH (2018) An overview of *Bacillus* proteases: from production to application. Crit Rev Biotechnol 38(3):321–334. https://doi.org/10.1080/07388551.2017.1354354

da Silva TRF, Bandeira SR (2025) Solid waste and environmental impacts: a study on local perceptions and geographic distribution of reject points in the flores neighborhood, Manaus/Amazonas. Contrib Cienc Soc 18(3):1–24. https://doi.org/10.55905/revconv.18n.3-021

de Lima JFP, Marinho RP, do Nascimento EC, Barros VH d O (2021) Análise dos impactos ambientais decorrentes da disposição irregular dos resíduos de construção e demolição no bairro Alto Bom Jesus em Serra Talhada-PE. Holos Environ 21(3):338–349. https://doi.org/10.14295/holos.v21i3.12454

de Oliveira Simas AL, de Alencar Guimarães NC, Glienke NN, Galeano RMS, de Sá Teles JS, Kiefer C, de Souza Nascimento KMR, Masui DC, Zanoelo FF, Giannesi GC (2024) Production of phytase, protease and xylanase by *Aspergillus niveus* with rice husk as a carbon source and application of the enzymes in animal feed. Waste Biomass Valori 15(7):3939–3951. https://doi.org/10.1007/s12649-024-02455-x

Dalev PG (1994) Utilization of waste feathers from poultry slaughter for production of a protein concentrate. Bioresour Technol 48:265–267. https://doi.org/10.1016/0960-8524(94)90156-2

Damare S, Raghukumar C, Muraleedharan UD, Raghukumar S (2006) Deep-sea fungi as a source of alkaline and cold-tolerant proteases. Enzym Microb Technol 39:172–181. https://doi.org/10.1016/j.enzmictec.2006.03.032

Das K, Mukherjee AK (2007) Comparison of lipopeptide biosurfactants production by *Bacillus subtilis* strains in submerged and solid state fermentation systems using a cheap carbon source: some industrial applications of biosurfactants. Process Biochem 42:1191–1199. https://doi.org/10.1016/j.procbio.2007.05.011

Deivasigamani B, Alagappan KM (2008) Industrial application of keratinase and soluble proteins from feather keratins. J Environ Biol 29(6):933–936

Doan CT, Tran TN, Nguyen VB, Vo TPK, Nguyen AD, Wang SL (2019) Chitin extraction from shrimp waste by liquid fermentation using an alkaline protease-producing strain, *Brevibacillus parabrevis*. Int J Biol Macromol 131:706–715. https://doi.org/10.1016/j.ijbiomac.2019.03.117

Dumas A, Spicer CD, Gao Z, Takehana T, Lin YA, Yasukohchi T, Davis BG (2013) Self-liganded Suzuki–Miyaura coupling for site-selective protein PEGylation. Angew Chem Int Ed 52:3916–3921. https://doi.org/10.1002/anie.201208626

Ellaiah P, Adinarayana K, Rajyalaxmi P, Srinivasulu B (2003) Optimization of process parameters for alkaline protease production under solid state fermentation by alkalophilic *Bacillus* sp. Asian J Microbial Biotechnol Environ Sc 5:49–54

Fulzele R (2011) Characterization of novel extracellular protease produced by marine bacterial isolate from the Indian Ocean. Braz J Microbiol 42:1364–1373. https://doi.org/10.1590/s1517-83822011000400018

Germano S, Pandey A, Osaku CA, Rocha SN, Soccol CR (2003) Characterization and stability of proteases from *Penicillium* sp. produced by solid-state fermentation. Enzym Microb Technol 32:246–251. https://doi.org/10.1016/s0141-0229(02)00283-1

Ghatage A, Chavan OD, Jadhav A (2024) Isolation and detailed characterization of protease-producing bacteria derived from dairy waste. BPAS-Zool 43B:344

Gilan I, Sivan A (2013) Effect of proteases on biofilm formation of the plastic-degrading actinomycete *Rhodococcusruber* C208. FEMS Microbiol Lett 342(1):18–23. https://doi.org/10.1111/1574-6968.12114

Gonzalez-Rabade N, Badillo-Corona JA, ArandaBarradas JS, del Carmen O-SM (2011) Production of plant proteases in vivo and in vitro-a review. Biotechnol Adv 29:983–996. https://doi.org/10.1016/j.biotechadv.2011.08.017

Gupta R, Beg QK, Lorenz P (2002) Bacterial alkaline proteases: molecular approaches and industrial applications. Appl Microbiol Biotechnol 59:15–32. https://doi.org/10.1007/s00253-002-0975-y

Gurumallesh P, Alagu K, Ramakrishnan B, Muthusamy S (2019) A systematic reconsideration on proteases. Int J Biol Macromol 128:254–267. https://doi.org/10.1016/j.ijbiomac.2019.01.081

Hoffmeister D, Keller NP (2007) Natural products of filamentous fungi: enzymes, genes, and their regulation. Nat Prod Rep 24:393–416. https://doi.org/10.1039/b603084j

Isaksen MF, Cowieson AJ, Kragh KM (2010) Starch- and protein-degrading enzymes: biochemistry, enzymology and characteristics relevant to animal feed use. In: Bedford MR, Partridge GG (eds) Enzymes in farm animal nutrition, 2nd edn. CAB International, Oxfordshire, pp 85–95

Islam S, Islam S, Islam SA, Eaton DW (2016) Total and dissolved metals in the industrial wastewater: a case study from Dhaka metropolitan, Bangladesh. Environ Nanotechnol Monit Manag 5:74–80. https://doi.org/10.1016/j.enmm.2016.04.001

Jamal GA, Jahangirian E, Hamblin MR, Mirzaei H, Tarrahimofrad H, Alikowsarzadeh N (2025) Proteases, a powerful biochemical tool in the service of medicine, clinical and pharmaceutical. Prep Biochem Biotechnol 55(1):1–25. https://doi.org/10.1080/10826068.2024.2364234

Jellouli K, Bayoudh A, Manni L, Agrebi R, Nasri M (2008) Purification, biochemical and molecular characterization of a metalloprotease from *Pseudomonas aeruginosa* MN7 grown on shrimp wastes. Appl Microbiol Biotechnol 79:989–999. https://doi.org/10.1007/s00253-008-1517-z

Jellouli K, Ghorbel-Bellaaj O, Ayed HB, Manni L, Agrebi R, Nasri M (2011) Alkaline-protease from *Bacillus licheniformis* MP1: purification, characterization and potential application as a detergent additive and for shrimp waste deproteinization. Process Biochem 46(6):1248–1256. https://doi.org/10.1016/j.procbio.2011.02.012

Jin SH, Corless A, Sell SL (1998) Digestive system development in post-hatch poultry. Worlds Poult Sci J 54:335–345. https://doi.org/10.1079/wps19980023

Jisha VN, Smitha RB, Pradeep S, Sreedevi S, Unni KN, Sajith S, Priji P, Josh MS, Benjamin S (2013) Versatility of microbial proteases. Adv Enzy Res 1:39–51. https://doi.org/10.4236/aer.2013.13005

Jo GH, Jung WJ, Kuk JH, Oh KT, Kim YJ, Park RD (2008) Screening of protease-producing *Serratia marcescens* FS-3 and its application to deproteinization of crab shell waste for chitin extraction. Carbohydr Polym 74(3):504–508. https://doi.org/10.1016/j.carbpol.2008.04.019

Karak T, Bhagat RM, Bhattacharyya P (2012) Municipal solid waste generation, composition, and management: the world scenario. Crit Rev Environ Sci Technol 42(15):1509–1630. https://doi.org/10.1080/10643389.2011.569871

Karigar CS, Rao SS (2011) Role of microbial enzymes in the bioremediation of pollutants: a review. Enzyme Res 1:1–11. https://doi.org/10.4061/2011/805187

Kennedy J, Flemer B, Jackson SA, Lejon DP, Morrissey JP, O'gara F, Dobson AD (2010) Marine metagenomics: new tools for the study and exploitation of marine microbial metabolism. Mar Drugs 8:608–628. https://doi.org/10.3390/md8030608

Kitole FA, Ojo TO, Emenike CU, Khumalo NZ, Elhindi KM, Kassem HS (2024) The impact of poor waste management on public health initiatives in shanty towns in Tanzania. Sustainability 16(24):10873. https://doi.org/10.3390/su162410873

Kumar A (2020) *Aspergillus nidulans*: a potential resource of the production of the native and heterologous enzymes for industrial applications. Int J Microbiol 2020(1):1–11. https://doi.org/10.1155/2020/8894215

Kalisz HM (1988) Microbial proteinases. In: Fiechter, A. (ed) Adv Biochem Eng/Biotechnol 36: 1–65

Lad BC, Coleman SM, Alper HS (2022) Microbial valorization of underutilized and nonconventional waste streams. J Ind Microbiol Biotechnol 49(2):kuab056. https://doi.org/10.1093/jimb/kuab056

Lange L (2017) Fungal enzymes and yeasts for conversion of plant biomass to bioenergy and high-value products. Microbiol Spectrum 2017:1–19. https://doi.org/10.1128/microbiolspec.funk-0007-2016

Lei XG, Stahl CH (2001) Biotechnological development of effective phytases for mineral nutrition and environmental protection. Appl Microbiol Biotechnol 257:474–481. https://doi.org/10.1007/s002530100795

Liao CH, McCallus DE (1998) Biochemical and genetic characterization of an extracellular protease from *Pseudomonas fluorescens* CY091. Appl Environ Microbiol 64(3):914–921. https://doi.org/10.1128/aem.64.3.914-921.1998

Limkhada JR, Mander P, Cho SS, Yoo JC (2010) A novel fibrinolytic protease from *Streptomyces* sp. CS684. Process Biochem 45:88–93. https://doi.org/10.1016/j.procbio.2009.08.010

Lin H, Ying Y (2009) Theory and application of near infrared spectroscopy in assessment of fruit quality: a review. Sens & Instrumen Food Qual 3:130–141. https://doi.org/10.1007/s11694-009-9079-z

Liu CH, Chiu CS, Ho PL, Wang SW (2009) Improvement in the growth performance of white shrimp, *Litopenaeus vannamei*, by a protease-producing probiotic, *Bacillus subtilis* E20, from natto. J Appl Microbiol 107(3):1031–1041. https://doi.org/10.1111/j.1365-2672.2009.04284.x
Lu M, Zhou C, Wang C, Jackson RB, Kempes CP (2024) Worldwide scaling of waste generation in urban systems. Nature Cities 1(2):126–135. https://doi.org/10.1038/s44284-023-00021-5
Mahmood T, Mirza MA, Nawaz H, Shahid M (2017) Effect of difference exogenous proteases on growth performance, nutrient digestibility, and carcass response in broiler chickens fed poultry by-product meal-based diets. Livest Sci 200:71–75. https://doi.org/10.1016/j.livsci.2017.04.009
Majumder R, Banik SP, Ramrakhiani L, Khowala S (2015) Bioremediation by alkaline protease (AkP) from edible mushroom *Termitomyces clypeatus*: optimization approach based on statistical design and characterization for diverse applications. J Chem Technol Biotechnol 90:1886–1896. https://doi.org/10.1002/jctb.4500
Mane P, Tale V (2015) Overview of microbial therapeutic enzymes. Int J Curr Microbiol App Sci 4(4):17–26
Mienda BS, Yahya A (2011) Engineering of microbial proteases: improving stability and catalytic performances. IIOAB J 2:10–15
Mienda BS, Yahya A, Galadima IA, Shamsir MS (2014) An overview of microbial proteases for industrial applications. Res J Pharm Biol Chem Sci 5(1):388–396
Motyan JA, Toth F, Tozser J (2013) Research applications of proteolytic enzymes in molecular biology. Biomolecules 3:923–942. https://doi.org/10.3390/biom3040923
Mukhtar S, Ishaq A, Hassan SARA, Mehnaz SAMINA, Mirza MS, Malik KA (2017) Comparison of microbial communities associated with halophyte (*Salsola stocksii*) and nonhalophyte (*Triticum aestivum*) using culture-independent approaches. Pol J Microbiol 66:375–386. https://doi.org/10.5604/01.3001.0010.4866
Naeem M, Manzoor S, Abid MUH, Tareen MBK, Asad M, Mushtaq S, Ehsan N, Amna D, Xu B, Hazafa A (2022) Fungal proteases as emerging biocatalysts to meet the current challenges and recent developments in biomedical therapies: an updated review. J Fungi 8(2):1–32. https://doi.org/10.3390/jof8020109
Nair P, Kuriakose T, Das B (2025) Pharmaceutical applications of microbial proteases. In: Francis D, Abdulhameed S (eds) Microbial enzymes as potential biotherapeutics in human healthcare, 1st edn. CRC Press, Floride, pp 130–146
Nascimento WCA, Martins MLL (2004) Production and properties of an extracellular protease from thermophilic *Bacillus* sp. Braz J Microbiol 35:1–2. https://doi.org/10.1590/s1517-83822004000100015
Neidleman SL (1984) Applications of biocatalysis to biotechnology. Biotechnol Gen Eng Rev 1:1–38. https://doi.org/10.1080/02648725.1984.10647779
Okafor N (2007) Modern industrial microbiology and biotechnology. Science Publishers, Enfield
Olsen M, Iverson B, Georgiou G (2000) High-throughput screening of enzyme libraries. Curr Opin Biotechnol 11:331–337. https://doi.org/10.1016/s0958-1669(00)00108-7
Palleroni NJ (1984) Pseudomonadaceae. (Wilson, Broadhurst, Buchanan, Krumwide, Rogers and Smith 1917). In: Krieg NR, Holt JG (eds) Bergey's manual of systematic bacteriology, vol 1. Baltimore, The Williams & Wilkins Co, pp 143–213
Palsaniya P, Mishra R, Beejawat N, Sethi S, Gupta BL (2012) Optimization of alkaline protease production from bacteria isolated from soil. J Microbiol Biotechnol Res 2:695–701
Pandey D, Singh R, Chand D (2011) An improved bioprocess for synthesis of acetohydroxamic acids using DTT (dithiothreitol) treated resting cells of *Bacillus* sp. APB-6. Bioresour Technol 102(11):6579–6586. https://doi.org/10.1016/j.biortech.2011.03.071
Park S, Lee JJ, Yang BM, Cho JH, Kim S, Kang J, Oh S, Park DJ, Perez-Maldonado R, Cho JY, Park IH, Kim HB, Song M (2020) Dietary protease improves growth performance, nutrient digestibility, and intestinal morphology of weaned pigs. J Anim Sci Technol 62:21–30. https://doi.org/10.5187/jast.2020.62.1.21

Phadatare SU, Srinivasan MC, Deshpande VV (1993) High activity alkaline protease from *Conidiobolus coronatus* (NCL 86.8. 20): enzyme production and compatibility with commercial detergents. Enz Microbial Technol 15:72–76. https://doi.org/10.1016/0141-0229(93)90119-m

Pham VHT, Kim J, Shim J, Chang S, Chung W (2021) Purification and characterization of strong simultaneous enzyme production of protease and α-amylase from an extremophile-*Bacillus* sp. FW2 and its possibility in food waste degradation. Fermentation 8(1):1–13. https://doi.org/10.3390/fermentation8010012

Pudova D, Koryagina A, Rudakova N, Mardanova A, Sharipova M (2020) Effect of *Bacillus pumilus* 3–19 protease on growth parameters and gut microbiome of broiler chickens. In: E3S Web of conferences, vol 222. EDP Sciences, p 02051

Raj J, Prasad S, Bhalla TC (2006) *Rhodococcus rhodochrous* PA-34: a potential biocatalyst for acrylamide synthesis. Process Biochem 41:1359–1363. https://doi.org/10.1016/j.procbio.2006.01.022

Rao MB, Tanksale AM, Ghatge MS, Deshpande VV (1998) Molecular and biotechnological aspects of microbial proteases. Microbiol Mol Biol R 62:597–635. https://doi.org/10.1128/MMBR.62.3.597-635.1998

Raval KM, Vaswani PS, Majumder DR (2012) Microbial biotransformation: pharmaceutical peptides. J Biol Food Sci Res 1(1):1–14

Rawlings ND, Barrett AJ, Thomas PD, Huang X, Bateman A, Finn RD (2017) The MEROPS database of proteolytic enzymes, their substrates and inhibitors in 2017 and a comparison with peptidases in the PANTHER database. Nucleic Acids Res 46:D624–D632. https://doi.org/10.1093/nar/gkx1134

Razzaq A, Shamsi S, Ali A, Ali Q, Sajjad M, Malik A, Ashraf M (2019) Microbial proteases applications. Front Bioeng Biotechnol 7:1–20. https://doi.org/10.3389/fbioe.2019.00110

Rehman R, Ahmed M, Siddique A, Hasan F, Hameed A, Jamal A (2017) Catalytic role of thermostable metalloproteases from *Bacillus subtilis* KT004404 as dehairing and destaining agent. Appl Biochem Biotechnol 181:434–450. https://doi.org/10.1007/s12010-016-2222-5

Rubilar O, Diez MC, Gianfreda L (2008) Transformation of chlorinated phenolic compounds by white rot fungi. Crit Rev Environ Sci Technol 38:227–268. https://doi.org/10.1080/10643380701413351

Sandhya C, Madhavan Nampoothiri K, Pandey A (2005) Microbial protease, microbial enzyme and biotransformation, 1st edn. Humana Press Inc, Totowa

Saran S, Isar J, Saxena RK (2007) A modified method for the detection of microbial proteases on agar plates using tannic acid. J Biochem Biophys Methods 70:697–699. https://doi.org/10.1016/j.jbbm.2007.03.005

Sathishkumar R, Ananthan G, Arun J (2015) Production, purification and characterization of alkaline protease by ascidian associated *Bacillus subtilis* GA CAS8 using agricultural wastes. Biocatal Agric Biotechnol 4:214–220. https://doi.org/10.1016/j.bcab.2014.12.003

Savitha S, Sadhasivam S, Swaminathan K, Lin FH (2011) Fungal protease: production, purification and compatibility with laundry detergents and their wash performance. J Taiwan Inst Chem Eng 42:298–304. https://doi.org/10.1016/j.jtice.2010.05.012

Schallmey M, Singh A, Ward OP (2004) Developments in the use of *Bacillus* species for industrial production. Can J Microbiol 50(1):1–17. https://doi.org/10.1139/w03-076

Sharma KM, Kumar R, Panwar S, Kumar A (2017) Microbial alkaline proteases: optimization of production parameters and their properties. J Genet Eng Biotech 15:115–126. https://doi.org/10.1016/j.jgeb.2017.02.001

Singh CJ (2003) Optimization of an extracellular protease of *Chrysosporiumkeratinophilum* and its potential in bioremediation of keratinic wastes. Mycopathologia 156:151–156. https://doi.org/10.1023/a:1023395409746

Singh R, Kumar M, Mittal A, Mehta PK (2016) Microbial enzymes: industrial progress in 21st century. 3 Biotech 6:1–15. https://doi.org/10.1007/s13205-016-0485-8

Singhal P, Nigam V, Vidyarthi A (2012) Studies on production, characterization and applications of microbial alkaline proteases. Int J Adv Biotechnol Res 3:653–669
Smitha MS, Singh S, Singh R (2017) Microbial biotransformation: a process for chemical alterations. J Bacteriol Mycol Open Access 4(2):47–51. https://doi.org/10.15406/jbmoa.2017.04.00085
Song P, Zhang X, Wang S, Xu W, Wang F, Fu R, Wei F (2023) Microbial proteases and their applications. Front Microbiol 14:1–24. https://doi.org/10.3389/fmicb.2023.1236368
Souza PMD, Bittencourt MLDA, Caprara CC, de Freitas M, de Almeida RPC, Silveira D, Fonseca YM, Filho EXF, Junior AP, Magalhães PO (2015) A biotechnology perspective of fungal proteases. Braz J Microbiol 46:337–346. https://doi.org/10.1590/s1517-838246220140359
Sun SY, Xu Y (2009) Membrane-bound 'synthetic lipase' specifically cultured under solid-state fermentation and submerged fermentation by *Rhizopus chinensis*: a comparative investigation. Bioresour Technol 100:1336–1342. https://doi.org/10.1016/j.biortech.2008.07.051
Suthindhiran K, Jayasri MA, Dipali D, Prasar A (2014) Screening and characterization of protease producing actinomycetes from marine saltern. J Basic Microbiol 54(10):1098–1109. https://doi.org/10.1002/jobm.201300563
Tahri N, Bahafid W, Sayel H, El Ghachtouli N (2013) Biodegradation: involved microorganisms and genetically engineered microorganisms. Biodegradation-Life Sci 1:289–320. https://doi.org/10.5772/56194
Tanwar M, Debnath M, Debnath S, Sharma P, Mukhopadhay A, Kakar N, Ramakrishna S (2022) Exploring the utility of nanoprotease as environmentally friendly benign laundry detergent fabric cleaner. J Clean Prod 334:130243. https://doi.org/10.1016/j.jclepro.2021.130243
Theron LW, Divol B (2014) Microbial aspartic proteases: current and potential applications in industry. Appl Microbiol Biotechnol 98:8853–8868. https://doi.org/10.1007/s00253-014-6035-6
Verma A (2014) Alkaline protease from *Thermoactinomyces* sp. RS1 mitigates. J Med Plants 1:11–15
Verma A, Singh H, Anwar MS, Ansari MW, Agrawal S (2014) Production of alkaline protease from a haloalkaliphilic soil thermoactinomycete and its application in feather fibril disintegration. Afr J Microbiol Res 8(27):2565–2573. https://doi.org/10.5897/ajmr2014.6682
Verma A, Singh H, Anwar S, Chattopadhyay A, Tiwari KK, Kaur S, Dhilon GS (2017) Microbial keratinases: industrial enzymes with waste management potential. Crit Rev Biotechnol 37(4):476–491. https://doi.org/10.1080/07388551.2016.1185388
Ward OP (2011) Proteases. In: Moo-Young M (ed) Comprehensive biotechnology, 2nd edn, Waterloo, Canada, pp 571–582
Wietzorrek A, Bibb M (1997) A novel family of proteins that regulates antibiotic production in Streptomycetes appears to contain an OmpR-like DNA-binding fold. Mol Microbiol 25:1181–1184. https://doi.org/10.1046/j.1365-2958.1997.5421903.x
Wojtowicz-Krawiec A, Sokolowska I, Smorawinska M, Chojnacka-Puchta L, Mikiewicz D, Lukasiewicz N, Marciniak-Rusek A, Wolinowska R, Bierczynska-Krzysik A, Porebska AJ, Kuthan-Styczen J, Gurba L, Borowicz P, Mazurkiewicz A, Plucienniczak G, Plucienniczak A (2014) Use of Ubp1 protease analog to produce recombinant human growth hormone in *Escherichia coli*. Microb Cell Factories 13:1–12. https://doi.org/10.1186/s12934-014-0113-4
Wilson DS, Pollock GB, Dugatkin LA (1992) Can altruism evolve in purely viscous populations? In: Evol Ecol 6:331–341. https://doi.org/10.1007/BF02270969
Xue B, Dunker AK, Uversky VN (2012) Orderly order in protein intrinsic disorder distribution: disorder in 3500 proteomes from viruses and the three domains of life. J Biomol Struct Dyn 30:137–149. https://doi.org/10.1080/07391102.2012.675145
Yegin S, Dekker P (2013) Progress in the field of aspartic proteinases in cheese manufacturing: structures, functions, catalytic mechanism, inhibition, and engineering. Dairy Sci Technol 93(6):565–594. https://doi.org/10.1007/s13594-013-0137-2. 2-s2.0-84885587642

Zambare VP, Nilegaonkar SS (2016) Proteases in leather processing. In: Thangadurai D, Sangeetha J (eds) Industrial biotechnology, sustainable production and bioresource utilization. Apple Academic Press, CRC Press., (e-Book ISBN-13: 978-1-77188-262-0), pp 209–238

Zambare V, Nilegaonkar S, Kanekar P (2011) A novel extracellular protease from Pseudomonas aeruginosa MCM B-327: enzyme production and its partial characterization. New Biotechnol 28:173–181. https://doi.org/10.1016/j.nbt.2010.10.002

Zhang S, Li H, Uluko H, Liu L, Pang X, Lv J (2015) Investigation of protease production by *Pseudomonas fluorescens* BJ-10 and degradation on milk proteins. J Food Process Preserv 39(6):2466–2472. https://doi.org/10.1111/jfpp.12496

Zhang C, Bijl E, Svensson B, Hettinga K (2019) The extracellular protease AprX from *Pseudomonas* and its spoilage potential for UHT milk: a review. Compr Rev Food Sci Food Saf 18(4): 834–852. https://doi.org/10.1111/1541-4337.12452

Zhu D, Yang S, Zhan S, Yu H, Yang Y, Li Y (2014) Waste management in a typical Chinese gravure plate-making industry. Environ Nanotechnol Monit Manag 1:50–56. https://doi.org/10.1016/j.enmm.2014.07.002

Chapter 14
From Tradition to Innovation: Next-Generation Microbial Proteases Transforming Dairy Processing

Latika Shendre, Anshu Kumari, Vasudeo Zambare, and Amit Kumar Singh

Abstract Protease utilization in the milk processing sector is changing significantly as a result of new process technologies, changing customer preferences, and sustainability requirements. This chapter examines how protease uses in milk processing are evolving, from conventional cheese manufacturing to cutting-edge functional milk products. Extremophile enzymes and precision fermentation technology are two new sources of proteases that provide more accurate and efficient protein modification while meeting clean-label requirements. These sophisticated uses include the development of medicinal components, the elimination of allergies, and the production of bioactive peptides. Important technological advancements include real-time enzymatic analysis, AI-optimized process improvement, and enzyme immobilization technologies that reduce costs and increase sustainability. Proteases targeted at rapid cheese ripening technologies, plant-based dairy alternatives, and animal rennet substitutes are all making notable advancements in the sector. Regulations are changing to allow for new enzyme sources while maintaining consumer safety and acceptance. Shorter processing times, more ecological friendly manufacturing techniques, and an increase in the demand for functional dairy products are some of the current industry trends. Among the market obstacle are scaling up innovative protease technologies, managing intricate international regulatory frameworks, and reaching cost parity with current solutions. The chapter combines market data, industry case studies, and new research discoveries to provide dairy processors with strategic advice for utilizing modern protease technology. Due to ongoing developments in enzyme engineering and their convergence with digital

L. Shendre (✉) · A. Kumari
Dr. D. Y. Patil Biotechnology and Bioinformatics Institute, Dr. D. Y. Patil Vidyapeeth, Pune, Maharashtra, India
e-mail: latika.shendre@dpu.edu.in

V. Zambare (✉)
R&D Department, Balaji Enzyme & Chemical Pvt Ltd, Mumbai, Maharashtra, India

A. K. Singh
Center of Excellence in Tissue Engineering, Department of Biotechnology and Medical Engineering, National Institute of Technology, Rourkela, Odisha, India

V. Zambare et al. (eds.), *Biotechnological Advances and Applications in Microbial Proteases*, Interdisciplinary Biotechnological Advances,
https://doi.org/10.1007/978-981-95-9011-7_14

manufacturing system, proteases are anticipated to play a significant part in the creation of the future generation of dairy products.

Keywords Microbial proteases · Dairy biotechnology · Enzyme applications · Food processing innovation · Industrial fermentation

14.1 Introduction

The quality of milk stored is greatly impacted by enzymes, which have an impact on a number of biochemical processes. Among these enzymes, proteases and lipases are special because they affect flavor, texture, and stability over time. This study thoroughly examines the enzymatic activity and modes of action of the proteases found in milk during preservation (Khan et al. 2025). Proteases are a broad and varied class of hydrolytic enzymes that are categorized according to their unique structure, site of action, and reaction mechanism. Proteases are widely distributed in biology and affect many aspects of cell and organism function, including protein turnover, growth, adaptability, control, sporulation and germination, disease, and death. Numerous physiological functions, including the essential cellular functions of differentiation, motility, division, and cell death, are regulated by proteases (Ward 2011). Enzymes have been used extensively in industrial applications since the nineteenth century. Enzymes have long been used in the dairy industry for a variety of purposes, including making yogurt, milk, Cheddar cheese, and milk products. A variety of exogenous (made commercially) and endogenous (found in milk) enzymes are crucial for the digestion of dairy products. In order to preserve milk and enhance its quality, commercial enzymes are added at various stages of the dairy manufacturing process. The manufacture of dairy enzymes has a sizable worldwide market. Rennet–chymosin, lipases, protease, and β-galactosidase are among the enzymes that mostly dominate it; these will be covered in detail in this chapter. The majority of these exogenous enzymes come from genetically modified organisms, plants, and microorganisms (Annapure and Gaur 2022). Proteases and peptidases are used to produce dairy protein hydrolysates for clinical, sports, or newborn nutrition in order to decrease the risk of allergic responses and increase the protein's digestibility. Proteases will break down the substrate into bigger peptides and fewer amino acids, changing the protein substrate's structure (Dekker 2019). Proteases are enzymes that cleave peptide bonds in proteins to release smaller peptides or amino acids. Casein is the primary protein found in dairy products, and proteases are important for the following processes: (1) cheesemaking (milk coagulation and curd production), (2) cheese develops its flavor and texture as it ripens, (3) manufacturing of protein hydrolysates for functional meals and baby formulae, (4) quality monitoring, as milk's inherent proteases can occasionally lead to spoiling (Naveed et al. 2021). Zambare et al. (2011) reported the production optimization, purification, and biochemical characterization of a novel extracellular microbial protease from *Pseudomonas aeruginosa* MCM B-327. The study demonstrated that the optimization of cultural and nutritional parameter significantly enhanced

Table 14.1 Importance of proteases in dairy industry

Function	Protease role	Impact on dairy products	References
Curd firmness	Cleaves k-casein to form firm curds	Determines yield and texture	Asioli et al. (2017)
Cheese ripening rate	Breaks down casein to peptides and amino acids	Speeds up aging, reduces storage cost	Li et al. (2023)
Flavor profile	Generates flavor precursors	Creates nutty, savory, or tangy cheese notes	Liu et al. (2020)
Nutritional properties	Produces easily digestible peptides	Improves digestibility and reduces allergenicity	Parmar et al. (2021)

protease yield, highlighting the importance of process engineering in the development of industrially viable enzymes. The purified protease exhibited desirable catalytic efficiency and stability under a broad range of operational conditions, emphasizing its suitability for large-scale applications. Such systematic optimization and purification strategies represent a critical advancement in next-generation microbial protease development.

Proteases are very important in the dairy industry because they directly affect curds formation, cheese aging, dairy products taste, and nutritional value (Table 14.1). Proteins like chymosin break down k-casein during milk coagulation, which makes curds with the right firmness and yield. Proteolytic enzymes break down casein into peptides and free amino acids over time, speeding up the maturation processes and lowering the cost and time needed to store cheese. These proteolysis product act as key precursors for flavor development, contributing to characteristic nutty, savory, and tangy notes in different cheese varieties. Additionally, controlled protease activity enhances the nutritional profile of dairy products by generating bioactive and easily digestible peptides, thereby improving protein digestibility and reducing allergenic potential. Overall, proteases are essential for optimizing the functional, sensory, and nutritional attributes of dairy products (McSweeney 2004).

14.1.1 Historical Context of Enzyme Uses in Milk Processing

Since rennet from animal stomachs was first used to make cheese in antiquity, enzymes use in the dairy industry has developed through microbial and recombinant innovation. Today enzymes such as chymosin, lipase, and lactase are essential for large-scale, effective, and specialized dairy production globally (McSweeney 2004).

14.1.2 Overview of Current Market Drivers and Regulatory Landscape

Figure 14.1 illustrates the global distribution of proteases in the dairy industry. About 65% proteases contributes in cheese production (milk coagulation). Cheese ripening and flavor development accounts for 20%, while 10% of protease is for protein hydrolysates for infant and sports nutrition. A relatively smaller proportion, about 5%, is applied in other dairy processes. This highlights the critical role of proteases in cheese manufacturing compared to other applications.

14.1.3 Novel Protease Sources and Engineering

Diverse microbiological environments are now being considered in the hunt for novel proteases, in addition to the conventional animal and plant sources. The discovery of enzymes with distinct catalytic qualities appropriate for industrial applications has been made possible by developments in protein engineering and

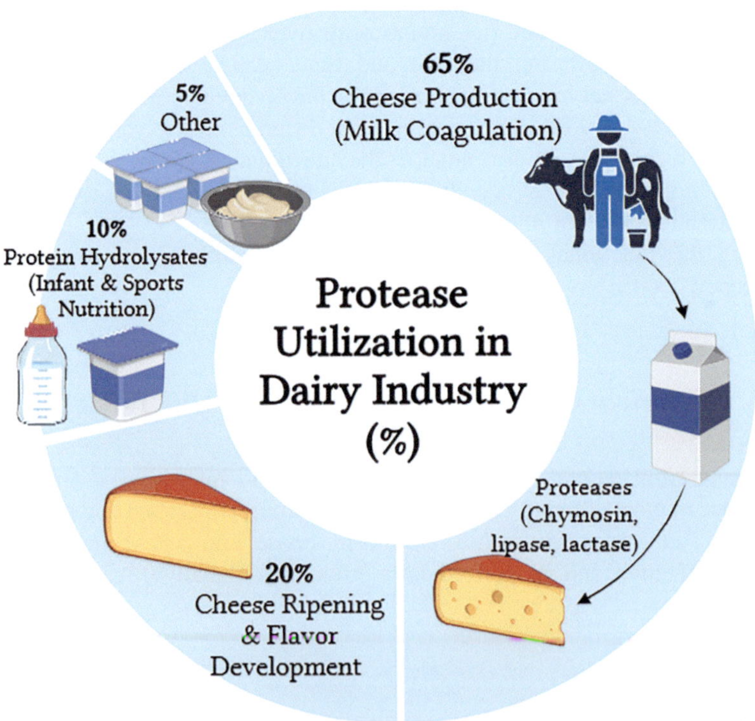

Fig. 14.1 Global usage of proteases in dairy industry

metagenomics, and the cost-effective manufacture, scalability, and adaptation to recombinant expression systems of microbial proteases make them especially appealing. In an effort to improve enzyme stability, specificity, and activity in difficult circumstances, recent research has focused on investigating unusual microbiological sources, such as extremophiles, and using engineering techniques (Yao et al. 2023).

14.1.4 Microbial Proteases from Extremophile Organisms

Novel proteases can be obtained from extremophile microbes that thrive under conditions of high pressure, high temperature, high salt, or extreme pH. When exposed to severe industrial settings that denature traditional enzymes, these enzymes—known as extremozymes—display remarkable catalytic stability (Song et al. 2023). The detergent, textile, and dairy sectors can benefit from thermophilic proteases, which are found in genera such as *Geobacillus* and *Thermus*. Between 60 and 100 °C, these proteases continue to function. Because they are active at low temperatures, psychrophilic proteases from organisms such as *Pseudoalteromonas* or *Shewanella* allow for energy-saving uses in seafood processing (Yang et al. 2023). Microbes such as *Halobacterium* and *Salinivibrio* produce halophilic proteases, which are especially important for high-salt food fermentations and maintain their functionality in hypersaline environments. Alkaliphilic and acidophilic proteases, on the other hand, are useful in the biocleaning, waste treatment, and leather processing industries (Lai et al. 2023). New protein engineering and metagenomic techniques have improved our access to new extremophile proteases and allowed us to maximize their stability and specificity. To increase keratinolytic activity and stability in high heat and salt, for example, a polyextremotolerant subtilisin-like protease from *Thermoactinomyces vulgaris* CDF was created (Contesini et al. 2018). Recombinant strains such as *Lactobacillus helveticus* CP790, *L. rhamnosus* GG, *L. bulgaricus* SS1, and *L. lactis* subsp. *cremoris* FT4 have high efficiency of producing dairy products that contain bioactive peptides (Feijoo-Siota et al. 2014).

14.2 Recombinant and Genetically Modified Proteases

Recombinant and genetically modified proteases have been developed in response to the need for highly effective and application-specific proteases. High-level production, simpler purification, and enhanced stability are made possible by recombinant DNA technology, which permits the heterologous expression of protease genes in hosts such as *Escherichia coli*, *Bacillus subtilis*, *Pichia pastoris*, and *Filamentous fungi* (Liu et al. 2024). Site-directed mutagenesis, directed evolution, and rational design are examples of protein engineering techniques that are frequently employed to improve solvent or pH tolerance, substrate specificity, thermal stability, and catalytic efficiency. Site-directed mutagenesis of subtilisin-like proteases, for instance, has produced variations with improved keratinolytic and detergent-stable qualities that are appropriate for the leather and textile industries. Surface display systems and fusion protein techniques further facilitate the production of multifunctional proteases for use in biomedical and industrial settings (Lai et al. 2023). Furthermore, new protease variants with customized characteristics are being designed more and more through the use of synthetic biology and CRISPR-Cas-based editing. These genetically modified proteases, which offer great specificity, sustainability, and cost-efficiency, are revolutionizing industries such as pharmaceutical manufacture, food processing, waste management, and detergent formulation (Yao et al. 2023).

14.3 Microbial Proteases in Value-Added Dairy Products

In the dairy sector, the microbial enzyme plays a vital significance, where they are used to improve and enhance organoleptic qualities such as fragrance, color, and flavor, as well as increasing yield of milk products (Table 14.2). In the dairy

Table 14.2 Microbial source of protease, their type, and dairy application

Enzyme type	Microbial source	Dairy application	Value-added result	References
Aspaetic proteases	*Rhizomucoe miehei*, *Rhizomucor pusillus*	Cheese manufacturing	Efficient for milk clotting and curd formation	Aljammas et al. (2022)
Fermentation-produced chymosin	*Aspergillus niger*, *Kluyveromyces lactis*	Cheese manufacturing	Highly specific cleavage of casein, resulting in high quality, consistent curd suitable for hard cheeses	Hellmuth (2006)
Endopeptidases	*Bacillus species*	Accelerated cheese ripening	Breaks down the casein matrix into small peptides, dramatically enhancing flavor intensity and softening texture	Gupta et al. (2002)
Exopeptidase	*Rhizopus arrhizus*	Protein hydrolysates/ debittering	Cleaves peptides into free amino acids; removes bitter peptides that can develop during ripening; improves solubility and reduces allergenicity	Soliman et al. (2022)
Alkaline proteases	*Bacillus licheniformis*	Enzyme-modified cheese	Extensive protein breakdown creates concentrated, strong cheese flavors for use in food seasonings and sauces	Gupta et al. (2002)

industries, microbial enzymes have been applied to make varied goods, such as yogurt, cheese, syrup, bread, and so on to enhance their qualities (Abada 2019).

Microorganism are a key source of industrial protease because of their consistent enzyme activity, high production, and ease of cultivation. *Aspergillus niger*, *Rhizomucor miehei*, *Rhizomucor pusillus*, and *Rhizopus arrhizus* are among the fungi that produce proteases with a high milk clotting specificity, which makes them ideal for protein breakdown and cheese production. High purity, high activity recombinant chymosin is frequently produced using yeast such as *Kluyveromyces lactis*. Alkaline proteases and endopeptides are best produced by bacterial species, particularly *Bacillus* strains such as *Bacillus licheniformis*, because of their quick growth, thermal stability, and appropriateness for large-scale fermentation (Hellmuth 2006).

14.4 Precision Fermentation Approaches for Protease Production

The development of precision fermentation is a revolutionary method for producing industrial enzymes, such as proteases. Precision fermentation combines synthetic biology, metabolic engineering, and recombinant DNA technology to create high-yield, customized protease producers, in contrast to conventional fermentation, which depends on native microbial strains (Liang et al. 2024).

14.4.1 Key Principles of Precision Fermentation

Strain engineering for high-productivity engineered microbial chassis such as *Escherichia coli*, *Bacillus subtilis*, *Corynebacterium glutamicum*, *Saccharomyces cerevisiae*, and *Pichia pastoris* are used for targeted protease expression. Gene expression, protein folding, and secretion efficiency are improved by combining CRISPR-Cas gene editing, codon optimization, and promoter engineering (Shea et al. 2024). Optimization of metabolism and pathways in order to enhance enzyme yield while avoiding by-products, precision fermentation systems optimize carbon flux and energy balance. For instance, rewiring the secretory routs and metabolism of amino acids in *Bacillus* and *Pichia* has increased the output of papain and subtilisin on an industrial scale (Ndochinwa et al. 2024).

14.5 Advanced Cheese Manufacturing Applications

Modern enzymatic, microbiological, and engineering methods have been applied in cheese production advancements to transform antiquated artisanal methods into extremely productive industrial operations. Modern cheese manufacturing is increasingly focused on improving yield flavor, texture, and shelf life in order to meet clean-label and sustainable production criteria (Alhelli et al. 2021).

14.5.1 Site-Specific Proteases to Speed Up the Ripening of Cheese

The primary biochemical mechanism underlying the intricate biochemical process of cheese ripening is proteolysis, which hydrolyzes casein proteins into peptides and free amino acids. Flavor and texture are then formed with the help of these peptides and free amino acids. Conventional ripening methods might take weeks or months, which lowers factory efficiency and market reaction. Site-specific proteases offer a tailored way to speed up ripening and improve taste development while preventing bitterness or textural defects by hydrolyzing caseins at predetermined cleavage sites (Alhelli et al. 2021). There are casein specific and sequencing proteolysis mechanisms, which are represented in Fig. 14.2. Casein specific mechanism showed

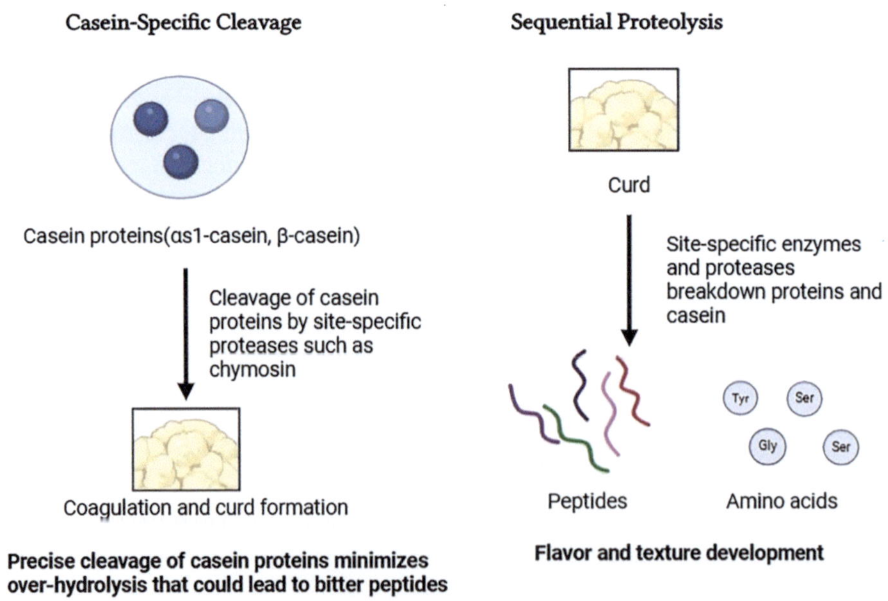

Fig. 14.2 The mechanism of action of both casein-specific cleavage and sequential proteolysis

curding of milk with bitter peptides and bitter taste. However, the sequential proteolysis showed formation of peptide and amino acids, which enhance flavor and texture.

14.5.2 Proteases to Enhance Taste and Alter Texture

During the cheesemaking process, proteases help with texture modification and flavor complexity in addition to curd development and ripening. Cheese producers can enhance mouthfeel, speed up maturity, and add distinctive flavors without sacrificing product quality by carefully choosing their proteases (West 2007).

14.5.3 Targeted Proteolysis for the Manufacture of Low-Sodium Cheese

Reducing cheese's salt level is an essential first step in producing healthier dairy products, but doing so poses serious issue with flavor, texture, and microbiological stability. Sodium chloride (NaCl) has several functions in cheese, including regulating water activity and curd hardness, enhancing flavor perception, and acting as a preservative. However, reducing NaCl sometimes leads to some microbiological spoilage risks as well as softer texture, bitterness, and reduced flavor (Fernandes et al. 2025). Targeted proteolysis, the controlled enzymatic destruction of casein proteins, has been found to effectively offset many of the quality issues caused by salt reduction. Specifically, site-specific proteases hydrolyze as αs1- and β-caseins to create peptides and free amino acids that are used as flavoring building blocks. This controlled hydrolysis improves texture and restores umami flavor even in low salt recipes. Because of the increased water activity and decreased ionic strength caused by salt reduction, proteolysis tends to accelerate on its own in low-sodium cheese. This effect can be focused and enhanced by selecting proteolytically active cultures, such as *Lactobacillus helveticus*, which has a large number of cells envelop proteinases (CEP). These enzymes initiate and control the breakdown of caseins, producing small peptides that contribute desired textural qualities and improve flavor (Choi et al. 2020). Fragrance molecules, including butter-like or sardine-like volatiles, can enhance flavor perception without increasing sodium levels. These volatiles can mimic the taste-enhancing qualities of salt and cooperate with the outcomes of certain proteolysis (Wang et al. 2022). Furthermore, salt reduction often alters the microbial communities in cheese. An increase in proteolytic and lipolytic metabolism was found in a recent microbiome investigation on half-salt Cheddar varieties (1.0% NaCl), indicating that low-sodium environments may encourage advantageous microbial pathways that improve flavor production (Hao et al. 2025).

14.5.4 Artisanal and Specialty Cheese Applications

In order to enhance texture, quicken ripening, and create complex flavors in artisanal and specialty cheeses without sacrificing regional or traditional identities, proteases are being utilized more and more in their manufacturing. Numerous microbial and fungal strains with considerable protease and lipase activity were found during research into handmade raw ewe's milk cheeses. These strains included *Brevibacterium antiquum*, *Micrococcus luteus*, and *Yarrowia lipolytica*. Since they are aspartic proteases, several of these enzymes offer a viable substitute for rennet-like action in conventional cheesemaking (Ozturkoglu-Budak et al. 2016).

14.6 Milk Protein Modification and Functionality

In the manufacture of dairy products, milk proteins—mostly caseins (80%) and whey proteins (20%)—are essential. The main milk protein components are presented in Table 14.3. Their solubility, emulsification, foaming, gelation, and nutritional value are all improved by altering their structure by physical, chemical, or enzymatic methods.

Caseins, the main protein component of milk, are heat-stable micellar structures that are necessary for the coagulation of milk and cheese production. On the other hand, whey proteins are globular, heat sensitive, and have a high nutritional and biological value because of their functional qualities and rich essential amino acid treatment (Kuddus and Ramteke 2012).

14.6.1 Modification Techniques

Enzymatic modification: Certain peptide bonds are hydrolyzed by enzymes such as subtilisin, chymosin, or trypsin. Increases bioactivity, decreases allergenicity, and improves solubility. Uses include cheese ripening, protein drinks, and baby formula.

Heat treatment: Increases reactivity by denaturing whey proteins, particularly β-lactoglobulin. Enhances gelation and water-holding ability by encouraging contact with caseins. Used in the production of UHT milk and yogurt.

Table 14.3 Types of milk proteins (Kuddus and Ramteke 2012)

Category	Main proteins	Key characteristics
Caseins	αs1-, αs2-, β-, k-casein	Heat stable, form micelles, good for coagulation
Whey	β-Lactoglobulin, α-lactalbumin, immunoglobulin	Heat sensitive, globular, high nutritional value

pH adjustment: Alters protein charge and solubility. Useful in acid coagulation for paneer, yogurt, and ricotta cheese.

High-pressure processing (HPP) and ultrasonication: Nonthermal methods used to modify protein structure and improve techno-functionality while preserving nutrition (Nayik et al. 2024).

14.7 Selective Proteolysis for Improved Protein Solubility

The targeted enzymatic breakdown of particular peptide bonds in milk proteins, especially caseins and whey proteins, by highly selective proteases is known as selective proteolysis. Proteins are partially hydrolyzed without undergoing excessive breakdown due to a carefully planned method that improves their solubility, digestibility, and nutritional value (Fan et al. 2018).

Mechanism of Action The precise selection of enzymes is essential for this method. Trypsin, chymotrypsin, and subtilisin are examples of endoproteases that specifically target interior hydrophobic areas and start hydrolysis. Terminal amino acids are then cut by exopeptidases, such as aminopeptidases and carboxypeptidases, to improve solubility. In order to prevent the formation of bitterness, hydrolysis is intentionally kept at a low level—typically 2–10%—disturbing aggregated protein structures and revealing hydrophilic regions that improve water–protein interactions. According to research, the solubility of milk protein components, such as milk protein concentrates (MPC), may be enhanced by such regulated enzymatic action. For example, utilizing proteases such as trypsin, pepsin, chymotrypsin, or papain to enzymatically hydrolyze MPC80 resulted in significant increases in nitrogen solubility and decreased viscosity, which improved dispersion in aqueous environments (Kaur et al. 2023). Furthermore, research has found that physical–chemical changes, such as casein micelle aggregation and Maillard-induced cross-linking, are important causes of low solubility in MPCs and milk protein isolates (MPIs) (Khalesi and FitzGerald 2022).

14.8 Bioactive Peptide Generation Through Controlled Proteolysis

Short amino acid sequences, usually ranging from 2 to 20 residues, known as bioactive peptides, are liberated from parent milk proteins, including casein, β-lactoglobulin, and α-lactalbumin, by means of regulated proteolysis. After digestion or processing, these peptides can have particular physiological consequences, because enzymatic cleavage reveals their functional locations, even though they are inactive in the intact protein structure (Tang et al. 2023).

14.9 Protein Hydrolysates for Sports Nutrition and Infant Formulas

Protein hydrolysates are predigested proteins made from dairy, plant, or animal proteins that have been hydrolyzed by enzymes. Proteins are broken down into smaller peptides and free amino acids by hydrolysis, which improves digestibility, speeds up absorption, and lessens allergenicity. Because of these qualities, they are especially useful in infant formulae, where a lower possibility for allergies is essential for delicate or vulnerable infants, and in sports nutrition, where quick administration of amino acids aids in muscle repair (Rios-Mera et al. 2017). These properties make them highly valuable in infant formula and sports nutrition. In infant formula, protein hydrolysates reduce the allergic reactions by removing complete β-lactoglobulin and casein epitopes that triggers cow's milk protein allergy (CMPA). Moreover, their smaller peptides require less enzymatic breakdown, making them beneficial for infant's digestive system development (Sekkidou et al. 2021). These hydrolysates can additionally contain bioactive peptides with immune-modulatory or gut barrier-protective qualities (Dias et al. 2022). In sports nutrition, hydrolysate such as whey protein hydrolysate enables rapid absorption of amino acids by avoiding prolong digestion as well as early stimulation of muscle protein synthesis (MPS). After exercise, it also encourages faster recovery through enhanced glycogen replenishment and decreased muscle soreness (Tang et al. 2023). Additionally, hydrolysis produces peptides with anti-inflammatory and antioxidant properties, which may help lower oxidative stress during vigorous exercise (Horstman and Huppertz 2023). Production involves choosing the right enzymes—trypsin, alcalase, papain, or pancreatin for sports nutrition, and pepsin–pancreatin combinations or microbial proteases for infant formula—followed by regulating the degree of hydrolysis (DH) that influences bitterness and digestibility (Vandenplas et al. 2025). In contrast to sports supplements, which balance DH for palatability and bioavailability, infant formulae frequently call for considerable hydrolysis (>80% peptide fraction <3 kDa). The final purification steps, including filtration and spray-drying, ensure microbiological safety and yield a stable powder product (Vandenplas et al. 2025).

14.10 Allergenicity Reduction in Milk Proteins

The goal of allergenicity reduction in milk proteins is to change or get rid of the epitopes that cause immunological reactions, especially in people who have cow's milk protein allergy (CMPA). Caseins (αs1-, αs2-, β-, and κ-casein) and whey proteins (β-lactoglobulin and α-lactalbumin) are the main allergenic proteins found in milk (Park et al. 2023). One of the most popular methods to lessen allergenicity is enzymatic hydrolysis. When peptide bonds in allergenic proteins are broken by proteases such trypsin, pepsin, alkaline, and bromelain, conformational and linear epitopes are disrupted (Mirzapour-Kouhdasht and Garcia-Vaquero 2022). Commercially available partially hydrolyzed formulas (PHF) and extensively hydrolyzed protein formulas (EHF) are utilized in baby nutrition for the management of CMPA (Lozano-Ojalvo and López-Fandiño 2018). To guarantee adequate allergenicity reduction while preserving nutritional and sensory qualities, the degree of hydrolysis is meticulously regulated (D'Auria et al. 2021). Lactic acid bacteria fermentation can also lessen allergenicity by changing the immunogenicity of allergenic proteins and breaking them down proteolytically. While simultaneously generating bioactive peptides with health-promoting properties, some strains of *Bifidobacterium* and *Lactobacillus* can produce peptidases that target allergic sequences. Enzymatic hydrolysis is frequently used in conjunction with physical processing methods including heat treatment, ultrasound, and high-pressure processing (HPP) to further denature proteins and increase the accessibility of epitopes for enzymatic cleavage. For instance, it has been demonstrated that HPP at 400–600 MPa, followed by protease treatment, dramatically lowers β-lactoglobulin immunoreactivity (Lozano-Ojalvo and López-Fandiño 2018).

14.11 Sustainable and Clean-Label Innovations

14.11.1 Animal Rennet Alternatives and Vegetarian Cheese Production

Animal rennet, an enzyme extract from the abomasum of unweaned calves that mostly contains chymosin, is used in the traditional process of making cheese. However, the hunt for animal rennet substitutes has been fueled by growing ethical concerns, dietary preferences (vegetarianism, veganism), and sustainability objectives (Madende et al. 2018). Some fungus, such as *Rhizomucor miehei* and *Rhizomucor pusillus*, produce chymosin-like milk-clotting proteases. Because they are nonanimal-derived, scalable, and reasonably priced, microbial rennets are frequently employed in the production of vegetarian cheese (Mokrani et al. 2024). Nevertheless, excessive proteolysis brought on by some microbial enzymes might result in bitterness when cheese ripens. Enzyme engineering and modern strain selection have enhanced clotting efficiency and flavor characteristics (Kashid and Joshi 2022).

14.11.2 Enzyme Recycling and Immobilization Technologies

Production economics in dairy and functional food processing can be greatly impacted by the stability and cost of enzymes. The goals of enzyme recycling and immobilization technologies are to decrease waste, increase enzyme reusability, and sustain activity throughout several processing cycles (Sheldon and van Pelt 2013). Enzyme immobilization involves physically confining or attaching enzymes to solid supports while retaining their catalytic activity. Common methods include adsorption (i.e., physical binding to carriers such as silica, activated carbon, or ion-exchange resins (Ishak et al. 2025)) and covalent bonding where enzyme molecules are chemically linked to functionalized supports, enhancing stability but often reducing flexibility (Karakashev and Manev 2015). Entrapment or encapsulation traps enzymes in polymer gels, nanofibers, or microcapsules to prevent leaching whereas cross-linking in which cross-linked enzyme aggregates are formed without an external carrier, offering high enzyme density (Barua et al. 2024). Recycling of enzymes while membrane retention system uses ultrafiltration or nanofiltration to keep enzymes in the reaction vessel while eliminating small product molecules, magnetic immobilization based on nanoparticles uses magnetic separation to enable quick enzyme recovery and redeployment, automated sequential batch systems, and enzyme recycling techniques to optimize operational efficiency (Gong et al. 2019).

14.12 Demand from Customers for Natural and Organic Enzyme Solutions

Consumer health consciousness, environmental concerns, and regulatory pressures are driving a transition in the global food business toward clean-label, natural, and organic products. Enzymes from non-GMO, plant-based, and microbial sources that satisfy organic and natural certification standards are becoming more and more popular in the dairy and functional food industries (Asioli et al. 2017). The clean-label movement, in which customers seek out products devoid of artificial chemicals and additives and choose enzymes derived from plant sources or natural fermentation, is driving the increasing demand for natural and organic enzyme solutions. Enzymes and other processing aids must originate from non-synthetic, non-GMO sources, according to laws such as the USDA National Organic Program and the EU Organic Standards. Furthermore, in comparison, animal-derived enzymes are thought to be healthier, more sustainable, and morally superior (Gandhewar and Shende 2025). This aligns with what customers want in terms of transparency, animal welfare, and environmental sustainability. The global market for food enzymes, especially in the natural and organic sector, is increasingly quickly due to rising consumer confidence, the possibility of premium pricing, and its support of developments in organic goods (Mujeeb and Miller 2023).

14.13 Quality Control and Process Optimization

Real-time proteolytic activity monitoring ensures optimal product quality and functional performance by enabling precise control of enzymatic processes. Advanced analytical techniques such as surface plasmon resonance (SPR), isothermal titration calorimetry (ITC), and microfluidic biosensors provide high sensitivity continuous detection of peptide bond cleavage (Ong and Yang 2017). Fluorescence resonance energy transfer (FRET)-based substrate are highly helpful for identifying protease activity in situ because they immediately provide a detectable signal upon cleavage. Additionally, mass spectrophotometry (MS) and Fourier-transform infrared spectroscopy (FTIR) provide the rapid profiling of peptide release during hydrolysis (Chauhan et al. 2025). Industrial applications can prevent over hydrolysis, conserve energy, and maintain product consistency through feedback-controlled enzymatic processing by incorporating these methods into processes analytical technology (PAT) frameworks. Through remote, real-time monitoring, and data logging, fully automated enzyme-based processing systems are now feasible due to the integration of IoT-enabled sensor platforms (Rathod and Sherje 2025).

14.14 AI-Driven Enzyme Condition and Dosage Optimization

Artificial intelligence (AI) is transforming enzyme-based food processing by providing predictive and adaptive management of enzymatic processes. Machine learning (ML) algorithms can analyze past and present process data, such as temperature, pH, substrate concentration, and enzyme kinetics, to determine the optimal dosage schedules and reaction parameters. This procedure reduces processing time, ensures consistent product quality, and minimizes the misuse of costly enzyme (Ge et al. 2023). Reinforcement learning models are increasingly being utilized to dynamically change conditions in response to process differences in order to continually optimize without the need for human intervention. AI-based predictive models in dairy processing enable closed-loop control that balances product functionality and efficiency by utilizing data from inline sensors and process analytical technology (PAT) frameworks (Rao and Neethirajan 2025).

14.15 Consistency of Protease Activity among Production Batches

Consistent protease activity is required for dairy products to have reproducible functional and sensory properties. The quantity of enzyme activity that fluctuates between production batches can be influenced by microbial strain performance, fermentation conditions, raw material quality, and downstream purification techniques (Dai et al. 2024). Calibration of enzyme assays were performed using globally accepted units on various standard substrtaes (e.g., International Milk Clotting Units, Anson Units). Enzyme batches are blended to produce a consistent activity profile prior to packaging and real-time activity verification were checked using inline monitoring and process analytical technology (PAT). Formulation using protective carriers (such glycerol or maltodextrin) to maintain action during storage and transit are examples of stabilization strategies. In addition to ensuring product quality and regulatory compliance, standardization encourages process efficiency, reduces waste, and boosts scalability. For dairy proteases such as chymosin, microbial rennet, and plant-based alternatives, strict batch-to-batch quality control is necessary for consistent curd formation, flavor development, and texture outcomes (Saito et al. 2022).

14.16 Integration with Current Dairy Processing Machinery

The successful application of cutting-edge proteolytic technologies in the dairy industry depends on their seamless interaction with the existing processing infrastructure. New enzyme-based solutions must be compatible with standard dairy equipment such as pasteurizers, ultrafiltration units, fermenters, and cheese vats in order to achieve targeted proteolysis, produce bioactive peptides, or decrease allergenicity (Zhu et al. 2020). Enzyme addition is one of the essential components of integration. Determining when to dosage enzymes (pre-pasteurization, post-homogenization, fermentation, etc.) is to maximize efficacy and preserve product safety. Process flow compatibility ensures that without requiring significant equipment changes, enzyme activity synchronizes with processing schedules, pH changes, and heat profiles. Using enzyme formulations that do not conflict with automated cleaning procedures is known as CIP (clean-in-place) system adaption (Lu et al. 2019).

14.17 Emerging Applications Beyond Traditional Uses

14.17.1 Lactose-Free Product Development Support

Due to worldwide market trends toward digestive wellness, dietary choices, and growing knowledge of lactose intolerance, the demand for lactose-free dairy products has increased dramatically. Development of lactose-free products without sacrificing flavor, texture, or nutritional value is made possible by enzyme technology, especially β-galactosidase (lactase) applications (Sznajder-Katarzyńska et al.

2019). β-Galactosidases from *Aspergillus oryzae*, *Kluyveromyces lactis*, and other GRAS (generally recognized as safe) microorganisms provide high activity at a variety of pH and temperature levels. *Integration into processing*: Depending on the kind of product (milk, yogurt, ice cream), lactase may be introduced prior to pasteurization, during fermentation, or in a postproduction phase. Lactose is broken down into glucose and galactose by controlled enzymatic hydrolysis, which naturally increases sweetness and gets rid of the digestive pain that comes with lactose. *Impact on nutrition and taste*: Appropriate dosage avoids too sweet flavors while preserving a preferred mouthfeel and viscosity, which is essential for high-end dairy products. *Optimization of process*: Adherence to legal limitations (less than 0.1 g lactose/100 g in many countries) is ensured by real-time assessments of residual lactose levels by inline monitoring equipment (Shepherd et al. 2012).

14.17.2 Applications of Dairy

Lactose-free fluid milk is clear and tastes good, and it has a long shelf life. Lactose-free fermented products and yogurt enhanced digestibility and fermentation efficiency. *Lactose-free cheese*: New techniques allow for the removal of lactose without changing the dynamics of ripening. Without the addition of sugar, lactose-free ice cream maintains its smooth texture and enhances its sweetness profile. *Waste stream valorization and protein recovery*: Significant by-products from dairy production, such as whey, buttermilk, and cheese fines, have generally been regarded as waste or low-value streams. By converting these by-products into high-value protein ingredients, bioactive chemicals, and functional food additives, waste stream valorization is made possible by contemporary enzyme-based technology, which helps the dairy industry achieve its circular economy objectives (Smithers 2015).

14.17.3 Important Process Elements

Whey protein recovery: When whey proteins (β-lactoglobulin and α-lactalbumin) are hydrolyzed by enzymes, peptides with antihypertensive, antibacterial, and antioxidant qualities are produced. Yield and purity are increased when enzymatic pretreatment is coupled with membrane filtration (ultrafiltration, nanofiltration).

Use of buttermilk and cheese fines: For use in dairy products, nutritious powders, or baked goods, targeted proteolysis releases soluble protein fractions.

Conversion of lactose: What was once a disposal problem can be improved by enzymatically hydrolyzing and fermenting the residual lactose in whey permeate into lactic acid, ethanol, or galacto-oligosaccharides (GOS).

Recovering fat: Milk fat globule membrane components with emulsifying and nutraceutical potential are recovered by lipase-assisted procedures (Huang et al. 2019).

14.18 Functional Dairy Ingredients for Food Fortification

For nutritional, physiological, and technological reasons, functional dairy ingredients such as caseinates, whey protein hydrolysates (WPH), milk protein isolates (MPI), milk protein concentrates (MPC), and lactose-derived prebiotics are being used more and more in fortified meals (Dass 2010). *Improved water-binding and emulsification qualities* have a balanced amino acid profile and a high protein content (>80%). *Better digestion and solubility* include bioactive peptides that have immune-modulatory, antihypertensive, and antioxidant properties. Offer slow-releasing amino acids to promote muscle protein synthesis and sustained fullness. Galacto-oligosaccharides (GOS) are lactose derivatives that are prebiotics that improve the absorption of minerals and intestinal bacteria. Enzymatic processing ensures minimal denaturation and maximal retention of bioactivity in a variety of applications, including sports nutrition, newborn formulae, clinical nutrition for the aged, and functional beverages (Korhonen and Pihlanto 2006).

14.19 Therapeutic and Nutraceutical Applications

The bioactive characteristics of dairy peptides and hydrolysates formed from proteolysis provide great promise for the creation of medicinal and nutraceutical products. Numerous peptides have antihypertensive, antioxidant, antibacterial, immunomodulatory, and mineral-binding properties, which may help prevent or treat chronic illnesses (Pelletier 2005).

Antihypertensive effects: Tripeptides produced from casein, such as Val-Pro-Pro (VPP) and Ile-Pro-Pro (IPP), help control blood pressure by inhibiting the angiotensin-converting enzyme (ACE).

Antioxidant activity: Whey protein hydrolysates contain peptides that can reduce oxidative stress and scavenge-free radicals, which is good for cardiovascular health and aging.

Immunomodulation: Peptides produced from β-lactoglobulin and α-lactalbumin help regulate inflammatory responses and boost immunological function; these properties are helpful in therapeutic nutrition.

Improvement of mineral absorption: By binding and solubilizing calcium and other minerals, casein phosphopeptides (CPPs) improve their bioavailability. For vitamins that promote bone health, this is essential.

Metabolic health: Dairy-derived peptides may improve insulin sensitivity and lipid metabolism, supporting metabolic syndrome treatment (Hartmann and Meisel 2007). Dairy proteins can now be used not just in food applications but also in the pharmaceutical and nutraceutical industries by adding these bifunctional molecules to capsules, powders, fortified drinks, and medical nutrition formulations. For medical substances to be approved by regulators, regulated enzymatic hydrolysis and purification are necessary to provide reproducible bioactivity and safety profiles (Phelan et al. 2009).

14.20 Regulatory Considerations and Market Trends

Various regulatory frameworks that control the evaluation, manufacturing, and use of enzyme safety have an impact on the worldwide dairy enzyme market. Proteases for targeted proteolysis and other enzymes used in dairy processing are subject to different food safety and labeling regulations in different countries (Silano et al. 2016). Novel proteases must undergo a thorough safety assessment before being used in dairy processing in order to protect consumers and adhere to global standards. A systematic, scientific approach is usually followed by these protocols, which include source characterization, production process control, toxicological evaluation, and allergenicity assessment (Tang et al. 2023). The market for dairy enzymes is expected to expand at a compound annual growth rate (CAGR) of approximately 7.1%, from USD 1.01 billion in 2025 to USD 2.01 billion by 2035 (Jenkins 2024).

14.21 Conclusion

The dairy industry's global proteases market is changing as a result of developments in biotechnology, digitization, and sustainability. While microbial enzymes enable clean-label and environmentally friendly production, precision enzyme engineering, AI, and the Internet of things are improving process efficiency and product quality. In addition to using digital monitoring and concentrating on high-value functional products, dairy processors should diversify their enzyme supply. Enzyme recycling and the utilization of renewable energy are examples of sustainable methods that improve both commercial and environmental performance. Finding new proteases, optimizing using AI and omics, and confirming the advantages of bioactive peptides are among the top research goals. Collaboration between academic, industry, and regulators will drive enzyme-based dairy technology innovation, standardization, and wider use.

References

Abada EA (2019) Application of microbial enzymes in the dairy industry. In: Enzymes in food biotechnology. Academic, pp 61–72. https://doi.org/10.1016/B978-0-12-813280-7.00005

Alhelli AM, Mohammed NK, Khalil ES, Hussin ASM (2021) Optimizing the acceleration of cheddar cheese ripening using response surface methodology by microbial protease without altering its quality features. AMB Express 11(1):45. https://doi.org/10.1186/s13568-021-01205-9

Aljammas HA, Yazji S, Azizieh A (2022) Optimization of protease production from *Rhizomucor miehei* Rm4 isolate under solid-state fermentation. J Genet Eng Biotechnol 20(1):82. https://doi.org/10.1186/s43141-022-00358-9

Annapure US, Gaur SS (2022) Commercial enzymes in dairy processing. In: Value-addition in food products and processing through enzyme technology. Academic Press, pp 205–219. https://doi.org/10.1016/B978-0-323-89929-1.00015-9

Asioli D, Aschemann-Witzel J, Caputo V, Vecchio R, Annunziata A, Næs T, Varela P (2017) Making sense of the "clean label" trends: a review of consumer food choice behavior and discussion of industry implications. Food Res Int 99:58–71. https://doi.org/10.1016/j.foodres.2017.07.022

Barua RC, Coniglio RO, Molina MA, Díaz GV, Fonseca MI (2024) Fungi as biotechnological allies: exploring contributions of edible and medicinal mushrooms. J Food Sci 89(11):6888–6915. https://doi.org/10.1111/1750-3841.17390

Chauhan B, Singh DP, Sharma P (2025) Bioactive pigments in functional foods: insights into their diversity, extraction, and applications. Food Sci Biotechnol:1–21. https://doi.org/10.1007/s10068-025-01896-x

Choi J, In Lee S, Rackerby B, Frojen R, Goddik L, Ha SD, Park SH (2020) Assessment of overall microbial community shift during Cheddar cheese production from raw milk to aging. Appl Microbiol Biotechnol 104(14):6249–6260. https://doi.org/10.1007/s00253-020-10651-7

Contesini FJ, Melo RRD, Sato HH (2018) An overview of *bacillus* proteases: from production to application. Crit Rev Biotechnol 38(3):321–334. https://doi.org/10.1080/07388551.2017.1354354

D'Auria E, Salvatore S, Acunzo M, Peroni D, Pendezza E, Di Profio E, Pedezza E, Di Profio E, Fiore G, Zuccotti GV, Verduci E (2021) Hydrolysed formulas in the management of cow's milk allergy: new insights, pitfalls and tips. Nutrients 13(8):2762. https://doi.org/10.3390/nu13082762

Dai Y, Chen Y, Lin X, Zhang S (2024) Recent applications and prospects of enzymes in quality and safety control of fermented foods. Foods 13(23):3804. https://doi.org/10.3390/foods13233804

Dass SC (2010) Ecology and molecular typing of *L. monocytogenes* in a processing plant for cold-smoked salmon in the Republic of Ireland. Food Res Int. https://doi.org/10.1016/j.foodres.2010.04.030

Dekker P (2019) Dairy enzymes. In: Industrial enzyme applications, pp 143–166. https://doi.org/10.1002/9783527813780.ch2_3

Dias JA, Santos E, Asseiceira I, Jacob S, Koninckx CR (2022) The role of infant formulas in the primary prevention of allergies in non-breastfed infants at risk of developing allergies—recommendations from a multidisciplinary group of experts. Nutrients 14(19):4016. https://doi.org/10.3390/nu14194016

Fan F, Liu M, Shi P, Xu X, Lu W, Wang Z, Du M (2018) Protein cross-linking and the Maillard reaction decrease the solubility of milk protein concentrates. Food Sci Nutr 6(5):1196–1203. https://doi.org/10.1002/fsn3.657

Feijoo-Siota L, Blasco L, Luis Rodriguez-Rama J, Barros-Velázquez J, de Miguel T, Sánchez-Pérez A, G. Villa T (2014) Recent patents on microbial proteases for the dairy industry. Recent Adv DNA Gene Seq 8(1):44–55. https://doi.org/10.2174/2352092208666141013231720

Fernandes J, Gomes S, Reboredo FH, Pintado ME, Amaral O, Dias J, Alvarenga N (2025) Clean label approaches in cheese production: where are we? Foods 14(5):805. https://doi.org/10.3390/foods14050805

Gandhewar N, Shende P (2025) Gold nanoparticle embedded color changing film for theranostics application in wound management. Naunyn Schmiedebergs Arch Pharmacol:1–13. https://doi.org/10.1007/s00210-025-04374-7

Ge F, Chen G, Qian M, Xu C, Liu J, Cao J, Li X, Hu D, Xu Y, Xin Y, Wang D, Zhou J, Shi H, Tan Z (2023) Artificial intelligence aided lipase production and engineering for enzymatic performance improvement. J Agric Food Chem 71(41):14911–14930. https://doi.org/10.1021/acs.jafc.3c05029

Gong K, Chen L, Xia H, Dai H, Li X, Sun L et al (2019) Driving forces of disaggregation and reaggregation of peanut protein isolates in aqueous dispersion induced by high-pressure microfluidization. Int J Biol Macromol 130:915–921. https://doi.org/10.1016/j.ijbiomac.2019.02.123

Gupta R, Beg Q, Lorenz P (2002) Bacterial alkaline proteases: molecular approaches and industrial applications. Appl Microbiol Biotechnol 59(1):15–32. https://doi.org/10.1007/s00253-002-0975-y

Hao Y, Han J, Cao Q, Yu Y, Jiang C, Wu L et al (2025) Application of electrospun biomimetic packaging materials for prolonged preservation and quality retention of Mongolian cheese. Int Dairy J 167:106263. https://doi.org/10.1016/j.idairyj.2025.106263

Hartmann R, Meisel H (2007) Food-derived peptides with biological activity: from research to food applications. Curr Opin Biotechnol 18(2):163–169. https://doi.org/10.1016/j.copbio.2007.01.013

Hellmuth K (2006) Industrial scale production of chymosin with *Aspergillus niger*. Microb Cell Factories 5(Suppl 1):S31. https://doi.org/10.1186/1475-2859-5-S1-S31

Horstman AM, Huppertz T (2023) Milk proteins: processing, gastric coagulation, amino acid availability and muscle protein synthesis. Crit Rev Food Sci Nutr 63(30):10267–10282. https://doi.org/10.1080/10408398.2022.2078782

Huang C, Lin W, Lai C, Li X, Jin Y, Yong Q (2019) Coupling the post-extraction process to remove residual lignin and alter the recalcitrant structures for improving the enzymatic digestibility of acid-pretreated bamboo residues. Bioresour Technol 285:121355. https://doi.org/10.1016/j.biortech.2019.121355

Ishak SNH, Saad AHM, Latip W, Rahman RNZRA, Salleh AB, Kamarudin NHA et al (2025) Enhancing industrial biocatalyst performance and cost-efficiency through adsorption-based enzyme immobilization: a review. Int J Biol Macromol:144278. https://doi.org/10.1016/j.ijbiomac.2025.144278

Jenkins V 2024 EU agricultural outlook 2024–2035: projected trends and challenges. https://doi.org/10.2762/2329210

Karakashev SI, Manev ED (2015) Hydrodynamics of thin liquid films: retrospective and perspectives. Adv Colloid Interf Sci 222:398–412. https://doi.org/10.1016/j.cis.2014.07.010

Kashid S, Joshi K (2022) Biopolymer-based nanomaterials: synthesis, characterization, and applications. In: Innovations in green nanoscience and nanotechnology. CRC Press, pp. 15–59. https://doi.org/10.1201/9781003319153-3

Kaur S, Vasiljevic T, Huppertz T (2023) Influence of actinidin-induced hydrolysis on the functional properties of milk protein and whey protein concentrates. Foods 12(20):3806. https://doi.org/10.3390/foods12203806

Khalesi M, FitzGerald RJ (2022) Insolubility in milk protein concentrates: potential causes and strategies to minimize its occurrence. Crit Rev Food Sci Nutr 62(25):6973–6989. https://doi.org/10.1080/10408398.2021.1908955

Khan MU, Yu P, Wu Y, Chen Z, Kong L, Farid A, Cui J, Yang J (2025) Comprehensive review of enzymes (protease, lipase) in milk: impact on storage quality, detection methods, and control strategies. Compr Rev Food Sci Food Saf 24(3):e70164. https://doi.org/10.1111/1541-4337.70164

Korhonen H, Pihlanto A (2006) Bioactive peptides: production and functionality. Int Dairy J 16(9):945–960. https://doi.org/10.1016/j.idairyj.2005.10.012

Kuddus M, Ramteke PW (2012) Recent developments in production and biotechnological applications of cold-active microbial proteases. Crit Rev Microbiol 38(4):330–338. https://doi.org/10.3109/1040841X.2012.678477

Lai R, Lin M, Yan Y, Jiang S, Zhou Z, Wang J (2023) Comparative genomic analysis of a thermophilic protease-producing strain *geobacillus stearothermophilus* H6. Genes 14(2):466. https://doi.org/10.3390/genes14020466

Li B, Kelly AL, Mcsweeney PL (2023) Cheese manufacture. In: Tunick MC (ed) Handbook of cheese chemistry. Royal Society of Chemistry, pp 62–86. https://doi.org/10.1039/BK9781839169908-00062

Liang X, Dai N, Yang F, Zhu H, Zhang G, Wang Y (2024) Molecular identification and safety assessment of the potential probiotic strain *Bacillus paralicheniformis* HMPM220325 isolated from artisanal fruit dairy products. Food Funct 15(2):747–765. https://doi.org/10.1039/D3FO04625G

Liu Y, Zhang W, Zhang L, Hettinga K, Zhou P (2020) Characterizing the changes of bovine milk serum proteins after simulated industrial processing. LWT 133:110101. https://doi.org/10.1016/j.lwt.2020.110101

Liu X, Lian M, Zhao M, Huang M (2024) Advances in recombinant protease production: current state and perspectives. World J Microbiol Biotechnol 40(5):144. https://dx.doi.org/. https://doi.org/10.1007/s11274-024-03957-5

Lozano-Ojalvo D, López-Fandiño R (2018) Immunomodulating peptides for food allergy prevention and treatment. Crit Rev Food Sci Nutr 58(10):1629–1649. https://doi.org/10.1080/10408398.2016.1275519

Lu X, Chen J, Guo Z, Zheng Y, Rea MC, Su H et al (2019) Using polysaccharides for the enhancement of functionality of foods: a review. Trends Food Sci Technol 86:311–327. https://doi.org/10.1016/j.tifs.2019.02.024

Madende M, Kemp G, Stoychev S, Osthoff G (2018) Characterisation of African elephant beta casein and its relevance to the chemistry of caseins and casein micelles. Int Dairy J 85:112–120. https://doi.org/10.1016/j.idairyj.2018.05.013

McSweeney PL (2004) Biochemistry of cheese ripening. Int J Dairy Technol 57(2–3):127–144. https://doi.org/10.1111/j.1471-0307.2004.00147.x

Mirzapour-Kouhdasht A, Garcia-Vaquero M (2022) Cardioprotective peptides from milk processing and dairy products: from bioactivity to final products including commercialization and legislation. Foods 11(9):1270. https://doi.org/10.3390/foods11091270

Mokrani S, Nabti EH, Mokrani S, Nabti EH (2024) Recent status in production, biotechnological applications, commercial aspects, and future prospects of microbial enzymes: a comprehensive review. Int J Agric Sci Food Technol 10:006–020. https://doi.org/10.17352/2455-815X.000202

Mujeeb A, Miller AF (2023) Peptide bionanomaterials global market: the future of emerging industry. In: Peptide bionanomaterials: from design to application. Springer International Publishing, pp 539–555. https://doi.org/10.1007/978-3-031-29360-3_16

Naveed M, Nadeem F, Mehmood T, Bilal M, Anwar Z, Amjad F (2021) Protease—a versatile and ecofriendly biocatalyst with multi-industrial applications: an updated review. Catal Lett 151(2):307–323. https://doi.org/10.1007/s10562-020-03316-7

Nayik GA, Gull A, Masoodi L, Navaf M, Sunooj KV, Ucak İ et al (2024) Milk proteins: chemistry, functionality and diverse industrial applications. Cogent Food Agric 10(1):2377686. https://doi.org/10.1080/23311932.2024.2377686

Ndochinwa OG, Wang QY, Amadi OC, Nwagu TN, Nnamchi CI, Okeke ES, Moneke AN (2024) Current status and emerging frontiers in enzyme engineering: an industrial perspective. Heliyon 10(11). https://doi.org/10.1016/j.heliyon.2024.e32673

Ong ILH, Yang KL (2017) Recent developments in protease activity assays and sensors. Analyst 142(11):1867–1881. https://doi.org/10.1039/C6AN02647H

Ozturkoglu-Budak S, Wiebenga A, Bron PA, de Vries RP (2016) Protease and lipase activities of fungal and bacterial strains derived from an artisanal raw ewe's milk cheese. Int J Food Microbiol 237:17–27. https://doi.org/10.1016/j.ijfoodmicro.2016.08.007

Park W, Park J, Im S, Choi SJ (2023) Influence of the type and concentration of hydrocolloids on Ostwald ripening of emulsions stabilized with small molecular and non-ionic surfactants. Food Chem 411:135504. https://doi.org/10.1016/j.foodchem.2023.135504

Parmar P, Lopez-Villalobos N, Tobin JT, Murphy E, Buckley F, Crowley SV, Kelly AL, Shalloo L (2021) Effect of temperature on raw whole milk density and its potential impact on milk payment in the dairy industry. Int J Food Sci Technol 56(5):2415–2422. https://doi.org/10.1111/ijfs.14869

Pelletier DL (2005) Science, law, and politics in FDA's genetically engineered foods policy: scientific concerns and uncertainties. Nutr Rev 63(6):210–223. https://doi.org/10.1111/j.1753-4887.2005.tb00139.x

Phelan M, Aherne A, FitzGerald RJ, O'Brien NM (2009) Casein-derived bioactive peptides: biological effects, industrial uses, safety aspects and regulatory status. Int Dairy J 19(11):643–654. https://doi.org/10.1016/j.idairyj.2009.06.001

Rao S, Neethirajan S (2025) Computational architectures for precision dairy nutrition digital twins: a technical review and implementation framework. Sensors 25(16):4899. https://doi.org/10.3390/s25164899

Rathod J, Sherje A (2025) FRET-based ratiometric fluorescent probes for enzyme detection: current insight. Curr Anal Chem 22(2):236–255. https://doi.org/10.2174/0115734110348463241121191421

Rios-Mera JD, da Silva Pinto JS, Contreras-Castillo CJ (2017) Effect of ultimate pH and ageing on thermal denaturation of bovine muscle proteins. Meat Sci 131:25–27. https://doi.org/10.1016/j.meatsci.2017.04.017

Saito M, Nakaya M, Kondo T, Nakazawa M, Ueda M, Naganawa S et al (2022) Gelation of konjac glucomannan by acetylmannan esterases from *Aspergillus oryzae*. Enzym Microb Technol 160:110075. https://doi.org/10.1016/j.enzmictec.2022.110075

Sekkidou M, Muhardi L, Constantinou C, Kudla U, Vandenplas Y, Nicolaou N (2021) Nutritional management with a casein-based extensively hydrolysed formula in infants with clinical manifestations of non-IgE-mediated CMPA enteropathies and constipation. Front Allergy 2:676075. https://doi.org/10.3389/falgy.2021.676075

Shea Z, Ogando do Granja M, Fletcher EB, Zheng Y, Bewick P, Wang Z et al (2024) A review of bioactive compound effects from primary legume protein sources in human and animal health. Curr Issues Mol Biol 46(5):4203–4233. https://doi.org/10.3390/cimb46050257

Sheldon RA, van Pelt S (2013) Enzyme immobilisation in biocatalysis: why, what and how. Chem Soc Rev 42(15):6223–6235. https://doi.org/10.1039/C3CS60075K

Shepherd R, Dean M, Lampila P, Arvola A, Saba A, Vassallo M, Claupein E, Winkelmann M, Lähteenmäki L (2012) Communicating the benefits of wholegrain and functional grain products to European consumers. Trends Food Sci Technol 25(2):63–69. https://doi.org/10.1016/j.tifs.2012.01.002

Silano V, Bolognesi C, Castle L, Cravedi JP, Fowler P et al (2016) EFSA panel on food contact materials, enzymes, flavourings and processing aids (CEF), exposure assessment of food enzymes. EFSA J 14(11):e04581. https://doi.org/10.2903/j.efsa.2016.4581

Smithers GW (2015) Whey-ing up the options–yesterday, today and tomorrow. Int Dairy J 48:2–14. https://doi.org/10.1016/j.idairyj.2015.01.011

Soliman SS, El-Labbad EM, Abu-Qiyas A, Fayed B, Hamoda AM, Al-Rawi AM, Dakalbab S, El-Shorbagi AN, Hamad M, Ibrahim AS, Mohammad MG (2022) Novel secreted peptides from *Rhizopus arrhizus* var. delemar with immunomodulatory effects that enhance fungal pathogenesis. Front Microbiol 13:863133. https://doi.org/10.3389/fmicb.2022.863133

Song P, Zhang X, Wang S, Xu W, Wang F, Fu R, Wei F (2023) Microbial proteases and their applications. Front Microbiol 14:1236368. https://doi.org/10.3389/fmicb.2023.1236368

Sznajder-Katarzyńska K, Surma M, Wiczkowski W, Cieślik E (2019) The perfluoroalkyl substance (PFAS) contamination level in milk and milk products in Poland. Int Dairy J 96:73–84. https://doi.org/10.1016/j.idairyj.2019.04.008

Tang L, Wu A, Li S, Tuerdimaimaiti M, Zhang G (2023) Impacts of climate change on rice grain: a literature review on what is happening, and how should we proceed? Foods 12(3):536. https://doi.org/10.3390/foods12030536

Vandenplas Y, Meyer R, Hoffman I, Alliet P, De Greef E, Hauser B, Huysentruyt K (2025) Discrepancies in the diagnosis, treatment and prevention of cow Milk allergy among Paediatricians. Acta Paediatr 114(11):3000–3012. https://doi.org/10.1111/apa.70209

Wang M, Wan Y, Liu T, Zeng X, Liang X, Wu X, Fu G (2022) Effect of refining degree on the quality changes and lipid oxidation of camellia (*Camellia oleifera*) oil during heating. Foods 11(15):2232. https://doi.org/10.3390/foods11152232

Ward OP (2011) Proteases. In: Moo-Young M (ed) Comprehensive biotechnology, 3rd edn. Pergamon, pp 604–615. https://doi.org/10.1016/B978-0-444-64046-8.00187-7

West S (2007) Production of flavours, flavour enhancers and other protein-based speciality products. In: Novel enzyme technology for food applications. Woodhead Publishing, pp 183–204. https://doi.org/10.1533/9781845693718.2.183

Yang Z, Huang Z, Wu Q, Tang X, Huang Z (2023) Cold-adapted proteases: an efficient and energy-saving biocatalyst. Int J Mol Sci 24(10):8532. https://doi.org/10.3390/ijms24108532

Yao H, Liu S, Liu T, Ren D, Zhou Z, Yang Q, Mao J (2023) Microbial-derived salt-tolerant proteases and their applications in high-salt traditional soybean fermented foods: a review. Bioresour Bioprocess 10(1):82. https://doi.org/10.1186/s40643-023-00704-w

Zambare V, Nilegaonkar S, Kanekar P (2011) A novel extracellular protease from Pseudomonas aeruginosa MCM B-327: enzyme production and its partial characterization. New Biotechnol 28(2):173–181. https://doi.org/10.1016/j.nbt.2010.10.002

Zhu Y, Bhandari B, Prakash S (2020) Relating the tribo-rheological properties of chocolate flavoured milk to temporal aspects of texture. Int Dairy J 110:104794. https://doi.org/10.1016/j.idairyj.2020.104794

Chapter 15
Role of Microbial Proteases in Meat and Poultry Industries

U. Subhalakshmi, Pallavi Singh, K. Kesavardhini, Abdel Razzaq Mohammad Al-Tawaha, M. Manigandan, C. Mabel Joshaline, and G. Vithiya

Abstract Microbial proteases constitute a critically important class of industrial enzymes, accounting for a dominant share of the global enzyme market due to their versatile biocatalytic properties. In the meat and poultry industries, these enzymes, derived predominantly from bacterial and fungal sources, have transitioned from simple processing aids to indispensable tools for enhancing quality, efficiency, and sustainability. The core discussion details their multifaceted applications: in post-mortem tenderization to import meat texture through controlled hydrolysis of myofibrillar and connective tissue proteins; in flavor enhancement via the generation of umami peptides and savory compounds; and in the production of bioactive protein hydrolysates for functional foods and nutraceuticals. A significant focus is placed on by-product valorization, where proteases are employed to recover proteins and nutrients from low value streams such as bones, trimmings, blood, and feathers, converting waste into high-value ingredients such as protein powders, collagen peptides, and animal feed, thereby advancing circular economy principles within the industry. This further examines the enzymes' role in accelerating fermentation and ripening of dry-cured products and in detergent formulations for cleaning-in-place systems. Critically, the article highlights the transformative potential of emerging technologies, specifically enzyme engineering and immobilization. Engineering techniques such as rational design and directed evolution are creating proteases with tailored stability, specificity, and activity, while immobilization on

U. Subhalakshmi (✉) · K. Kesavardhini · M. Manigandan · G. Vithiya
PG and Research Department of Microbiology, Sacred Heart College (Autonomous), Tirupattur, Tamil Nadu, India

P. Singh
Department of Biotechnology, Graphic Era Deemed to be University, Dehradun, Uttarakhand, India

A. R. M. Al-Tawaha
National Agricultural Research Center (NARC), Baq'a, Jordan

C. M. Joshaline
Department of Rural Development Science, Arul Anandar College, Madurai, Tamil Nadu, India

V. Zambare et al. (eds.), *Biotechnological Advances and Applications in Microbial Proteases*, Interdisciplinary Biotechnological Advances,
https://doi.org/10.1007/978-981-95-9011-7_15

various supports enables enzyme reuse, enhances stability, and facilitates continuous processing in bioreactors. Looking forward, the future perspective underscores the convergence of synthetic biology, digital process control, and sustainability metrics. The integration of engineered proteases with nonthermal technologies (e.g., high-pressure processing, pulsed electric fields), their application in developing plant-based and hybrid meat analogs, and their role in enabling radical sodium reduction strategies are identified as key frontiers. Ultimately, the strategic deployment of microbial proteases is poised to drive the meat and poultry industries toward more precise, sustainable, and innovative production paradigms, addressing core challenges of resource efficiency, environmental impact, and consumer demand for high-quality, clean-label products.

Keywords Microbial proteases · Meat industry · Poultry processing · Enzyme applications · Tenderization · By-product valorization · Protein hydrolysates · Enzyme Engineering · Immobilization · Sustainability · Food biotechnology

15.1 Introduction

Enzymes are macromolecules made up of amide bonds connecting amino acid subunits. They are extremely selective biocatalysts that, without changing their structural makeup, lower the activation energy of biological reactions to increase their pace and specificity. These macromolecules' selectivity for substrate is determined by the hydrophobic pocket that contains their active site. Living things secrete enzymes, which are necessary for life. In many biotechnological applications, enzymes are essential (Rawlings et al. 2017). Hydrolases, an enzyme family that catalyzes the process of hydrolysis of different natural compounds, are currently the most widely utilized (more than 75%) organisms for commercial uses. Versatility in biotechnology, proteases are acknowledged as the leading enzyme. The largest and most intricate class of enzymes, proteases, break down proteins by cleaving the peptide bonds that hold amino acid residues together in a chain of polyamino acids. They make up about 65% of the industrial enzyme market, making them one of the most significant categories of enzymes. They come from a diverse range of sources that include plants, animals, and microorganisms, and are found throughout nature. Since the very beginning of enzymology, microbial proteases have been one of the most significant and thoroughly researched hydrolytic enzymes (Masi et al. 2021). They make up over 40% of the enzymes produced globally. Numerous microorganisms, such as bacteria, fungi, and yeasts, create them. Because of their quick growth, wide biochemical variety, simplicity of genetic modification, and small cultivation space requirements, microbes are an excellent source of proteases. Additionally, an established fermentation procedure for the mass manufacturing of the enzymes can cultivate the microorganisms in enormous quantities in a comparatively short length of time (Shaikh et al. 2023).

The generating organisms secrete microbial proteases directly into the microbial growth medium, which shortens the enzyme's downstream processing. They may be stored for a long time without experiencing a noticeable loss of activity, and they have an extended lifespan. However, because of their broad substrate specificity, high catalytic activity, and stability at ideal pH and temperature, bacterial proteases are particularly interesting among all microbial sources. All living things contain enzymes: biological catalysts made of proteins that aid in accelerating chemical processes and metabolism. Because of their ability to carry out reactions with great specificity and efficiency, enzymes have become popular instruments in chemistry and are being utilized more frequently in industrial processes. Only around 5% of the more than 3000 catalysts that have been discovered are used in industry (Rathod and Pathak 2014).

Many businesses' energy requirements have significantly decreased the industrial use of enzymes, and the waste products produced by this process are environmentally friendly, biodegradable, and nontoxic. Additionally, the use of enzymes for industrial purposes is more economical, and it is possible to integrate genetically modified bacteria to create better and more stable enzymes on an industrial scale (de Matos et al. 2022). Because using enzymes in industries is more environmentally friendly than using chemicals, which produce a lot of greenhouse gases, there is a growing need for industrial enzymes as a substitute for both conventional and synthetic compounds in many industrial processes. Since the early days of enzymology, microbial proteases have been one of the most significant and thoroughly researched hydrolytic enzymes (Mortezaei et al. 2021). They make up over 40% of the enzymes produced globally. Numerous microorganisms, such as bacteria, fungi, and yeasts, create them. Because of their quick growth, wide biochemical variety, simplicity of genetic modification, and small cultivation space requirements, microbes are an excellent source of proteases. Additionally, an established fermentation procedure for the mass manufacturing of the enzymes can cultivate the microorganisms in enormous quantities in a comparatively short length of time (Pereira et al. 2020).

The generating organisms secrete microbial proteases immediately into the fermentation medium, which shortens the enzyme's downstream processing. They may be stored for a long time without experiencing a noticeable loss of activity, and they have an extended lifespan. However, because of their broad substrate specificity, high catalytic activity, and stability at ideal pH and temperature, bacterial proteases are particularly interesting among all microbial sources. Additionally, microbial proteases are used in a variety of biotechnological applications, such as waste management and silver recovery, the detergent, engineering, healthcare, textile, food and feed, and leather sectors (Nnolim et al. 2020).

15.2 Classification of Microbial Proteases

Proteases produced by microbial species can be classified into several categories based on their substrate specificity, similarity to known enzymes, and catalytic mechanisms. Based on substrate specificity, proteases include enzymes such as collagenases, elastases, and keratinases, which hydrolyze specific protein substrates. Based on similarity to well-characterized proteolytic enzymes, they include enzymes resembling trypsin and pepsin (Prabhu et al. 2021). Furthermore, based on their catalytic mechanisms, proteases are classified into aspartic proteases, cysteine proteases, metalloproteases, and serine proteases, depending on the amino acid residues present at their active sites. Additionally, based on the site of action on polypeptide chains, proteases are categorized as endopeptidases, which cleave internal peptide bonds, and exopeptidases, which act near or at the terminal ends of polypeptide chains (Anis Ahamed et al. 2022).

Different ending and exopeptidase subtypes have been identified by the Enzyme Committee (EC). According to the International Organization of Biochemistry and Molecular Biology's Nomenclature Committee, proteases belong to subgroup 4 of category 3 (hydrolase). Growing understanding of the catalyst's mechanism and structure has led to a comprehensive classification system. Proteases are divided into various clans and families based on their amino acid sequences and evolutionary ties (Zins et al. 2001). A clan, or collection of families, has an evolutionary link but no discernible sequence similarities. Families from various catalytic classes can also be included in clans because their enzymatic-site residues have comparable tertiary folds and follow the same order. The presence of homologous proteolytic enzymes within a family is indicated by a high degree of amino acid sequence similarity. They can be recognized by the kind of enzyme that belongs to the family or by a protein that is homologous to the type of enzyme. The MEROPS database offers thorough information about several proteases based on this classification. All proteases in clans and families can be categorized into the following groups: asparagine proteases, aspartic proteases, which include cysteine protein proteases, glutamic proteases, which include mixed proteases, the metalloproteases, threonine proteases, also known as serine proteases, and unknown proteases based on these ancestral relationships and mechanisms of action (Wang et al. 2013).

15.3 Detection of Microbial Proteases

15.3.1 Endopeptidase Detection

15.3.1.1 Observation of Halos

The development of transparent halos surrounding populations that have grown on protein-containing agar plates is a sign of protease synthesis. This happens when microorganisms in solid media create extracellular endopeptidases. After adding growth media, the protein substrates were transferred to Petri dishes. Skim milk agar, casein agar, bovine serum albumin (BSA) agar, collagen agar, keratin agar, fibrin agar, and elastin agar are among the frequently utilized substrates. The final product of fermentation broth (for extracellular proteases) or cell lysate (for intracellular proteases) containing protease was collected when the protease was generated in liquid media (Yang et al. 2016). In order to see the halo, either an Oxfordshire cup has been set on the plate for the enzyme liquid container, or a well was constructed in the same agar plates (containing protein substrates) as stated for solid media. The most straightforward and intuitive technique for identifying proteases is the observation halo, but it is only appropriate for endopeptidases in particular and proteases that are active enough to produce distinct halos (Cejudo-Bastante et al. 2022). Measuring the hydrolysate or the decrease in substrate brought on by protease hydrolysis is a typical method of determining the activity of proteases. Different substrates and techniques are required to identify the various types of protean enzymes that exhibit distinct activity, employ distinct hydrolysis mechanisms, and produce hydrolysis products with distinct properties. To date, the substrates used to detect proteases are roughly divided into native substrates and modified substrates (De Azeredo et al. 2001).

15.3.1.2 Detection by Natural Protein Substrates

Plant, animal, microbial, and other naturally occurring proteins are examples of natural protein substrates. Casein is the substrate that is most frequently used to measure protease activity. Under specific pH and temperature circumstances, protease hydrolyzes casein to create peptides or protein amino acids that are accessible in an alkaline solution. The newly created product dissolved in the greater acid solution following acid deposition, whereas the unhydrolyzed protein precipitated. Centrifugation is used to collect the supernatant, and Folin reagent, ninhydrin, 2,4,6-trinitrobenzenesulfonic acid (TNBS), or *ortho*-phthaldialdehyde (OPA), all of which have pros and cons, are used to examine the resultant peptides or amino acids to determine the protease's activity (Marson et al. 2022).

15.3.1.3 Detection by Modified Protein Substrates

15.3.1.3.1 Detection by Chromogenic Substrates

In chromogenic substrates, some approaches to detect protease activity involve a modified protein substrate, which should produce a colored end-product after proteolysis or a product that may be transformed into a colored complex, in order to increase substrate solubility and detection sensitivity. Azocasein, a casein dyed with aminobenzenesulfonic acid, is one example. After being broken down by proteases, it forms a colorful complex that is soluble in trichloroacetic acid, a common solvent, and exhibits absorbance at 440 nm (Ding et al. 2022). Casein, succinyl casein, has chemically modified amino groups and dissolves readily at pH levels higher than 4. The type of protease examined, such as Tosyl-Gly-Pro-Arg-pNA for trypsin, Suc-Ala-Ala-Pro-Phe-pNA for chymotrypsin, and Suc-Ala-Ala-Pro-Val-pNA for elastase, can be used to identify the synthetic substrate based on substrate specificity. However, N-Cbz-Ala-Ala-Leu-pNA and N-Cbz-Gly-Gly-Leu-pNA are good substrates for subtilisin. The principle underlying the assay is that proteases hydrolyze the amide bond connecting *p*-nitroaniline (pNA) to the neighboring amino acid residue, and released pNA exhibits specific absorption at a 405 nm wavelength (enzyme activity is proportional to fluorescence intensity) (Taylor et al. 2022).

15.3.1.3.2 Detection by Fluorogenic Substrates

The sensitive fluorescent peptide substrates are available, the amount or activity of protease enzymes is low, and more sensitive techniques are required, allowing the limit of detection to be reached at the nanogram (ng) level. Protease substrate modification via fluorescent labeling falls into one of the following categories: single fluorescence-based labeling, in which a single type of fluorescent dye binds to the substrate protein to provide fluorescent labeling; double fluorescence labeling, in which the peptides are labeled using two distinct fluorescent dyes (Feng et al. 2022). The labeled peptide, which is activated by protease hydrolysis, does not exhibit fluorescence because one dye is an energy provider and the other is an energy acceptor. The third method is homotransfer fluorescence labeling, which uses a specific type of fluorescence labeling substrate protein and fluorescence resonance energy transfer (FRET) between the fluorescent molecules that are labeled. The fluorescent molecules do not exhibit fluorescence, but they undergo hydrolysis by proteases to activate fluorescence (Sumantha et al. 2006).

15.3.1.3.3 Single Fluorescence-Based Labeling

Fluorophores linked to side chain amino acids, such as the N-terminus, C-terminus, Glu, Lys, or Cys of a peptide, are introduced during single fluorescent dye labeling. Different kinds of fluorescence dyes have been created. Carboxyfluorescein, fluorescein isothiocyanate, dansyl chloride, 2,4-dinitrophenylhydrazine, 7-amino-4-methylcoumarin (AMC), 7-amino-4-trifluoromethyl coumarin, carboxyrhodamine 110, Texas Red, pentamethine cyanine (Cy5), and heptamethine cyanine (Cy7) are the more commonly used colors. An increased emission produced an enzyme cleaves a peptide and releases it from the fluorophore is used to detect protease activity. For proteases, single fluorescence-based labeling has detection limits as low as the ng level. To improve the detection signal, the product and substrate must be split, and the pH must be changed when utilizing a single fluorescently labeled protease substrate. The detection procedures are still somewhat intricate (Yoo and Han 2021).

15.3.1.3.4 Double Labeling Using Fluorescent Dye

The FRET idea, double fluorescence labeling offers a more accurate and convenient technique than single fluorescence-based labeling, such as the widely used

FTC-casein assay. Assay methods that do not require separation stages can be used to determine the kinetics of exo- and endopeptidases across a broad pH range. By the addition of an enzyme, the total substrate turnover can be evaluated at a predetermined period. FRET-based measurement can detect decreased fluorescence quenching (i.e., higher total fluorescence), which happens when peptides (labeled proteins) are broken down into smaller fluorescein-labeled fragments. Two fluorophores, an energy donor and an energy acceptor, transfer electron energy in traditional FRET (Oliveira et al. 2001).

15.3.1.3.5 Labeling with Homotransfer Fluorescence

As previously stated, classical FRET involves the transfer of energy between two distinct fluorophores. However, fluorescence homotransfer can also result in FRET events, in which the same fluorophore (e.g., fluorescein) acts as both the energy donor and the energy acceptor. This phenomenon is known as homotransfer fluorescence labeling. These tests are 100 times more sensitive than the FTC-casein assay and are also simpler to conduct (Austin et al. 2022). The boron-dipyrromethene (BODIPY) dye is a common single-fluorescence dye used in FRET; casein conjugates of BODIPY dyes are created by attaching BODIPY dye molecules to casein. For effective protein quenching, the dyes in these conjugates are labeled. Nonfluorescent substrate molecules are produced by this method. During proteolysis, these fluorogenic substrates produce highly fluorescent peptides labeled with BODIPY dye, increasing fluorescence in relation to enzyme activity. This activity can be quantified with fluorescence microplate readers, filter fluorometers, or ordinary fluorometers. BODIPY casein hydrolysis can be measured using the excitation and emission wavelengths of fluorescein. A highly labeled casein derivative is included in ThermoFisher Scientific's EnzChek™ Protease Assay kits. For this purpose, green-fluorescent BODIPY FL dye and red-fluorescent BODIPY TR-X dye are frequently utilized (Legare et al. 2022).

15.3.2 Identification of Exopeptidase

Since exopeptidases hydrolyze proteins and release free amino acids, while endopeptidases hydrolyze proteins and mostly release peptides, the methods used to detect endopeptidases and exopeptidases must be different, and the method used to detect endopeptidases is not very sensitive to exopeptidases. Synthetic peptide is typically utilized as the substrate in assays that evaluate exopeptidase activity; in the case of the aminopeptidase test, a peptide with two or three amino acid residues is synthesized in order to detect aminopeptidases (Elston et al. 2007). Peptides can be linked to their carboxyl terminus with *p*-nitroaniline or 7-methoxycoumarin-4-acetic acid for more sensitive detection; upon hydrolysis, a free pNA or MCA molecule is liberated into the reaction solution. This technique can identify the activity of aminopeptidases by detecting a particular absorbance value. The most popular substrate for the carboxypeptidase assay is a peptide whose amino terminus is inhibited by benzoyl or benzyloxycarbonyl. Amino acids can only be released from the carboxyl terminal by carboxypeptidase. Following hydrolysis, the synthetic peptide releases free amino acids, which can be identified using the OPA reagent or ninhydrin. Because of their high sensitivity, fluorescent substrates can also be used to identify exopeptidases, such as aminopeptidases and carboxypeptidases. There are also several protease detection techniques, such as ELISAs and assays using ultrasonic resolver technology. For instance, ELISA-based assays require prior knowledge of the enzyme's structure. A distinct analytical technique is required for ultrasonic resolver technology, and it must first be correlated with the relevant ultrasonic velocity signals (Thompson et al. 2000).

15.4 Protease in the Industrial Domain

In industries such as paper mills, leather processing, food processing, liquid detergent manufacture, and medicine, microbial enzymes, particularly proteases, play several significant roles. Alkaline and heat-stable serine proteases are thought to be the most significant of the many kinds. The following highlights the role that proteases play in several industrial areas (Mathew et al. 2000).

15.4.1 Industries that Process Leather

Protease is used extensively in the leather industry for the processing of raw leather, which includes soaking, liming, dehairing, deliming, bating, degreasing, and pickling. Microorganisms produce alkaline proteases, which are highly helpful for hair removal and far safer than the conventional method that uses dangerous chemicals such as sodium sulfide, which pollutes the environment. Alkaline proteases hydrolyze the non-collagenous components of skin and eliminate non-fibrillar proteins, such as albumin and globulin, which are essential for the production of leather. Alkaline proteases produced by microorganisms are a better option for tanning raw leather in high alkaline conditions (pH 12.0), according to the use of some industrially significant protease-producing microbes, such as *Streptomyces* sp., *A. flavus*, *Bacillus amyloliquefaciens*, and *Bacillus subtilis*. However, in order to achieve superior outcomes, genetically modified strains are utilized at random and have a higher importance in enterprises. For instance, site-directed mutagenesis in *Bacillus amyloliquefaciens*, the inclusion of the pBX96 plasmid in *Bacillus pumilus* c172-14, and the alteration of the enzyme-producing gene in *B. licheniformis*, all improve performance in the leather industry (Ayhan et al. 2001).

15.4.2 Management of Waste

The broad substrate specificity of keratinases derived from microbial sources makes them highly valuable in waste management. These enzymes degrade keratin-rich wastes such as feathers, scales, hair, and wool into simpler, utilizable products, including amino acids and peptides, which can be further processed into protein-rich feed components for the aquaculture and poultry industries. Microbially derived keratinases have diverse industrial applications due to their efficiency and versatility. Despite the fact that keratin is an insoluble protein, it can be made soluble by submerged fermentation with keratinases produced by bacteria. Owing to their unique properties, keratinases have been employed extensively in the synthesis of fertilizer from poultry waste, biofuel, and nematicidal agents. *B. subtilis* produces an alkaline keratinase that may be helpful in the poultry industry's feather processing. Chicken farms can use RS1 to prepare feed. Other keratin-degrading proteases have also been described for a variety of uses, including the synthesis of peptides, waste recycling, food processing, and wastewater treatment (Gu and Walling 2002).

Microbial proteases have been sporadically employed in the food industry in recent years for tasks such as cheesemaking and meat tenderization. The first step in making cheese is milk protein coagulation, which is carried out with the aid of the protease enzyme. The microbiological source of proteases used in the cheesemaking industry has two disadvantages: low yield and bitterness after prolonged storage (Nurhidayat et al. 2018). Aspartic proteases are thought to be the most suitable for the cheese industry out of all the different groups of proteases. A variety of nonspecific proteases produced by microbes give stored food a bitter taste. Alcalase is an alkaline protease that makes food sweeter by eliminating bitterness. On the other hand, *B. amyloliquefaciens* protease can be used to hydrolyze chickpea protein into a methionine-rich protein. Microbially derived proteases are used to prepare protein-rich therapeutic diets and to fortify fruit juices and soft beverages. The main ingredient in baking is wheat flour, which contains gluten, an insoluble protein that controls the characteristics of bakery dough. *Aspergillus oryzae* produces protease, which hydrolyzes gluten and improves the product's color, texture, and softness (Geeganage and Frey 2002).

15.4.3 Recovery of Silver

A valuable metal that is utilized in photography, MRIs, CT scans, and X-rays, silver is a significant component of film. Silver concentration in the gelatin layer of X-ray films ranges from 1.5 to 2%. Silver recovery from discarded X-ray film is a challenging and time-consuming process. There are a number of environmentally unfriendly chemical techniques for recovering silver, including oxidizing metallic silver and burning film. For this specific aim, alkaline proteases are employed, which actively break down the film's gelatin layer and liberate the bound silver.

Additionally, they reported that this protease hydrolyzed silver at pH 10.0 in 6 min. They also examined the silver recovery efficiency of a protease derived from *Bacillus sphaericus*, which hydrolyzed silver at pH 11.0–12.0 in 8 min. It has also been observed that *Bacillus coagulans* and *Bacillus* sp. 21-2 are significant bacterial strains in the businesses that recover silver (Babalola et al. 2023).

15.4.4 Industries That Use Detergents

The detergent industry uses a variety of enzymes, including lipase, amylase, and protease. Proteases make up about 60.0% of the detergent industry's total enzyme sales. In 1913, Otto Rohm found that pancreatic extract could be used to wash textiles and break down proteins. Subsequently, a detergent containing bacterial enzymes was released into the market in 1956 under the brand name Bio-40. Numerous harmful compounds, including polycarboxylates, chlorine, phosphonates, and others, are included in household detergents. These chemicals are bad for the environment and can lead to skin conditions. Several variables, including water hardness, stain kind, washing temperature, and washing technique, affect how well an enzyme in detergent works (Yang et al. 2017). Two qualities should be present in the perfect protease for usage in the detergent industry: stability at alkaline pH and detergent compatibility. In the production of laundry detergents such as Era-plus, Dynamo, and Tide (Colgate Palmolive), proteases derived from *Bacillus* sp. are highly helpful. Protease generated from two putative bacterial strains, *Bacillus* sp. KSM-KP43 and *B. clausii* KSM-K16, have been used in the laundry industry in the past years. The screening of bacteria that produce protease is a continual process since there is a significant demand for detergent-compatible protease. In terms of the environment, using microbial proteases to make detergent is beneficial for both the environment and human health (Clapp et al. 2004).

15.4.5 Chemical Sectors

As biocatalysts, proteases are essential for the synthetic chemical industry. Numerous researchers have documented how alkaline proteases catalyze the synthesis of peptides in organic solvents. One such alkaline protease, called proleather, is produced by *Bacillus* sp. and is particularly helpful when transesterifying D-glucose into pyridine and synthesizing sucrose polymers in anhydrous conditions. However, *C. coronatus* in alkaline proteases may be able to take the place of subtilisin in different reaction mixtures (Table 15.1).

Table 15.1 Major sources and industrial applications of microbial proteases

Microbial source (genus)	Protease type (common)	Optimal pH	Industrial applications	Relevance to meat/poultry industry	References
Bacillus sp. (e.g., *B. licheniformis*, *B. amyloliquefaciens*)	Alkaline serine protease (subtilisin)	8.0–11.0	Detergents, leather processing, food processing, and silver recovery	Primary source for detergent-based cleaning (CIP systems), with potential for protein hydrolysis	Qazi et al. (2008)
Aspergillus sp. (e.g., *A. oryzae*, *A. niger*)	Acidic/aspartic and neutral proteases	3.0–7.0	Food processing (baking, cheese, soy sauce), digestive aids	Flavor enhancement, fermentation acceleration, and bioactive peptide production	Geeganage and Frey (2002)
Rhizopus sp.	Acidic protease	3.0–5.0	Food fermentation, dairy	Tenderization (specific to certain applications)	Alipour et al. (2016)
Streptomyces sp.	Neutral/ metalloprotease	6.0–8.0	Leather bating, therapeutic agents	Potential for controlled tenderization and collagen hydrolysis	Sharma et al. (2017)
Pseudomonas sp.	Alkaline serine protease	8.0–10.0	Detergents, waste processing	By-product valorization (keratin hydrolysis from feathers)	Gamblin et al. (2008)
Bacillus sp. (keratinolytic strains)	Keratinase (serine/metallo)	7.0–9.0	Waste management, feed production	Crucial: Valorization of poultry feathers into feed-grade protein hydrolysates	Elston et al. (2007)
Mucor sp.	Aspartic protease (e.g., rennet-like)	3.0–5.0	Dairy (cheese coagulation)	Analogous processes in certain fermented meat product textures	Cruz et al. (2020)

15.5 Enzymes in the Meat Industry

Even before recorded history, enzymes were in use for a very long time. Enzymes have long been used in the processing of traditional foods, including bread, cheese, and wine. As an enzyme have been used in a variety of food processes, this method is still used in contemporary food technology, but with greater accuracy. Enzymes have been used in the food business under strict control to create high-quality food products. Due to customer awareness and quality consciousness, the meat business is undergoing a revolution where quality is the top priority (Hafid et al. 2020). The main draw has been the affordability and quality of meat products. The primary problem with meat products is the postmortem alterations in muscle, which impact texture and softness and are significantly influenced by enzymes.

Enzyme-induced postmortem alterations include glycolytic activity, lowers muscle pH by metabolizing glycogen to produce lactic acid. Meat quality may be impacted by the rate at muscle pH drops, and the care of animals before slaughter may also have an impact. Changes in membranes, sarcoplasm, sarcolemma, and muscle fiber follow the depletion of energy. Simultaneously, the contractile protein complexes (myosin and actin) become irrevocably bonded together, causing muscles to stiffen and develop rigor mortis, which makes meat tough (Gagaoua et al. 2014). A variety of factors, some of which may be preslaughter or postslaughter in nature, affect quality metrics like meat softness. To produce tender meat, this hardness can be broken down by the enzymes found in the muscle cell or by external sources. Enzymes are used in the meat business to enhance subpar meat and enhance the production process. The two main methods for producing high-quality meat products are restructuring low-value meat pieces into higher-value ones and tenderizing tough slices. According to the T index, tenderness is the most important attribute of high-quality beef. Three factors determine a meat product's tenderness: the residue left over after chewing, the ease of tooth penetration, and the ease of fragment breakdown. Myofibrillar components give meat from young animals their toughness because actomyosin is formed during postmortem transformations (Neath et al. 2007). Meat's actomyosin-induced myofibrillar roughness is what makes meat and meat products more tender. It has been discovered that the endogenous calpain system is crucial for the postmortem proteolysis of proteins, which results in a more delicate end-product. Proteins are attacked by calpain, which breaks them up into huge pieces, making them more vulnerable to other enzymes. Significant variations in meat tenderness have been associated with the activities of calpain and calpastatin in postmortem muscle. Although the lysosomal enzymes, also known as cathepsins, and other enzymatic components have been linked to meat softness, their functions are still unclear and require more research (Yeo et al. 2011). It has been discovered that in addition to meat with high hardness, exogenous enzymes are also necessary. Proteolytic enzymes found in many fruits and vegetables have the ability to make tough meat softer. Papain, zingibain, and ficin are the three plant proteolytic enzymes that are most frequently mentioned. Likewise, it has been discovered that certain microbial enzymes have a tenderizing effect.

Numerous bioactive peptides have been identified in meat, and scientists have documented various health-promoting uses for these peptides. They are produced following the deliberate application of enzymes to the meat as a substrate. In the meat protein hydrolysates, bioactive peptides with antihypertensive, antioxidant, and antibacterial properties have been found. To increase the market value of small bits of meat, the enzymes can also be utilized to bond them. Additionally, this improves carcass utilization efficiency. Transglutaminases (TGase) and other protein cross-linking enzymes have been widely employed in food products to enhance texture. Because the enzymes catalyze the cross-linking of glutamine and lysine residues, they can produce regular shapes and sizes in raw meat products. Enzymes also have a significant role in enhancing the flavor of meat products, particularly during the ripening or ageing phase (Qazi et al. 2008).

15.6 Enzymes in the Poultry Industry

Enzymes have substantial benefits to nutrition, growth performance, product quality, and environmental sustainability, and enzymes have emerged as essential instruments in the contemporary poultry business. The primary economic input in poultry production systems is feed, and profitability is directly impacted by how birds use nutrients. Complex carbohydrates, non-starch polysaccharides, and antinutritional factors that restrict nutrient digestion are included in a large number of plant-based feed ingredients frequently used in poultry diets. The digestive system of poultry is supported with enzymes such as lipases, carbohydrates, proteases, and phytases, which break down substances that indigenous enzymes would otherwise find difficult to access (Su et al. 2017). The presence of non-starch polysaccharides (NSPs) in grains such as wheat, barley, rye, and corn, carbohydrates are among the most often utilized enzymes in chicken feed. Poor growth performance is caused by NSPs because they increase intestinal viscosity and obstruct nutrient absorption. By breaking down arabinoxylans and β-glucans, xylanases and β-glucanases improve energy utilization and decrease digesta viscosity. According to research, broiler diets supplemented with xylanase improve total weight gain, feed intake efficiency, and metabolizable energy (Sathiya 2013). Although multienzyme complexes have also demonstrated efficacy in corn–soy diets by targeting residual NSP fractions, these advantages are more significant in wheat- and barley-based diets. Carbohydrates not only affect gut viscosity but also encourage beneficial microbial fermentation in the hindgut, which inhibits the growth of harmful bacteria and improves gut health (Padmapriya et al. 2012).

Important class of enzymes used to increase the digestibility of proteins is proteases. Endogenous protease activity is naturally low in young chicks, particularly in the first few weeks of life. Exogenous proteases added to feed improve nutrient retention, increase amino acid availability, and lessen the demand for high-protein diets. There is less of an influence on the environment, lower feed costs, and less nitrogen excretion. Additionally, trypsin inhibitors found in soybean meal and other antinutritional proteins are broken down by proteases, increasing the biological value of plant-based proteins. Protease supplementation improves feed conversion ratio, decreases flock-to-flock performance variability, and increases ileal amino acid digestibility, according to numerous studies (Firouzbakht et al. 2015). Because it plays a crucial part in releasing phosphorus from phytic acid, the main type of phosphorus storage found in plant-based feed materials, phytotase is the most commonly utilized enzyme in poultry nutrition. Due to their low intrinsic phytase activity, poultry have limited access to dietary phosphorus without supplementation.

Microbial phytase enhances skeletal development, decreases the need for inorganic phosphate supplementation, and promotes the absorption of calcium and phosphorus in poultry diets. In addition to utilizing minerals, phytase also improves energy and amino acid digestibility, a process known as the "extra-phosphoric effect." Significant cost-savings and a decrease in environmental phosphorus excretion are the outcomes of this widespread improvement in nutrient availability, which is crucial for sustainable chicken production (Sugumaran et al. 2012). The use of lipases to increase fat digestion is growing, especially in diets that contain lower-quality lipid sources or saturated fats. Pancreatic lipase activity is low in young birds, which reduces the efficiency of dietary fat digestion in the early stages of growth. Lipase supplementation promotes improved feed efficiency, increases energy utilization, and improves fat emulsification. This is very helpful for creating inexpensive diets with different types of fat (Sharma et al. 2017). Enzymes are crucial for the processing of poultry flesh in addition to their nutritional value. Papain, bromelain, ficin, and fungal proteases are examples of proteolytic enzymes that are frequently employed to tenderize poultry meat. Tenderness, juiciness, and general sensory qualities are enhanced by these enzymes' hydrolysis of myofibrillar and connective tissue proteins. In order to produce processed poultry goods, including marinated pieces, ready-to-cook items, and value-added meat formulations, controlled enzymatic tenderization minimizes cooking losses and improves water-holding capacity.

Enzymes aid in the formation of flavors by hydrolyzing proteins to produce amino acids and peptides that take part in Maillard reactions during cooking, which enhance flavors in a desired way (Duman and Tekin 2020). Enzymes also contribute to waste reduction and the use of poultry by-products. Feathers, blood, viscera, and processing residues are all produced in considerable quantities by the chicken industry. Feathers and other protein-rich wastes are hydrolyzed by keratinases, proteases, and lipases to create feather meal, amino acid supplements, pet food ingredients, and organic fertilizers. Enzymatic hydrolysis enhances the nutritional content and digestibility of by-products while providing a sustainable substitute for harsh chemical treatments (Nadeem et al. 2013).

15.7 Applications in the Meat and Poultry Industries

Strategic applications of microbial proteases in meat and poultry processing along with its source and application-wise mechanism of action are summarized in Table 15.2.

Table 15.2 Strategic applications of microbial proteases in meat and poultry processing

Application target	Primary objective	Mechanism of action	Example enzymes (microbial source)	References
Tenderization	Improve texture by reducing hardness	Hydrolysis of myofibrillar proteins (myosin, actin) and connective tissue collagen	Fungal proteases (*Aspergillus* spp.), bacterial neutral proteases (*Bacillus* spp.)	Vaid and Bajaj (2017)
Flavor enhancement	Develop umami, savory notes; reduce bitterness	Release of free amino acids (glutamate) and savory peptides Debittering via exopeptidase activity	*Bacillus*-derived proteases (e.g., Neutrase, Alcalase) Complex blends (e.g., Flavourzyme from *A. oryzae*)	Koak et al. (2011)
Bioactive peptide production	Generate health-promoting protein fragments	Selective hydrolysis to release encrypted peptide sequences with ACE-inhibitory, antioxidant activities	Specific microbial proteases (e.g., from *Bacillus* or *Aspergillus*) with known cleavage specificity	Zanutto-Elgui et al. (2019)
By-product valorization	Recover proteins/nutrients from low-value streams	Extensive hydrolysis of insoluble/structural proteins (collagen, keratin)	Keratinases (*Bacillus*, *Streptomyces*), collagenases (*Bacillus*, *Pseudomonas*)	Marson et al. (2022).
Process acceleration	Shorten fermentation/ripening time for cured products	Accelerated breakdown of proteins and lipids to form flavor precursors	Proteases and lipases from mold starters (e.g., *Penicillium*) or bacterial cultures	Ravindran et al. (2011)
Cleaning-in-place (CIP)	Degrade proteinaceous soils on equipment	Hydrolysis of adhered meat proteins, blood, and fats into soluble peptides	Thermostable alkaline proteases (*Bacillus* spp.)	Vijayaraghavan and Vincent (2015)

15.7.1 *Tenderization*

15.7.1.1 Postmortem Application Methods

The ultimate textural quality, consumer acceptance, and total economic worth of meat cuts, postmortem tenderization is a crucial step in the meat processing process. Following slaughter, rigor mortis, structural stiffening, and a drop in pH are caused by biochemical alterations in muscle. Specific techniques, including physical processing, enzymatic treatments, and regulated biological processes, can improve tenderness. Enzymatic marinating is one of the most effective methods among them (Ghobadi Nejad et al. 2014). Collagen, elastin, and myofibrillar proteins are commonly broken down by natural proteases such as papain (found in papayas), bromelain (found in pineapples), ficin (found in figs), and microbial proteases originating from *Bacillus* and *Aspergillus* species. Usually, injection, marination, or surface application are used to introduce these enzymes. In order to ensure uniform penetration and quick reactivity, injection pumping distributes brine containing enzymes straight into the musculature. This is improved by vacuum tumbling, which opens the muscle fibers under less pressure, enabling more thorough and reliable enzyme diffusion. Another important factor in postmortem tenderization is physical interventions (Alipour et al. 2016). Immediately upon slaughter, electrical stimulation is used to boost early activation of endogenous calpains, speed up ATP depletion, and accelerate glycolysis. The condition is known as "cold shortening," in which quick cooling before rigor mortis results in toughening and tightness of the muscles. By changing the ultrastructure of muscle fibers, partially unfolding proteins, and boosting the activity of endogenous proteases, high-pressure processing (HPP), a nonthermal technique, also increases tenderness. Meat treated with HPP has reduced ageing time, decreased shear force, and increased juiciness (Bose et al. 2014). Calpain and cathepsin-induced natural enzymatic degradation is the basis for traditional ageing methods, including dry and wet ageing. By keeping meat in vacuum-sealed packaging at low temperatures, wet ageing enables proteolysis to generate mild flavor notes and softness. Meat is kept under controlled humidity and ventilation during dry ageing, which produces a stronger flavor but takes longer and requires more resources (Anbu 2013).

15.7.1.2 Management of Texture Uniformity and Overtenderization

Excessive breakdown of muscle proteins unwanted softness, mushiness, and loss of structural integrity, even though tenderization is essential for improving the quality of meat. Thus, managing overtenderization is a significant technological challenge. Enzyme concentration, exposure duration, and environmental factors (pH, temperature, and ionic strength) are all regulated as part of the first line of control. If not carefully regulated, powerful enzymes such as papain and bromelain can rapidly overhydrolyze proteins (Chou et al. 2003). In order to estimate the best tenderization times based on muscle type, connective tissue density, and fiber composition, modern processors employ mathematical modeling and kinetic investigations. Using enzyme inhibitors or thermal inactivation once the required softness is attained is another important tactic. Proteases can be denatured, and their activity stopped by heating meat to a certain temperature (55–60 °C) (Gamblin et al. 2008). This is especially helpful for marinated meats since, following controlled proteolysis, cooking directly stabilizes the structure. Incorporating mechanical tenderization, such as blade tenderizers, tumblers, and massagers, which accomplish structural softening without excessive proteolysis, also improves texture consistency (Huang et al. 2017). Because high-pressure processing denatures proteins uniformly, regardless of muscle variations, it can also aid in uniformity control. Processors are able to ensure quality consistency between manufacturing batches through continuous monitoring using instruments such as instrumental texture profiling, NIR spectroscopy, and shear force measurement. The right enzymes for the various tissues are also essential to preventing overtenderization. For instance, fungal collagenases preferentially degrade connective tissues, providing a more regulated tenderizing action, whereas bromelain aggressively hydrolyzes both collagen and myofibrillar proteins. By combining particular enzymes, processors can adjust the level of tenderization (Kloetzel 2004).

15.7.2 Enhancement and Development of Flavor

15.7.2.1 Manufacturing of Savory Flavors and Umami Peptides

One of the most important factors influencing consumers' acceptance of meat products is flavor. A potent method for creating savory flavor molecules and umami-enhancing peptides is enzymatic hydrolysis. The umami taste is primarily produced by free amino acids such as glutamate and aspartate, which are released when proteases break down muscle proteins. Furthermore, the savory sense is reinforced by short-chain peptides that include glutamic acid residues (Puente et al. 2003). To produce these taste-active ingredients from the proteins of beef, chicken, and pork, enzymes such as alkalase, flavorzyme, and neutrase are frequently employed. These peptides contribute kokumi qualities, such as mouthfulness, depth, and lingering flavor, in addition to enhancing umami. Additionally, umami peptides complement nucleotides such as guanosine monophosphate (GMP) and inosine monophosphate (IMP). These nucleotides frequently rise as flesh is broken down by enzymes or as it ages after death through autolysis. Umami intensity can be increased several times when free glutamate and IMP are combined (Tavano 2013).

Soups, broths, bouillons, processed meats, quick spices, and plant-based meat substitutes are all products that use umami-rich hydrolysates. Umami peptides are essential for preserving palatability while making up for flavor loss in reduced sodium meat compositions. Additionally, they improve the flavor of thermally processed foods, cured goods, and restructured meats. Complex savory scents are produced during cooking when umami peptides mix with Maillard reaction products made from reducing sugars and amino acids. The meaty overtones found in grilled, roasted, or stewed meat are synergy. Proteolysis enhances aroma profiles in addition to taste, making enzymatic flavor enhancement a crucial method in contemporary meat preparation (Simkhada et al. 2010).

15.7.2.2 Decrease in Bitterness

One of the main problems with the enzymatic hydrolysis of animal proteins is bitterness. Hydrophobic peptides comprising amino acids such as leucine, valine, phenylalanine, and tryptophan are produced proteases thoroughly degrade proteins.

The intense bitter taste of these peptides may make them less palatable to consumers. By removing hydrophobic residues from the ends of peptides, exopeptidases, in particular, aminopeptidases and carboxypeptidases, help to lessen bitterness. Because they have both endopeptidase and exopeptidase activity, enzymes like Flavourzyme are useful for creating hydrolysates with a mild flavor. Another technique that reduces the production of bitter fractions is selective hydrolysis utilizing proteases with low hydrophobic specificity (Vijayaraghavan and Vincent 2015). To extract bitter peptides from hydrolysates, fractionation methods such as activated carbon adsorption, nanofiltration, and ultrafiltration are used. Smaller, less bitter peptides can pass through ultrafiltration membranes with certain molecular weight cutoffs, but bigger, bitter-tasting peptides are retained. This creates hydrolysates with a cleaner flavor that can be added to baby food, soups, and dietary supplements. By altering the hydrolysis conditions, debittering can also be enhanced. The development of bitterness can be considerably decreased by lowering the degree of hydrolysis, modifying pH, and managing temperature. Bitter peptides are further broken down into non-bitter components by combining enzymatic and heat treatments (Xue et al. 2012).

15.7.3 Hydrolyzed Meat Proteins and Useful Substances

15.7.3.1 Bioactive Peptide Production

The health benefits of bioactive peptides made from animal proteins have drawn a lot of interest. These peptides have immunomodulatory, antihypertensive, antioxidant, antibacterial, and anti-inflammatory properties. Usually, they are made by enzymatic hydrolysis with the aid of microbial enzymes or proteases such as trypsin, pepsin, and alkaline proteases (Nakiboglu et al. 2003). Certain peptide sequences are liberated from parent proteins, including collagen, myosin, actin, troponin, and sarcoplasmic proteins, during hydrolysis. By blocking the angiotensin-converting enzyme (ACE), antihypertensive peptides cause vasodilation and a drop in blood pressure. Proline, tyrosine, and phenylalanine are examples of hydrophobic and aromatic residues that are frequently found in these peptides (Orgaz et al. 2006).

Antioxidant peptides strengthen cellular defenses, bind metal ions, and scavenge free radicals. They contain a lot of hydrophobic amino acids, histidine, and cysteine, which interact with reactive oxygen species. By rupturing bacterial cell membranes or interfering with metabolic processes, antimicrobial peptides produce their effects. There may be uses for these peptides in the natural preservation of animal products. Bioactive peptides produced from meat find extensive use in sports nutrition, functional foods, nutraceuticals, and therapeutic diets. Clinical nutrition formulations, drinks, protein powders, and fortified foods all contain them. They are promising components for upcoming health-conscious food developments because of their great digestibility and potent biological activity (Madhavi et al. 2011).

15.7.3.2 Soups, Seasonings, and Flavors

Rich amino acid content, solubility, and ability to enhance flavor, meat protein hydrolysates are useful ingredients in the creation of flavorings, soups, broths, and spices. Peptides and amino acids produced by hydrolysis add to the mouthfeel, savory taste, and overall flavor profile. By intensifying the umami, kokumi, and meaty overtones, hydrolysates function as organic flavor enhancers. They are frequently found in canned soups, processed meats, instant noodles, bouillon cubes, ready-to-eat meals, and snack spices (Radnedge et al. 2003). Hydrolysates enhance the functional qualities of processed foods by improving emulsification, water-binding, and protein solubility in addition to improving flavor. They are particularly useful in low-sodium formulations where adding flavor without adding more salt is required. Hydrolysates also help plant-based meat substitutes taste better by simulating the nuanced flavor of real meat. Deep, roasted, and grilled fragrances are produced by the interaction of hydrolysates with Maillard reaction products, which are enhanced by thermal processing. Meat hydrolysates are being used more often in high-end food products because they provide natural, clean-label substitutes for artificial flavor enhancers (Yoshioka et al. 2007).

15.7.4 By-product Appraisal

15.7.4.1 Protein Recovery from Feathers, Bones, and Trimmings

Bones, trimmings, skin, blood, and feathers are among the many by-products produced by the meat and poultry industries. In addition to lessening the impact on the environment, valorizing these materials through protein recovery produces important chemicals. Significant amounts of collagen, gelatin, minerals, and leftover meat can be found in bones and trimmings (Vaid and Bajaj 2017). These components can be extracted in a gentle and environmentally friendly enzymatic hydrolysis. Because of its gelling, emulsifying, and structural qualities, collagen and gelatin are widely employed in the food, pharmaceutical, and cosmetic sectors. Because of substantial disulfide cross-linking, keratin, a high-sulfur, fibrous protein found in poultry feathers, is difficult to break down. This stiff protein can be broken down into soluble keratin hydrolysates by keratinases that are generated by *Bacillus*, *Streptomyces*, and fungal species. These hydrolysates are used as components in cosmetic formulations, biomaterials, animal feed, and fertilizers. By-products are also rendered and enzymatically processed into ingredients for pet food, fish meal, and meat-and-bone meal (MBM) (Wang et al. 2011). Restructured meats, surimi-style products, and emulsified sausages can all benefit from the application of protein-rich extracts from trimmings. By turning waste into valuable functional materials, value addition not only increases sustainability but also boosts economic benefits.

15.7.4.2 Manufacturing of Nutritional Supplements and Protein Powders

Protein powders made from leftover meat are becoming more and more well-liked because of their high nutritional value, ease of digestion, and vital amino acid composition. Usually, enzymatic hydrolysis is used to create these powders, which are then filtered, concentrated, and spray-dried (Pillai et al. 2011). Hydrolysates offer easily absorbed smaller peptides, which makes them perfect for therapeutic diets, sports nutrition, and baby food. Protein powders made from meat hydrolysates, collagen, gelatin, and poultry proteins have practical advantages such as better intestinal health, joint support, and muscle recuperation. They are frequently found in weight-management products, protein shakes, dietary supplements, and fortified foods. Meat hydrolysate powders enhance nutrient absorption and palatability in animal nutrition. These powders lessen dependency on traditional protein sources such as dairy or soy and promote the circular economy since they are made from underutilized by-products. The increasing demand for high-protein foods, clean-label goods, and sustainable ingredients is what is driving their growing use (Ravindran et al. 2011).

15.7.5 Quicker Fermentation and Ripening of Meat Products

15.7.5.1 Dry-Cured Hams and Sausages

During fermentation and ripening, dry-cured sausages and hams go through intricate biochemical changes. Proteolysis, lipolysis, and microbial metabolism create distinctive flavors, textures, and scents during these processes, which often take weeks or months. Both endogenous and microbial enzymes are essential. Amino acids and peptides that are released during proteolysis act as building blocks for volatile substances such as alcohols, sulfur-containing chemicals, and aldehydes. Free fatty acids produced by lipolysis undergo additional oxidation to produce aromatic molecules that contribute to notes of cheese, nutty, smoky, or fermented flavors. Fermentation is frequently controlled with starter cultures that contain *Lactobacillus*, *Pediococcus*, *Staphylococcus*, and *Micrococcus* species. These microorganisms speed up flavor creation, improve safety, improve color development, and lower pH.

In sausages and hams, moisture loss during ripening concentrates flavors and helps create a solid texture. Controlled ventilation, humidity, and temperature allow for quicker ripening without sacrificing quality. In order to accelerate biological operations, industrial processors frequently add microbial proteases to natural proteolysis (Maqsood et al. 2018).

15.7.5.2 Shortening the Ripening Period

Long ripening times needed for traditional beef fermentation and curing raise expenses, energy consumption, and microbiological hazards. Industrial processing is heavily focused on technologies that reduce ripening time. Flavor precursors are produced more quickly when muscle proteins and lipids are broken down by enzymatic treatments utilizing lipases and proteases. Further reducing processing time without sacrificing quality are starter cultures that have been developed or chosen for their high lipolytic and proteolytic activity (Richardson et al. 2018). High-pressure processing, ultrasonic treatment, pulsed electric fields, and regulated fermentation chambers are examples of advanced curing technologies that provide ideal conditions for quick biological reactions. Ultrasound speeds up salt transport and boosts mass transfer in meat tissues, whereas HPP improves proteolysis and microbiological safety. Ripening is predictable and quick due to modern fermentation chambers' exact control over temperature, humidity, and ventilation. These chambers greatly shorten the timetable while simulating conventional settings. Without sacrificing taste qualities, accelerated ripening methods lower costs, enhance product uniformity, and facilitate large-scale commercial manufacturing (Zhu et al. 2018a, b).

15.8 Enzyme Engineering and Immobilization for Advanced Meat and Poultry Applications

The global meat and poultry industries face mounting pressure to enhance efficiency, sustainability, and product quality while meeting clean-label consumer demands. Enzymes, particularly proteases, have long been indispensable tools for processes such as tenderization, flavor development, and by-product valorization. However, the native forms of these biocatalysts often possess inherent limitations, including narrow pH/thermal stability, poor reusability, difficulty in process control, and high costs that restrict their optimal industrial application (Rao et al. 2009). To overcome these challenges, the synergistic convergence of enzyme engineering and immobilization has emerged as a transformative technological frontier. This paradigm involves the deliberate molecular redesign of enzymes to confer superior functional attributes, followed by their strategic attachment to solid supports to enhance stability, enable recovery, and facilitate continuous processing. Together, these approaches are forging a new generation of robust, efficient, and tailored biocatalysts poised to revolutionize meat and poultry processing (Kakash et al. 2019).

Enzyme engineering employs rational design, directed evolution, and semi-rational methods to reprogram enzyme molecules. In meat science, the goals are precise: to expand the working pH and temperature range of proteases to match harsh processing conditions (e.g., hot-boning or high-temperature cooking); to increase catalytic activity on specific substrates such as collagen or myofibrillar proteins while minimizing nonspecific hydrolysis that causes mushiness; and to enhance stability against autolysis, oxidizing agents, or denaturants present in food matrices (Afshar-Mohammadian et al. 2011). For instance, site-directed mutagenesis of subtilisin (a common bacterial protease) has been used to introduce disulfide bonds, improving its thermostability for applications in precooked meat products. Similarly, engineering the substrate-binding pockets of proteases can shift their specificity, enabling the selective release of savory umami peptides (glutamate-rich) or bioactive sequences (e.g., antihypertensive peptides) from meat protein hydrolysates, while avoiding the generation of bitter-tasting hydrophobic peptides. This level of precision moves enzyme application from a blunt instrument for general tenderization to a scalpel for customizing texture, flavor, and functionality (Zhu et al. 2018a, b).

Complementing molecular engineering, enzyme immobilization addresses practical operational constraints. Immobilization involves attaching engineered enzymes onto inert carriers such as chitosan beads, silica nanoparticles, magnetic particles, or food-grade polymeric resins through adsorption, covalent bonding, entrapment, or cross-linking (Lees et al. 2019). This confinement offers multifaceted advantages for meat and poultry processing. First, it significantly improves enzyme stability against thermal denaturation and pH fluctuations, allowing reuse over multiple production cycles, dramatically reducing cost-per-unit operations. Second, immobilization enables easy separation of the biocatalyst from the product stream, which is critical for producing clear protein hydrolysates, flavorings, or bioactive peptide fractions without residual enzyme contamination, aligning with clean-label requirements. Third, immobilized enzyme systems can be configured in packed-bed or fluidized-bed reactors for continuous processing of meat slurries or by-product streams (e.g., feather or blood protein hydrolysates), enhancing throughput and scalability (Biffin et al. 2020).

A compelling application lies in controlled tenderization. Traditional methods using free papain or bromelain can lead to inconsistent or overtenderized meat due to uncontrolled diffusion and activity. Immobilizing these proteases on a macroscopic support allows them to act only on the meat surface during a controlled tumbling or dipping process, enabling precise, uniform, and shallow proteolysis that improves tenderness without compromising structural integrity. Moreover, the reaction can be instantly terminated by simply removing the immobilized enzyme carrier, offering unprecedented control (Oyeleke et al. 2010). In by-product valorization, immobilized engineered enzymes are game changers. For example, keratinolytic proteases engineered for higher oxidative stability can be immobilized on magnetic nanoparticles. These can be used to hydrolyze tough poultry feathers in a stirred-tank reactor and then easily recovered with a magnet for repeated use, transforming waste into valuable protein hydrolysates for feed or fertilizers with high process economy. Similarly, immobilized transglutaminase (TGase) is used in restructuring lower-value meat trimmings into uniform steaks or nuggets; the immobilization allows the enzyme to be contained and reused, making the process more economically viable (Page and Di Cera 2008).

Future perspectives point toward multifunctional and "smart" immobilized systems. Research is advancing into co-immobilizing complementary enzymes, such as a protease for protein hydrolysis and a lipase for flavor development, on the same carrier for simultaneous reactions in fermented sausage ripening. "Smart" carriers that respond to stimuli (e.g., pH, temperature) could allow on-demand activation or deactivation of enzymes, providing ultimate process control. Furthermore, the integration of engineered enzymes with novel immobilization supports such as 3D-printed biocatalytic scaffolds or nano-fibrous membranes opens avenues for innovative continuous processing reactor designs tailored for meat and poultry streams (Nam et al. 2016).

15.9 Future Perspective

The industrial value of this enzyme has been realized; researchers and engineers are searching for new and robust bacterial enzymes. Protein engineering will be the main method used in the future to create proteases with novel characteristics. Because of their potential, alkaline bacterial proteases are essential in a variety of industries, and their use is only expected to grow in the future (Koak et al. 2011). To create better strains of protease, the researchers are using cutting-edge techniques such as protein/genetic engineering, molecular biology, and computational biology. In vitro evolutionary modifications to the fundamental structure of proteins will be used to create bacterial strains with desired traits. Developing bacterial proteases with traits such as increased yield, altered substrate specificity, improved heat stability, altered optimal pH, and prevention of auto-proteolytic inactivation is one of scientists' main objectives. The integration of proteases in the meat and poultry sectors is poised for a transformative leap, driven by converging trends in sustainability, precision biotechnology, digitalization, and evolving consumer demands. The future will see enzymes transition from mere processing aids to central tools for creating value, ensuring quality, and fundamentally reshaping industry practices.

Future advancements will move beyond broad-spectrum tenderization to precision proteolysis. This involves engineering proteases with exquisite substrate specificity enzymes designed to cleave only specific peptide bonds in target proteins (e.g., collagen in connective tissue, or specific actomyosin cross-links) while leaving others intact. This will eliminate overtenderization, enable the customization of texture (e.g., "al dente" vs. "melt-in-the-mouth"), and allow for the targeted release of specific bioactive or flavor-enhancing peptides (Zanutto-Elgui et al. 2019). Computational protein design, machine learning models of enzyme–substrate interactions, and directed evolution will be standard tools for developing these "designer enzymes." The focus on by-product valorization will intensify. Future biorefineries will employ tailored multienzyme cocktails (proteases combined with lipases, carbohydrates) in fully optimized, continuous flow systems to perform "cell-free" biotransformation of low-value streams (feathers, blood, bones, trimmings). The goal will be the holistic fractionation of these materials into a spectrum of high-purity, high-value products: not just generic protein hydrolysates but isolated functional peptides (for nutraceuticals), chondroitin/glucosamine (for joint health), iron–peptide complexes (for nutrition), and bio-based polymers (for packaging). This represents a shift from waste management to purpose-driven molecular upcycling (Cruz et al. 2020).

Protease applications will be synergistically combined with nonthermal technologies to create efficient, green processing lines. For example, high-pressure processing (HPP) will be used to precondition meat structure, making it more susceptible to subsequent low-dose enzymatic treatment, drastically reducing enzyme requirements and processing time. Pulsed electric fields (PEF) could be used to create temporary micropores in cell membranes, allowing for the ultra-rapid and uniform infusion of protease solutions for marination or curing. Ultrasound will enhance enzyme kinetics and mass transfer during brining and hydrolysis processes. Furthermore, inline biosensors and real-time spectroscopy (e.g., NIR, Raman) will be deployed to monitor the degree of proteolysis in situ, feeding data to AI-controlled systems that automatically adjust enzyme dosage, temperature, or process duration to achieve a perfect, consistent product (Gagaoua et al. 2015).

Consumer demand for natural, healthy products will be a primary driver. Proteases will be key enablers of "radical sodium reduction": By generating potent umami peptides and kokumi compounds, proteases will allow for 50% + sodium reduction in processed meats without compromising taste. The mitigation of allergens and anti-nutritional factors has been widely studied (Moon 2018). Specific proteases could be designed to hydrolyze and deactivate potential allergenic epitopes in meat-based ingredients or residual soy/wheat proteins in composite products. *Next-generation functional foods*: The controlled hydrolysis of meat and collagen will produce standardized, clinically validated bioactive peptide ingredients for foods targeting hypertension, oxidative stress, and muscle-wasting conditions. The environmental benefits of protease use, reducing chemical waste, lowering energy consumption (e.g., in tenderization vs. mechanical methods), and valorizing by-products will be rigorously quantified. Full lifecycle assessments (LCAs) will become mandatory to demonstrate the carbon footprint reduction and circular economy contribution of enzymatic processes. This data will not only guide internal process optimization but will also become a powerful marketing tool for sustainability-conscious brands (Machado et al. 2016). The role of proteases will expand beyond traditional meat into the alternative protein sector. In plant-based meats, proteases will be crucial for modifying the texture of plant proteins (e.g., soy, pea, wheat gluten) to better mimic the fibrous structure of muscle tissue. They will also be used to hydrolyze plant proteins to eliminate "beany" off-flavors and generate savory, meaty flavor precursors. In cultured (cell-based) meat production, proteases (like trypsin) are already used in the harvesting phase. Future research will focus on finding animal-free, food-grade proteases for this step and potentially using proteases to structure and texture the final cultivated product (Singh et al. 2012).

15.10 Conclusion

Microbial proteases have firmly established themselves as indispensable biocatalysts within the modern meat and poultry industries, driving a paradigm shift from traditional, often inefficient, chemical and physical processes toward precise, sustainable, and value-adding biotechnological solutions. This review has detailed their multifaceted roles, extending far beyond simple protein degradation to become central tools for quality enhancement, process efficiency, and waste minimization. The strategic application of these enzymes enables significant advancements in core product attributes. Through controlled postmortem tenderization, proteases address the critical textural demands of consumers, improving palatability and consistency. In flavor development, they act as natural bio-factories, generating savory umami peptides and kokumi compounds that enhance taste profiles while enabling critical strategies like sodium reduction. Furthermore, the hydrolysis of meat proteins unlocks a treasure trove of bioactive peptides, transforming ordinary meat into a source of functional ingredients for the nutraceutical and health-food sectors. Perhaps one of the most significant contributions lies in the realm of sustainability through by-product valorization. Microbial proteases, particularly specialized keratinases and collagenases, are the key to unlocking the latent value in low-value streams—bones, trimmings, blood, and feathers. By converting these potential wastes into high-value products such as protein hydrolysates, collagen peptides, animal feed, and fertilizers, the industry moves decisively toward a circular economy model, reducing environmental impact and improving overall resource efficiency. The advent of advanced technologies such as enzyme engineering and immobilization marks the beginning of a new era. The ability to design proteases with tailored specificity, stability, and activity, and to deploy them in reusable, controlled systems, addresses the historical limitations of enzyme use. This precision allows for unprecedented control over processes like tenderization and hydrolysis, minimizes unwanted side-effects like bitterness, and paves the way for continuous, cost-effective bioreactor operations.

References

Afshar-Mohammadian M, Rahimi-Koldeh J, Sajedi RH (2011) The comparison of protease activity and total protein in three cultivars of kiwifruit of northern Iran during fruit development. Acta Physiol Plant 33(2):343–348. https://doi.org/10.1007/s11738-010-0554-2

Alipour H, Raz A, Zakeri S, Dinparast Djadid N (2016) Therapeutic applications of collagenase (metalloproteases): a review. Asian Pac J Trop Biomed 6(11):975–981. https://doi.org/10.1016/j.apjtb.2016.10.003

Anbu P (2013) Characterization of solvent stable extracellular protease from *Bacillus koreensis* (BK-P21A). Int J Biol Macromol 56:162–168. https://doi.org/10.1016/j.ijbiomac.2013.02.014

Anis Ahamed N, Arif IA, Al-Rashed S, Panneerselvam A, Ambikapathy V (2022) *In vitro* thrombolytic potential of fibrinolytic enzyme from *Brevibacterium* sp. isolated from the root of the plant, *aloe castellorum*. J King Saud Univ Sci 34(4):101868. https://doi.org/10.1016/j.jksus.2022.101868

Austin MJ, Schunk H, Watkins C, Ling N, Chauvin J, Morton L (2022) Fluorescent peptomer substrates for differential degradation by metalloproteases. Biomacromolecules 23(12): 4909–4923. https://doi.org/10.1021/acs.biomac.2c01077

Ayhan F, Çelebi SS, Tanyolaç A (2001) The effect of fermentation parameters on the production of *Mucor miehei* acid protease in a chemically defined medium. J Chem Technol Biotechnol 76(2): 153–160. https://doi.org/10.1002/jctb.364

Babalola BA, Akinwande AI, Gboyega AE, Otunba AA (2023) Extraction, purification and characterization of papain cysteine proteases from the leaves of *Carica papaya*. Sci Afr 19: e01538. https://doi.org/10.1016/j.sciaf.2022.e01538

Biffin TE, Smith MA, Bush RD, Morris S, Hopkins DL (2020) The effect of whole carcase medium voltage electrical stimulation, tenderstretching and longissimus infusion with actinidin on alpaca meat quality. Meat Sci 164:108107. https://doi.org/10.1016/j.meatsci.2020.108107

Bose A, Chawdhary V, Keharia H, Subramanian RB (2014) Production and characterization of a solvent-tolerant protease from a novel marine isolate *Bacillus tequilensis* P15. Ann Microbiol 64(1):343–354. https://doi.org/10.1007/s13213-013-0668-z

Cejudo-Bastante MJ, Oliva-Sobrado M, González-Miret ML, Heredia FJ (2022) Optimization of the methodology for obtaining enzymatic protein hydrolysates from an industrial grape seed meal residue. Food Chem 370:131078. https://doi.org/10.1016/j.foodchem.2021.131078

Chou K-C, Wei D-Q, Zhong W-Z (2003) Binding mechanism of coronavirus main proteinase with ligands and its implication to drug design against SARS. Biochem Biophys Res Commun 308(1):148–151. https://doi.org/10.1016/S0006-291X(03)01342-1

Clapp AR, Medintz IL, Mauro JM, Fisher BR, Bawendi MG, Mattoussi H (2004) Fluorescence resonance energy transfer between quantum dot donors and dye-labeled protein acceptors. J Am Chem Soc 126(1):301–310. https://doi.org/10.1021/ja037088b

Cruz PL, Panno PHC, Giannotti JDG, de Carvalho RV, Roberto CD (2020) Effect of proteases from ginger rhizome on the fragmentation of myofibrils and tenderness of chicken breast. LWT Food Sci Technol 120:108921. https://doi.org/10.1016/j.lwt.2019.108921

De Azeredo LA, Leite SG, Freire DM, Benchetrit L, Coelho RR (2001) Proteases from actinomycetes interfere in solid media plate assays of hyaluronidase activity. J Microbiol Methods 45(3): 207–212. https://doi.org/10.1016/s0167-7012(01)00251-2

de Matos FM, de Lacerda J, Gomes TJ, Zanetti G, de Castro R, Soares J (2022) Production of black cricket protein hydrolysates with α-amylase, α-glucosidase and angiotensin I-converting enzyme inhibitory activities using a mixture of proteases. Biocatal Agric Biotechnol 39: 102276. https://doi.org/10.1016/j.bcab.2022.102276

Ding S, Mao B, Lu X, Zhuge B, Zong H (2022) Efficient production and biochemical characterization of a thermostable carboxypeptidase from *Bacillus megaterium* and its application on flavor improvement of soy isolate protein hydrolysates. Eur Food Res Technol 248(9): 2135–2143. https://doi.org/10.1007/s00217-022-04036-5

Duman YA, Tekin N (2020) Kinetic and thermodynamic properties of purified alkaline protease from *Bacillus pumilus* Y7 and non-covalent immobilization to poly(vinylimidazole)/clay hydrogel. Eng Life Sci 20(1–2):36–49. https://doi.org/10.1002/elsc.201900094

Elston C, Wallach J, Saulnier J (2007) New continuous and specific fluorometric assays for *Pseudomonas aeruginosa* elastase and LasA protease. Anal Biochem 368(1):87–94. https://doi.org/10.1016/j.ab.2007.04.041

Feng T, Yan S, Hou S, Fan X (2022) Novel fluorescence biosensor custommade for protein tyrosine phosphatase 1B detection based on titanium dioxidedecorated single-walled carbon nanohorn nanocomposite. Sens Actuators B Chem 369:132329. https://doi.org/10.1016/j.snb.2022.132329

Firouzbakht H, Zibaee A, Hoda H, Sohani MM (2015) Purification and characterization of the cuticle-degrading proteases produced by an isolate of *Beauveria bassiana* using the cuticle of the predatory bug, *Andrallus spinidens* Fabricius (Hemiptera: Pentatomidae). J Plant Prot Res 55(2):179–186. https://doi.org/10.1515/jppr-2015-0025

Gagaoua M, Boucherba N, Bouanane-Darenfed A, Ziane F, Nait-Rabah S, Hafid K, Boudechicha HR (2014) Three-phase partitioning as an efficient method for the purification and recovery of ficin from Mediterranean fig (*Ficus carica* L.) latex. Sep Purif Technol 132:461–467. https://doi.org/10.1016/j.seppur.2014.06.001

Gagaoua M, Hoggas N, Hafid K (2015) Three phase partitioning of Zingibain, a milk-clotting enzyme from *Zingiber officinale* roscoe rhizomes. Int J Biol Macromol 73:245–252. https://doi.org/10.1016/j.ijbiomac.2014.11.015

Gamblin DP, Scanlan EM, Davis BG (2008) Glycoprotein synthesis: an update. Chem Rev 109(1): 131–163. https://doi.org/10.1021/cr078291i

Geeganage S, Frey PA (2002) Galactose-1-phosphate uridylyltransferase: kinetics of formation and reaction of uridylyl-enzyme intermediate in wild-type and specifically mutated uridylyltransferases. Methods Enzymol 354:134–148

Ghobadi Nejad Z, Yaghmaei S, Moghadam N, Sadeghein B (2014) Some investigations on protease enzyme production kinetics using *Bacillus licheniformis* BBRC 100053 and effects of inhibitors on protease activity. Int J Chem Eng 2014:394860. https://doi.org/10.1155/2014/394860

Gu YQ, Walling LL (2002) Identification of residues critical for activity of the wound-induced leucine aminopeptidase (LAP-A) of tomato. Eur J Biochem 269(6):1630–1640. https://doi.org/10.1046/j.1432-1327.2002.02795.x

Hafid K, John J, Sayah TM, Domínguez R, Becila S, Lamri M, Dib AL, Lorenzo JM, Gagaoua M (2020) One-step recovery of latex papain from *Carica papaya* using three phase partitioning and its use as milk-clotting and meat-tenderizing agent. Int J Biol Macromol 146:798–810. https://doi.org/10.1016/j.ijbiomac.2019.10.043

Huang M, Chen R, Ren G (2017) Secretory expression and purification of *Bacillus licheniformis* keratinase in insect cells. PLoS One 12(8):e0183764. https://doi.org/10.1371/journal.pone.0183764

Kakash SBAGHERI, Hojjatoleslamy M, Babaei G, Molavi H (2019) Kinetic study of the effect of kiwi fruit actinidin on various proteins of chicken meat. Food Sci Technol 39(4):980–992. https://doi.org/10.1590/fst.14218

Kloetzel PM (2004) Generation of major histocompatibility complex class I antigens: functional interplay between proteasomes and TPPII. Nat Immunol 5(7):661–669. https://doi.org/10.1038/ni1090

Koak J-H, Kim H-S, Choi YJ, Baik M-Y, Kim B-Y (2011) Characterization of a protease from over-matured fruits and development of a tenderizer using an optimization technique. Food Sci Biotechnol 20(2):485–490. https://doi.org/10.1007/s10068-011-0067-9

Lees A, Konarska M, Tarr G, Polkinghorne R, McGilchrist P (2019) Influence of kiwifruit extract infusion on consumer sensory outcomes of striploin (*M. longissimus lumborum*) and outside flat (*M. biceps femoris*) from beef carcasses. Foods 8(8):332. https://doi.org/10.3390/foods8080332

Legare S, Heide F, Bailey-Elkin BA, Stetefeld J (2022) Improved SARSCoV-2 main protease high-throughput screening assay using a 5-carboxyfluorescein substrate. J Biol Chem 298(4):101739. https://doi.org/10.1016/j.jbc.2022.101739

Machado ARGM, Teixeira MFS, de Souza Kirsch L, Campelo M d CL, de Aguiar Oliveira IM (2016) Nutritional value and proteases of *Lentinus citrinus* produced by solid state fermentation of lignocellulosic waste from the tropical region. Saudi J Biol Sci 23(5):621–627. https://doi.org/10.1016/j.sjbs.2015.07.002

Madhavi J, Srilakshmi J, Raghavendra Rao MV, Sambasiva Rao KRS (2011) Efficient leather dehairing by bacterial thermostable protease. Int J Bio-Sci Bio-Technol 3(4):11–26

Maqsood S, Manheem K, Gani A, Abushelaibi A (2018) Degradation of myofibrillar, sarcoplasmic and connective tissue proteins by plant proteolytic enzymes and their impact on camel meat tenderness. J Food Sci Technol 55(9):3427–3438. https://doi.org/10.1007/s13197-018-3256-1

Marson GV, Lacour S, Hubinger MD, Belleville MP (2022) Serial fractionation of spent brewer's yeast protein hydrolysate by ultrafiltration: a peptide-rich product with low RNA content. J Food Eng 312:110737. https://doi.org/10.1016/j.jfoodeng.2021.110737

Masi C, Gemechu G, Tafesse M (2021) Isolation, screening, characterization, and identification of alkaline protease-producing bacteria from leather industry effluent. Ann Microbiol 71(1):1–11. https://doi.org/10.1186/s13213-021-01631-x

Mathew Z, Knox TM, Miller CG (2000) *Salmonella enterica* serovar typhimurium peptidase B is a leucyl aminopeptidase with specificity for acidic amino acids. J Bacteriol 182(12):3383–3393. https://doi.org/10.1128/JB.182.12.3383-3393.2000

Moon SS (2018) Effect of proteolytic enzymes and ginger extract on tenderization of M. *pectoralis profundus* from Holstein steer. Korean J Food Sci Anim Resour 38(6):2234–2246. https://doi.org/10.5851/kosfa.2018.e80

Mortezaei M, Dadmehr M, Korouzhdehi B, Hakimi M, Ramshini H (2021) Colorimetric and label free detection of gelatinase positive bacteria and gelatinase activity based on aggregation and dissolution of gold nanoparticles. J Microbiol Methods 191:106349. https://doi.org/10.1016/j.mimet.2021.106349

Nadeem M, Qazi JI, Syed Q, Gulsher M (2013) Purification and characterization of an alkaline protease from *Bacillus licheniformis* UV-9 for detergent formulations. Songklanakarin J Sci Technol 35(2):187–195

Nakiboglu N, Toscali D, Nisli G (2003) A novel silver recovery method from waste photographic films with NaOH stripping. Turk J Chem 27(1):127–133

Nam S-H, Kim Y-M, Walsh MK, Yim S-H, Eun J-B (2016) Functional characterization of purified pear protease and its proteolytic activities with casein and myofibrillar proteins. Food Sci Biotechnol 25(1):31–39. https://doi.org/10.1007/s10068-016-0005-y

Neath KE, Del-Barrio AN, Lapitan RM, Herrera JRV, Cruz LC, Fujihara T, Muroya S, Chikuni K, Hirabayashi M, Kanai Y (2007) Protease activity higher in postmortem water buffalo meat than Brahman beef. Meat Sci 77(3):389–396. https://doi.org/10.1016/j.meatsci.2007.04.001

Nnolim NE, Okoh AI, Nwodo UU (2020) Proteolytic bacteria isolated from agro-waste dumpsites produced keratinolytic enzymes. Biotechnol Rep 27:e00483. https://doi.org/10.1016/j.btre.2020.e00483

Nurhidayat I, Setiasih S, Handayani S, Hudiyono S (2018) Kinetic studies of bromelain purified from Palembang pineapple (*Ananas comosus* [L.] Merr) using gel filtration chromatography and its activity as antiplatelet aggregation. AIP Conference Proceedings 2021(1):020068. https://doi.org/10.1063/1.5062722

Oliveira V, Campos M, Melo RL, Ferro ES, Camargo ACM, Juliano MA et al (2001) Substrate specificity characterization of recombinant metallo oligopeptidases thimet oligopeptidase and neurolysin. Biochemistry 40(14):4417–4425. https://doi.org/10.1021/bi002715k

Orgaz B, Kives J, Pedregosa AM et al (2006) Bacterial biofilm removal using fungal enzymes. Enzym Microb Technol 40(1):51–56. https://doi.org/10.1016/j.enzmictec.2005.10.037

Oyeleke S, Egwim EC, Auta S (2010) Screening of *Aspergillus flavus* and *Aspergillus fumigatus* strains for extracellular protease enzyme production. J Microbiol Antimicrob 2(7):83–87

Padmapriya B, Rajeswari T, Nandita R, Raj F (2012) Production and purification of alkaline serine protease from marine *Bacillus* species and its application in detergent industry. Eur J Appl Sci 4(1):21–26

Page MJ, Di Cera E (2008) Evolution of peptidase diversity. J Biol Chem 283(44):30010–30014. https://doi.org/10.1074/jbc.M804650200

Pereira WES, da Silva RR, de Amo GS, Ruller R, Kishi LT, Boscolo M et al (2020) A collagenolytic aspartic protease from *Thermomucor indicae-seudaticae* expressed in *Escherichia coli* and *Pichia pastoris*. Appl Biochem Biotechnol 191(3):1258–1270. https://doi.org/10.1007/s12010-020-03292-z

Pillai P, Mandge S, Archana G (2011) Statistical optimization of production and tannery applications of a keratinolytic serine protease from *Bacillus subtilis* P13. Process Biochem 46(5):1110–1117. https://doi.org/10.1016/j.procbio.2011.01.030

Prabhu N, Gajendran T, Karthikadevi S, Archana A, Arthe R (2021) Utilization of sugarcane bagasse for enhancement production of fibrinolytic enzyme using statistical approach. Clean Eng Technol 5:100269. https://doi.org/10.1016/j.clet.2021.100269

Puente XS, Sánchez LM, Overall CM, López-Otín C (2003) Human and mouse proteases: a comparative genomic approach. Nat Rev Genet 4(7):544–558. https://doi.org/10.1038/nrg1111

Qazi JI, Jamshaid H, Nadeem M, Ali SS (2008) Production of proteases by *Aspergillus niger*, through solid state fermentation. Punjab Univ J Zool 23:037–046

Radnedge L, Agron PG, Hill KK et al (2003) Genome differences that distinguish *Bacillus anthracis* from *Bacillus cereus* and *Bacillus thuringiensis*. Appl Environ Microbiol 69(5): 2755–2764. https://doi.org/10.1128/AEM.69.5.2755-2764.2003

Rao CS, Sathish T, Ravichandra P, Prakasham R (2009) Characterization of thermo-and detergent stable serine protease from isolated *Bacillus circulans* and evaluation of eco-friendly applications. Process Biochem 44(3):262–268. https://doi.org/10.1016/j.procbio.2008.10.022

Rathod MG, Pathak AP (2014) Wealth from waste: optimized alkaline protease production from agro-industrial residues by *Bacillus alcalophilus* LW8 and its biotechnological applications. J Taibah Univ Sci 8(4):307–314. https://doi.org/10.1016/j.jtusci.2014.04.002

Ravindran B, Kumar AG, Aruna Bhavani PS, Ganesan Sekaran G (2011) Solid-state fermentation for the production of alkaline protease by *Bacillus cereus* 1173900 using proteinaceous tannery solid waste. Curr Sci 100(5):726–730

Rawlings ND, Barrett AJ, Thomas PD, Huang X, Bateman A, Finn RD (2017) The MEROPS database of proteolytic enzymes, their substrates and inhibitors in 2017 and a comparison with peptidases in the PANTHER database. Nucleic Acids Res 46(D1):D624–D632. https://doi.org/10.1093/nar/gkx1134

Richardson DP, Ansell J, Drummond LN (2018) The nutritional and health attributes of kiwifruit: a review. Eur J Nutr 57(8):2659–2676. https://doi.org/10.1007/s00394-018-1627-z

Sathiya G (2013) Production of protease from *Bacillus subtilis* and its application in leather making process. Int J Res Biotechnol Biochem 3(4):7–10

Shaikh IA, Turakani B, Malpani J, Goudar SV, Mahnashi MH, Hamed Al-Serwi R et al (2023) Extracellular protease production, optimization, and partial purification from *Bacillus nakamurai* PL4 and its applications. J King Saud Univ Sci 35(1):102429. https://doi.org/10.1016/j.jksus.2022.102429

Sharma KM, Kumar R, Panwar S, Kumar A (2017) Microbial alkaline proteases: optimization of production parameters and their properties. J Genet Eng Biotechnol 15(1):115–126. https://doi.org/10.1016/j.jgeb.2017.02.001

Simkhada JR, Mander P, Cho SS, Yoo JC (2010) A novel fibrinolytic protease from *Streptomyces* sp. CS684. Process Biochem 45(1):88–93. https://doi.org/10.1016/j.procbio.2009.08.010

Singh PK, Kumar S, Kumar P, Bhat ZF (2012) Pulsed light and pulsed electric field-emerging non thermal decontamination of meat. Am J Food Technol 7(9):506–516. https://doi.org/10.3923/ajft.2012.506.516

Su C, Gong JS, Zhang RX et al (2017) A novel alkaline surfactantstable keratinase with superior feather-degrading potential based on library screening strategy. Int J Biol Macromol 95:404–411. https://doi.org/10.1016/j.ijbiomac.2016.11.089

Sugumaran KR, Ponnusami V, Gowdhaman D, Gunasekar V, Srivastava SN (2012) Thermostable alkaline protease production from *Bacillus thuringiensis* MTCC 1953: optimization and kinetic studies. Int J ChemTech Res 4(1):198–202

Sumantha A, Larroche C, Pandey A (2006) Microbiology and industrial biotechnology of food-grade proteases: a perspective. Food Technol Biotechnol 44(2):211–220. https://doi.org/10.1177/1082013206063838

Tavano OL (2013) Protein hydrolysis using proteases: an important tool for food biotechnology. J Mol Catal B Enzym 90:1–11. https://doi.org/10.1016/j.molcatb.2013.01.011

Taylor G, Frommherz Y, Katikaridis P, Layer D, Sinning I, Carroni M et al (2022) Antibacterial peptide CyclomarinA creates toxicity by deregulating the *mycobacterium tuberculosis* ClpC1–ClpP1P2 protease. J Biol Chem 298(4):102202. https://doi.org/10.1016/j.jbc.2022.102202

Thompson VF, Saldaña S, Cong J, Goll DE (2000) A BODIPY fluorescent microplate assay for measuring activity of calpains and other proteases. Anal Biochem 279(2):170–178. https://doi.org/10.1006/abio.1999.4475

Vaid S, Bajaj BK (2017) Production of ionic liquid tolerant cellulase from *Bacillus subtilis* G2 using agroindustrial residues with application potential for saccharification of biomass under one pot consolidated bioprocess. Waste Biomass Valoriz. 8(4):949–964. https://doi.org/10.1007/s12649-016-9642-x

Vijayaraghavan P, Vincent SP (2015) A low cost fermentation medium for potential fibrinolytic enzyme production by a newly isolated marine bacterium, *Shewanella* sp. IND20. Biotechnol Rep 7:135–142. https://doi.org/10.1016/j.btre.2015.06.005

Wang SL, Wu YY, Liang TW (2011) Purification and biochemical characterization of a nattokinase by conversion of shrimp shell with *Bacillus subtilis* TKU007. New Biotechnol 28(2):196–202. https://doi.org/10.1016/j.nbt.2010.10.002

Wang J, Xu A, Wan Y, Li Q (2013) Purification and characterization of a new metallo-neutral protease for beer brewing from *Bacillus amyloliquefaciens* SYB-001. Appl Biochem Biotechnol 170(8):2021–2033. https://doi.org/10.1007/s12010-013-0350-8

Xue B, Dunker AK, Uversky VN (2012) Orderly order in protein intrinsic disorder distribution: disorder in 3500 proteomes from viruses and the three domains of life. J Biomol Struct Dyn 30(2):137–149. https://doi.org/10.1080/07391102.2012.675145

Yang H, Zhu Q, Zhou N, Tian Y (2016) Optimized expression of prolyl aminopeptidase in *Pichia pastoris* and its characteristics after glycosylation. World J Microbiol Biotechnol 32(11):176. https://doi.org/10.1007/s11274-016-2135-z

Yang Y, Shen D, Long Y, Xie Z, Zheng H (2017) Intrinsic peroxidase-like activity of ficin. Sci Rep 7(1):43141. https://doi.org/10.1038/srep43141

Yeo WS, Seo MJ, Kim MJ et al (2011) Biochemical analysis of a fibrinolytic enzyme purified from *Bacillus subtilis* strain A1. J Microbiol 49(3):376–380. https://doi.org/10.1007/s12275-011-1017-1

Yoo S, Han MS (2021) New strategy to design fluorescent substrates of carboxypeptidases using a combination of dansylated peptides and albumin. Dyes Pigments 196:109804. https://doi.org/10.1016/j.dyepig.2021.109804

Yoshioka M, Miwa T, Horii H et al (2007) Characterization of a proteolytic enzyme derived from a *Bacillus* strain that effectively degrades prion protein. J Appl Microbiol 102(2):509–515. https://doi.org/10.1111/j.1365-2672.2006.03080.x

Zanutto-Elgui MR, Vieira JCS, do Prado DZ, Buzalaf MAR, de Magalhães PP (2019) Production of milk peptides with antimicrobial and antioxidant properties through fungal proteases. Food Chem 278:823–831. https://doi.org/10.1016/j.foodchem.2018.11.119

Zhu X, Kaur L, Boland M (2018a) Thermal inactivation of actinidin as affected by meat matrix. Meat Sci 145:238–244. https://doi.org/10.1016/j.meatsci.2018.06.032

Zhu X, Kaur L, Staincliffe M, Boland M (2018b) Actinidin pretreatment and sous vide cooking of beef brisket: effects on meat microstructure, texture and in vitro protein digestibility. Meat Sci 145:256–265. https://doi.org/10.1016/j.meatsci.2018.06.035

Zins MM, Zimprich CA, Petermann SR, Rust L (2001) Expression and partial characterization of an elastase from *Chromobacterium violaceum*. Vet Microbiol 80(1):63–74. https://doi.org/10.1016/s0378-1135(00)00370-9

Chapter 16
Microbial Proteases for Detergent and Cleaning Applications

Latika Shendre and Vasudeo Zambare

Abstract Proteases have revolutionized the detergent industry by offering eco-efficient and high-performance alternatives to traditional chemical-based cleaners. This chapter presents a comprehensive overview of the historical evolution, biochemical mechanisms, and industrial applications of proteases in detergents. Microbial alkaline proteases, especially those from *Bacillus* species, are widely used for their ability to break down protein-based stains under alkaline and high-temperature conditions. The evolution of enzyme technology—from natural extracts to engineered enzymes—has enhanced protease stability, substrate specificity, and performance in detergent formulations. Key advancements in protein engineering, including site-directed mutagenesis and directed evolution, have enabled the development of enzymes that withstand harsh detergent environments. Proteases not only improve stain removal at lower temperatures but also reduce energy consumption and chemical usage, contributing to environmental sustainability. This chapter also explores formulation strategies, regulatory frameworks, market dynamics, and emerging trends in enzyme-based cleaning solutions. Ultimately, the integration of biotechnological innovations with detergent formulations signifies a major step toward safer, efficient, and sustainable cleaning products for both household and industrial use.

Keywords Protease · Detergents · Enzyme engineering · Alkaline protease · Stability · Sustainable cleaning · *Bacillus* · Protein hydrolysis · Enzymatic detergents

L. Shendre (✉)
Dr. D. Y. Patil Biotechnology and Bioinformatics Institute, Dr. D. Y. Patil Vidyapeeth, Pune, Maharashtra, India
e-mail: latika.shendre@dpu.edu.in

V. Zambare (✉)
R&D Department, Balaji Enzyme & Chemical Pvt Ltd, Mumbai, Maharashtra, India

V. Zambare et al. (eds.), *Biotechnological Advances and Applications in Microbial Proteases*, Interdisciplinary Biotechnological Advances,
https://doi.org/10.1007/978-981-95-9011-7_16

16.1 Introduction

Proteases (EC 3.4) have become an important part of modern detergents due to their ability to break down proteins into smaller fragments such as peptides and amino acids. This action makes them especially effective against stubborn stains caused by substances such as blood, milk, eggs, sweat, meat, and grass—types of stains that are often resistant to ordinary washing methods (Gupta et al. 2002a; Nilegaonkar et al. 2006; Zambare et al. 2014). Unlike typical detergent ingredients like surfactants or builders, which act more generally, proteases specifically target the proteins that cling stubbornly to fabric fibers. As a result, they have become an essential component of today's cleaning formulations. Among the various types of proteases, those produced by microbes especially alkaline serine proteases from *Bacillus* subtilis, *Bacillus pumilus*, *B. licheniformis*, and *B. amyloliquefaciens*, are widely preferred in detergent formulations. These enzymes work best in alkaline conditions (pH 8–11), which suits the environment of most washing processes (Aruna et al. 2023). They're also stable in the presence of common detergent ingredients like oxidants, surfactants, and chelating agents (Maurer 2004; Saeki et al. 2007). Thanks to these traits, they're used in both powdered and liquid detergent varieties, as well as in specialized products like prewash stain removers and automatic dishwashing detergents. Proteases do more than clean stains—they also help take care of fabrics over time. By keeping clothes bright, stopping dirt from settling back during the wash, and making low-temperature cleaning more effective, they help save energy and extend the life of garments. These practical benefits make them valuable for everyday use and for the detergent industry as a whole (Kumar and Takagi 1999).

The use of enzymes in detergents began in the early 1960s, marking a significant shift in how cleaning products were formulated. One of the earliest examples was Burnus, introduced by the German company Rohm, which contained a crude protease extract (Horikoshi 1999). This innovation signaled a move away from relying entirely on harsh chemical ingredients, toward using biologically active compounds capable of breaking down stains more precisely. A major advancement came with the adoption of microbial alkaline proteases, produced through submerged fermentation of *Bacillus* species, particularly *Bacillus licheniformis*. These enzymes performed reliably under the alkaline and high-temperature conditions typical of laundry processes and soon became a standard component in detergent manufacturing (Gupta et al. 2002b). Enzyme technology has since evolved with the help of protein engineering, site-directed mutagenesis, and directed evolution. These improvements in enzyme technology have made it possible to develop proteases that remain stable at higher temperatures, withstand exposure to oxidants, and work more precisely on protein-based stains. Companies such as Novozymes, DuPont (formerly Genencor), BASF, and AB Enzymes have played a leading role in these developments, producing enzymes suited to a wide range of detergent formulations and regional washing practices (Godfrey 1996) Proteases now make up nearly two-thirds of the enzymes used in detergents and account for over 60% of total enzyme volume in this industry (Anwar and Saleemuddin 1998). Enzyme-based detergents are widely adopted in Europe, North America, and many developing countries, where demand is increasing for cleaning products that are both efficient and environmentally responsible.

With increasing attention on environmental impact and energy use, enzymatic detergents have taken on a more significant role in the cleaning sector. Earlier detergent formulations often relied on synthetic surfactants and phosphate-based builders, which are known to contribute to water pollution, eutrophication, and harm to aquatic ecosystems (Kirk et al. 2002). Proteases offer a more sustainable option—they are biodegradable, work well in small amounts, and reduce the need for strong chemical additives. A key advantage of protease-based detergents is their ability to clean effectively at lower temperatures, usually between 30 °C and 40 °C, unlike conventional products that often require washing at 60 °C or higher. Washing at lower temperatures can reduce energy use by up to 60%, lowering electricity costs and cutting down on carbon emissions. Proteases play a crucial role in making this energy-saving shift possible without compromising cleaning results. For manufacturers, the inclusion of proteases improves cleaning performance while reducing reliance on more expensive chemical ingredients. This makes detergent formulations more cost-effective and supports the creation of compact and concentrated products that are easier to store, handle, and distribute (Cherry and Fidantsef 2003). The global detergent enzyme market, now worth over USD 1 billion, continues to grow in response to rising interest in sustainable consumer products. Proteases remain the most widely used enzymes in this field, reflecting their proven value in both product effectiveness and user satisfaction. In institutional and commercial settings—such as hospitals, hotels, and food processing facilities, protease-based detergents offer practical benefits. They help reduce water consumption, shorten rinse cycles, and work well with automated dosing systems, all of which support greater efficiency and lower operating costs.

This chapter aims to provide a comprehensive understanding of the role of proteases in the detergent industry, encompassing their biochemical principles, historical evolution, and industrial applications. The objectives of this chapter are to explore the biochemical properties of proteases, including their classification, catalytic mechanisms, substrate specificity, and molecular structure–function relationships that make them suitable for detergent formulations. The chapter then traces the historical development of enzyme-based detergents, highlighting key milestones in the commercial use of proteases and the transformation of detergent formulations over time. In addition, it examines the industrial and economic significance of proteolytic enzymes, including their contribution to enhanced cleaning performance, energy efficiency, and environmental sustainability. Furthermore, this chapter analyzes protein engineering strategies and formulation techniques used to optimize enzyme activity under harsh detergent conditions such as high pH, elevated temperatures, and oxidative stress. Finally, it assesses the current challenges, innovations, and future trends in protease development and their integration into advanced, sustainable cleaning technologies.

The scope of this chapter includes an interdisciplinary overview, integrating molecular biology, enzymology, protein engineering, and industrial bioprocessing.

It is intended for students, researchers, and professionals in biotechnology, biochemical engineering, and industrial enzyme development who seek to understand the foundational and applied aspects of proteases in the detergent sector.

16.2 Detergent Protease: Development and Evolution

The introduction of proteolytic enzymes into detergent formulas brought a major shift in how cleaning products were designed and used. This development began in the early 1960s with the launch of Biotex by Novo Industry A/S (now Novozymes). Released in 1963, Biotex was among the first detergents to include microbial proteases. It was sold as a presoak powder aimed at breaking down stubborn stains like blood, milk, and egg. Although the protease used—extracted from *Bacillus licheniformis*—was relatively unrefined and less stable than those used today, it was still effective against protein-based stains and helped establish a new class of biologically active cleaning products (Horikoshi 1999; Godfrey 1996; Maurer 2004). The success of Biotex sparked interest among detergent manufacturers in Europe and North America, prompting many to explore the potential of enzyme-based cleaning products. As more households adopted washing machines, the need for effective stain removal became a priority. Major brands like Procter & Gamble, Unilever, and Henkel began adding enzymes to their detergents to improve cleaning results and distinguish their products in a competitive market. One clear advantage was that enzymes worked well even at lower temperatures, appealing to consumers aiming to reduce electricity use (Kirk et al. 2002). Despite this early momentum, the adoption of enzymes was not without challenges. In the late 1960s and 1970s, reports began to surface linking enzyme dust from powdered detergents to respiratory health problems among factory workers. Cases of asthma and hypersensitivity pneumonitis raised concerns about occupational safety. As a result, regulatory agencies took notice, and some manufacturers reduced or temporarily halted the use of enzymes in their products. In response, both detergent companies and enzyme suppliers worked to make enzyme handling safer. New formulation techniques were introduced to convert enzymes into granulated or encapsulated forms. Using methods like prilling and microencapsulation, enzymes were turned into larger, dust-free particles that greatly reduced the risk of inhalation (Cherry and Fidantsef 2003). These changes not only improved safety for workers handling enzymes but also made enzyme formulations more stable and longer-lasting. With regulatory bodies like OSHA in the United States and the European Economic Community setting clear exposure limits, manufacturers were better equipped to create safer production environments. These steps helped restore trust in enzyme-based detergents and encouraged their continued use. Once these safety issues were addressed, efforts shifted back to improving how well the enzymes worked. By the 1970s and 1980s, subtilisin-type proteases—especially Subtilisin Carlsberg and Subtilisin BPN'—had become widely used in laundry detergents. Extracted from *Bacillus licheniformis* and *Bacillus amyloliquefaciens*, these enzymes worked efficiently in the alkaline conditions typical of wash cycles and proved especially effective at breaking down protein-based stains like blood and food residues (Gupta et al. 2002a; Kumar and Takagi 1999). The late 1980s marked another turning point. Advances in molecular biology allowed scientists to fine-tune enzymes for better performance using genetic techniques. Through site-directed mutagenesis and recombinant DNA methods, researchers were able to enhance enzyme resistance to heat and oxidation and improve their ability to target specific stain types. One such development was Subtilisin 309, introduced by Novozymes, which featured structural changes that helped the enzyme maintain activity under the demanding conditions of modern laundry. In the years that followed, the use of directed evolution—a process that mimics natural selection—enabled the creation of enzymes suited to cold-water and quick-wash settings, meeting changing consumer preferences for faster, energy-saving laundry (Al-Ghanayem and Joseph 2020). These newer enzymes were more versatile than earlier versions. They could be used in both liquid and powder detergents, performed well in hard or soft water, and worked effectively in hand and machine washing. Their compatibility with a wide range of ingredients—like surfactants, chelators, and fragrance agents—allowed manufacturers to create more specialized products tailored to different consumer needs and local conditions.

By the early 2000s, enzymatic detergents were no longer a niche product. They had become widely used in global markets. In many parts of Asia, Latin America, and Africa, growing access to washing machines and rising incomes supported demand for high-performance detergents. Detergent manufacturers such as Novozymes, DuPont (formerly Genencor), BASF, and AB Enzymes began adapting their enzyme formulations to suit local washing conditions. These adjustments helped ensure reliable performance across a range of environments—whether dealing with hard or soft water, high humidity, or stain types common to specific regions (Lund et al. 2012). As safety practices improved, public trust in enzyme-based detergents began to rise. Earlier concerns about allergic reactions and skin irritation were addressed through clearer product labels, thorough dermatological testing, and better methods of incorporating enzymes into detergents. These changes helped reduce direct skin contact and limited airborne exposure, making these products safer for both consumers and factory workers. At the same time, advertising began to emphasize the benefits that mattered most to modern consumers—like natural origins, biodegradability, and energy savings—which made enzyme-based detergents more appealing to those seeking safer and more sustainable choices (Nguyen et al. 2025). Regulatory systems also evolved alongside these changes. In the European Union, enzymes in detergents were brought under the REACH (Registration, Evaluation, Authorization and Restriction of Chemicals) legislation, requiring thorough testing for human and environmental safety. In the United States, the FDA and EPA took on oversight roles to ensure that enzymes used in household and food-related products met safety standards. Commonly used proteases, including variants of subtilisin, received GRAS (Generally Recognized as Safe) classification. Industry groups like the Enzyme Technical Association (ETA) and the Association of Manufacturers and Formulators of Enzyme Products (AMFEP) continued to collaborate with both manufacturers and regulators to maintain high safety and sustainability standards (Kirk et al. 2002). In short, the journey of proteases in detergents tells a story of steady innovation and adaptation. What began with basic microbial extracts has grown into a well-regulated and consumer-trusted technology—bringing together effective stain removal, better environmental outcomes, and improved safety for both workers and users.

16.3 Biochemistry and Molecular Mechanisms

16.3.1 Fundamental Properties of Proteases

Proteases, or proteolytic enzymes, are essential biological catalysts responsible for cleaving peptide bonds within proteins and peptides. By facilitating this hydrolysis, they convert large polypeptide chains into smaller peptides and eventually into free amino acids (Rao et al.,1998). This function is not only crucial for maintaining protein homeostasis in living organisms but also underpins a range of physiological processes such as digestion, immune defense, blood coagulation, tissue remodeling, and apoptosis. Their presence is universal, spanning all forms of life—from bacteria and archaea to plants, animals, and humans. Because of their biological importance and catalytic versatility, proteases have also become indispensable in several industries, including food processing, leather treatment, pharmaceuticals, and household detergents (Nilegaonkar et al. 2007a, b).

Proteases are commonly defined as enzymes that hydrolyze the amide bonds linking amino acids in polypeptides. While this reaction is thermodynamically favorable, it is extremely slow under physiological conditions due to a high activation energy barrier. Proteases overcome this barrier by stabilizing the transition state and efficiently catalyzing peptide bond cleavage (Maurer 2004). They are broadly classified based on the position at which they cleave the polypeptide chain. Endopeptidases act within the interior of the peptide sequence, cleaving internal peptide bonds, whereas exopeptidases function near the terminal ends, releasing single amino acids from either the amino (N-) or carboxyl (C-) terminus (Maurer 2004). A more detailed classification emerges when proteases are grouped according to the nature of the functional group at their active site that participates in catalysis. This biochemical classification recognizes six major types: serine, cysteine, aspartic, metalloproteases, threonine, and glutamic proteases. Each class employs a distinct catalytic mechanism. In serine proteases, a serine residue plays the central role in nucleophilic attack, typically supported by a histidine and aspartate residue forming a catalytic triad (Lund et al. 2012). Cysteine proteases function via a mechanism involving a cysteine thiol group and a histidine, forming a catalytic dyad. Aspartic proteases utilize two aspartate residues to activate a water molecule that then directly attacks the peptide bond, without forming a covalent enzyme-substrate intermediate. Metalloproteases require a divalent metal ion, usually zinc, which activates water and stabilizes the transition state. Threonine proteases depend on an N-terminal threonine, while glutamic proteases use a carboxyl group of a glutamate residue, though this latter class is relatively rare (Polgar 2005).

Another axis of classification considers the pH range at which the protease functions optimally. Acid proteases, such as pepsin, are most active in highly acidic environments like the stomach. Neutral proteases operate effectively at physiological pH, and alkaline proteases show optimum activity in basic conditions. Alkaline proteases, especially those derived from *Bacillus* species, have become the cornerstone of many detergent formulations due to their robust activity and stability in high-pH and high-temperature environments (Rao et al. 1998; Zambare 2010a).

To provide a standardized reference framework, the MEROPS database systematically categorizes proteases based on sequence similarity, structure, and evolutionary relationships. This classification groups enzymes into families and clans, offering an informative overview that facilitates comparative studies and predictions about unknown or newly discovered enzymes (Barrett et al. 2001.

Proteolysis, while chemically straightforward in principle, involves complex molecular mechanisms that vary among the different types of proteases. At the heart of the reaction is the nucleophilic attack on the carbonyl carbon of the peptide bond, leading to the formation of a tetrahedral intermediate. These intermediate collapses break the peptide bond and release the cleaved fragments. In serine proteases, the catalytic triad composed of serine, histidine, and aspartate residues orchestrates this process. Histidine abstracts a proton from the hydroxyl group of the serine, making it a stronger nucleophile, while aspartate stabilizes the positive charge on histidine (Rajagopalan et al. 2014). The serine then attacks the carbonyl carbon of the peptide bond, forming a transient tetrahedral intermediate that is stabilized by an oxyanion hole—a structural feature comprising backbone amide groups that donate hydrogen bonds to the negatively charged oxygen atom. The intermediate collapses to form an acyl-enzyme complex, which is then hydrolyzed by a water molecule to release the product and regenerate the free enzyme (Polgar 2005; Zambare 2010b).

Cysteine proteases follow a similar catalytic pathway, though they employ a thiol group from a cysteine residue as the nucleophile. The thiol is activated by a histidine residue, forming a cysteine-histidine dyad. Aspartic proteases, on the other hand, do not form a covalent intermediate. Instead, they use two aspartate residues to polarize a water molecule, which then directly attacks the peptide bond. This mechanism is particularly suited to acidic environments, such as the stomach or lysosomes, where aspartic proteases are commonly found (Verma et al. 2016).

Metalloproteases take advantage of a coordinated zinc ion to activate a water molecule and facilitate peptide bond hydrolysis. The metal ion not only activates the water molecule for nucleophilic attack but also helps stabilize the negative charges that develop during the transition state. A canonical example is thermolysin, which has been extensively studied for both its natural function and its industrial applications (Klein and Bischoff 2011).

From a kinetic perspective, proteases typically follow Michaelis-Menten kinetics, characterized by parameters such as Km, Vmax, and kcat. The Km value reflects the enzyme's affinity for its substrate; a lower Km indicates higher affinity. The Vmax represents the maximum rate of the enzymatic reaction when the enzyme is saturated with substrate. The turnover number, or kcat, denotes the number of substrate molecules converted to product per unit of time by a single enzyme molecule. The catalytic efficiency of a protease is often represented by the kcat/Km ratio, which combines information about both substrate affinity and catalytic turnover. For industrial applications, particularly in detergents, high catalytic efficiency under alkaline conditions, along with stability at elevated temperatures and resistance to harsh chemicals, is essential for effective enzyme performance (Das and Ghosh 2023).

Understanding the function of proteases also requires a detailed examination of their structure. Most proteases are compact, globular proteins, with their catalytic residues nestled within a cleft or cavity that serves as the active site. This active site is not only the locus of chemical reactivity but also the region where substrate recognition occurs. Substrate-binding regions, referred to as subsites, are located around the scissile bond and are often designated S1, S2, S3, and so forth, each interacting with specific residues of the substrate. The arrangement and chemical characteristics of these subsites largely determine the specificity of the protease. For example, trypsin selectively cleaves peptide bonds after positively charged residues like lysine and arginine due to the presence of a negatively charged aspartate residue in its S1 pocket (Ochoa et al. 2020).

In addition to the catalytic and substrate-binding regions, other structural features contribute to the enzyme's overall stability and functionality. Disulfide bridges help maintain the integrity of the protein under denaturing conditions. Salt bridges and hydrogen bonds reinforce the tertiary structure, especially at elevated temperatures. These structural elements are particularly important for proteases used in detergent formulations, where the enzyme must withstand alkaline pH, high temperatures, surfactants, and oxidizing agents (Yao et al. 2023).

Insights into the structure-function relationship of proteases have been significantly advanced through techniques such as X-ray crystallography and nuclear magnetic resonance spectroscopy, as well as through computational modelling. Experimental approaches like site-directed mutagenesis have enabled the engineering of proteases with enhanced properties. For instance, modifications in subtilisin, a widely used bacterial serine protease, have led to variants with greater thermostability, altered substrate specificity, and improved resistance to chemical denaturation. These engineered proteases—such as Savinase and Alcalase—are now routinely used in commercial detergent products, where they contribute to efficient protein stain removal under challenging washing conditions (Von der Osten et al. 1993).

16.3.2 Mechanism of Protein Hydrolysis

The hydrolysis of peptide bonds by proteolytic enzymes is a fundamental biochemical process that supports critical cellular functions such as protein turnover, cell signaling, apoptosis, and nutrient assimilation. This process, while chemically simple in principle, unfolds through a precise and highly regulated catalytic cycle involving substrate recognition, bond cleavage, and product release. Each stage of this cycle is intricately shaped by the structural features of the enzyme and influenced by the physicochemical environment in which it operates (Hedstrom 2002).

16.3.2.1 Enzyme–Substrate Interaction and Binding

The first step in proteolysis is the interaction between the protease and its substrate. This interaction is not a random event but is governed by molecular complementarity between the substrate's peptide backbone and the binding site of the enzyme. The substrate typically binds within a cleft or groove on the enzyme's surface that is shaped to accommodate specific amino acid sequences. This interaction is stabilized by multiple non-covalent forces, including hydrogen bonds, van der Waals interactions, hydrophobic effects, and electrostatic attractions (Hedstrom 2002). In many proteases, substrate recognition is mediated through defined substrate-binding subsites located around the active site. These subsites are designated S1, S2, S3, etc., on the enzyme and interact with corresponding P1, P2, P3 residues on the substrate. The specificity of a protease is largely determined by the chemical environment of these binding pockets. For instance, the S1 pocket of trypsin contains a negatively charged aspartate residue that strongly favors binding of positively charged side chains such as lysine and arginine (Bryan 2000). Furthermore, binding often induces minor conformational changes in the enzyme, a phenomenon described as the "induced fit" mechanism, which optimally positions the substrate for catalysis (Koshland Jr 1958).

16.3.2.2 Catalytic Cycle and Product Release

Once the substrate is bound, the enzyme proceeds through a series of catalytic steps that culminate in peptide bond cleavage. Although the exact mechanism varies by enzyme class, all proteases act by lowering the activation energy required for hydrolysis, typically by stabilizing a high-energy transition state.

In serine proteases, such as subtilisin and chymotrypsin, the catalytic triad—serine, histidine, and aspartate—plays a central role in facilitating nucleophilic attack on the peptide bond. The hydroxyl group of serine, activated by histidine (which in turn is stabilized by aspartate), attacks the carbonyl carbon of the peptide bond, forming a tetrahedral intermediate. This intermediate is stabilized by the enzyme's oxyanion hole, a structural feature formed by backbone amide groups that hydrogen-bond with the negatively charged oxygen atom of the intermediate (Hedstrom 2002; Bryan 2000).

The intermediate collapses, resulting in cleavage of the peptide bond and the release of the amino-terminal product. A covalent acyl-enzyme complex is formed with the carboxyl-terminal part of the substrate. In the next phase, a water molecule—again activated by the catalytic triad—attacks this complex, forming a second tetrahedral intermediate that then collapses to release the second product and regenerate the free enzyme. This two-step process, involving acylation and deacylation, is a hallmark of serine and cysteine proteases (Basketter et al. 2008).

Cysteine proteases follow a similar mechanism but use a thiolate anion from cysteine as the nucleophile. Aspartic proteases, such as pepsin and HIV protease, do not form covalent intermediates. Instead, two aspartate residues polarize and activate a water molecule, which then directly attacks the peptide bond (Turk et al. 2012). Metalloproteases also employ activated water, coordinated by a metal ion (usually Zn^{2+}), to act as the nucleophile. This coordination not only stabilizes the negative charge that develops during the reaction but also contributes to the proper orientation of substrate and water (Stocker et al. 1995).

The final step—product release—is often the rate-limiting stage of the cycle. After hydrolysis, the two resulting peptide fragments must dissociate from the enzyme's active site. This may require a subtle rearrangement of enzyme conformation or a decrease in binding affinity once catalysis is complete. Efficient product release is essential to allow the enzyme to engage in further catalytic cycles, especially in industrial or physiological settings where high turnover is required (Rao et al. 1998).

16.3.2.3 Factors Affecting Enzymatic Activity

The efficiency of proteolytic hydrolysis depends on a variety of internal and external factors. Among the most critical are temperature, pH, substrate concentration, and the presence of inhibitors or stabilizing agents.

Temperature influences both the rate of reaction and the stability of the enzyme. While increased temperatures generally accelerate enzymatic activity by enhancing molecular motion, excessively high temperatures can destabilize the enzyme's tertiary structure, leading to denaturation and loss of function. Proteases from thermophilic organisms, such as *Bacillus licheniformis*, exhibit enhanced thermal stability and are well suited for high-temperature applications such as in detergent and food industries (Gupta et al. 2002b).

pH affects the ionization state of catalytic residues and substrate-binding sites. Each protease has an optimal pH range that reflects the protonation requirements of its active site. For example, pepsin functions optimally at pH 1.5–2.0 in the stomach, while subtilisin and other alkaline proteases perform best at pH 9–11, conditions often found in laundry detergents (Maurer 2004). Deviations from the optimal pH can lead to reduced substrate binding or impaired catalytic turnover.

Substrate concentration, a core variable in enzyme kinetics, follows the principles of Michaelis–Menten behavior. At low concentrations, the rate of hydrolysis increases nearly linearly with substrate levels. However, as the enzyme becomes saturated, the rate approaches a maximum velocity (V_{max}). The Km value reflects the substrate concentration at which the reaction proceeds at half of Vmax, and the kcat/Km ratio is widely used as a measure of catalytic efficiency under physiological or industrial conditions (Rao et al. 1998).

Inhibitors—both natural and synthetic—can profoundly affect protease activity. Natural inhibitors such as serpins (serine protease inhibitors) serve important regulatory roles in biological systems. Synthetic inhibitors are often designed to bind irreversibly to the active site or to mimic the transition state, thereby blocking enzymatic function. Competitive inhibitors resemble the natural substrate and bind the active site, while noncompetitive inhibitors interact with the enzyme at an allosteric site, altering its conformation and reducing activity (Turk et al. 2012).

Finally, environmental components such as surfactants, oxidizing agents, chelators, and organic solvents can also influence protease activity. For industrial applications, especially in the detergent industry, proteases must be stable and active under harsh conditions. Enzymes are often engineered or selected for traits such as resistance to oxidative stress, tolerance to high alkalinity, and stability in the presence of chelating agents like EDTA (Maurer 2004; Gupta et al. 2002a).

Taken together, the mechanism of protein hydrolysis reflects a finely tuned series of molecular events that are both robust and adaptable. From substrate recognition and catalytic turnover to the influence of environmental parameters, each aspect of the hydrolytic process has been shaped by evolutionary pressures and, more recently, by human intervention for biotechnological purposes. Understanding these mechanisms at a detailed level is essential for both basic biological research and the rational design of improved protease-based technologies.

16.3.3 Substrate Specificity and Soil Targeting

The success of proteases in detergent formulations lies in their ability to selectively and efficiently degrade proteinaceous soils bound to textile fibers. These protein soils are chemically complex and often firmly attached to fabric surfaces, requiring precise enzymatic action to break them down into soluble fragments. Understanding substrate specificity and targeting is therefore central to the optimization of enzymatic detergent systems. Enzymes must recognize a wide range of protein substrates, cleave them under harsh washing conditions, and do so without damaging the fabric or reducing enzyme stability. This section explores the nature of protein soils, the molecular determinants of protease specificity, and how these enzymes are optimized for practical laundry applications (Saeki et al. 2007).

16.3.3.1 Proteinaceous Soils in Laundry Applications

In domestic and industrial laundering, proteinaceous stains represent a persistent challenge due to their chemical diversity and strong adherence to textiles. These stains are typically derived from organic matter such as blood (rich in hemoglobin and serum proteins), sweat (containing albumin, urea, and salts), milk (which includes caseins and whey proteins), egg yolk (rich in lipovitellin and phosvitin), meat juices, body secretions, and food residues. Each type of protein has its own physicochemical characteristics—such as isoelectric point, hydrophobicity, and secondary structure—that influence its interaction with textile fibers (Rao et al. 1998; Gupta et al. 2002a).

Moreover, environmental factors during use—such as heat from ironing or drying, UV exposure, and microbial action—can cause protein stains to undergo cross-linking, oxidation, or denaturation, making them more resistant to removal. For instance, denatured albumin adheres tightly to cotton fibers, forming hydrophobic interactions that are resistant to both mechanical agitation and surfactant action alone (Maurer 2004). These proteinaceous layers often act as binding matrices for other stain components such as lipids, pigments, or dirt particles, thereby creating a multilayered soil structure.

Proteases used in laundry detergents target the peptide bonds within these protein layers, breaking them into smaller oligopeptides and amino acids. This enzymatic hydrolysis not only weakens the structural integrity of the stain but also enhances the action of surfactants, which can now access the hydrophobic interior more effectively. The use of proteases, therefore, not only improves stain removal but also reduces the need for harsh chemicals or high-temperature washes, aligning with environmental and energy-saving goals (Gupta et al. 2002b).

16.3.3.2 Enzyme Selectivity and Substrate Recognition

Protease activity depends critically on the enzyme's ability to recognize and bind to specific peptide bonds within a diverse range of protein substrates. This recognition is mediated by the architecture of the enzyme's active site and its surrounding substrate-binding subsites. The central catalytic cleft of most proteases is flanked by a series of amino acid pockets—commonly referred to as S1, S2, S3, etc.—which interact with side chains of the substrate's residues designated P1, P2, P3, and so on (Abramowitz et al. 1967). The specificity of a protease is defined by the chemical compatibility between these enzyme pockets and the substrate residues.

For instance, subtilisin, a serine protease widely used in detergents, possesses a deep S1 pocket with a preference for bulky hydrophobic residues such as phenylalanine, leucine, or tyrosine. This allows subtilisin to effectively cleave within hydrophobic regions of proteins such as caseins or meat-derived albumins (Bryan 2000). However, unlike therapeutic proteases, which often require precise substrate targeting to avoid off-target effects, detergent proteases must exhibit broad substrate tolerance. This functional promiscuity allows them to act on a wide variety of proteins from different stain types.

The structural plasticity of these enzymes contributes to this adaptability. Studies using crystallography and computational modeling have shown that subtilisin and its engineered variants, such as Savinase and Alcalase, possess flexible surface loops near the substrate-binding cleft, enabling them to accommodate substrates of varying size and shape Hedstrom 2002). The oxyanion hole—a stabilizing feature that supports transition state formation—also plays a key role in catalytic efficiency and is conserved among many detergent-grade proteases.

Advanced protein engineering has further expanded the substrate range of commercial enzymes. By employing techniques such as site-directed mutagenesis and directed evolution, researchers have altered amino acid residues near the active site to improve substrate affinity or catalytic turnover. For example, replacing methionine residues near the active site with oxidation-resistant amino acids like leucine has yielded proteases that retain activity in bleach-containing detergents (Maurer 2004). Furthermore, chimeric proteases have been developed by combining domains from different enzymes, thereby broadening their pH tolerance and substrate range without compromising stability.

Beyond individual peptide bonds, protease selectivity is also influenced by the secondary and tertiary structure of the substrate. Native proteins with tightly folded domains or extensive disulfide bonding—such as keratins in hair or fibrillar proteins in dried blood—can limit protease accessibility. In such cases, detergent formulations may include reducing agents or chaotropes to help unfold the protein substrate and expose cleavage sites for enzymatic attack (Gupta et al. 2002a).

16.3.3.3 Optimization of Cleaning Performance

To maximize cleaning performance in real-world conditions, detergent proteases must retain their catalytic efficiency under chemically and physically challenging environments. Modern laundry systems expose enzymes to high alkalinity (pH 9–11), elevated temperatures (up to 60 °C), surfactants, oxidizing agents (like sodium percarbonate or perborate), and mechanical stress during washing. The optimization of proteases for these conditions involves both biological and chemical strategies.

One primary goal is enhancing enzyme stability at alkaline pH. Alkaline proteases sourced from *Bacillus licheniformis*, *Bacillus subtilis*, or *Thermomyces lanuginosus* have naturally evolved to function in high-pH niches and serve as starting points for further engineering. Subtilisin Carlsberg, Savinase, and Esperase are examples of such enzymes that maintain catalytic integrity in alkaline conditions while exhibiting resistance to denaturation (Maurer 2004). Directed mutagenesis has been used to strengthen intramolecular interactions, such as salt bridges and hydrogen bonds, that stabilize the enzyme fold at high pH (Bryan 2000).

Thermal stability is another important factor, especially in regions where high-temperature washing is standard. Enzymes for such markets are designed to retain structure and activity at temperatures exceeding 60 °C. Strategies to achieve this include the introduction of additional disulfide bonds, replacement of thermolabile residues, and surface charge modifications to improve heat resistance (Rao et al. 1998). In contrast, cold-active proteases are also gaining attention due to rising demand for energy-efficient, low-temperature wash cycles. Enzymes derived from psychrophilic organisms are adapted for activity at 15–30 °C, but often require trade-offs between stability and flexibility (Gupta et al. 2002a).

Resistance to detergent additives is another optimization target. Oxidants present in detergents, intended to enhance whitening and microbial control, can oxidize sensitive residues (e.g., methionine, cysteine) near the active site. Engineering enzymes with substitutions that eliminate these reactive residues has led to bleach-stable protease variants. For example, the replacement of methionine-222 in subtilisin with alanine or leucine drastically improves oxidative stability (Bryan et al. 1992).

In addition, formulation chemistry plays a vital role in enzyme performance. Proteases must remain active throughout the product's shelf life, whether in powder or liquid form. Stability during storage is maintained through protective coatings, enzyme granulation, or encapsulation with inert carriers. This protects the enzyme from premature hydrolysis, moisture, and interactions with reactive detergent components (Maurer 2004).

The performance of these optimized enzymes is evaluated through laboratory-scale and field tests using standardized stain swatches (e.g., EMPA 101 blood/milk/ink or WFK-10 fabrics). Cleaning efficiency is measured in terms of stain removal percentage, enzyme residual activity, fabric damage potential, and compatibility with other detergent components. These tests help manufacturers fine-tune enzyme dosage, formulation composition, and washing protocols to deliver effective, sustainable laundry performance across a range of water qualities and user preferences (Gupta et al. 2002b).

16.4 Classification and Sources of Detergent Proteases

16.4.1 Enzyme Classification Systems

Proteases used in detergent formulations are remarkably diverse, both in their catalytic mechanisms and in their environmental preferences. To better understand and categorize this diversity, proteolytic enzymes are classified based on several complementary systems—most notably by their mechanism of action, optimal pH and temperature ranges, and substrate specificity profiles. These classifications not only aid in enzyme identification and selection but also help manufacturers design formulations that are more effective, stable, and suited to specific washing conditions.

16.4.1.1 Mechanistic Classification (Serine, Cysteine, Aspartic, Metallo Proteases)

One of the most fundamental classification systems divides proteases according to the nature of their catalytic mechanism—specifically, the identity of the active-site residue or cofactor that acts as the nucleophile during peptide bond hydrolysis. This mechanistic framework groups proteases into four major types: serine proteases, cysteine proteases, aspartic proteases, and metalloproteases (Hedstrom 2002).

(A) *Serine proteases*: Serine proteases are the most widely used class in detergent applications. These enzymes utilize a nucleophilic serine residue, usually in conjunction with a histidine and an aspartate residue, to catalyze peptide bond cleavage via a charge relay system. The active site forms a transient acyl-enzyme intermediate, which is subsequently hydrolyzed by a water molecule. Members of this group include subtilisin, Savinase, and Alcalase derived primarily from *Bacillus* species (Bryan 2000). Their high activity, robustness, and broad substrate range make them ideal for removing proteinaceous stains in alkaline washing environments.

(B) *Cysteine proteases*: Cysteine proteases use a thiol group from a cysteine residue as the nucleophile, usually aided by a histidine base. They are more sensitive to oxidative inactivation due to the susceptibility of the thiol group to oxidation. As such, their utility in detergent formulations—particularly those containing bleaching agents—is limited, although they are commonly used in food and pharmaceutical applications (Turk et al. 2012).

(C) *Aspartic proteases*: Aspartic proteases rely on a pair of aspartate residues to activate a water molecule for direct nucleophilic attack on the peptide bond. These enzymes function optimally at acidic pH and are not generally favored in laundry detergents, which are usually alkaline. However, they find application in specific niche formulations such as those for enzyme presoaks or dishwashing aids.

(D) *Metalloproteases*: Metalloproteases employ a divalent metal ion, usually Zn^{2+}, at the catalytic center to polarize a water molecule that performs the nucleophilic attack. While some metalloproteases show promise in detergent applications, their dependency on metal ions and sensitivity to chelating agents like EDTA (commonly found in detergents) have limited their widespread adoption in laundry products (Stocker et al. 1995).

The MEROPS database provides a comprehensive classification of proteases based on these mechanistic categories and further organizes them into families and clans based on sequence homology and structure.

16.4.1.2 pH and Temperature Preferences

In addition to mechanistic type, proteases are often classified according to their optimal pH and temperature ranges, which reflect their native environmental origins and determine their suitability for detergent applications. Detergents, particularly those used in laundry, typically function under alkaline conditions (pH 9–11) and moderate to high temperatures (30–60 °C). Thus, detergent-compatible proteases must retain activity and structural stability in these environments (Maurer 2004).

(A) *Alkaline proteases*: Alkaline proteases especially those from *Bacillus licheniformis*, *Bacillus subtilis*, and *Bacillus clausii*, have evolved to function under highly basic conditions. These enzymes typically have a pH optimum around 9–11 and are stable in the presence of anionic surfactants and oxidizing agents—two common detergent additives. The ability to remain active at elevated pH is often linked to amino acid substitutions that stabilize charged interactions within the enzyme and prevent denaturation (Gupta et al. 2002a).

(B) *Thermostable proteases*: Thermostable proteases are derived from thermophilic organisms such as *Thermus aquaticus*, *Thermococcus* spp., and some alkaliphilic *Bacillus* strains. These enzymes maintain catalytic efficiency at elevated temperatures (up to 70 °C), making them suitable for hot-water laundry cycles. Engineering efforts have enhanced thermal stability through techniques like disulfide bond insertion and modification of surface loop regions that typically unfold at high temperatures (Rao et al. 1998).

Conversely, cold-active proteases from psychrophilic organisms are gaining attention due to the growing demand for energy-efficient, low-temperature washing. These enzymes remain active at 10–30 °C but tend to be less stable in storage and under chemical stress. Balancing catalytic flexibility with structural resilience remains a key challenge in their commercial deployment (Maurer 2004).

Classifying proteases based on pH and temperature preferences allows formulators to tailor enzyme systems for specific consumer markets—ranging from high-efficiency machines in colder climates to industrial laundering in hot, alkaline environments.

16.4.1.3 Substrate Specificity Patterns

Proteases can also be classified by their substrate specificity—that is, their ability to recognize and hydrolyze particular peptide sequences or protein structures. This classification is especially important in the context of detergents, where enzymes must target a wide variety of natural protein soils, including blood proteins (e.g., hemoglobin), food residues (e.g., casein, egg albumin), and body secretions (e.g., keratin, mucins). Proteases such as subtilisin are often described as having broad substrate specificity, meaning they can cleave peptide bonds adjacent to multiple types of amino acid residues. This non-specificity is advantageous for detergents, as the enzyme does not require a well-defined sequence for activity and can hydrolyze a wide range of stain proteins regardless of their source (Bryan 2000). Nevertheless, even broad-specificity enzymes exhibit preferences. Subtilisin, for example, prefers large, hydrophobic residues at the P1 site, such as leucine, phenylalanine, or tyrosine. Engineering of the S1 binding pocket can shift this preference, allowing for better performance against target stains (Bryan et al. 1992). On the other hand, narrow-specificity proteases—such as trypsin or chymotrypsin—are less suitable for detergent use because their action is limited to specific cleavage sites, such as after lysine/arginine or aromatic residues, respectively. However, their specificity can be harnessed for specialized cleaning applications, such as in medical textiles or sensitive fabrics. Classification by substrate specificity also considers the structural accessibility of peptide bonds in folded proteins. Many protein stains, particularly those that are dried or heat-denatured, have hydrophobic cores and rigid tertiary structures. Detergent proteases must therefore exhibit high structural flexibility to bind, unfold, and hydrolyze these inaccessible regions. Some advanced formulations incorporate chaotropic agents or nonionic surfactants that help expose buried peptide bonds, improving enzyme access and stain degradation (Gupta et al. 2002b).

In summary, proteases are classified not only by the chemistry of their catalytic sites but also by their ability to tolerate environmental conditions and their effectiveness against practical targets. These classification systems are indispensable for selecting and engineering enzymes that meet the rigorous demands of modern detergent applications.

16.4.2 *Microbial Sources and Production*

The large-scale production of proteases for detergent applications relies almost entirely on microbial systems. Microorganisms provide an ideal source of proteolytic enzymes due to their rapid growth, amenability to genetic modification, scalability of cultivation, and capacity to secrete enzymes directly into the culture medium. Among these, bacteria, particularly species from the *Bacillus* genus, dominate commercial production, while fungi and actinomycetes have also contributed significantly to the diversity and versatility of industrial enzymes. This section explores the primary microbial sources of detergent proteases, alternative producers of growing relevance, and the fermentation technologies used to produce them at industrial scale (Solanki et al. 2021).

16.4.2.1 *Bacillus* Species as Primary Sources

Species within the *Bacillus* genus are by far the most important microbial sources of detergent proteases. These Gram-positive, endospore-forming bacteria are valued for their ability to thrive under harsh environmental conditions, including high pH, elevated temperatures, and nutrient limitations—traits that align well with the functional requirements of detergent proteases (Gupta et al. 2002a). One of the most studied proteases is subtilisin Carlsberg, an alkaline serine protease originally isolated from *Bacillus subtilis*. This enzyme and its engineered derivatives, such as Savinase and Alcalase produced by *Bacillus lentus* and *Bacillus licheniformis*, are standard components in modern laundry detergents (Maurer 2004). These enzymes are prized for their high catalytic efficiency, broad substrate specificity, and stability in alkaline environments, where most household detergents operate (pH 9–11). *Bacillus licheniformis* is particularly favored in industrial enzyme production due to its high extracellular enzyme yield, nonpathogenic nature, and well-characterized genetics. It can produce large quantities of proteases that remain active in the presence of surfactants, bleaching agents, and chelators, making them ideal for modern formulations (Rao et al. 1998). Furthermore, *Bacillus clausii*, *Bacillus pumilus*, and *Bacillus amyloliquefaciens* have also been employed for the production of specialized or engineered variants with enhanced thermal or oxidative stability (Gupta et al. 2002b). These strains have naturally evolved or been engineered to produce enzymes with desirable traits such as cold-activity, rapid inactivation post-wash (to protect fabrics), or resistance to inhibitors.

16.4.2.2 Other Bacterial and Fungal Producers

While *Bacillus* species remain dominant, other microbial groups have contributed valuable protease variants with unique properties that may complement or enhance existing detergent formulations. Among bacteria, genera such as *Streptomyces*, *Pseudomonas*, and *Thermomonospora* have been studied for their production of thermostable and solvent-tolerant proteases. For example, *Pseudomonas aeruginosa* produces elastase-like enzymes with high activity at elevated temperatures, while *Streptomyces griseus* produces serine proteases similar to trypsin but with greater stability at low temperatures (Rao et al. 1998). Thermophilic bacteria like *Thermus aquaticus* and *Geobacillus stearothermophilus* are also noteworthy. These organisms, isolated from hot springs and geothermal environments, produce proteases that remain catalytically active at temperatures exceeding 70 °C. Such enzymes are valuable in industrial laundering where high-temperature cycles are common. Fungal proteases, while less frequently used in detergents, offer advantages in specific formulations (Boyce and Walsh 2012). Species such as *Aspergillus oryzae*, *Aspergillus niger*, and *Rhizopus* spp. produce acid or neutral proteases that are widely used in food and leather industries, and in some enzyme pretreatments for laundry (Gupta et al. 2002b). Fungal enzymes are particularly useful where mild reaction conditions are preferred, such as for delicate fabrics or wool. The main challenge with fungal proteases is their generally lower alkaline stability and slower production rates compared to bacterial systems. Nevertheless, advances in fungal strain engineering, such as heterologous expression of alkaline proteases in *Aspergillus* spp., are gradually bridging this gap (Al-Ghanayem and Joseph 2020).

16.4.2.3 Fermentation and Production Processes

The production of microbial proteases at industrial scale is carried out through controlled fermentation, typically employing submerged fermentation (SmF) or, less frequently, solid-state fermentation (SSF), depending on the microbial strain and enzyme properties desired.

(a) *Submerged fermentation (SmF)*: SMF is the most widely used approach for detergent proteases, especially in large-scale bioreactors. In this method, microorganisms are grown in liquid nutrient media under carefully monitored conditions of pH, temperature, aeration, and agitation. The protease is secreted extracellularly into the culture broth, from which it is later harvested and purified. SmF allows for consistent product quality, ease of scaling, and real-time monitoring of growth and enzyme production (Rao et al. 1998).

(b) *Solid-state fermentation (SSF)*: SSF is primarily used for fungal enzymes or thermotolerant strains. It involves the cultivation of microbes on solid materials such as wheat bran, rice husk, or soybean meal without free-flowing water. SSF systems can yield high enzyme concentrations with lower energy and water input, though they are harder to automate and control compared to SmF (Pandey 2003).

Following fermentation, the protease-rich broth undergoes downstream processing, which includes filtration, ultrafiltration, precipitation, and spray-drying or granulation. These steps are crucial for producing stable enzyme formulations that retain activity during storage and washing. In some cases, enzymes are encapsulated or coated with inert polymers or waxes to protect them from oxidation and premature activation during shelf life (Maurer 2004). Recent innovations in fermentation include the use of genetically modified strains to increase enzyme yields, reduce by-product formation, and enhance enzyme properties. Techniques such as CRISPR/Cas9 genome editing, plasmid-based overexpression, and synthetic promoter engineering are increasingly being applied to protease-producing microbes to achieve more efficient and sustainable production pipelines (Al-Ghanayem and Joseph 2020). Moreover, the use of waste substrates—such as dairy whey, poultry feathers, or fish waste—as cost-effective feedstocks for protease production has gained traction in recent years, supporting the development of circular and eco-friendly bioprocesses (Gupta et al. 2002b).

16.4.3 Commercial Enzyme Families

16.4.3.1 Subtilisin Variants and Derivatives

Proteases have played an indispensable role in revolutionizing the detergent industry by enhancing the removal of proteinaceous stains under a wide range of washing conditions. Among the diverse categories of proteolytic enzymes explored for such applications, serine proteases—especially those derived from the subtilisin family-remain the most prominent. The development of commercial enzyme families has involved not only the discovery of naturally robust proteases but also the extensive use of protein engineering to tailor their stability, specificity, and compatibility with detergent formulations. This section elaborates on the major commercial protease families used in detergents, with a particular focus on subtilisin variants, other industrially relevant serine proteases, and the comparative advantages and limitations of each group in the context of real-world detergent performance (Wang et al. 2023).

Subtilisins represent the largest and most industrially significant group of serine proteases in commercial detergent formulations. These are alkaline endopeptidases produced primarily by *Bacillus* species, and their widespread adoption is a direct result of their high catalytic efficiency, broad substrate specificity, and remarkable stability under harsh washing conditions. The parent molecule, subtilisin Carlsberg, was originally isolated from *Bacillus subtilis*, and it provided the foundation for the development of numerous industrial variants through both classical mutagenesis and modern protein engineering. One of the most notable derivatives is Savinase, an engineered subtilisin produced by *Bacillus lentus*, known for its high resistance to oxidants and surfactants typically present in laundry detergents. Savinase and similar variants such as Alcalase and Esperase have become mainstays in enzyme-based formulations due to their robust performance in alkaline conditions and their ability to degrade a wide spectrum of protein stains including blood, milk, egg yolk, and sweat residues (Azrin et al. 2022).

The structure–function relationship in subtilisins has been well studied, allowing for rational modifications that enhance their resistance to destabilizing agents. For example, the substitution of methionine-222—a residue prone to oxidative damage—with more stable amino acids like alanine or leucine has significantly improved bleach resistance in variants such as Savinase. Similarly, modifications in the surface loops near the active site have helped increase the flexibility and binding efficiency of these enzymes, allowing them to function more effectively at lower temperatures. Advances in directed evolution have enabled the generation of subtilisins that remain active even in cold-water washes, aligning with modern consumer demands for energy-efficient and eco-friendly laundry practices (Gron et al. 1990).

16.4.3.2 Other Serine Protease Families

In addition to subtilisins, other families of serine proteases have also found commercial utility, although to a lesser extent. Elastase-like enzymes derived from species such as *Pseudomonas aeruginosa* and *Streptomyces griseus* have been investigated for their ability to degrade more recalcitrant or fibrous proteins, such as those found in dried blood, meat residues, or keratinous soils. While these enzymes generally exhibit narrower specificity and less stability in alkaline or oxidizing conditions, their inclusion in detergent blends can provide complementary action by targeting protein substrates that are less susceptible to subtilisin hydrolysis (Fujii et al. 2020).

Another interesting group of non-subtilisin serine proteases includes engineered hybrids or chimeric enzymes, which are developed by fusing structural domains from different protease families. These hybrid enzymes are designed to combine the desirable properties of multiple parent enzymes—such as high thermal stability, broad substrate tolerance, and oxidative resistance—into a single molecule. Though still under development in many cases, such engineered proteases have the potential to outperform natural enzymes by delivering tailored activity profiles suited to specialized detergent applications (Ji et al. 2023).

16.4.3.3 Comparative Performance Characteristics

The comparative performance of these commercial enzyme families is typically evaluated based on several criteria, including catalytic activity over a range of pH and temperature conditions, stability in the presence of detergent components, resistance to oxidation and autolysis, and substrate specificity. Subtilisin variants consistently demonstrate superior overall performance in most of these categories. Their optimal activity typically lies within the pH range of 9 to 11, making them well suited for modern alkaline detergent systems. Furthermore, they maintain substantial activity at temperatures ranging from 30 °C to 60 °C, with engineered forms capable of functioning at both lower and higher extremes. In contrast, many of the non-subtilisin enzymes, while potentially valuable for specific stains or applications, may be limited by their sensitivity to high pH or their narrower substrate recognition patterns (Gulmez et al. 2018).

Nevertheless, each enzyme family brings unique strengths that contribute to improved washing performance when used in combination. For example, while a subtilisin derivative may rapidly degrade the bulk of a protein stain, an elastase-like protease can complement this action by targeting the remaining insoluble fragments, thereby enhancing overall stain removal. Similarly, the inclusion of cold-active proteases from psychrophilic organisms in low-temperature formulations has enabled effective protein degradation without the need for energy-intensive hot washes (Vojcic et al. 2015). In commercial practice, detergent manufacturers often formulate enzyme cocktails that include multiple proteases alongside other hydrolases such as amylases, lipases, and mannanases. This multienzyme approach ensures that a wide array of organic soils can be addressed simultaneously, delivering a more thorough and consistent clean across various fabric types and washing conditions.

In summary, subtilisin and its engineered variants remain the cornerstone of commercial detergent protease families, owing to their high activity, stability, and broad applicability. Other serine proteases and hybrid enzymes serve as valuable complements in more specialized or extreme washing environments. The continued evolution of these enzyme families through recombinant DNA technology, protein engineering, and high-throughput screening promises to further expand their functional range and performance in next-generation detergents (Gupta et al. 2002a).

16.5 Protein Engineering and Enzyme Optimization

16.5.1 Challenges in Detergent Environments

The detergent industry imposes a unique set of environmental stresses on enzymes, markedly different from those encountered in natural biological systems. While enzymes are inherently evolved to function under tightly regulated physiological conditions—typically neutral pH, moderate temperatures, and minimal chemical stress—commercial detergents present a more aggressive milieu. These products often contain surfactants, oxidizing agents, chelators, and builders, and they operate under alkaline pH and variable temperatures (Rodrigues et al. 2013). Proteases used in detergents must not only remain catalytically active under such extremes but also resist inactivation during formulation, storage, and use. This complex requirement has propelled the advancement of protein engineering and optimization technologies aimed at enhancing enzyme resilience, while maintaining or improving their catalytic efficacy. The most critical challenges include maintaining enzyme stability at high pH, preserving activity across a range of temperatures, and ensuring compatibility with oxidizing agents such as bleach (Gupta et al. 2002b).

16.5.1.1 High pH and Alkaline Stability Requirements

One of the foremost challenges for detergent enzymes is alkaline stability. The majority of laundry detergents are designed to work best in alkaline environments, typically between pH 9 and 11. Alkalinity helps in breaking down fatty acids, suspending dirt particles, and improving the action of surfactants. However, enzymes are often sensitive to such conditions. High pH values can disrupt the network of electrostatic interactions, salt bridges, and hydrogen bonds that stabilize the protein's tertiary and quaternary structures. In proteases, the ionization states of key residues can also shift, potentially affecting catalytic activity and folding equilibrium (Gupta et al. 2002b).

Naturally occurring alkaline proteases—particularly those secreted by alkaliphilic *Bacillus* strains such as *Bacillus licheniformis*, *Bacillus clausii*, and *Bacillus lentus*—already exhibit some resistance to these destabilizing effects. However, commercial demands for improved shelf-life and wash performance have necessitated further improvements. Protein engineers have addressed these needs by selectively substituting labile amino acids with residues that can tolerate or stabilize charged environments. For example, replacing surface-exposed acidic residues (such as aspartate or glutamate) with neutral or basic counterparts can help reduce repulsive forces and enhance structural rigidity at high pH (Maurer 2004).

In addition, stabilizing salt bridges are sometimes introduced at strategic positions to reinforce the electrostatic scaffold of the protein. Engineering disulfide bridges between cysteine residues is another common strategy to "lock in" local structural conformations. These covalent links can effectively constrain flexible loops or vulnerable helices, thereby reducing the entropy of unfolding. Such changes have been shown to significantly increase the half-life of proteases at alkaline pH, often without compromising their substrate affinity or turnover rate (Rao et al. 1998). Directed evolution has emerged as a particularly powerful approach to improve alkaline tolerance. This technique involves generating large libraries of enzyme variants through random or targeted mutagenesis, followed by high-throughput screening for improved activity under alkaline conditions. Using this strategy, researchers have successfully developed subtilisin variants that retain over 80% activity after prolonged exposure to pH 10.5 or higher levels at which most wild-type enzymes are rapidly inactivated (Al-Ghanayem and Joseph 2020).

16.5.1.2 Temperature Stability and Thermal Resistance

Temperature presents another crucial challenge. Washing temperatures can vary significantly depending on the application: from cold-water cycles in domestic laundry (15–30 °C), to high-temperature cycles in industrial laundering

(60–90 °C). Enzymes must function efficiently across this broad range. However, elevated temperatures can accelerate protein denaturation by increasing molecular motion and disrupting the non-covalent forces that stabilize the folded state. Denatured proteins tend to aggregate irreversibly, particularly in the presence of surfactants and salts common in detergents, leading to a loss of function (Maurer 2004).

Thermostability, the capacity of a protein to remain folded and functional at high temperatures, can be enhanced by several mechanisms. Natural proteases from thermophilic organisms such as *Thermus aquaticus* or *Geobacillus stearothermophilus* exhibit exceptional thermal resistance due to intrinsic structural features. These include an increased number of ionic interactions, greater hydrophobic core packing, reduced solvent-exposed loops, and shorter inter-domain linkers. These enzymes provide templates for designing detergent-compatible proteases that function at 60 °C and beyond (Vieille and Zeikus 2001). For enzymes like subtilisin, rational design has enabled improvements in thermal resistance through targeted substitutions. Proline residues, which restrict backbone flexibility, are often introduced into loop regions to reduce the entropic cost of folding. Hydrophobic residues can be incorporated into the protein's core to enhance van der Waals interactions and reduce the likelihood of unfolding. Increasing the polarity of solvent-exposed residues has also proven effective in improving aqueous stability under heat stress (Eijsink et al. 2004).

Enzyme stability during detergent processing and storage is another important consideration. Enzymes may be exposed to high temperatures during drying (such as spray drying at 80–100 °C) or to thermal fluctuations during shipment and storage. Proteins must therefore resist aggregation and retain their active conformation even in the absence of water. This has led to the use of protective excipients—such as polyols, sugars (like trehalose), and certain polymers—that form stabilizing glassy matrices around enzyme molecules. However, protein engineering still plays a primary role in developing intrinsically stable enzyme variants capable of surviving these processing steps without loss of functionality (Carpenter and Crowe, 1989).

16.5.1.3 Oxidative Stability and Bleach Compatibility

The presence of oxidizing agents such as hydrogen peroxide, percarbonate, perborate, or peracids in detergent formulations represents one of the most formidable challenges to enzyme survival. These agents are essential for stain oxidation, whitening, and disinfection, but they also produce reactive oxygen species (ROS) that can damage sensitive amino acid residues. Sulfur-containing residues such as methionine and cysteine are especially vulnerable to oxidation, which can lead to side chain modification, disruption of the enzyme's active site geometry, and irreversible deactivation (Bryan 2000).

Historically, subtilisin Carlsberg was quickly inactivated by bleach due to the presence of methionine-222 near the catalytic center. This discovery led to one of the most celebrated examples of site-directed mutagenesis in industrial biotechnology. By substituting methionine-222 with leucine, researchers created an oxidant-stable variant that became the basis for Savinase—one of the most widely used detergent proteases today. The success of this single substitution spurred further efforts to identify and modify other oxidation-sensitive residues on the enzyme surface and near functional domains (Bryan et al. 1992). In addition to amino acid substitutions, surface engineering techniques have been employed to reduce the overall hydrophobicity and solvent exposure of susceptible regions. Incorporating charged or polar residues at the solvent interface can sometimes prevent the accumulation of oxidants near sensitive side chains. Likewise, rigidifying the enzyme structure—by shortening surface loops or enhancing intra-molecular bonding—can help prevent ROS-induced unfolding (Fujii et al. 2020). To further enhance oxidative resistance, enzyme encapsulation has become an important formulation strategy. Enzymes can be coated with waxes, polysaccharides, or polymeric films that physically isolate them from oxidants during storage. These coatings dissolve upon contact with water during the wash cycle, releasing the active enzyme at the point of use. Encapsulation not only protects enzymes from oxidative attack but also improves handling, dosing, and shelf life (Lund et al. 2012). Despite these advances, achieving complete bleach compatibility remains a work in progress. Ongoing research is focused on integrating computational design with machine learning to predict oxidation hotspots, model interaction pathways with detergent components, and guide the development of ultra-stable enzyme variants. Moreover, the use of synergistic stabilizers—such as calcium ions, borates, and polyols—can be optimized in combination with engineered proteases to achieve long-term oxidative resilience (Yao et al. 2023).

16.5.2 Protein Engineering Approaches

The adaptation of proteolytic enzymes to the demanding conditions of detergent formulations has been one of the most significant achievements in industrial biotechnology. While nature provides a diverse set of enzymes with varying degrees of stability and specificity, few possess the exact combination of traits required for consistent performance in the chemically aggressive and physically dynamic environment of modern detergents. As a result, protein engineering has become an essential strategy, allowing scientists to systematically modify and improve the structure and function of proteases to meet these industrial standards. Over the past several decades, the field has evolved to incorporate increasingly sophisticated methodologies, from early rational mutagenesis to modern high-throughput directed evolution and structure-based computational design (Maurer 2024).

16.5.2.1 Site-Directed Mutagenesis and Rational Design

Site-directed mutagenesis was one of the earliest tools in protein engineering and remains fundamental for precise alterations of specific amino acids. This method involves the intentional substitution, insertion, or deletion of targeted residues within the enzyme's coding sequence based on prior knowledge of its structure, function, or conserved motifs. Rational design, often used in conjunction with site-directed mutagenesis, relies on understanding the three-dimensional structure of the protein, its catalytic mechanism, and the physicochemical properties of amino acids to hypothesize and introduce beneficial changes (Bryan 2000).

In the context of detergent proteases, rational design has been instrumental in enhancing pH and thermal stability, as well as oxidative resistance. A classic example is the substitution of methionine-222 in subtilisin Carlsberg, which sits adjacent to the active site. This residue was particularly vulnerable to oxidation by bleach components, and replacing it with leucine or alanine significantly improved the enzyme's resistance to oxidative deactivation while maintaining its catalytic efficiency (Bryan et al. 1992). Other successful rational modifications have included introducing disulfide bonds to stabilize flexible loop regions, increasing the number of salt bridges to support alkaline stability, and replacing thermolabile residues to enhance temperature tolerance. The effectiveness of these changes is often validated through kinetic studies, thermal shift assays, and structural analysis via X-ray crystallography or NMR spectroscopy. While rational design is powerful, it is limited by the extent of available structural data and the complexity of predicting the effects of mutations in highly dynamic protein environments (Gron et al. 1990).

16.5.2.2 Directed Evolution and Screening Methods

Where rational design reaches its limits, especially in the absence of high-resolution structural data or in proteins with unpredictable mutational responses, directed evolution offers a robust alternative. This technique mimics the process of natural selection in the laboratory by generating large libraries of enzyme variants through random or semi-random mutagenesis, followed by iterative rounds of screening or selection for improved traits (Al-Ghanayem and Joseph 2020).

The process typically begins with error-prone PCR, DNA shuffling, or saturation mutagenesis to introduce genetic diversity. These libraries can contain tens of thousands to millions of unique variants. The next step involves high-throughput screening or selection protocols that allow researchers to identify and isolate the best-performing mutants based on desired criteria such as activity, stability, or resistance to inhibitors under detergent-relevant conditions (Chen and Arnold 1991). For detergent proteases, this means screening variants for activity at high pH, resistance to surfactants and oxidants, or effectiveness at low temperatures. One successful example of this approach was the development of improved Savinase variants using directed evolution to enhance performance at lower wash temperatures, responding to consumer demand for energy-efficient detergents (Maurer 2004).

Modern screening systems employ microplate-based assays, fluorogenic or chromogenic substrates, and even in vivo selection platforms using genetically engineered reporter strains. These technologies have vastly increased the throughput and sensitivity of the process, making it possible to optimize multiple traits simultaneously—a process known as "multi-parameter" or "multiplex" evolution.

Directed evolution has become a staple in industrial enzyme development because it does not require detailed knowledge of protein structure and can uncover beneficial mutations that may not be intuitive from a theoretical standpoint. However, it does rely on having effective screening strategies, which can sometimes be limiting depending on the complexity of the trait being optimized (Polgar 2005).

16.5.2.3 Computational Design and Modeling

The latest frontier in enzyme engineering lies in the integration of computational tools to predict, model, and design protein structures and functions. Computational protein design enables researchers to simulate the effects of mutations, identify hotspots for stabilization, and model enzyme–substrate interactions in silico before moving to laboratory experiments. This approach has become especially powerful with the advent of accurate structure prediction tools like AlphaFold and advanced molecular dynamics simulations (Gonzalez et al. 2022).

Computational modeling allows for the identification of structurally weak or flexible regions in the enzyme that may be prone to unfolding under detergent conditions. Once identified, these regions can be engineered through targeted mutations that enhance local stability or reduce entropy. Additionally, enzyme-substrate docking simulations can help refine the geometry of the active site to improve catalytic efficiency or substrate selectivity. For subtilisin variants and other detergent proteases, computational tools have been used to guide the introduction of stabilizing salt bridges, improve packing density in the hydrophobic core, and redesign surface residues to reduce susceptibility to oxidation or surfactant interference. Algorithms such as FoldX, Rosetta, and molecular mechanics packages have become essential tools for prescreening mutations and narrowing down the most promising candidates for experimental testing (Kortemme and Baker 2002).

Moreover, machine learning approaches are now being used to predict functional outcomes of mutations based on training data from large-scale mutagenesis studies. These tools can help design focused libraries for directed evolution, reducing the need for excessive randomization and improving the efficiency of the discovery process. The combination of computational design with laboratory-based methods forms a powerful hybrid approach known as semi-rational design, where computational predictions are used to define mutation sites or residue positions for focused mutagenesis and subsequent high-throughput screening. This integration has accelerated the development of robust, high-performance proteases tailored for a wide range of detergent formulations and wash conditions (Ming et al. 2023).

16.5.3 Enhanced Enzyme Properties

The cumulative goal of protein engineering strategies in the detergent industry is not simply to produce enzymes that “survive” in harsh washing conditions, but rather to generate high-performance biocatalysts that retain optimal functionality, stability, and safety. Advances in site-directed mutagenesis, directed evolution, and computational modelling have enabled the development of protease variants with vastly improved properties tailored to industrial needs. These improvements span multiple dimensions, including enhanced structural stability under stress, greater catalytic efficiency against diverse protein substrates, and reduced allergenic potential to support consumer safety. The following sections explore each of these enhanced characteristics and their significance in modern enzyme design for detergents (Maurer 2024).

16.5.3.1 Improved Stability Profiles

Enzyme stability in detergent formulations is central to their efficacy and commercial viability. Proteases are required to remain active not only during the actual washing cycle—where they face extreme pH, variable temperatures, surfactants, and oxidants—but also throughout the entire product shelf life, which can extend for months under fluctuating storage conditions. Thus, improving stability across a spectrum of physical and chemical parameters has been a core focus of enzyme optimization efforts (Das and Ghosh 2023).

Enhanced pH stability is particularly critical, as most modern detergents operate in highly alkaline conditions. Wild-type proteases, which are typically adapted to near-neutral pH environments, often show limited activity or rapid degradation at pH levels above 9. Through targeted mutations, such as the introduction of stabilizing salt bridges or the replacement of pH-sensitive residues, proteases have been engineered to remain active at pH values up to 11. Some engineered subtilisin variants, for example, can retain more than 80% of their activity after prolonged exposure to high-pH environments—an outcome rarely achieved in unmodified enzymes (Gupta et al. 2002b).

Thermal stability is another crucial parameter. Many industrial wash cycles involve temperatures between 40 °C and 60 °C, with even higher temperatures used in commercial laundering. Thermostability can be enhanced through amino acid substitutions that increase hydrophobic interactions in the protein core, introduce rigidifying proline residues in loop regions, or eliminate entropy-promoting glycine residues at key positions. Additionally, improved oxidative stability—through the elimination or shielding of oxidizable residues such as methionine and cysteine—has greatly expanded the utility of proteases in bleach-containing formulations (Bryan 2000).

Stabilization strategies also include formulation enhancements, such as encapsulation or co-formulation with polyols and salts. However, enzymes with intrinsic resistance to thermal denaturation, alkaline hydrolysis, and oxidative attack reduce the need for such additives and simplify manufacturing and storage logistics (Ming et al. 2023).

16.5.3.2 Enhanced Catalytic Efficiency

In addition to surviving detergent conditions, engineered proteases must exhibit high catalytic turnover and effective substrate recognition to be commercially useful. Catalytic efficiency is typically defined by the enzyme's kcat/K_M value—a ratio reflecting how effectively the enzyme converts substrate to product at low concentrations. Proteases used in detergents must quickly hydrolyze a wide variety of proteinaceous stains, including blood, milk, grass, egg, and sweat, often under dilute, nonoptimal conditions (Zhang et al. 2021).

Protein engineering has been used to refine the active site geometry of proteases, tailoring them to bind more effectively to typical laundry soils. Substitutions in the S1 and S2 subsites of the protease binding cleft, for instance, can influence how tightly and specifically the enzyme binds to different amino acid residues within protein substrates. This allows engineers to broaden or narrow the enzyme's specificity profile as needed for particular detergent formulations (Yao et al. 2023). In some cases, modifications enhance the ability of proteases to function at lower temperatures, aligning with consumer preferences for energy-saving cold washes. Cold-adapted enzymes, sourced from psychrophilic organisms or engineered via directed evolution, have been shown to exhibit high activity even at 20–30 °C, thereby facilitating effective stain removal while conserving energy (Maurer 2004). Enhanced catalytic efficiency also supports enzyme cost-effectiveness. Enzymes with higher specific activity require smaller dosages, which not only reduces formulation costs but also minimizes the amount of biological material introduced into the environment. This advantage becomes particularly relevant in large-scale commercial and industrial detergent applications, where dosage precision is tightly regulated (Ochoa et al. 2020).

16.5.3.3 Reduced Allergenicity and Safety Improvements

While enzyme performance has traditionally been the primary focus of optimization, growing awareness of consumer health and safety has pushed allergenicity and environmental impact to the forefront of detergent enzyme design. Proteases, by nature of being foreign proteins, can sometimes act as allergens if inhaled or if they come into prolonged contact with the skin. This concern is especially relevant for workers in enzyme production and formulation plants, as well as for sensitive consumers (Azrin et al. 2022). Protein engineering offers several ways to reduce allergenic potential. One approach involves modifying or eliminating known IgE-binding epitopes—small peptide regions recognized by the human immune system. By substituting surface-exposed amino acids or altering conformational epitopes through loop shortening or domain deletion, it is possible to reduce immunogenicity while preserving catalytic function (Baur et al. 1998).

Another strategy is the reduction of dust formation during enzyme formulation. While not a protein-level engineering challenge per se, it complements safety improvements. Encapsulation technologies and granulation methods, such as the use of core–shell structures and inert coatings, significantly reduce airborne particles, thereby lowering the risk of respiratory sensitization (Eijsink et al. 2004). Environmental safety is also an important consideration. Engineered enzymes are typically biodegradable and can be rapidly broken down by native proteases in soil and aquatic environments. Improvements in catalytic efficiency, stability, and dosage control ensure that less enzyme is released into the environment overall. Moreover, enzymes allow detergents to be formulated with fewer harsh chemicals, such as phosphates or strong alkalis, thus contributing to broader environmental sustainability goals. The convergence of these safety improvements—reduced allergenicity, better formulation control, and environmental compatibility—has positioned engineered proteases as not only more effective but also more responsible components of modern detergent technology (Naganthran et al. 2017).

16.6 Formulation Science and Product Development

16.6.1 Detergent Formulation Principles

The successful inclusion of enzymes in detergent formulations is not merely a matter of mixing active proteins with surfactants and builders. Instead, it demands a carefully orchestrated approach—grounded in the principles of formulation science—that balances the chemical environment, physical form, and delivery system—to preserve enzyme activity from production through storage to use. Proteases must not only retain their functional integrity in the presence of aggressive detergent ingredients but also be stable across varying temperatures, humidity levels, and shelf-life durations. This section outlines the core formulation strategies used in modern enzyme-based detergents, highlighting enzyme–component compatibility, stabilization techniques, and long-term activity preservation (Ji et al. 2023).

16.6.1.1 Enzyme Compatibility with Detergent Components

One of the most fundamental challenges in enzyme-based detergent development lies in ensuring compatibility between the enzyme and the numerous other ingredients present in the formulation. A typical detergent contains a complex mixture of surfactants (anionic, nonionic, and sometimes cationic), builders (such as sodium carbonate, zeolites, or phosphates), bleaching agents (e.g., hydrogen peroxide, sodium percarbonate), chelating agents, optical brighteners, fragrances, and sometimes fabric softeners. Many of these chemicals can interact with the enzyme's structure, either stabilizing or destabilizing it depending on their concentration and chemical nature (Gupta et al. 2002b).

For proteases, surfactants are particularly significant. While low concentrations of certain nonionic surfactants can help solubilize enzymes and reduce aggregation, anionic surfactants like sodium dodecyl sulfate (SDS), commonly used in laundry products, can denature proteins by binding to their hydrophobic regions and disrupting the native conformation. Modern detergent-compatible proteases have been engineered to resist such denaturation by introducing mutations that strengthen internal hydrogen bonds, enhance surface hydrophilicity, or improve compactness of the folded structure (Maurer 2004). 1Bleaching agents also pose a serious challenge to enzyme compatibility. Reactive oxygen species generated from perborates or percarbonates during the wash cycle can oxidize sensitive residues in enzymes, especially methionine, cysteine, and tryptophan. Proteases must either be designed to exclude these labile residues near active or structural regions, or they must be physically separated from oxidants through microencapsulation or coated granules that prevent premature exposure before the detergent dissolves in water (Bryan et al. 1992).

Metal ions are another important factor in enzyme compatibility. While divalent cations like calcium can help stabilize certain proteases (e.g., by reinforcing loop conformations in subtilisin), excessive levels of chelators like EDTA or phosphonates can remove these stabilizing ions, leading to destabilization. To address this, some enzymes have been engineered to either reduce their calcium dependence or increase their affinity for residual trace metals in detergent formulations (Barrett et al. 2001).

16.6.1.2 Stabilization Strategies and Additives

Beyond compatibility, long-term stability is a critical requirement for detergent enzymes, especially those incorporated into powder or liquid products with extended shelf lives. Stabilization strategies can be chemical, physical, or both, and often involve the inclusion of protective excipients and formulation additives.

In powdered detergents, enzymes are commonly stabilized through granulation and coating techniques. The enzyme is usually embedded in a core material—such as starch or salt—and coated with multiple protective layers composed of waxes, polymers, or nonionic surfactants. These coatings prevent direct contact with incompatible ingredients like bleaches or high-pH builders and delay enzyme activation until the wash cycle begins. Additionally, these coatings reduce the generation of airborne dust particles, thereby enhancing user safety (Al-Ghanayem and Joseph 2020).

Liquid detergents pose a greater formulation challenge because they contain water and surfactants that can lead to gradual enzyme denaturation or proteolytic autolysis (self-digestion). In such formulations, stabilizing additives are essential. Polyols like glycerol or sorbitol, sugars like trehalose, and stabilizing salts are used to create a microenvironment that preserves enzyme structure. Chelating agents like borate can reduce autolysis by forming reversible complexes with the enzyme, thereby reducing its proteolytic activity until dilution in water. Some liquid detergents also use reversible inhibitors or buffer systems that control pH to reduce enzymatic degradation during storage (Basketter et al. 2008).

In both powder and liquid formulations, preservatives are often added to prevent microbial contamination, which can degrade both the enzyme and the overall product integrity. However, these biocides must be selected carefully to avoid unintended interactions with the enzyme protein. Modern stabilization methods also include enzyme immobilization on solid supports and encapsulation using liposomes or polymer matrices. These advanced systems not only protect the enzyme but also allow for controlled release, improving overall performance during the wash cycle (Basketter et al. 2008).

16.6.1.3 Activity Preservation During Storage

Preserving enzyme activity over time is essential for maintaining consistent product performance and consumer satisfaction. Factors such as humidity, temperature fluctuations, exposure to light, and oxygen all contribute to the gradual decline in enzyme activity during storage. Understanding and mitigating these effects requires a multilayered approach involving formulation, packaging, and enzyme engineering (Abramowitz et al. 1967).

One of the most critical parameters for activity preservation is moisture control. In powder formulations, even small amounts of absorbed moisture can trigger enzyme activation, leading to self-digestion or interaction with other detergent components. Therefore, packaging must be designed to minimize moisture ingress—often using foil-lined bags or plastic containers with desiccant sachets. Some formulations also include drying agents, such as sodium sulfate, to passively absorb ambient moisture (Gupta et al. 2002b). Temperature stability is equally important. Proteins, especially in aqueous formulations, are sensitive to thermal denaturation. During transportation or storage in warm climates, product temperatures may rise significantly. Engineered enzymes with increased thermal tolerance (as described in earlier chapters) help address this challenge, but additional formulation strategies—such as the use of stabilizing buffers or antioxidants—are also employed to limit temperature-induced degradation (Maurer 2004). In liquid formulations, oxygen exposure is a major concern. Oxidative stress can cause irreversible damage to the enzyme's active site, leading to loss of function. To counter this, oxygen scavengers such as ascorbic acid or sodium bisulfite are sometimes added, and headspace oxygen in containers is minimized through inert gas flushing during packaging. Shelf-life studies using accelerated ageing tests (e.g., storing product at 40 °C for 12 weeks) are routinely performed to predict enzyme stability over time. These studies inform both formulation adjustments and expiration date labelling, ensuring that enzymes remain active until the product's end of use (Klein and Bischoff 2011).

Altogether, activity preservation is not solely the responsibility of the enzyme itself—it is a collaborative outcome of formulation science, storage conditions, packaging technology, and continuous monitoring. Advances in these areas have allowed enzyme-based detergents to reach markets across climates and demographics, all while maintaining high cleaning performance and product consistency (Solanki et al. 2021).

16.6.2 Physical Forms and Delivery Systems

The success of enzyme-based detergents depends not only on the inherent stability and catalytic efficiency of the enzymes themselves, but also on the formulation technologies used to deliver them in a controlled and effective manner. Enzymes in detergents are formulated into different physical forms—including powders, liquids, and tablets—each with its own advantages, challenges, and engineering solutions. Beyond simply choosing a form, developers must also consider how to protect enzymes from premature inactivation and ensure that their release during the wash cycle is timed for optimal performance. Recent advances in encapsulation and controlled release systems have greatly expanded the versatility and reliability of enzyme delivery across a wide range of detergent products and usage conditions (Fleißner and Dersch 2010).

16.6.2.1 Liquid, Powder, and Tablet Formulations

Proteases are incorporated into detergent products in various physical formats, with each format presenting its own constraints and opportunities. Powder formulations have long dominated the detergent industry due to their ease of handling, stability, and cost-effectiveness. In these products, enzymes are typically added as granules—either as dry-blended particles or integrated into the base powder during the manufacturing process. To protect enzyme activity in powdered detergents, they are often granulated with stabilizing excipients and coated with protective materials. These coatings prevent direct contact with bleach, alkaline builders, and moisture until the detergent dissolves in water (Maurer 2024).

Liquid detergents, though increasingly popular for their convenience and ability to dissolve quickly in cold water, pose a more challenging environment for enzymes. Unlike dry powders, liquid formulations contain water, which can accelerate enzyme degradation and increase the risk of proteolytic self-digestion. Stabilization in liquid products is achieved using a combination of pH control, chelating agents (to limit metal-catalyzed oxidation), reversible inhibitors, and polyols or sugars that help preserve the enzyme's folded structure. In many cases, enzyme concentration must be carefully optimized to balance cleaning performance with formulation stability (Gupta et al. 2002a).

One tablet formulations—used in both laundry and dishwashing applications—represent a more recent innovation. Tablets can contain compacted detergent along with discrete layers or compartments that house enzymes separately from other reactive components like bleaches. This physical separation allows for better protection during storage and more controlled release during use. Multichamber designs in dishwashing tablets, for example, may release enzymes during the prewash phase and bleaches during the main wash, optimizing the efficacy of each component without causing mutual deactivation (Das and Ghosh 2023).

16.6.2.2 Encapsulation and Protection Technologies

To enhance enzyme stability and longevity, encapsulation has emerged as a key technological innovation in detergent enzyme formulation. Encapsulation refers to the process of enclosing enzymes within a physical barrier or matrix that isolates them from harsh environmental components until they are needed during the wash cycle.

In powdered detergents, encapsulation typically takes the form of coated granules. The enzyme is first adsorbed or embedded into a solid core material—often composed of salts, sugars, or starch—and then layered with one or more protective coatings. These coatings are usually made from waxes, fatty acids, polyethylene glycol (PEG), or nonionic surfactants. Each coating serves a specific purpose: the inner layer protects the enzyme from interaction with the detergent matrix, while the outer layer controls dust formation and improves flowability. This multilayered protection enables enzymes to retain their activity over months of storage, even in warm and humid climates (Rao et al. 1998).

In liquid formulations, encapsulation must allow enzymes to remain suspended in solution while also being protected from proteolytic or oxidative degradation. This is often achieved through microencapsulation techniques using polymeric shells, liposomes, or reverse micelle systems. These delivery vehicles not only shield the enzyme but can also release it in response to specific triggers such as dilution, pH shift, or temperature change during the wash cycle (Abramowitz et al. 1967).

More advanced systems have explored the use of spray-dried microcapsules and coacervate gels, which offer tunable permeability and release kinetics. These approaches allow for more precise control over when and how much enzyme is delivered, reducing waste and improving overall cleaning efficiency. Importantly, these systems must be designed to disintegrate rapidly and completely during the wash, ensuring full enzyme activation without leaving behind residue on fabrics or dishes (Gulmez et al. 2018).

16.6.2.3 Controlled Release Mechanisms

Controlled release is central to maximizing the efficiency of enzymes in detergent applications. Rather than having all components interact simultaneously—an approach that often results in antagonism or early degradation—modern detergents use delivery systems that stage the release of enzymes in sync with other cleaning phases (Yang et al. 2024).

In laundry detergents, for example, it is advantageous to release proteases during the early part of the wash cycle, when protein-based stains such as blood, sweat, and food residues are most responsive to enzymatic breakdown. If released too late, after bleach activation or surfactant saturation, enzyme performance can be significantly impaired. Time-release coatings, pH-sensitive matrices, and moisture-activated granules are commonly used to achieve this early phase delivery (Zhang et al. 2021).

In dishwashing applications, where enzyme activity must precede bleach activation, multichamber tablet technology enables temporal separation of incompatible ingredients. Each compartment is designed to dissolve at a specific point during the wash, guided by water exposure, temperature, or mechanical agitation (Naganthran et al. 2017). Even in liquid detergents, where the opportunity for physical separation is more limited, designers use inhibitors and protective additives to suppress enzyme activity during storage, followed by rapid activation upon dilution in the wash water. This type of chemical "release" control helps extend shelf life and maintain consistent product performance (Lund et al. 2012). Future advancements in controlled release mechanisms are likely to integrate responsive materials that can sense and react to wash conditions in real-time. For instance, materials that respond to temperature, fabric type, or stain load could optimize enzyme release dynamically, providing a more intelligent and efficient wash (Yang et al. 2024).

16.6.3 Multienzyme Systems

In recent years, the formulation of detergent products has evolved significantly from using single-enzyme systems to incorporating multienzyme blends that offer broader substrate coverage and enhanced cleaning efficacy. This transition has been driven by the increasing complexity of modern stains—which often contain a combination of proteins, fats, carbohydrates, and synthetic polymers—as well as growing consumer expectations for effective stain removal under mild wash conditions. Multienzyme systems, when designed carefully, provide synergistic benefits that exceed the capabilities of individual enzymes, making them a cornerstone of high-performance detergent formulations (Bié et al. 2022).

16.6.3.1 Synergistic Enzyme Combinations

The cornerstone of a successful multienzyme detergent system is synergy—the phenomenon where the combined action of multiple enzymes leads to greater stain removal than the sum of their individual effects. For example, combining a protease with a lipase and an amylase allows for the simultaneous breakdown of protein-, fat-, and starch-based stains, respectively. Such combinations are particularly useful when dealing with composite soils, like food residues (e.g., egg yolk, gravy, or chocolate), that contain multiple stain components embedded in fabric fibers (Kirk et al. 2002).

Synergy may also arise through sequential or cooperative action. Proteases can cleave structural proteins that trap fats or polysaccharides within stains, allowing lipases and amylases to access their substrates more efficiently. In other cases, enzymes may act on separate regions of a complex molecule, leading to more complete degradation than would occur with a single enzyme. This interplay is particularly beneficial in cold or quick wash cycles, where limited time and lower temperatures reduce the effectiveness of individual enzyme action (Maurer 2024). When carefully balanced, these synergistic blends allow detergents to achieve high performance across a wide range of stain types and washing conditions, all while using lower enzyme concentrations. This reduces formulation cost and environmental impact, making synergy a key target in modern detergent design (Ming et al. 2023).

16.6.3.2 Protease Interactions with Other Enzymes

While combining enzymes has clear advantages, it also presents challenges—particularly when proteases are involved. Proteases are inherently capable of hydrolyzing other proteins, including fellow enzymes in the formulation. If not properly stabilized or protected, enzymes such as amylases, lipases, or cellulases can be degraded by proteases, leading to diminished effectiveness of the entire blend.

To address this, several strategies have been developed. One approach is to engineer proteases with reduced activity against other detergent enzymes. This may involve altering substrate-binding sites to limit their access to nontarget proteins or adjusting protease specificity so they preferentially attack loosely folded, soil-derived proteins rather than structured globular enzymes (Von der Osten et al. 1993). Another common strategy is physical or temporal separation. In powder detergents, enzymes can be granulated separately with different coatings that delay or stagger their release. This allows proteases to act early in the wash cycle, degrading stain proteins, while other enzymes are activated later—after the risk of proteolytic degradation has passed. In liquid detergents, reversible inhibitors or chelating agents may be included to suppress protease activity during storage, only releasing full functionality upon dilution in the wash water (Gupta et al. 2002b). It's also possible to co-formulate enzymes with stabilizing agents such as polyols, calcium ions, or borates, which preserve enzyme conformation and resist denaturation caused by protease interaction. This approach not only protects enzyme activity but also improves overall formulation stability and shelf life. The successful integration of proteases with other enzyme classes therefore depends on both biochemical compatibility and innovative formulation design—ensuring that all enzymes remain functional and effective throughout the product's use cycle (Vojcic et al. 2015).

16.6.3.3 Formulation Optimization Strategies

Optimizing multienzyme detergent formulations requires a fine balance between performance, stability, cost, and user safety. This process begins with selecting enzyme types and variants that complement each other in terms of substrate specificity, optimal pH and temperature range, and resistance to inactivation by other formulation components. Enzymes that operate effectively within overlapping conditions—such as alkaline pH and moderate to high temperatures—are ideal candidates for combined use (Patel and Makampara 2023). Once the enzyme set is chosen, their relative concentrations must be optimized. This is typically done through a series of factorial design experiments, where the ratios of enzymes are systematically varied and tested against standardized stain panels. These tests help determine the minimal enzyme load required for maximum stain removal, ensuring cost-effective use of high-value biocatalysts (Vicaria et al. 2022). Formulators must also consider interactions with surfactants, builders, and bleaches. For example, some enzymes may be sensitive to anionic surfactants or hydrogen peroxide; therefore, selecting oxidant-stable or surfactant-tolerant variants is crucial. Encapsulation technologies can also be employed to control the release profile of individual enzymes, allowing for time-staged activation during the wash cycle. In addition to experimental optimization, computational modelling and predictive analytics are increasingly used to simulate enzyme interactions, predict synergistic potential, and streamline formulation development. These tools can identify nonobvious synergies and minimize trial-and-error testing, speeding up product development and innovation (Vega et al. 2025).

Finally, modern multienzyme formulations often include additional performance enhancers—such as soil-release polymers, anti-redeposition agents, and optical brighteners—that work in concert with enzymes to deliver a superior cleaning experience. These add-ons must be tested for compatibility with enzymes to ensure they do not inhibit enzyme activity or stability over time. In sum, the design of multienzyme systems is a multidimensional challenge that blends enzyme biochemistry with advanced formulation science. The result is a new generation of high-performance detergents capable of tackling complex soils, delivering deep cleaning in short cycles, and supporting sustainability goals through reduced chemical and energy use (Abramowitz et al. 1967).

16.7 Industrial Applications and Market Analysis

16.7.1 Commercial Market Segments

The commercial success of proteases in detergents is not only a reflection of their biochemical versatility but also of their integration into diverse market sectors that range from domestic laundry to industrial-scale cleaning. As detergent formulations have become more sophisticated and environmentally conscious, enzymes—particularly proteases—have found a prominent role in enhancing cleaning efficiency, reducing the need for harsh chemicals, and supporting sustainability goals. This section explores the primary commercial segments where proteases are employed, including household laundry detergents, industrial and institutional cleaning solutions, and emerging niche markets.

16.7.1.1 Household Laundry Detergents

The household laundry segment represents the largest and most established market for detergent proteases. Enzymes have been a mainstay in this sector since the 1960s, when early subtilisin-type proteases were first introduced to improve the removal of protein-based stains such as blood, sweat, milk, grass, and food residues. Since then, the role of proteases in household detergents has expanded significantly, both in powdered and liquid formats (Wang et al. 2017).

Modern consumers expect detergents that can deliver high performance under a variety of wash conditions—including cold temperatures, quick cycles, and eco-friendly settings. This shift has placed greater demand on enzymes to work effectively at lower temperatures and alkaline pH, often in the presence of reduced surfactant loads or without strong bleaches. Proteases tailored for these conditions have enabled a new generation of enzyme-rich detergents that maintain stain removal efficacy while supporting energy conservation and fabric care (Okino-Delgado et al. 2019).

Within household laundry, proteases are rarely used alone. Instead, they are formulated alongside amylases, lipases, and cellulases to address the multicomponent nature of common stains. Nevertheless, proteases often take the lead role due to the prevalence of protein-rich soils in everyday laundering. In liquid detergents especially, where chemical stability is more challenging, engineered proteases with high oxidative and thermal stability have become critical to maintaining product integrity and consumer trust over extended shelf lives. (Barrett et al. 2001). Brands such as Ariel, Tide, Persil, and Surf have all prominently featured enzyme technologies in their flagship products, leveraging the cleaning power of proteases as part of their value proposition. As the global market for laundry detergents continues to grow—particularly in emerging economies—so does the demand for robust, low-cost, and environmentally sound enzyme solutions (Lund et al. 2012).

16.7.1.2 Industrial and Institutional Cleaning

Beyond the household, proteases play a vital role in the industrial and institutional (I&I) cleaning segment, which includes applications in healthcare, hospitality, food service, and textile processing. Cleaning challenges in these sectors often involve high soil loads, stringent hygiene requirements, and the need for cost-effective, large-scale performance (Verma et al. 2016). In healthcare settings, for instance, enzyme-based detergents are used to clean surgical instruments and hospital linens that are contaminated with blood, tissue, and other proteinaceous materials. Proteases are prized in these environments for their ability to degrade organic matter without damaging sensitive surfaces or requiring extremely high temperatures or corrosive chemicals. Their use supports both material longevity and compliance with strict sanitization standards (Rao et al. 1998). In the hospitality and food service industries, proteases contribute to cleaning linens, uniforms, and surfaces that accumulate stubborn food and grease stains. Their inclusion in commercial laundry systems allows for effective cleaning at lower temperatures, reducing energy costs and extending the life of fabrics. Similarly, in the food and beverage industry, proteases are used to remove proteinaceous residues from production equipment, storage tanks, and processing surfaces—tasks where traditional chemical cleaning can be insufficient or damaging (Gulmez et al. 2018).

The textile industry also utilizes proteases in processing applications, such as the bio-polishing of cotton and the degumming of silk, where selective hydrolysis of protein-based materials improves texture and finish. In all these settings, the scalability, specificity, and environmentally friendly profile of proteases make them an attractive alternative to purely chemical cleaning agents (Polgar 2005). Market trends indicate that the I&I enzyme detergent segment is growing steadily, driven by regulatory pressure to reduce water usage, energy consumption, and hazardous chemicals in cleaning operations. This growth opens new opportunities for the application of engineered proteases designed for high stability, low-dose efficiency, and targeted performance (Basketter et al. 2008).

16.7.1.3 Specialty Applications and Niche Markets

In addition to mainstream household and industrial cleaning, proteases have found a foothold in several specialized applications that rely on their unique ability to break down protein substrates with precision and efficiency. These niche markets, though smaller in scale, often demand high-performance enzymes tailored for specific substrates, surfaces, or operational environments (Saeki et al. 2007). One such area is personal care and cosmetic products, where proteases are occasionally used in exfoliating scrubs, enzymatic peels, and acne treatments. In these formulations, mild proteases such as papain or subtilisin derivatives are used to gently remove dead skin cells or excess keratin without causing irritation. While the regulatory and safety requirements in this sector are stringent, advances in enzyme formulation and stabilization have made cosmetic proteases increasingly viable (Rajagopalan et al. 2014).

Another niche market lies in environmental bioremediation and household drain cleaners, where enzymes help degrade organic matter that clogs pipes or contributes to odors. Proteases, in conjunction with lipases and amylases, are employed in enzymatic drain treatment products to digest complex organic blockages in plumbing systems. Their biodegradable nature and low toxicity make them preferable to traditional caustic or acid-based cleaners.

Additionally, the pet care and veterinary industry uses protease-containing products for cleaning stains and odors from pet bedding, grooming tools, and accident spots. The selective breakdown of protein-based stains such as urine, feces, and saliva allows for more effective deodorization and sanitization (Zhang et al. 2021). Specialized enzyme cleaning systems are also used in sensitive electronics and optical equipment cleaning, where residues must be removed without mechanical abrasion or harsh solvents. Proteases are valued here for their substrate specificity and gentle action. While these applications constitute a smaller portion of the overall market, they represent areas of high growth potential and technical innovation. Customization, sustainability, and performance are the key drivers for enzyme deployment in these niche areas, and advances in protein engineering continue to expand the scope of what enzymes, particularly proteases, can achieve outside traditional detergent contexts (Ji et al. 2023).

16.7.2 Performance Evaluation and Testing

Understanding how well proteases perform in detergent formulations requires more than just theoretical predictions or enzyme activity assays in isolation. In the highly competitive detergent industry, performance must be demonstrated through rigorous testing—both under standardized laboratory conditions and in real-world consumer contexts. Protease performance is judged not only by how effectively it removes protein-based stains but also by its compatibility with formulation ingredients, stability during storage, and safety across diverse washing conditions. This section explores the various methods used to evaluate protease efficacy in detergent products, including standardized test protocols, real-use simulations, and comparison with traditional chemical-based cleaning agents (Das and Ghosh 2023).

16.7.2.1 Standardized Testing Protocols

To ensure consistency and reliability in evaluating detergent enzyme performance, manufacturers and regulatory bodies use a range of standardized testing methods. These protocols are designed to create repeatable laboratory conditions that mimic key elements of the wash process, such as water hardness, fabric type, stain load, temperature, and detergent concentration. Such methods allow for quantitative comparisons of enzyme-containing detergents across different product batches, formulations, or brands (Carpenter and Crowe 1989).

One of the most widely used frameworks comes from the International Association for Soaps, Detergents and Maintenance Products (A.I.S.E.) and the American Society for Testing and Materials (ASTM). These bodies provide detailed protocols for preparing standardized stained fabrics (often called "swatches") with common soils like blood, milk, egg yolk, grass, and cocoa. These swatches are washed under controlled conditions using test detergents with and without enzymes, and the degree of stain removal is then measured spectrophotometrically, gravimetrically, or visually (Gupta et al. 2002b).

Enzyme activity is typically quantified through reduction in stain intensity or release of chromogenic substrates. For proteases, azocasein and hide powder azure assays are often employed to track proteolytic degradation. Results are standardized by enzyme unit definitions (e.g., one unit liberating 1 μmol of product per minute) and normalized to per-gram-of-detergent or per-wash-load metrics to ensure industrial relevance (Gupta et al. 2002a). Such protocols are indispensable for initial formulation screening, quality control, and product benchmarking. They also form the basis for label claims (e.g., "removes tough stains" or "active even in cold water") and play an important role in regulatory compliance and consumer safety validation.

16.7.2.2 Real-World Performance Assessment

While laboratory testing offers precision and control, it is often limited in its ability to capture the full complexity of real-world washing environments. As a result, manufacturers also conduct field tests and in-use simulations to evaluate how protease-based detergents perform in actual consumer scenarios. These assessments are critical because factors such as wash load composition, fabric types, water quality, washing machine technology, and user behavior can vary widely and significantly influence detergent efficacy (Zhang et al. 2021). In typical field tests, enzyme detergents are distributed to households or institutional laundry services, where volunteers use them over a fixed period while recording washing conditions and outcomes. Collected garments are then analyzed for residual staining, fabric wear, color fading, and odor retention. This approach helps reveal how proteases perform under variable and uncontrolled conditions—particularly with respect to stain variety, wash temperature, and soil load (Basketter et al. 2008). Real-world testing also accounts for consumer preferences, such as the shift toward lower-temperature washes and quicker cycles. Proteases that perform well in controlled laboratory tests may not necessarily exhibit sufficient stain removal at 20–30 °C in modern high-efficiency washing machines. Thus, enzyme variants designed for cold activity and fast action are increasingly favored in performance assessment. Additionally, consumer surveys and blind testing panels provide qualitative data that help manufacturers understand how enzyme-based cleaning is perceived in terms of freshness, softness, and overall satisfaction. These real-world assessments not only validate laboratory findings but also guide interactive formulation improvements and marketing strategies (Zhang et al. 2021).

16.7.2.3 Comparative Analysis with Chemical Alternatives

A key aspect of performance evaluation is comparing enzyme-based cleaning with traditional chemical approaches. Historically, proteinaceous stains were removed using strong alkalis, bleaches, and surfactants, which often required high temperatures and extended wash times. While effective, these chemicals can damage fabrics, consume high energy, and contribute to environmental pollution. Proteases offer a compelling alternative, working under milder conditions and targeting protein stains with high specificity (Shankar et al. 2024).

Comparative testing involves side-by-side analyses of enzyme-containing and enzyme-free formulations, as well as those with high concentrations of chemicals such as sodium hypochlorite or perborate bleach. The cleaning efficacy, fabric care, environmental load (e.g., Chemical Oxygen Demand—COD), and energy use are measured across identical stain and wash conditions. These comparisons consistently show that enzyme-based detergents can match or exceed the performance of chemical-intensive products—particularly in removing biological stains—while requiring less energy and reducing environmental burden. Moreover, proteases reduce the need for repeated washes, which translates into lower water and detergent consumption. Engineered proteases with high activity at low temperatures have been particularly instrumental in enabling cold-wash detergents that meet both consumer performance expectations and sustainability goals (Maurer 2024).

Cost-performance analysis is also an important dimension. While enzymes are biologically derived and can be costlier to produce than some commodity chemicals, their high specificity and low dosage requirements make them economically viable. As enzyme production becomes more efficient through fermentation, purification, and formulation innovations, the cost gap between biological and chemical solutions continues to narrow. In summary, performance evaluation of detergent proteases spans rigorous laboratory tests, real-world usage assessments, and comparative benchmarks against chemical agents. Together, these methodologies provide a comprehensive picture of how proteases contribute to effective, safe, and sustainable cleaning solutions (Carpenter and Crowe 1989).

16.7.3 Economic Considerations

The adoption and continued use of proteases in detergent formulations are influenced not only by performance but also by their economic viability. For manufacturers and consumers alike, enzymes must demonstrate measurable benefits in terms of cost-effectiveness, production scalability, and market competitiveness. Although enzyme production involves sophisticated biotechnological processes, the value they offer—through enhanced cleaning, environmental compliance, and consumer appeal—has secured their place in both premium and mass-market detergent brands. This section explores the economic landscape of detergent proteases by comparing enzymatic and traditional chemical cleaning, assessing manufacturing economics, and analyzing pricing strategies and competition (Zhang et al. 2021).

16.7.3.1 Cost–Benefit Analysis of Enzymatic vs. Chemical Cleaning

From a formulation standpoint, enzymes offer a different economic value proposition compared to conventional cleaning chemicals. Chemical agents such as sodium carbonate, perborates, or surfactants are typically inexpensive and readily available in bulk. However, they often require high dosages and elevated wash temperatures to achieve desired cleaning effects, leading to higher overall energy and environmental costs. In contrast, proteases operate catalytically and at lower concentrations, which translates into fewer resources used per wash cycle (Khan 2025).

For example, a well-optimized protease can significantly improve protein stain removal even at 30 °C, eliminating the need for hot water cycles that contribute to high electricity bills in domestic settings. This capability not only enhances consumer experience but also reduces household energy consumption—a critical consideration in markets where energy efficiency is a key selling point. Studies have shown that the inclusion of enzymes in detergents can reduce total energy consumption per load by up to 30%, especially when used in cold wash formulations (Maurer 2024; Nielsen et al. 2007).

On the manufacturing side, the higher unit cost of enzymes is often offset by the lower quantities required and the broader functional benefits they provide. For instance, by integrating a protease, a manufacturer may reduce the need for expensive surfactants or additional bleaching agents, effectively balancing the total cost of goods while improving the product's cleaning profile. This form of cost-offsetting makes enzymes particularly attractive in eco-labelled or high-performance detergents. Furthermore, enzymes support brand differentiation. Detergents marketed as "enzyme-based" or "biological" are often perceived by consumers as more advanced, safer, or more environmentally friendly, justifying a price premium in competitive retail markets (Ming et al. 2023).

16.7.3.2 Manufacturing Economics and Scaling

The large-scale production of detergent proteases has become increasingly cost-efficient due to improvements in microbial fermentation, downstream processing, and formulation technologies (Gimenes et al. 2021). Most commercial detergent proteases are produced using recombinant *Bacillus* strains in industrial bioreactors. These microorganisms are selected for their ability to secrete high titers of extracellular proteases, which simplifies purification and reduces production costs (Rao et al. 1998).

Modern fermentation technologies allow for batch or continuous production with optimized growth media and controlled conditions, leading to high enzyme yields per liter of culture. The use of genetically engineered strains with enhanced productivity and resistance to proteolysis has further lowered manufacturing costs. For instance, yield improvements through promoter engineering, metabolic pathway tuning, and feedback inhibition control have made it possible to produce kilograms of purified enzyme per cubic meter of fermentation broth (Gupta et al. 2002a).

Downstream processing—typically including filtration, precipitation, and drying—has also evolved to become more energy- and time-efficient. Spray drying, granulation, and encapsulation systems now allow for rapid conversion of liquid enzyme concentrate into stable solid forms suitable for powdered detergent products. In parallel, the development of stabilizing excipients for liquid formulations has enabled longer shelf life and easier logistics (Basketter et al. 2008).

The economies of scale achieved through large-volume fermentation and streamlined processing have significantly reduced the cost per active unit of enzyme, making proteases accessible not only to premium brands but also to mid-range and even budget product lines. Additionally, enzyme producers often work closely with detergent manufacturers under long-term supply agreements, offering bulk pricing, formulation support, and proprietary enzyme variants tailored to specific brand needs (Abramowitz et al. 1967).

16.7.3.3 Market Pricing and Competitive Dynamics

In today's detergent market, enzyme pricing reflects a balance between technological sophistication and mass-market accessibility. While enzymes remain one of the costlier components in a detergent formulation on a per-gram basis, their high activity and catalytic mode of action allow them to deliver value at very low inclusion rates—often between 0.2% and 2% of the total product weight. This means that even premium protease variants represent a small fraction of the overall cost structure but contribute disproportionately to product performance and brand value (Saeki et al. 2007).

Pricing strategies in the enzyme market are shaped by multiple factors, including supply chain maturity, patent landscapes, production location, and formulation compatibility. Large enzyme producers such as Novozymes, DuPont (now part of IFF), and BASF dominate the global market and invest heavily in R&D to develop proprietary variants with improved stability, performance, or cost-efficiency. These innovations are often protected by intellectual property rights, which allow for differentiation and market segmentation (Beniwal et al. 2024).

In terms of competitive dynamics, the detergent enzyme market has seen increasing competition from regional players and smaller biotech firms, especially in emerging markets like India, China, and Southeast Asia. These companies often focus on localized production and cost-effective fermentation systems, targeting value-oriented detergent manufacturers with tailored enzyme solutions (Saeki et al. 2007).

At the retail level, enzyme-rich detergents generally command a premium price due to their performance advantages and environmentally friendly appeal. However, the rise of private-label and economy brands with enzyme-inclusive formulations indicates that the technology is no longer exclusive to high-end products. Consumer demand for greener, more efficient detergents is pushing manufacturers to include enzymes as standard, regardless of price point (Basketter et al. 2008). In conclusion, the economic success of proteases in detergents is underpinned by their catalytic efficiency, energy-saving potential, and contribution to sustainable cleaning—all of which support a favorable cost-benefit profile. Continued advances in protein engineering, process optimization, and formulation science will further enhance their economic competitiveness in the global detergent market.

16.8 Environmental Impact and Sustainability

16.8.1 Environmental Benefits of Enzymatic Cleaning

Proteases have long been recognized not only for their effectiveness in removing stains but also for their role in advancing environmentally friendly cleaning practices. Their integration into detergent formulations represents a significant step toward more sustainable consumer products. By operating under mild conditions and targeting stains with high specificity, proteases help reduce the need for energy-intensive processes and environmentally damaging chemicals. In this context, enzymatic detergents offer three major environmental benefits: biodegradability, energy savings through cold-water washing, and a substantial reduction in the overall chemical footprint (Abramowitz et al. 1967).

One of the primary advantages of enzymes like proteases is their inherent biodegradability and ecological safety. Being proteins themselves, enzymes are readily broken down by microbial activity in natural ecosystems after use. They do not persist in the environment, nor do they accumulate in aquatic or terrestrial organisms, which significantly lowers the risk of long-term ecological harm. Proteases degrade into amino acids and peptides that re-enter the nitrogen cycle, unlike certain synthetic surfactants or bleaching agents that may cause bioaccumulation or toxicity in aquatic fauna (Maurer 2024). Regulatory authorities such as the EU and US EPA consider enzymatic ingredients as Generally Recognized As Safe (GRAS) when used appropriately, further reinforcing their environmental credibility.

Another substantial ecological advantage comes from energy savings enabled by low-temperature washing. Traditional detergents often require wash temperatures of 60 °C or higher to achieve acceptable stain removal, leading to increased household energy consumption and greenhouse gas emissions. Enzyme-based detergents, particularly those containing cold-active proteases, are effective even at temperatures as low as 20–30 °C. By facilitating effective cleaning at these temperatures, enzymes contribute directly to reduced electricity use during laundry cycles. For example, studies estimate that a shift to low-temperature washing enabled by enzymes could reduce carbon emissions by up to 30% in developed countries, provided wide consumer adoption (Nielsen et al. 2007).

Finally, enzymatic detergents support reduced chemical usage and environmental footprint. Proteases enhance the cleaning power of detergents by breaking down protein-based soils that might otherwise require high levels of alkaline builders, phosphates, or oxidizing agents. This allows formulators to reduce the total load of harsh chemicals, many of which contribute to eutrophication, water toxicity, and chemical oxygen demand (COD) in wastewater effluents. By replacing or reducing these components, enzymes lower the detergent's ecological burden while maintaining or improving cleaning performance. This shift also supports the development of eco-labelled and biodegradable detergent brands, which are increasingly favored by environmentally conscious consumers (Carpenter and Crowe 1989).

16.8.2 Life Cycle Assessment of Protease Production

While the use-phase benefits of proteases are well recognized, a complete understanding of their sustainability must also include the life cycle assessment (LCA) of protease production, which evaluates the environmental impact across all stages—from raw material sourcing and manufacturing to transport, usage, and end-of-life disposal.

In terms of manufacturing environmental impact, enzyme production is relatively efficient, especially compared to the synthesis of petrochemical-based surfactants or chelators. Proteases are typically produced through microbial fermentation, using genetically optimized *Bacillus* strains in large-scale bioreactors. This process consumes renewable feedstocks such as glucose, derived from agricultural sources like corn or sugarcane, and operates under ambient or moderately heated conditions, which significantly reduces energy inputs. Furthermore, closed-loop fermentation systems help capture waste gases and minimize water discharge, further lowering the carbon and water footprints (Gupta et al. 2002b). However, downstream processing—including purification, concentration, drying, and granulation—can be resource-intensive. Spray drying and encapsulation steps may consume thermal and electrical energy, and stabilizers or encapsulating agents sometimes involve petroleum-derived materials. As a result, enzyme manufacturers are increasingly exploring bio-based alternatives for formulation additives and are optimizing processing conditions to reduce heat and solvent use (Carpenter and Crowe 1989).

The transportation and packaging considerations also influence the overall environmental profile. Due to the high potency of enzymes, even small volumes of protease concentrate can treat large quantities of detergent—reducing bulk, shipping emissions, and packaging waste compared to traditional ingredients. Furthermore, enzymes formulated in solid forms such as granules or powders often have a longer shelf life and are less sensitive to temperature fluctuations, decreasing the need for refrigeration or specialized transport. Recent efforts in compact packaging and refillable formats further reduce the environmental impact of logistics and retail distribution (Gray 2022).

At the end of their life cycle, enzymes offer favorable disposal and biodegradation characteristics. Once released into domestic wastewater, they are rapidly denatured by temperature and dilution and then digested by microbial communities in sewage systems or natural water bodies. Unlike persistent cleaning agents such as alkylphenol ethoxylates or phosphate builders, proteases do not contribute significantly to chemical loading or pollution in aquatic environments. Their natural origin, low toxicity, and fast decomposition make them particularly well-suited for integration into circular economy models (Shakilanishi and Shanthi 2024).

16.8.3 Sustainable Production Practices

As environmental accountability becomes central to corporate strategy, enzyme manufacturers are increasingly adopting sustainable production practices guided by green chemistry principles, renewable inputs, and waste minimization.

Green chemistry in enzyme production involves designing processes that minimize waste, avoid hazardous reagents, and maximize atom economy. In the context of protease manufacturing, this includes optimizing fermentation media to reduce by-products, eliminating the need for organic solvents in purification, and using energy-efficient drying techniques. Enzyme stability improvements have also contributed to greener formulations—by extending shelf life and reducing formulation wastage, enzymes reduce the environmental cost per wash load (Leśniewski et al. 2023). One of the most impactful developments is the use of renewable feedstocks. Carbohydrate sources like molasses, starch hydrolysates, or lignocellulosic biomass serve as carbon and energy sources for microbial fermentation. Nitrogen inputs can also be sourced from agricultural waste streams, further supporting a bio-based economy. By shifting from fossil fuel-derived inputs to renewable agricultural residues, protease production becomes more carbon-neutral and aligns with long-term sustainability goals (Ottone et al. 2021). Additionally, leading producers are embracing waste minimization and circular economy approaches. Fermentation residues are increasingly valorized as fertilizers, biogas feedstock, or animal feed, reducing solid waste. Process water is recycled internally using filtration and ultrafiltration technologies, cutting down water consumption by up to 90% in some facilities. By creating value from waste and designing enzyme production as a closed-loop system, companies can dramatically reduce their environmental impact while boosting operational efficiency. As consumer demand and regulatory pressure continue to favor greener products, the detergent industry's move toward enzymes—particularly sustainably produced proteases—offers a compelling model of innovation that is both economically and ecologically viable. These developments are not just beneficial for manufacturers and consumers; they represent a meaningful contribution to global sustainability goals in chemical manufacturing and domestic product use (Bommarius and Paye 2013).

16.9 Regulatory Framework and Safety Considerations

As biologically active ingredients, proteases used in detergent formulations are subject to strict regulatory oversight to ensure their safety for consumers, workers, and the environment. While enzymes offer clear benefits in terms of biodegradability and cleaning efficiency, their proteinaceous nature also carries potential risks, such as allergenicity or respiratory sensitization, particularly in occupational settings. The global enzyme industry must therefore comply with a wide array of safety, quality, and labeling regulations—each guided by scientific risk assessment, international harmonization efforts, and consumer protection principles. This section explores the international regulatory landscape, the safety evaluation processes for protease-containing products, and the quality control mechanisms that ensure consistent, safe use in commercial detergent applications (Espejo 2021).

16.9.1 International Regulatory Landscape

Regulations governing the use of enzymes such as proteases vary across jurisdictions, but they share a common goal: to ensure that enzymes are safe for their intended use and do not pose undue risks to human health or the environment. The enzyme industry has benefited from increasing regulatory harmonization, particularly in safety evaluation protocols, which has facilitated the global distribution of detergent enzyme products (Basketter et al. 2008).

In the United States, regulatory oversight is primarily administered by the Food and Drug Administration (FDA) and the Environmental Protection Agency (EPA). The FDA regulates enzymes used in food processing, while the EPA governs the safety of industrial and cleaning agents, including those containing enzymes. Proteases used in consumer detergents must be assessed for potential skin and eye irritation, sensitization, and environmental toxicity under the EPA's Toxic Substances Control Act (TSCA). Many commonly used detergent proteases are granted exemptions or preapproved status, provided they meet safety thresholds and are manufactured under Good Manufacturing Practices (GMPs) (Espejo 2021).

In the European Union, enzymes fall under the jurisdiction of the European Chemicals Agency (ECHA) and are subject to the Registration, Evaluation, Authorization and Restriction of Chemicals (REACH) regulation. Under REACH, manufacturers and importers of enzymes must submit detailed safety dossiers, including toxicological data, production processes, and exposure risk assessments. Proteases used in cleaning products must also comply with the EU Detergents Regulation (EC No 648/2004), which mandates full ingredient disclosure and safety labelling for consumer protection. If a detergent is considered a biocidal product—such as those with antimicrobial enzyme components—it may also fall under the Biocidal Products Regulation (EU 528/2012), which imposes stricter evaluation criteria (Abramowitz et al. 1967).

In the Asia-Pacific region, regulations are diverse and vary significantly between countries. For instance, in Japan, enzymes used in industrial products are regulated by the Ministry of Economy, Trade and Industry (METI) under the Chemical Substances Control Law (CSCL), while China enforces safety standards via the Ministry of Ecology and Environment and the State Administration for Market Regulation. In India, enzymes used in cleaning products must comply with guidelines under the Central Insecticides Board and Registration Committee (CIBRC) if biocidal activity is involved. Across the region, there is growing emphasis on adopting OECD guidelines for safety assessment and participating in international mutual recognition frameworks (Azrin et al. 2022).

Together, these regulatory systems require enzyme manufacturers to provide robust data on safety, functionality, and environmental impact, thereby ensuring that proteases used in detergents meet global standards for responsible product stewardship (Carpenter and Crowe 1989).

16.9.2 Safety Assessment and Risk Management

A critical aspect of bringing enzyme-based detergents to market is ensuring that the included proteases are safe across the product's entire life cycle—from manufacturing and packaging to end use by consumers. Safety assessments are conducted through a combination of in vitro, in vivo, and computational studies, often guided by OECD testing guidelines and ISO standards (Rao et al. 1998).

The toxicological evaluation of enzyme products typically begins with acute and chronic toxicity tests to determine whether the protease causes harmful effects when ingested, inhaled, or absorbed through the skin. While enzymes generally exhibit low systemic toxicity, their protein structure can sometimes act as allergens or sensitizing agents. For this reason, proteases are routinely screened for dermal and ocular irritation, as well as for respiratory sensitization potential using animal-free assays and structure-activity relationship (SAR) modelling. If a protease is considered a novel substance or produced using genetically modified microorganisms (GMMs), additional toxicological scrutiny is required, including DNA analysis, endotoxin testing, and allergenicity prediction (Basketter et al. 2010).

In terms of occupational safety, the risk of enzyme exposure is most significant in manufacturing plants, formulation facilities, and packaging units, where workers may encounter enzyme dust or aerosols. To mitigate these risks, companies implement engineering controls (e.g., closed systems, HEPA filtration), personal protective equipment (PPE), and enzyme encapsulation technologies that limit airborne dispersal. The use of safety data sheets (SDS) and on-site training ensures that workers understand proper handling, spill response, and medical protocols. In some countries, occupational exposure limits (OELs) for enzymes like subtilisin have been set to minimize respiratory sensitization in the workplace (Yao et al. 2023).

From the consumer safety perspective, enzymes used in detergent formulations are usually encapsulated or stabilized in such a way that dermal or inhalation exposure during product use is extremely low. Proteases are denatured quickly in the wash cycle due to dilution, surfactants, and thermal conditions, and thus pose minimal residual risk on laundered fabrics. Nonetheless, detergents must comply with labelling requirements that inform consumers of potential allergens, proper use, and storage. In the EU, for example, enzyme-containing products must carry precautionary statements (e.g., "contains enzymes") and allergen warnings if needed, as per the Detergents Regulation (Kirk et al. 2002).

16.9.3 Quality Control and Standardization

Ensuring the consistent safety and effectiveness of detergent proteases also requires robust quality control (QC) systems. These systems are grounded in internationally recognized industry standards and specifications, including ISO 9001 for quality management and ISO 14001 (International Organization for Standardization) for environmental management.

Manufacturers define enzyme activity units and establish potency benchmarks for each protease batch, ensuring predictable performance in detergent formulations. Batch-to-batch consistency is maintained through validated fermentation, purification, and formulation protocols. Analytical tools such as high-performance liquid chromatography (HPLC), mass spectrometry, and protein electrophoresis are routinely used to confirm protease identity, purity, and activity levels (Bravi et al. 2020). In addition to enzyme potency, analytical methods and testing protocols are used to detect microbial contamination, endotoxins, heavy metals, and formulation stability. Enzymes are also subjected to thermal and oxidative stress tests to ensure they remain stable throughout shelf life and perform under expected washing conditions. Quality control laboratories often perform accelerated ageing studies, storage compatibility assessments, and cross-interference checks to validate enzyme stability within the detergent matrix (Al-Ghanayem and Joseph 2020). Maintaining batch-to-batch consistency is particularly important in global brands, where detergent formulations are produced across multiple sites but must meet the same performance and safety expectations. To this end, enzyme suppliers provide Certificates of Analysis (CoAs) for each production batch, documenting compliance with specification limits and safety thresholds. In cases where genetically modified production strains are used, traceability protocols are implemented to ensure regulatory compliance across jurisdictions. By adhering to these stringent quality standards, the detergent industry ensures that proteases continue to be safe, reliable, and effective ingredients—upholding both regulatory compliance and consumer confidence.

16.10 Current Challenges and Limitations

While proteases have significantly transformed modern detergents with their efficiency and eco-friendliness, several hurdles continue to restrict their full potential across global markets. These challenges span technical constraints, economic barriers, and complex regulatory landscapes. Addressing these limitations is essential for the sustained growth of enzyme-based detergents and for their successful integration into emerging markets and next-generation product formulations. This section examines the key challenges currently facing detergent proteases and offers insights into areas where further innovation and collaboration are needed (Maurer 2004).

16.10.1 Technical Limitations

Despite their many advantages, proteases must contend with several technical limitations that can impact their efficacy and shelf stability in commercial detergent products. These challenges are particularly relevant as detergent formulations become more diverse and as consumers demand higher performance across a broader range of wash conditions.

One of the most persistent technical issues is enzyme stability under extreme conditions. Proteases are inherently sensitive to factors such as high pH, oxidative stress, and elevated temperatures—conditions that are common in many detergent applications. While protein engineering has improved the robustness of many commercial proteases, maintaining activity and structural integrity in environments with high levels of surfactants, peroxides, and metal ions remains challenging. For instance, bleach-based detergents or multienzyme formulations often pose risks of denaturation or inactivation unless the protease is specially stabilized (Maurer 2004).

Closely tied to this is the problem of storage and shelf-life constraints, especially in liquid detergent formulations. Unlike solid detergents, where enzymes are encapsulated and relatively protected, liquid detergents expose proteases to continual interaction with water, solvents, and other active ingredients. Over time, this can lead to gradual loss of activity, particularly in hot climates or under poor storage conditions. The challenge is compounded when enzymes are stored alongside incompatible ingredients like proteolytic inhibitors or reactive chemicals. To address this, formulators rely on stabilizers, polymers, or sugar-based glassing agents—but these can add cost and complexity to the formulation process (Gupta et al. 2002b).

Another growing technical concern is compatibility with emerging detergent ingredients, such as bio-based surfactants, fragrance microcapsules, and fabric-softening agents. While many traditional proteases have been optimized for classical detergent systems, their interactions with new, environmentally friendly ingredients are still being studied. The risk of enzyme inhibition, precipitation, or unexpected chemical reactions increases as detergents become more multifunctional and consumer-oriented. Ensuring seamless performance in complex formulations remains a frontier of detergent enzyme technology (Al-Ghanayem and Joseph 2020).

16.10.2 Market and Economic Challenges

In addition to technical issues, protease adoption in detergents faces several economic and market-related barriers, particularly in cost-sensitive regions and highly competitive commercial segments.

One of the most significant challenges is cost competitiveness with synthetic alternatives. While enzymes offer compelling advantages in performance and sustainability, their production through fermentation, purification, and stabilization remains more expensive than many commodity chemicals used in traditional detergents. In premium product segments, this cost is often justifiable—but in mass-market or low-income regions, price constraints can limit enzyme inclusion. Enzymes must prove not just superior efficacy but cost-efficiency at scale to displace cheaper, chemical-based cleaning agents (Nielsen et al. 2007).

Compounding this issue is consumer education and awareness. Despite their long history of safe use, enzymes like proteases are still poorly understood by many end users. Misconceptions about allergens, synthetic origins, or "biological" cleaning can lead to consumer hesitation or mistrust, particularly in regions where environmental and ingredient transparency is not widely promoted. Many consumers still associate foaming and strong chemical scent with cleanliness, underestimating the invisible but powerful action of enzymes in removing stubborn soils. To gain broader acceptance, enzyme-based products must be accompanied by educational campaigns, clearer labelling, and credible environmental claims (Gron et al. 1990).

Another critical factor is market penetration in developing regions, where detergent use is increasing but cost and infrastructure remain limiting factors. In countries across South Asia, Africa, and Latin America, enzyme detergents may be seen as luxury items or incompatible with local washing practices—such as hand-washing in cold water or use of high-dust environments that may degrade enzyme formulations. Enzyme producers and detergent brands must work collaboratively to adapt formulations, packaging, and marketing strategies for these diverse contexts while maintaining efficacy and affordability (Zhang et al. 2021).

16.10.3 Regulatory and Safety Hurdles

The final category of challenges relates to regulatory complexities and safety frameworks, which are continually evolving in response to new scientific insights and public health concerns. One such hurdle is the rise of evolving safety requirements, particularly around allergenicity and occupational exposure. Although detergent proteases are generally considered safe for consumer use, concerns persist regarding their potential to cause respiratory sensitization in industrial workers or, less commonly, dermal irritation in sensitive individuals. Regulatory agencies in the US, EU, and other regions now require more comprehensive safety data, including in vitro allergenicity assays, risk models, and post-marketing surveillance. Meeting these demands adds cost, delays product launches, and can restrict access to certain markets unless all documentation and safety thresholds are met (Basketter et al. 2010). Another persistent challenge is international harmonization. Despite efforts to standardize enzyme safety and environmental evaluations through bodies like the OECD and ISO, regional variations in data requirements, terminology, and product classification persist. An enzyme approved for use in North America may require additional animal testing or administrative clearance for sale in Asia or South America. This lack of harmonization increases regulatory burden for global enzyme producers and complicates the introduction of novel protease variants (Lund et al. 2012).

Finally, the approval process for new enzyme products remains complex, especially for genetically modified enzyme strains. Regulatory approval often involves extensive dossiers covering genetic construction, allergenicity potential, environmental fate, and consumer exposure estimates. While these processes are essential for ensuring public safety, they can be prohibitively slow or opaque, particularly for small or medium-sized biotech firms hoping to enter the detergent enzyme market. Streamlining regulatory pathways—without compromising safety—will be crucial for fostering innovation and competition in this field (Basketter et al. 2008).

16.11 Emerging Trends and Future Innovations

As consumer demands evolve and sustainability becomes a defining principle in industrial chemistry, the detergent industry continues to look toward advanced biotechnology to meet new expectations. Proteases, having already revolutionized the cleaning sector through biological stain removal and eco-friendly performance, now stand at the cusp of another transformation—driven by cutting-edge science, novel delivery systems, and powerful computational tools. This chapter explores the next frontier of detergent proteases by highlighting the most promising trends and innovations that will shape their future in household and industrial cleaning applications (Wang et al. 2017).

16.11.1 Next-Generation Enzyme Technologies

The development of proteases with novel biochemical properties is central to future innovation in detergent design. Traditional proteases, though highly effective, are limited by narrow optimal working conditions. Expanding their functional ranges will allow for broader application in diverse formulations and use environments.

One exciting development is the engineering and discovery of cold-active and psychrophilic proteases—enzymes that retain high catalytic activity at low temperatures, even below 20 °C. These enzymes are typically derived from extremophiles found in Arctic and Antarctic habitats or engineered through site-directed mutagenesis to maintain flexibility in low-energy environments. Their use supports energy-efficient cold washing, aligning detergent formulations with global energy conservation goals. The challenge lies in balancing cold activity with adequate shelf-life and formulation stability, which continues to drive structural studies and protein engineering strategies (Carpenter and Crowe 1989).

Another promising innovation is the creation of proteases that function at extreme pH and salinity levels, making them more compatible with niche or industrial detergents. Enzymes stable in highly alkaline (pH 10–12) or even acidic environments, as well as those that withstand high ionic concentrations, are increasingly sought after for specialized cleaning tasks, such as textile processing or industrial equipment sanitation. These robust variants are often sourced from halotolerant bacteria (Ahmed et al. 2016) or haloalkaliphilic bacteria or developed through directed evolution to improve salt-bridge networks and internal hydrogen bonding (Klein and Bischoff 2011).

Further innovation is seen in multifunctional enzyme chimeras—engineered proteins that combine the domains of multiple enzymes into a single molecule. These chimeras can offer combined protease, lipase, and amylase activity, reducing the number of additives required in a formulation and enabling synergistic cleaning action. For instance, a bifunctional protease-amidase may simultaneously break down both protein and peptide-based stains, increasing efficacy across diverse soil types (Sharma et al. 2020). Such hybrid enzymes not only simplify formulations but also reduce packaging and transport burdens, contributing to sustainability.

16.11.2 Advanced Formulation Technologies

Alongside new enzymes, the success of future protease detergents will also depend on the innovation of formulation techniques that enhance stability, targeted delivery, and user experience.

Nanotechnology offers one of the most exciting avenues for enzyme encapsulation and protection. Encapsulation within nanogels, liposomes, or mesoporous silica particles can shield proteases from premature denaturation, enhance their solubility, and even facilitate time- or pH-triggered release during washing cycles. This kind of controlled release is particularly valuable in liquid detergents, where protease degradation by water or other active ingredients is a major limitation. Research into biocompatible nano-carriers is growing, with a focus on cost-effective, biodegradable materials suitable for large-scale use (Patel et al. 2025).

In tandem, smart delivery systems are being designed to enable more efficient and responsive detergent performance. These include pH-sensitive or temperature-responsive polymers that only activate enzyme release when ideal conditions are met—ensuring that proteases work precisely when and where needed during the wash cycle. Smart coatings may also allow sequential enzyme activation, starting with lipases for greasy soils and followed by proteases for proteinaceous residues. Such systems optimize cleaning efficacy while conserving enzyme resources (Rodrigues et al. 2013).

A broader innovation involves the shift toward personalized cleaning solutions, especially in high-end or digital-first detergent brands. With the help of consumer input (e.g., fabric types, water quality, stain profiles), protease blends may be customized or modular, delivered via cartridges or refills tailored to individual households. As machine learning and internet-of-things (IoT) capabilities integrate with smart washing appliances, detergent delivery systems may eventually become adaptive—automatically dispensing enzyme cocktails based on load sensing and historical usage patterns (Polgar 2005).

16.11.3 Biotechnology Innovations

At the heart of these technological advances is the field of biotechnology, which continues to redefine the landscape of enzyme discovery, engineering, and manufacturing.

Synthetic biology is unlocking new ways to design and express proteases from scratch. By assembling synthetic gene circuits, scientists can create novel proteases with nonnatural properties or even entirely artificial scaffolds. These enzymes may offer advantages like resistance to oxidation, rapid folding, or binding to synthetic polymers. Expression systems are also becoming more diverse, with cell-free platforms and nontraditional hosts like *Pichia pastoris*, microalgae, or even engineered viruses offering scalable, low-cost alternatives to traditional microbial fermentation (Abramowitz et al. 1967).

Artificial intelligence (AI)-driven enzyme design is another transformative tool reshaping the protease landscape. Machine learning algorithms, trained on vast structural and activity datasets, are now capable of predicting enzyme-substrate interactions, identifying mutation hotspots, and proposing optimal residue substitutions for improved performance. AI platforms like Alpha Fold have revolutionized structural biology by accurately predicting protein folding from sequence alone, significantly accelerating rational enzyme engineering. These tools are not only speeding up the discovery pipeline but also lowering R&D costs and increasing the predictability of enzyme improvements (Khan 2025).

Lastly, sustainable production platforms are being prioritized to make enzyme manufacturing eco-friendlier and socially responsible. Innovations include the use of waste feedstocks (e.g., agricultural residues or industrial byproducts) for fermentation, as well as the development of circular bioprocesses that valorize enzyme production waste streams. Life cycle analyses are being incorporated into the early stages of enzyme design, ensuring that proteases are not just effective and affordable, but also compatible with planetary boundaries (Beniwal et al. 2024).

16.12 Conclusion

Proteases are now the mainstay of contemporary detergent technology, spurring advancements in environmental responsibility, sustainability, and cleaning efficiency. Proteases are a prime example of how biotechnology may replace harsh chemical agents with precise, energy-efficient, and biodegradable substitutes, from their initial use as biological stain removers to their current role as multipurpose and environmentally friendly biocatalysts. Their catalytic adaptability contributes to lower energy usage and better fabric care by enabling efficient stain removal under mild washing settings.

The stability, activity, and specificity of detergent proteases have consistently been improved by developments in protein engineering, directed evolution, and computer modeling. These advancements are making it possible to create next-generation enzymes that are suited for a variety of washing conditions, such as low-water and cold-water systems, in conjunction with AI-driven enzyme discovery and structure prediction tools. The strategic use of proteases puts detergent makers at the forefront of green innovation as customer demand and environmental restrictions increasingly favor sustainable products.

Enzyme manufacturers, formulators, legislators, and researchers must work closely together to ensure that scientific advancements are in line with safety regulations, financial viability, and environmental objectives. Proteases are positioned to continue being essential catalysts, driving the detergent sector toward a cleaner, smarter, and more sustainable future with sustained investment in green chemistry, digital design, and international collaboration.

References

Abramowitz N, Schechter I, Berger A (1967) On the size of the active site in proteases II. carboxypeptidase-A. Biochem Biophys Res Commun 29(6):862–867. https://doi.org/10.1016/0006-291X(67)90299-9

Ahmed M, Rehman R, Siddique A, Hasan F, Ali N, Hameed A (2016) Production, purification and characterization of detergent-stable, halotolerant alkaline protease for eco-friendly application in detergents' industry. Int J Biosci 8(2):47–65. https://doi.org/10.12692/ijb/8.2.47-65

Al-Ghanayem AA, Joseph B (2020) Current prospective in using cold-active enzymes as eco-friendly detergent additive. Appl Microbiol Biotechnol 104(7):2871–2882. https://doi.org/10.1007/s00253-020-10429-x

Anwar A, Saleemuddin M (1998) Alkaline proteases: a review. Bioresour Technol 64(3):175–183. https://doi.org/10.1016/S0960-8524(97)00182-X

Aruna V, Chandrakala V, Angajala G, Nagarajan ER (2023) Proteases: an overview on its recent industrial developments and current scenario in the revolution of biocatalysis. Mater Today Proc 92:565–573. https://doi.org/10.1016/j.matpr.2023.03.806

Azrin NAM, Ali MSM, Rahman RNZRA, Oslan SN, Noor NDM (2022) Versatility of subtilisin: a review on structure, characteristics, and applications. Biotechnol Appl Biochem 69(6):2599–2616. https://doi.org/10.1002/bab.2309

Barrett AJ, Rawlings ND, O'Brien EA (2001) The MEROPS database as a protease information system. J Struct Biol 134(2–3):95–102. https://doi.org/10.1006/jsbi.2000.4332

Basketter DA, English JSC, Wakelin SH, White IR (2008) Enzymes, detergents and skin: facts and fantasies. Br J Dermatol 158(6):1177–1181. https://doi.org/10.1111/j.1365-2133.2008.08561.x

Basketter DA, Broekhuizen C, Fieldsend M, Kirkwood S, Mascarenhas R, Maurer K et al (2010) Defining occupational and consumer exposure limits for enzyme protein respiratory allergens under REACH. Toxicology 268(3):165–170. https://doi.org/10.1016/j.tox.2009.12.014

Baur X, Chen Z, Allmers H (1998) Can a threshold limit value for natural rubber latex airborne allergens be defined? J Allergy Clin Immunol 101(1):24–27. https://doi.org/10.1016/S0091-6749(98)70188-5

Beniwal R, Kumar V, Sharma V (2024) Metaheuristics approaches towards secure and optimized routing in IoT: a systematic literature review. In: 2024 international conference on electrical electronics and computing technologies (ICEECT), vol 1, pp 1–6. https://doi.org/10.1109/ICEECT61758.2024.10739076

Bié J, Sepodes B, Fernandes PC, Ribeiro MH (2022) Enzyme immobilization and co-immobilization: main framework, advances and some applications. PRO 10(3):494. https://doi.org/10.3390/pr10030494

Bommarius AS, Paye MF (2013) Stabilizing biocatalysts. Chem Soc Rev 42(15):6534–6565. https://doi.org/10.1039/C3CS60137D

Boyce A, Walsh G (2012) Identification of fungal proteases potentially suitable for environmentally friendly cleaning-in-place in the dairy industry. Chemosphere 88(2):211–218. https://doi.org/10.1016/j.chemosphere.2012.03.022

Bravi L, Santos G, Pagano A, Murmura F (2020) Environmental management system according to ISO 14001: 2015 as a driver to sustainable development. Corp Soc Responsib Environ Manag 27(6):2599–2614. https://doi.org/10.1002/csr.1985

Bryan PN (2000) Protein engineering of subtilisin. Biochimica et Biophysica Acta (BBA)-Protein Struct Mol Enzymol 1543(2):203–222. https://doi.org/10.1016/S0167-4838(00)00235-1

Bryan P, Alexander P, Strausberg S, Schwarz F, Lan W, Gilliland G, Gallagher DT (1992) Energetics of folding subtilisin BPN. Biochemistry 31(21):4937–4945. https://doi.org/10.1021/bi00136a003

Carpenter JF, Crowe JH (1989) An infrared spectroscopic study of the interactions of carbohydrates with dried proteins. Biochemistry 28(9):3916–3922. https://doi.org/10.1021/bi00435a044

Chen K, Arnold FH (1991) Enzyme engineering for nonaqueous solvents: random mutagenesis to enhance activity of subtilisin E in polar organic media. Bio/Technology 9(11):1073–1077. https://doi.org/10.1038/nbt1191-1073

Cherry JR, Fidantsef AL (2003) Directed evolution of industrial enzymes: an update. Curr Opin Biotechnol 14(4):438–443. https://doi.org/10.1016/S0958-1669(03)00099-5

Das M, Ghosh M (2023) Screening, characterization, and kinetic studies of a serine alkaline protease from kitchen wastewater bacteria P2S1An and evaluation of its application in nutraceutical production. Water Environ Res 95(3):e10848. https://doi.org/10.1002/wer.10848

Eijsink VG, Bjørk A, Gåseidnes S, Sirevåg R, Synstad B, van den Burg B, Vriend G (2004) Rational engineering of enzyme stability. J Biotechnol 113(1–3):105–120. https://doi.org/10.1016/j.jbiotec.2004.03.026

Espejo F (2021) Role of commercial enzymes in wine production: a critical review of recent research. J Food Sci Technol 58(1):9–21. https://doi.org/10.1007/s13197-020-04489-0

Fleißner A, Dersch P (2010) Expression and export: recombinant protein production systems for *Aspergillus*. Appl Microbiol Biotechnol 87(4):1255–1270. https://doi.org/10.1007/s00253-010-2672-6

Fujii T, Fukano K, Hirano K, Mimura A, Terauchi M, Etoh SI, Iida A (2020) A new serine protease family with elastase activity is produced by *Streptomyces* bacteria. Microbiology 166(3):253–261. https://doi.org/10.1099/mic.0.000880

Gimenes NC, Silveira E, Tambourgi EB (2021) An overview of proteases: production, downstream processes and industrial applications. Sep Purif Rev 50(3):223–243. https://doi.org/10.1080/15422119.2019.1677249

Godfrey JS (1996) The effect of the Indonesian throughflow on ocean circulation and heat exchange with the atmosphere: a review. J Geophys Res Oceans 101(C5):12217–12237. https://doi.org/10.1029/95JC03860

Gonzalez NA, Li BA, McCully ME (2022) The stability and dynamics of computationally designed proteins. Protein Eng Des Sel 35:gzac001. https://doi.org/10.1093/protein/gzac001

Gray M (2022) Packaging design in the circular economy: reuse models in fast-moving consumer goods. Doctoral dissertation, The University of Texas, Austin. https://doi.org/10.26153/tsw/45987

Gron H, Bech LM, Branner S, Breddam K (1990) A highly active and oxidation-resistant subtilisin-like enzyme produced by a combination of site-directed mutagenesis and chemical modification. Eur J Biochem 194(3):897–901. https://doi.org/10.1111/j.1432-1033.1990.tb19484.x

Gulmez C, Atakisi O, Dalginli KY, Atakisi E (2018) A novel detergent additive: organic solvent- and thermo-alkaline-stable recombinant subtilisin. Int J Biol Macromol 108:436–443. https://doi.org/10.1016/j.ijbiomac.2017.11.133

Gupta R, Beg Q, Khan S, Chauhan B (2002a) An overview on fermentation, downstream processing and properties of microbial alkaline proteases. Appl Microbiol Biotechnol 60(4):381–395. https://doi.org/10.1007/s00253-002-1142-1

Gupta R, Beg Q, Lorenz P (2002b) Bacterial alkaline proteases: molecular approaches and industrial applications. Appl Microbiol Biotechnol 59(1):15–32. https://doi.org/10.1007/s00253-002-0975-y

Hedstrom L (2002) Serine protease mechanism and specificity. Chem Rev 102(12):4501–4524. https://doi.org/10.1021/cr000033x

Horikoshi K (1999) Alkaliphiles: some applications of their products for biotechnology. Microbiol Mol Biol Rev 63(4):735–750. https://doi.org/10.1128/mmbr.63.4.735-750.1999

Ji X, Peng Z, Song J, Zhang G, Zhang J (2023) Fusion of substrate-binding domains enhances the catalytic capacity of keratinases and promotes enzymatic conversion of feather waste. J Agric Food Chem 71(30):11579–11586. https://doi.org/10.1021/acs.jafc.3c03064

Khan MF (2025) Recent advances in microbial enzyme applications for sustainable textile processing and waste management. Sci 7(2):46. https://doi.org/10.3390/sci7020046

Kirk O, Borchert TV, Fuglsang CC (2002) Industrial enzyme applications. Curr Opin Biotechnol 13(4):345–351. https://doi.org/10.1016/S0958-1669(02)00328-2

Klein T, Bischoff R (2011) Physiology and pathophysiology of matrix metalloproteases. Amino Acids 41(2):271–290. https://doi.org/10.1007/s00726-010-0689-x

Kortemme T, Baker D (2002) A simple physical model for binding energy hot spots in protein–protein complexes. Proc Natl Acad Sci 99(22):14116–14121. https://doi.org/10.1073/pnas.202485799

Koshland DE Jr (1958) Application of a theory of enzyme specificity to protein synthesis. Proc Natl Acad Sci 44(2):98–104. https://doi.org/10.1073/pnas.44.2.98

Kumar CG, Takagi H (1999) Microbial alkaline proteases: from a bioindustrial viewpoint. Biotechnol Adv 17(7):561–594. https://doi.org/10.1016/S0734-9750(99)00027-0

Leśniewski B, Kotula M, Kubiak A, Pajewska-Szmyt M (2023) Thermostability of selected biological materials. Lett Appl NanoBiosci 12(3):88. https://doi.org/10.33263/LIANBS123.088

Lund H, Kaasgaard SG, Skagerlind P, Jorgensen L, Jørgensen CI, Van de Weert M (2012) Protease and amylase stability in the presence of chelators used in laundry detergent applications: correlation between chelator properties and enzyme stability in liquid detergents. J Surfactant Deterg 15(3):265–276. https://doi.org/10.1007/s11743-011-1318-8

Maurer KH (2004) Detergent proteases. Curr Opin Biotechnol 15(4):330–334. https://doi.org/10.1016/j.copbio.2004.06.005

Maurer KH (2024) Enzymes in detergents and cleaning agents. In: Introduction to enzyme technology. Springer, Cham, pp 361–383. https://doi.org/10.1007/978-3-031-42999-6_17

Ming Y, Wang W, Yin R, Zeng M, Tang L, Tang S, Li M (2023) A review of enzyme design in catalytic stability by artificial intelligence. Brief Bioinform 24(3):bbad065. https://doi.org/10.1093/bib/bbad065

Naganthran A, Masomian M, Rahman RNZRA, Ali MSM, Nooh HM (2017) Improving the efficiency of new automatic dishwashing detergent formulation by addition of thermostable lipase, protease and amylase. Molecules 22(9):1577. https://doi.org/10.3390/molecules22091577

Nguyen VML, Ndao A, Peterson EC, Blais JF, Adjallé K (2025) *Bacillus* species: evolving roles in bio-based detergents. PRO 13(6):1885. https://doi.org/10.3390/pr13061885

Nielsen PH, Oxenbøll KM, Wenzel H (2007) Cradle-to-gate environmental assessment of enzyme products produced industrially in Denmark by novozymes A/S. Int J Life Cycle Assess 12(6): 432–438. https://doi.org/10.1065/lca2006.08.265.1

Nilegaonkar SS, Zambare VP, Kanekar PP, Dhakephalkar PK, Sarnaik SS, Chandrababu NK, Rajaram R, Ramanaiah B, Ramasami T, Saikumari YK, Balaram P (2006) A novel protease for industrial application. Indian Patent 2471DEL2006

Nilegaonkar SS, Zambare VP, Kanekar PP, Dhakephalkar PK, Sarnaik SS, Chandrababu NK, Rajaram R, Ramanaiah B, Ramasami T, Saikumari YK, Balaram P (2007a) A novel protease for industrial application. German Patent 102007013950:2

Nilegaonkar SS, Zambare VP, Kanekar PP, Dhakephalkar PK, Sarnaik SS, Chandrababu NK, Rajaram R, Ramanaiah B, Ramasami T, Saikumari YK, Balaram P (2007b) A novel protease for industrial application. US Patent 11/931537

Ochoa R, Magnitov M, Laskowski RA, Cossio P, Thornton JM (2020) An automated protocol for modelling peptide substrates to proteases. BMC bioinformatics 21(1):586. https://doi.org/10.1186/s12859-020-03931-6

Okino-Delgado CH, Zanutto-Elgui MR, do Prado DZ, Pereira MS, Fleuri LF (2019) Enzymatic bioremediation: current status, challenges of obtaining process, and applications. In: Microbial metabolism of xenobiotic compounds. Springer Singapore, Singapore, pp 79–101. https://doi.org/10.1007/978-981-13-7462-3_4

Ottone C, Romero O, Urrutia P, Bernal C, Illanes A, Wilson L (2021) Enzyme biocatalysis and sustainability. In: Nanostructured catalysts for environmental applications. Springer, Cham, pp 383–413. https://doi.org/10.1007/978-3-030-58934-9_14

Pandey A (2003) Solid-state fermentation. Biochem Eng J 13(2–3):81–84. https://doi.org/10.1016/S1369-703X(02)00121-3

Patel, H. K., & Makampara, R. A. (2023). Microbial enzymes: a futuristic tools for a sustainable environment in 21st century. In: Emerging technologies in applied and environmental microbiology (pp. 45–63). Academic Press. xhttps://doi.org/10.1016/B978-0-323-99895-6.00011-3

Patel V, Shah JN, Dalwadi S, Barot P, Vaghela D, Patel M, Sutariya V (2025) Injectables protein-based nanodiscs in cancer drug delivery: from bench to clinical potential. AAPS PharmSciTech 27(1):6. https://doi.org/10.1208/s12249-025-03247-9

Polgar L (2005) The catalytic triad of serine peptidases. Cell Mol Life Sci 62(19):2161–2172. https://doi.org/10.1007/s00018-005-5160-x

Rajagopalan S, Wang C, Yu K, Kuzin AP, Richter F, Lew S et al (2014) Design of activated serine–containing catalytic triads with atomic-level accuracy. Nat Chem Biol 10(5):386–391. https://doi.org/10.1038/nchembio.1498

Rao MB, Tanksale AM, Ghatge MS, Deshpande VV (1998) Molecular and biotechnological aspects of microbial proteases. Microbiol Mol Biol Rev 62(3):597–635. https://doi.org/10.1128/mmbr.62.3.597-635.1998

Rodrigues RC, Ortiz C, Berenguer-Murcia Á, Torres R, Fernández-Lafuente R (2013) Modifying enzyme activity and selectivity by immobilization. Chem Soc Rev 42(15):6290–6307. https://doi.org/10.1039/C2CS35231A

Saeki K, Ozaki K, Kobayashi T, Ito S (2007) Detergent alkaline proteases: enzymatic properties, genes, and crystal structures. J Biosci Bioeng 103(6):501–508. https://doi.org/10.1263/jbb.103.501

Shakilanishi S, Shanthi C (2024) An overview on preparation of enzymes for industrial use. Biocatal Biotransformation 42(4):485–496. https://doi.org/10.1080/10242422.2024.2311098

Shankar J, Verma P, Shah MP (2024) Microbial approaches for sustainable green technologies. CRC Press. https://doi.org/10.1201/9781003407683

Sharma AK, Kikani BA, Singh SP (2020) Biochemical, thermodynamic and structural characteristics of a biotechnologically compatible alkaline protease from a haloalkaliphilic, *Nocardiopsis dassonvillei* OK-18. Int J Biol Macromol 153:680–696. https://doi.org/10.1016/j.ijbiomac.2020.03.006

Solanki P, Putatunda C, Kumar A, Bhatia R, Walia A (2021) Microbial proteases: ubiquitous enzymes with innumerable uses. 3 Biotech 11(10):428. https://doi.org/10.1007/s13205-021-02928-z

Stocker W, Grams F, Reinemer P, Bode W, Baumann U, Gomis-Rüth FX, Mckay DB (1995) The metzincins—topological and sequential relations between the astacins, adamalysins, serralysins, and matrixins (collagenases) define a super family of zinc-peptidases. Protein Sci 4(5):823–840. https://doi.org/10.1002/pro.5560040502

Turk B, Turk D, Turk V (2012) Protease signalling: the cutting edge. EMBO J 31(7):1630–1643. https://doi.org/10.1038/emboj.2012.42

Vega A, Planas A, Biarnés X (2025) A practical guide to computational tools for engineering biocatalytic properties. Int J Mol Sci 26(3):980. https://doi.org/10.3390/ijms26030980

Verma S, Dixit R, Pandey KC (2016) Cysteine proteases: modes of activation and future prospects as pharmacological targets. Front Pharmacol 7:107. https://doi.org/10.3389/fphar.2016.00107

Vicaria JM, Herrera-Márquez O, Serrano-Haro M, Vidal A, Jurado E, Jiménez-Pérez JL (2022) Optimization of surfactants formulations to stabilise proteases and amylases. Chem Eng Sci 260:117858. https://doi.org/10.1016/j.ces.2022.117858

Vieille C, Zeikus GJ (2001) Hyperthermophilic enzymes: sources, uses, and molecular mechanisms for thermostability. Microbiol Mol Biol Rev 65(1):1–43. https://doi.org/10.1128/mmbr.65.1.1-43.2001

Vojcic L, Pitzler C, Körfer G, Jakob F, Martinez R, Maurer KH, Schwaneberg U (2015) Advances in protease engineering for laundry detergents. New Biotechnol 32(6):629–634. https://doi.org/10.1016/j.nbt.2014.12.010

Von der Osten C, Branner S, Hastrup S, Hedegaard L, Rasmussen MD, Bisgård-Frantzen H et al (1993) Protein engineering of subtilisins to improve stability in detergent formulations. J Biotechnol 28(1):55–68. https://doi.org/10.1016/0168-1656(93)90125-7

Wang SZ, Zhang YH, Ren H, Wang YL, Jiang W, Fang BS (2017) Strategies and perspectives of assembling multi-enzyme systems. Crit Rev Biotechnol 37(8):1024–1037. https://doi.org/10.1080/07388551.2017.1303803

Wang X, Qin X, Tong L, Zheng J, Dong T, Wang X et al (2023) Improving the catalytic activity of a detergent-compatible serine protease by rational design. Microb Biotechnol 16(5):947–960. https://doi.org/10.1111/1751-7915.14218

Yang H, Ren X, Zhao Y, Xu T, Xiao J, Chen H (2024) Enhancing alkaline protease stability through enzyme-catalyzed crosslinking and its application in detergents. PRO 12(3):624. https://doi.org/10.3390/pr12030624

Yao H, Liu S, Liu T, Ren D, Zhou Z, Yang Q, Mao J (2023) Microbial-derived salt-tolerant proteases and their applications in high-salt traditional soybean fermented foods: a review. Bioresour Bioprocess 10(1):82. https://doi.org/10.1186/s40643-023-00704-w

Zambare V (2010a) Strain improvement of alkaline protease from *Trichoderma reesei* MTCC-3929 by physical and chemical mutagen. IIOAB J 1(1):25–20

Zambare V (2010b) Purification and characterization of neutral serine protease from *Bacillus* sp. Asiat J Biotechnol Res 3:183–192

Zambare VP, Nilegaonkar SS, Kanekar PP (2014) Scale up production of protease using *Pseudomonas aeruginosa* MCM B-327 and its detergent compatibility. J Biochem Technol 5(2):698–707

Zhang W, Wu J, Xiao J, Zhu M, Yang H (2021) Compatibility and washing performance of compound protease detergent. Appl Sci 12(1):150. https://doi.org/10.3390/app12010150

Chapter 17
Exploring the Versatility of Microbial Proteases in Food Processing and Preservation

K. Kesavardhini, M. Charumathy, Ola Alnaddaf, U. Subhalakshmi, S. Dayana, C. Mabel Joshaline, and K. Gayathri

Abstract Microbial proteases are a frontier in modern industrial field of biotechnology, and their use in the food industry is the most comprehensive and diverse. These bacterial, fungal, and yeast proteases have mostly replaced the plant and animal proteases because these enzymes are widely biochemically diverse, have high yield, are economical to produce, and can easily be genetically engineered. This chapter explores the various applications of microbiological proteases in the food business, highlighting their significant impact on food quality, texture, flavor, and nutrition. The main uses raised are that they are indispensable in dairy processing as a curdling agent in cheese and in accelerating the ripening process, in baking to enhance the dough, and in the softening of meat. Moreover, the chapter also explores the new and emerging potential of these enzymes in food preservation, whereby certain antimicrobial peptides (Bacteriocins) produced by these enzymatic activities can prevent food spoilage and microbial pathogens, extending the shelf life and enhancing food safety. Their capability to modify the properties of proteases to match the requirements of certain food matrices highlights their possible usage in creating new food products and greener food processing technologies. Although widely used, nowadays, there are still challenges, including how to optimize production, how to have fine control of proteolytic activity, and regulatory approval of new applications. This discussion concludes that microbial proteases are essential

K. Kesavardhini (✉) · U. Subhalakshmi · S. Dayana · K. Gayathri
PG and Research Department of Microbiology, Sacred Heart College (Autonomous), Tirupattur, Tamil Nadu, India

M. Charumathy
PG and Research Department of Biochemistry, Marudhar Kesari Jain College for Women (Autonomous), Vaniyambadi, Tamil Nadu, India

O. Alnaddaf
Department of Horticultural Sciences, Damascus University, Swaida, Syria

C. M. Joshaline
Department of Rural Development Science, Arul Anandar College, Madurai, Tamil Nadu, India

V. Zambare et al. (eds.), *Biotechnological Advances and Applications in Microbial Proteases*, Interdisciplinary Biotechnological Advances,
https://doi.org/10.1007/978-981-95-9011-7_17

instruments, and they will lead to another breakthrough in food technology to support the changing needs of an increased world population.

Keywords Microbial Proteases · Food Biotechnology · Dairy Processing · Meat Tenderization · Bakery Improvement · Antimicrobial Peptides and Food Preservation

17.1 Introduction

Enzymes have been employed in food manufacturing, including the production of wine and beer, for a minimum of 6000 years ago (Mishra et al. 2017). The utilization of chymosin as well as calf stomach enzymes has been crucial for dairy products like cheese, dating back to 800 BC (Mehta and Sehgal 2019). The industrial manufacturing of chymosin, or rennin, commenced in 1874 (trademark-rennet). At present, rennin is predominantly generated through heterologous expression to devise efficient methods for rennet production utilizing Bacteria, yeast, or molds, including *Escherichia coli*, *Lactococcus lactis*, *Pichia pastoris*, *Kluyveromyces lactis*, and *Aspergillus niger*. Genetic engineering, initiated in the late 1970s, has led to substantial progress in biotechnology, especially in enzyme research (Gonzalez-Rabade et al. 2011).

Proteases, also known as peptidases and proteinases, are a specific class of enzymes that promote protein disintegration. Food peptidases may derive from organisms including plants, animals, or microorganisms. Moreover, it is significant that microbial peptidases account for 75% of the global commercial protease market. Microorganisms have proven to be prolific sources of novel peptidases, exhibiting several industrial uses. The evolutionary complexity of microbial metabolic processes generates a diverse array of different proteases, each defined by unique functions, structures, and specificities (Tavano 2013). Peptidases are presently utilized commercially in several applications, particularly in the leather goods, pharmaceutical, and food industries. In the latter, the principal applications of proteases include dairy products, meat, fish and their derivatives, alcoholic beverages, baked goods, juice clarification, and protein hydrolysates employed as functional ingredients, flavor enhancers, nutritional supplements, as well as allergen mitigation substances (Ismail et al. 2019).

The remarkable efficacy of microbial proteases that are is significantly dependent on robust and varied products produced by the microbial host. Bacteria, primarily *Bacillus* species including *B. licheniformis* and *B. subtilis*, are industrial stalwarts, known to secrete large amounts of highly stable alkaline serine proteases directly into fermentation broths, making it easy to extract downstream. In contrast, complex arsenals of aspartic and neutral proteins are produced by filamentous fungi like *Aspergillus oryzae* and *Rhizomucor miehei* and play a key role in traditional fermentations and dairy processing, respectively. Such microbial cell factories can be grown in large quantities using economical and sustainable substrates, such as agro-industrial residues, under optimized fermentation conditions, such as submerged fermentation (SmF) to produce in large volumes or using solid-state fermentation (SSF) to produce in increased quantities and with greater stability of the enzymes. More importantly, genetic plasticity of these organisms, along with state-of-the-art approaches in metabolic engineering and synthetic biology facilitate the rational design of strains that over-express native enzymes or produce novel, chimeric proteins with custom properties, such as high thermotolerance, pH optimum, or cleavage specificities, to suit particular process needs (Duarte et al. 2009).

Microbial proteases work within the multidimensional field of food processing, excelling in their multipurpose nature, which usurps all industries, as silent constructors of texture, flavor, nutrition, and yield. They are not just coagulants in dairy technology, particularly aspartic proteases catalyze the unique, definitive, the hydrolysis in the Phe105-Met106 connection in κ-casein, which initiates curd formation, while the synergistic activity of endo- and exo-peptidases orchestrates the proteolytic degradation of proteins through cheese maturation, leading to the release of peptides as well as amino acids underlying flavor complexity and texture development. Neutral proteins are used in cereal processes as advanced dough conditioners and selectively break down the gluten networks to optimize the rheology, make the dough machinable, enhance loaf volume, and slow staling. Meat industry applies their power to tenderize using the specific degradation of collagen and elastin, and at the same time valorizes low-value by-products into useful protein hydrolysates. Outside these structural functions, proteins are prominent biocatalysts in the emerging industry of functional foods, in which controlled hydrolysis releases encrypted bioactive peptides of parent proteins, which offer known health benefits of angiotensin-Inhibition of Angiotensin-Converting Enzyme (ACE), antioxidant properties, and antibacterial effects, transforming traditional food matrices into promoters of wellbeing (de Souza and de Bittencourt 2015).

Microbial proteases are increasingly significant in addressing two critical contemporary demands within the food industry: improving natural preservation and the switch to a circular bio-economy. They support a paradigm switch in food safety by in-situ production of natural antimicrobial peptides (Bacteriocins), which is a powerful, clean-label biological barrier to microorganisms that spoil and/or cause food poisoning, without the use of synthetic chemical preservatives. At the same time, they are the laborers of the waste valorization, providing a nice biochemical fix to the issue of protein-rich processing side-streams. Proteases convert environmental burdens such as whey, slaughterhouse offal and oilseed cakes into high value products such as protein concentrates, savory flavorants, and bioactive ingredients, thereby bridging nutrient loops, lowering greenhouse gas emissions generated by the decomposition of waste and also increasing the sustainability of the entire process. Herein, microbial proteases as biochemical catalysts are therefore transcended to be critically and multi-purposely relevant in the food innovation, food safety, nutrition, and environmental responsibility sectors, and their future evolution is therefore imperative in developing resilient and sustainable food systems of the future (Razzaq et al. 2019).

17.2 Microbial Proteases

Proteases were among the earliest enzymes to be identified and studied, with several animal-derived proteases, such as trypsin, and plant-derived proteases, such as papain, being discovered and characterized as early as the 1800s and early 1900s (Bond 2019). As a result, microbial proteases gained prominence and rapidly became popular due to their numerous benefits. Microorganisms, as a source of the proteases, have various intrinsic advantages, including rapid production rates, diminished area and time needs, and robustness to environmental variations. Furthermore, the varied ecosystems inhabited by microorganisms facilitate the identification of bacteria that produce enzymes with beneficial properties (Putatunda et al. 2019). For instance, if one seeks thermostable proteases, it is likely to obtain these enzymes from thermophilic bacteria found in hot springs. Furthermore, the diminished generation time and very straightforward genetics architecture facilitate the genetic regulation of microbes significantly more uncomplicated. This, in turn, expands opportunities for the construction of customized enzymes with specific characteristics. The ethical dilemmas faced by workers about animal-derived proteases are not present in the case of microbial-derived proteases. The capacity of bacteria to generate extracellular proteases facilitates the production process by streamlining downstream operations. Moreover, several microbes have been cultivated on more economical substrates or as waste materials to enhance the economic viability of the procedure (Limkar et al. 2019).

Bacterial and fungal origin proteases have garnered significant economic interest, prompting extensive exploration by researchers. Gurumallesh et al. (2019) state that microbiological variants of proteases comprise two-thirds of the variety of proteases employed in industry. The primary contributors are bacterial alkaline proteases and fungal acidic proteases; nevertheless, viral proteases also possess relevance. Microorganisms that produce proteases have been extracted from many settings include rhizosphere dirt, abattoir soil, floor washings, effluent, and food waste. These also occur in marine ecosystems, sediments, extreme salinity lakes, foods, and soda lakes. Bacillus species are essential for commercial protease production, while fungal sources include *Aspergillus*, *Trichoderma*, and *Penicillium*. Certain researchers have studied endophytic fungi for proteases for various reasons (Mandal and Banerjee 2019). Meshram and Saxena (2016) reported proteolytic and fibrinolytic capability in the endophytic fungi *Lasiodiplodia pseudotheobromae*, extracted from the stem of the Aegle marmelos, and proposed its potential use for the dissolution of blood clots. Numerous protein proteases were recently identified and characterized in various viruses; nevertheless, in a commercial standpoint, they have not achieved substantial significance and mainly function as targets for several antiviral drugs. Viral origin proteases include cysteine proteases associated with Adenovirus and Alphavirus, serine proteases related to Hepatitis C Virus, and aspartic proteases present in HIV, among other viruses (Sharma and Gupta 2017).

17.2.1 Alkaline Proteases

Alkaline and neutral proteases are the two main categories of bacterial proteases used in various commercial applications. Alkaline proteases have been used for a long time, particularly extracellular proteases from the *Bacillus* genus, which mostly contain serine proteases (Pathak and Rathod, 2018). Bacillus-derived enzymes have attracted considerable attention, leading to extensive understanding of their activity, size, arrangement, and various other attributes. Subtilisin Carlsberg and subtilisin novo represent two prominent types of alkaline proteases obtained from the *Bacillus* species. Moreover, the strains of the bacteria *B. subtilis* and associated *Bacillus* species employed for commercial enzyme manufacturing are designated as GRAS (Generally Regarded as Safe). A growing number of articles about alkaline protease derived from various *Bacillus* sp. strains, together with novel proteases from well-known strains, were emerging globally. The purification and identification of serine alkali protease from a distinct strain of *B. caseinilyticus* were carried out by Mothe and Sultanpuram (2016). According to Hammami et al. (2020), the characterization of a novel H_2O_2-tolerant alkaline protease extracted from *B. invictae* AH1. Zhou et al. (2020) executed genetic modifications onto *B. licheniformis* 2709, resulting in a 62.19% increase in the synthesis by mucopolysaccharides in addition to foaming components and a rise in the creation of alkaline protease. Although proteases produced by the *Bacillus* genus were the most thoroughly characterized and employed, numerous additional Bacteria that can produce alkaline proteases have recently been identified. Boughachiche et al. (2016) reported the synthesis of alkaline protease derived from *Streptomyces* sp. obtained from the saltpan within Algeria. Al-Dhabi et al. (2020) identified a new thermally resilient keratinolytic alkali protease-generating strain of *Streptomyces* sp. Masi et al. (2017) documented the production of alkaline protease through *Enterococcus hirae* and *Pseudomonas aeruginosa* sourced from dairy effluents. Individuals also synthesized a nanocomposite that included bacterial protease and even superparamagnetic nanoparticles, which improved the stability of the protease. Bacteria, including *Bacillus* and *Pseudomonas*, as well as specific fungi, have been reported to have the capability to produce alkaline proteases. According to Meshram et al. (2016), the endophytic fungus *Xylaria curta* produces a metalloprotease with a pH 8.0 optimum that has significant fibrinolytic activity, suggesting that it may have therapeutic value. A thermodynamic analysis of the serine-derived alkaline protease generated by the fungus *Aspergillus tamari* URM4634 was conducted by da Silva et al. (2018). Significant alkaline protease synthesis using *Chetomium globosum* was reported by Nadeem et al. (2019). The ideal pH range for alkaline bacterial proteases is often between 8 and 12. At higher temperatures, many enzymes show greater thermostability and activity (Naveed et al. 2021). Alkaline proteases, active at alkaline pH conditions, are often employed as additions in detergents to enhance stain removal effectiveness. Similarly, leather manufacturing transpires at high pH levels, rendering alkaline proteases suitable for the tanning sector (Hakim et al. 2018).

17.2.2 Acidic Proteases

Acidic proteases function mostly within the pH range from 3.8 to 5.6, with an optimum pH of 3 to 4 and also an isoelectric point that ranges between 3 and 4.5; they are categorized as aspartic proteases. Pepsin-like and rennin-like are the two main classifications for microbial acidic proteases. The most commonly mentioned species include *Aspergillus, Penicillium, Endothia*, and *Mucor*. Fungi are the primary manufacturers of acid proteases, in contrast to alkaline proteases, which are mostly produced by bacteria (Mamo and Assefa 2018).

Numerous studies indicate the production of new acid proteases by molds and certain yeasts. An aspartic protease from *Rhizomucor miehei* was described by Da Silva et al. (2016) and may serve as a substitute enzyme in milk coagulation in the making of cheese. According to Souza et al. (2017), *Aspergillus foetidus*, which was isolated within a soil sample, produced an acidic peptidase with a pH of 5.0 and an optimal normal temperature of 55 °C. In a similar vein, da Silva et al. (2017) identified an aspartic protease derived from *Phanerochaete chrysosporium* that has beneficial milk coagulation capabilities. In comparison to papain, a frequently used plant-based protease for this purpose, Sun et al. (2018) observed improved beef tenderization after cloning the aspartic protease gene with *Rhizomucor miehei* into *Pichia pastoris*. According to Deng et al. (2018), recombinant aspartic protease that was produced in *Pichia pastoris* and derived from *Trichoderma harzianum* demonstrated efficacy against a variety of pathogenic fungi, such as *Candida albicans, Rhizoctonia solani, Aspergillus fumigatus, Aspergillus flavus, Botrytis cinerea*, and *Mucor circinelloides*. In their biochemical characterization of both enzymes, Song et al. (2020) found that the optimal temperature was 50 °C, with a pH of 2.5 for PepAb and 3.0 for PepAc.

Acidic proteases have been reported to be produced by a variety of microorganisms. According to Putranto et al. (2020a, b), proteases from *Enterococcus faecium* strain 1.15 and *Lactobacillus plantarum* strain 1.13, which were isolated using Bakasam (a fermenting beef product), showed rennin-like activity. According to Putranto et al. (2020a, b), the protease generated by *L. paracasei* strain 2.12 from goat milk had similar activity. In the food and beverage sector, acidic proteases are crucial. Microbial-derived proteases are gradually replacing animal-derived enzymes like rennin, which were once used for coagulation of milk in the making of cheese. According to Singh and Bajaj (2017), acid proteases are used in baking to enhance dough properties and in the synthesis of protein hydrolysates. Acid proteases are crucial for clearing beer and juices of haze.

17.2.3 Neutral Proteases

Proteolytic enzymes that operate at neutral, moderately alkaline, or slightly acidic pH levels are known as neutral proteases, demonstrating optimal activity with the pH range from 5 to 8. Like alkaline proteases, neutral proteases are also generated by the genus Bacillus. Numerous neutral proteases are classified as metalloproteases and necessitate a divalent cation for their functionality. Neutral proteases have been widely employed in various applications, particularly in the baking and brewing sectors. A neutral protease produced from *Trichoderma harzianum* was characterized by Aissaoui et al. (2016) and hydrolyzed the viscera of *Scorpaena notata.* It was noted that the hydrolysate of the fish viscera produced antibacterial peptides. Ao et al. (2018) described A 45 kD protein with a preferred pH of 7.0 and a temperature of 55 °C, *Aspergillus oryzae* Y1 is a neutral protease that is obtained from naturally fermented broad beans. According to De Oliveira et al. (2020), the fungus *Moorella speciosa* produces a protease enzyme that is fairly thermostable and has an ideal pH of 6.5. Following immobilization on MAT540 microspheres, the ideal pH shifted to 5.74. The potential application of recombinant neutralized protease rNpI as an addition in fish feed to improve protein digestion and foster growth was recorded by Deng et al. (2021). Microbial proteases are highly versatile enzymes with applications spanning multiple industries. Their specificity and activity under different pH conditions make them suitable for uses such as detergent formulation, food processing, leather tanning, bioremediation, pharmaceuticals, and meat tenderization. Additionally, specialized proteases like Keratinases, Elastases, and Collagenases play important roles in feather degradation, medical research, and tissue engineering, highlighting their broad industrial significance (Table 17.1).

Table 17.1 Microbial proteases and their industrial applications

Protease type	Source microorganism	Optimum pH	Industrial application	References
Alkaline protease	*Bacillus subtilis*	8–10	Detergent industry	Boughachiche et al. (2016)
Acid protease	*Aspergillus Niger*	3–5	Food processing	Mamo and Assefa (2018)
Neutral protease	*Bacillus amyloliquefaciens*	6–7	Leather tanning	De Oliveira et al. (2020)
Metalloprotease	*Pseudomonas aeruginosa*	7–8	Bioremediation	Meshram et al. (2016)
Serine protease	*Streptomyces griseus*	8–9	Pharmaceuticals	Silva et al. (2018)
Cysteine protease	*Papaya latex microbes*	5–7	Meat tenderization	Roy (2021)
Keratinase	*Bacillus licheniformis*	7–9	Feather degradation	da Silva et al. (2017)
Elastase	*Pseudomonas* spp.	7–8	Medical research	Aissaoui et al. (2016)
Collagenase	*Clostridium histolyticum*	6–7	Tissue engineering	Deng et al. (2021)
Peptidase	*Lactobacillus* spp.	6–7	Fermentation	Bond (2019)

17.3 Technology for the Production of Microbial Proteases

17.3.1 Microorganisms and Origin

The isolation and selection of suitable bacterial candidates constitute the initial phase of microbial protease production. The microbes come from a variety of environments and are subsequently selected for their essential characteristics. The afore mentioned sources and ecosystems display distinctive or detrimental characteristics that correspond to the properties of the enzyme that needs to be synthesized; for instance, landfill sites, geothermal springs, saline lakes, soil samples from dairy processing facilities, waste from poultry sites, soil contaminated by meat-related waste, tannery wastewater, detergent manufacturing, leather production, and wood manufacturing (Ramakodi et al. 2020). Protease-producing bacteria, fungi, and actinomycetes have been found *Bacillus amyloliquefaciens*, *B. subtilis*, *B. lentus*, *Pseudomonas fluorescens*, *P. aeruginosa*, *B. circulans*, *Geobacillus* SBS-4S, *B. alkaloophilus*, *B. halodurans*, *B. pumilus*, *B. clausii*, *B. licheniformis*, and *B. safensis* are protease-producing bacteria (Ahmad et al. 2020).

Fungi represent the subsequent principal microbiological category recognized for protease synthesis. The fungi identified as protease producers include *Aspergillus sterreus*, *Aspergillus niger*, *Aspergillus clavatus*, *Fusarium* sp., *Aspergillus niger*, *Penicillium chrysogenium*, *Aspergillus flavus*, *Aspergillus fumigatus*, *Aspergillus nidulans*, *Aspergillus oryzae*, *Penicillium chrysogenum*, *Mucor* sp., *Cephalosporium* sp., *Rhizopus* sp., *Penicillium italicum*, and *Trichoderma* sp. (Abu-Tahon et al. 2020). Two common strains used for commercial protease production are *Aspergillus* and *Trichoderma*. Apart from fungus and bacteria, the bacterial species actinomycetes, which resembles fungi, is also known to produce proteases. An extracellular protease produced by *Streptomyces nogalator* aids in the depilation in goat skin. A *Streptomyces* sp. alkali-thermotolerant protease that demonstrated heat stability at 90 °C was discovered by Bajaj and Sharma (2011). To improve economic results, many microorganisms have been employed for protease production; nonetheless, efforts to discover cost-effective industrial-grade enzymes from microbial sources persist (Razzaq et al. 2019).

17.3.2 Fermentation

Both submerged fermentation (SmF) and solid-state fermentation (SSF) are utilized in the manufacture of microbial proteases. Submerged fermentation (SmF), employing either wild type or as genetically modified microorganisms, accounts for approximately 90% of total commercial production of enzymes. SmF is more conducive to bacterial growth due to microbial proliferation occurring in a liquid medium abundant in free water. Enhanced the recovery of the extracellular enzymes and improved the dispersion of medium components superior control over physicochemical variables are all advantages of this technology. However, the resulting enzymes are less stable and diluted. Fructose, Lactose, Sucrose, Yeast Extract, Peptone, and several complex industrial by-products such as molasses, residual biomass, wastewater, and maize steep liquor are utilized as substrates in microbial protease production via submerged fermentation (SmF) (Gimenes et al. 2019).

In solid-state fermentation (SSF), growth of microbes occurs on a solid substrate with minimal free water, yet ensuring sufficient moisture to facilitate metabolic processes and microbial proliferation. The SSF is more conducive to fungi because of the proximity of their natural environment in the medium and their lower moisture requirements, particularly 40–60% being sufficient for fungus. Compared to SmF, SSF has advantages such as reduced costs, enhanced simplicity, increased yield, improved product stability, and fewer sterility requirements of the potential to utilize various agro-industrial wastes as substrates. Nonetheless, a restricted variety of microbes can proliferate under diminished moisture conditions, presenting a constraint in Solid-State Fermentation (SSF). Furthermore, characteristics such as substrate particle size and moisture content must be examined meticulously. A range of inexpensive substrates, such as Wheat Bran, Straw made Wheat, Red Gramme Husk, Cow Dung, and Bagasse from sugarcane, have been utilized for protease synthesis using solid-state fermentation (SSF). The standard fermentation periods for bacterial and fungal protease synthesis are 2–4 days and 3–5 days, respectively (dos Santos Aguilar and Sato 2018).

Numorous physico-chemical and nutritional factors, including substrate, nitrogen, carbon, pH value, humidity, temperature, airflow, stirring, dissolved oxygen, and inorganic elements, have a significant impact on the synthesis of protease. Since every microbial strain has different nutritional and physical needs for optimum enzyme synthesis, there are no standardized conditions for cultivation or media for microbial protease production. Microbial protease synthesis has been documented by numerous researchers employing solid-state fermentation (SSF) and submerged fermentation (SmF). Using *A. oryzae* and *A. fumigatus*, da Silva et al. (2013) compared solid-state fermentation (SSF) with submerged fermentation (SmF). Using wheat bran as the substrate increased the protease yield in solid-state fermentation by 3.5 times in *A. oryzae* and 30 times in *A. fumigatus*. Using *Aspergillus fischeri* and *Penicillium citrinum* in a mixture containing 0.5% crushed feathers as a nitrogen source, Ida et al. (2017) carried out submerged fermentation (SmF), which produced two collagenolytic serine peptidases. The results of the enzyme assay showed that the fungus *A. fischeri* and *P. citrinum* produced 460 U/mL afterwards 72 hours and 760 U/mL after 168 hours, respectively. Protease production by bacteria has been documented in numerous studies. Alkaline metalloproteases were found by Anandharaj et al. (2016) using *B. alkalitelluris* TWI3; metalloprotease production was shown by Rehman et al. (2017) using *B. subtilis* KT004404; *Bacillus* sp. CL33A was used by de Oliveira et al. (2017); and alkaline proteases were found by Baweja et al. (2016) in submerged fermentation. One study explained how *Pediococcus acidilactici* SH produces acidic proteases during submerged fermentation (SmF). Using submerged fermentation, Jensen et al. (2010)

identified a novel microbial glutamine protease and showed its heterologous expression using a clone produced by *B. subtilis*. Furthermore, during solid-state fermentation (SSF), some bacterial strains belonging to the genus *Bacillus* have demonstrated significant protease production. While Vijayaraghavan et al. (2014) reported an activity of 4813 U/g using *B. cereus* in solid-state fermentation (SSF), Qureshi et al. (2016) reported a protease enzyme level of 12,200.0 U/g with *Bacillus* sp. BBXS-2.

17.3.3 Contemporary Methodologies to Enhance Protease Production

Researchers have utilized diverse techniques to augment the yield of enzymes that produce protease from microbial sources, including cloning as well as Feed-Batch, Chemostat, and Over-Expression fermentation methods, as well as the individual factor optimization of parameters (such as medium and developmental circumstances) and statistical techniques, such as response surface methodology. Current initiatives aim to improve protease enzyme production via conventional approaches, like ultraviolet light and chemicals, alongside contemporary techniques such as genetic modification, CRISPR gene editing, the engineering of proteins, and metagenomics (Thakur et al. 2020). Researchers are utilizing error minimization and yield optimization of enzymes to dominate the global biotechnology sector. Numerous bioengineered enzymes exhibiting enhanced stability are being generated via recombinant DNA technology in the detergent sector (Razzaq et al. 2019).

One essential technique for improving enzyme synthesis is protein engineering (Li et al. 2020). The effectiveness and longevity of enzymes often depend on their amino acid content and arrangement. Protein engineering modifies the protease's structure to improve its selectivity, activity, and stability. Protease technology had been thoroughly investigated for several industrial applications. Nowadays, the two main approaches used for protease engineering are directed evolution and random design. Additionally, the method known as computer-aided protein-directed evolution (CAPDE) amalgamates directed evolution approaches with computational design and has improved protein engineering methods. Database mining is a vital area of emphasis in the sequence-associated domain, as well as metagenomics functions as an optimal instrument for examining sequence-associated data derived from ambient microbes' DNA (Prayogo et al. 2020). To find the genes encoding a new enzyme, the isolated sample is next analyzed for open reading frames. Metagenomic screening has been used to isolate and define a wide range of enzymes with distinct biocatalytic activities.

Using metagenomic methods, Waschkowitz et al. (2009) identified a new protease enzyme. Verma et al. (2021) conducted an *in-silico* analysis of the metagenome of the solid tannery waste and provided a comprehensive structural characterization of protease. Devi et al. (2016) found and characterized the Prt1A gene, which encodes an alkaline serine protease, using a metagenomic library derived from tannery waste. At a pH of 11 and room temperature of roughly 55 °C, the Prt1A protease showed its peak activity. Pessoa et al. (2017) found a clone that synthesizes protease after screening a metagenomic library taken from mangrove sediments. After further purification, the enzyme yielded 77.27% and showed an activity in particular of about 8.5 U/μg. In order to find a novel protease generated from marine metagenomes that can hydrolyze insoluble zein under temperatures up to 50 °C and under certain pH values between 9 and 11, Garcia-Moyano et al. (2021) used a two-stage biological examination procedure. The enzyme came from a specific bacterial strain whose capacity to break down zein had not yet been recognized. CRISPR, a very powerful gene editing technique, is currently being used by researchers to modify genes quite well. The Cas9 enzyme and a Single Guide RNA (SgRNA), which both specifically target gene for DNA cleavage, are the two main parts of the CRISPR system. *B. subtilis* ATCC 6051 is a bacterial strain that secretes proteins effectively and is used in the production of enzymes for industrial uses. A significant challenge for researchers is the reduced transformation efficiency associated with this strain. In this case, Zhang et al. (2016) chose CRISPR/Cas9. By incorporating two CRISPR/Cas plasmids into a cloning process, Garcia-Moyano et al. (2020) altered *B. subtilis* to allow for flexible editing at any chosen genomic site or defined genomic locus. The protease enzyme was successfully synthesized by the genetically altered *B. subtilis*. Industrial requirement for enzymes is still high despite that of many contemporary and cutting-edge enzyme production techniques (Gurumallesh et al. 2019).

17.4 Fundamental Concepts of Microbial Proteases

17.4.1 Biochemical Classification

17.4.1.1 Serine Proteases

The active site of serine proteases contains a highly energetic serine atom, one of the three conserved enzymatic triads that typically include Aspartate, Histidine, and Serine. The trio works in unison to attack the substrate's peptide bond in a nucleophilic way, which causes the substrate to hydrolyze. These enzymes mostly operate in neutral to alkaline PH systems and are also one of the most prevalent and utilized proteases in the industry. A good example is subtilisin, which is secreted by the *Bacillus* species, and is famous due to its wide substrate specificity and exceptional stability in strong detergents and organic solvents. The high potency of serine proteases renders them essential in food industry in tasks like dough conditioning in baking as well as creating protein hydrolysates where they are effective in decomposing proteins into small peptides and amino acids. The fact that they do not need a metal cofactor also makes their use in different food matrices more straightforward because it is sure that they will produce the same effects and provide predictable results in the complex biochemical conditions (Niyonzima 2021).

The deprotonated serine protein residue is first the nucleophilic that strikes the carbonyl carbon for the scissile peptide bonds in the two-step reaction that serine proteases perform. This attack creates a transient, high-energy tetrahedron that is stabilized by oxyanion hole in the enzyme's active site. The acyl portion of the substrate is then acylated by the serine residues to produce an acyl-enzyme complex when the intermediate caves in, cleaving the peptide link. To generate the second product, the acyl–enzyme complex is hydrolyzed by a water molecule activated by the histidine residue, thereby releasing the product and regenerating the free enzyme. These accurate catalytic setup enables high turnover of rates and such enzymes are very efficient biocatalysts. Their specificity may however differ greatly; some are very specific to certain sequence of amino acids, whereas others such as subtilisin are more promiscuous. This flexibility is the basis of their applicability in a range of food processing procedures, with specific protein modification tasks to large-scale hydrolysis to form flavors and develop texture (Badoei-Dalfard and Sohrabi 2022).

17.4.1.2 Aspartic Proteases

Aspartic proteases, or acid proteases, employ two essential aspartic acid residues in their active site to activate a water molecule, which then acts as a nucleophile and then attacks the peptide bond. These enzymes work best in highly acidic conditions, which are normally between pH 3 and 4 and hence are particular food conditions. The most notable representatives of food biotechnology include pepsin-like enzymes, including those of fungi include *Aspergillus niger* and *Rhizomucor miehei*. They are also less promiscuous than serine proteases because of their high specificity to hydrophobic and aromatic amino acid residues especially phenylalanine and tyrosine. This property is important in the dairy industry where they are used as microbial rennets in the coagulation of cheese, which is the cleavage of the Phe105-Met106 link of κ-casein exhibits excellent specificity. Such a selective operation is necessary to produce the desired curd structure and texture of cheese without the undesirable result of excessive proteolysis that will result in bitterness and loss of yield (Mamo and Assefa 2021).

Aspartic proteases use catalytic mechanism of general acid-base catalysis in which one aspartate residue serves as a general base to activate the water molecule, whereas the other one serves as a general acid to protonate the carbonyl oxygen of the peptide. This leads to a tetrahedral gem-diol intermediate, which further collapses to cleave the peptide bond. An important structural aspect of most microbial aspartic proteins is that they are inhibited by pepstatin A, an element that is frequently employed to define this type of enzyme. This stability and low-pH activity is not only used in the manufacture of cheese but also in the manufacture of fermented drinks and soy sauce, which involves assistance in the breakdown of proteins during the fermentation process. Moreover, their specificity reduces the formation of bitter peptides that are likely to be produced using the less specific proteases, hence improving the entire flavor profile of the end product. This acidic stability coupled with targeted action entrenches it within the arsenal of the food technologist (Contesini et al. 2023).

17.4.1.3 Cysteine Proteases

The cysteine proteins are characterized by a catalytic mechanism which depends on deprotonated cysteine thiol group that is a strong nucleophilic group within the enzyme's active site. The thiol group is typically linked to a histidine residue, which functions as a general base to facilitate deprotonation process to create a thiolate-imidazolium ion pair, which is very reactive. From the neutral up to slightly reducing conditions, these enzymes usually work and are susceptible to oxidative inactivation. The most widely known example of a plant origin source is papain; microbial analogs are created by some bacteria, such as *Staphylococcus aureus*, and some fungi. Their primary application in the food industry for processing is in tenderizing of meat as they are able to break down the tough collagen and elastin fibers of the connective tissues in a process that results in a more palatable product. The fact that they can destroy certain muscle proteins without producing excessive mushiness should be carefully controlled in time and temperature when using it to achieve the maximum texture enhancement but to prevent over-processing (Roy 2021).

In the catalytic cycle of cysteine proteases, a covalent cysteine–substrate thioester complex is formed after the catalytic cysteine residue performs a nucleophilic attack on the substrate's carbonyl group. This initial step results in the liberation of the initial product, which is an amine fragment. This is then followed by hydrolysis of the thioester intermediated by a water molecule, leading to the production of the second product, a carboxylic acid fragment, as well as the restoration of the active enzyme. This mechanism is efficient and exposes the enzyme to inhibitors such as iodoacetate and heavy metals which attack the thiol group. In microbial systems, cysteine proteases may be involved in virulence and nutrient provision to the producing organism. In industrial food applications, they are appreciated due to their strong activity at intermediate temperatures, potentially beneficial in processes where heat-sensitive ingredients have to be maintained. Yet, they are susceptible to oxidation, and thus reducing agents or controlled atmospheres are used in most applications to preserve their catalyst properties across the food processing cycle (de Oliveira et al. 2022).

17.4.1.4 Metalloproteases

Metalloproteases constitute a vast and diverse class of enzymes that require a divalent metallic ion, typically zinc (Zn^{2+}), to enable their catalytic function. Histidine, Glutamate, or Aspartate residues typically coordinate the metallic ion in the active site and serve for polarizing the carbonyl group of the scissile peptide linkage and stabilize a negative charge that forms on transition state. This polarizing effect activates the hydrolysis reaction nucleophile, which is a polarity-charged water molecule. The metalloproteases exhibiting high thermal stability and wide substrate specificity are microbial metalloproteases, e.g., thermolysin of *Bacillus thermoproteolyticus*, which tends to hydrolyze peptide bonds at the amino terminus of the hydrophobic amino acid bonds. Such property renders them especially applicable in food processes, which have high temperatures, including preparation of protein hydrolysates used as flavorings or the rearrangement of wheat gluten during baking, where they can resist the initial stages of baking without being detected by immediate denaturation (Naveed et al. 2021).

The metalloproteases' mechanochemical attack is described as a base-assisted water attack, in which the metal-bound water molecule is deprotonated by another base, usually a glutamate residue, to give a strong hydroxide ion nucleophile. This ion interacts with the carbon atom that makes up the carbonyl group within the substrate, resulting in a tetrahedral intermediate that is coordinated to the Zn ion. The peptide bond is broken as a result of the intermediate rupturing. Among the primary advantages of several microbial metalloproteases is their resilience when subjected to chelating chemicals like EDTA, the agent is able to maintain or eliminate their functions by depriving the proteins of the essential metal cofactor. This property is a good control tool in food manufacturing. The flexibility in terms of operating at high temperature and over a very broad pH area makes them adaptable biocatalysts that can be employed in several biocatalytic applications, including the generation of savory flavors by controlled hydrolysis of plant and animal proteins. Their metal dependency, however, also should be handled with care so as not to be inactivated in food systems that can contain natural chelators (Li and Chen 2023).

17.4.2 Major Microbial Sources

17.4.2.1 Bacterial Proteases

The *Bacillus* genus of bacteria are the work-horses of industrial protease production, as they are noted to secrete high levels of extracellular enzymes with them being introduced into the process of fermentation broth, facilitating downstream processing significantly. Other species, including *Bacillus licheniformis, Bacillus amyloliquefaciens*, and *Bacillus subtilis*, are particularly notable for their diverse array of alkaline serine proteases, such as Subtilisin, and neutral metalloproteins. These enzymes exhibit remarkable robustness and demonstrate outstanding stability throughout a wide variety of pH and temperature ranges; in most instances, they can withstand the presence of surfactants, oxidants, and organic solvents. This resilience guarantees their suitability in the demanding conditions of food processing, where they are employed in baking to alter gluten and in the production of savory dish flavors, such as protein hydrolysates and in the extraction of useful substances out of meat and fish by-products. The genetic malleability of *Bacillus* species also enables the strain to be improved by metabolic engineering to achieve yield, specificity, and stability to provide an ever-present supply of customized enzymes to the food industry (Niyonzima 2021).

Highly optimized fermentation of those organisms through efficient large-scale submerged fermentation processes forms the basis of the commercial success of *Bacillus* proteases. These are Gram-positive bacteria that are not pathogenic and have received Generally Recognized as Safe (GRAS) status by food regulatory agencies, which is an essential procedure to obtain their application in food. Their proteases are frequently synthesized in a complex form, which could be advantageous in situations in which there is a requirement to have a synergistic hydrolytic action on complex protein substrates. An example is the dairy industry where certain proteins such as those of *Bacillus*-derived proteins are applied to hasten the ripening of the cheese by hydrolyzing the casein. Proteins are hydrolyzed into smaller peptides that contain amino acids, which contribute to flavor development. Moreover, their enzymes are also involved in the biorefinery concept in which they are employed to hydrolyze agricultural waste that contains a lot of proteins into useful feeds or biofertilizers, and this helps in the production of food systems that are less pollutant. High yield, strength, and security consolidate the leading position of bacterial proteases in the international enzyme market (Liu et al. 2023).

17.4.2.2 Fungal Proteases

Aspergillus fungi, especially *A. oryzae*, *A. niger,* and *A. flavus*, grow abundantly in a large range of proteases, including acid (Aspartic), neutral, and alkaline proteins. This enzymatic repertoire enables them to break down complex protein substrates in their natural habitat and is used widely in traditional and modern biotechnology of food. *Aspergillus oryzae* or Koji Mold is a fundamental component of Asian food fermentations, and has the GRAS status, meaning that it is a safe and safe source of enzymes used to prepare soy sauce, miso, and sake. Fungal proteases are normally expressed through solid-state fermentation (SSF), which recreates the natural environment of these organisms and, in many cases, leads to greater yields and stability of the enzymes than submerged fermentation. Produced enzymes tend to be more thermolabile than their bacterial equivalents, which can be an advantage in applications where tight control and termination of proteolytic activity is only needed by adding a small amount of heat (de Castro and Sato 2022).

The application of fungal proteases within the food business is extensive and varied. *Rhizomucor miehei* and *Cryphonectria parasitina* aspartic proteins are used extensively as microbial rennets in the production of cheeses, as a nonvegetarian and consistent substitute to animal rennet. The multifaceted secretion of a combination of endo- and exo-peptidase by fungi is of special interest to the generation of flavors in fermented foods and ripening of cheeses because these enzymes produce an equilibrated assortment of peptides and unbound amino acids, which are flavor precursors. Moreover, fungal proteases have been applied in baking sector, which is essential in altering the rheological characteristics of the dough by hydrolyzing gluten to enhance the extensibility and machinability, in addition to reducing the time of mix. They can further be seen to be versatile, by clarifying the fruit juices and alcoholic beverages, and by degrading haze-forming proteins. The co-production of proteases in conjunction with other hydrolytic enzymes, including amylases and lipases, is a characteristic of fungi, which offers a synergistic effect that is highly useful in complex food matrices, thus a necessity in multistep bioprocessing (Farha et al. 2023).

17.4.2.3 Yeast Proteases

Yeast proteases are proteins of diverse origin derived in part or entirely, utilized by *Saccharomyces cerevisiae, Kluyveromyces lactis*, and *Yarrowia lipolytica*, which are commonly employed as sources in food enzymology. They can be intracellular, cell-wall associated, or extracellular, and they are divided into several different classes, which are serine, metallo-, and aspartic proteases. Yeast proteases have essential roles in cellular activities in their natural environment such as protein turnover, nutrient uptakes, and maturation of other proteins. In food, their most important application is in fermentation, e.g., beer brewing and wine making, where they help in autolysis of the yeast to release amino acids and peptides that can affect the flavor, aroma, and mouthfeel of the end product. *Kluyveromyces lactis* proteases have been of special interest due to their milk-clotting property, which is yet another possible source of vegetarian rennet to the cheese sector, and the advantage of being a microorganism with a GRAS status (Görgens et al. 2021).

The research on exploiting yeast proteases in controlled food production is a field of increasing concern. Their relative enzyme activity and selectivity may be beneficial in processes where too much or nonselective proteolysis may be unfavorable, e.g., the selective proteolysis of proteins to produce bioactive peptide fragments without bitter flavors. Moreover, the genetic engineering of yeasts may be performed comparatively easily to over-express and secrete particular proteases, which opens opportunities of designing enzymes with new peculiarities and using them in specialized food production processes. As an example, the development of yeasts that can make proteases with increased thermostability or changed substrate specificity may provide an application in baking, or valorization of food industry side streams. Whole cell-based approach to intracellular enzymes also offers an opportunity of in-situ proteolysis in fermentations, making it simpler to design processes. With the current progress in the study of nonconventional yeasts, the diversity in protease profiles of the latter is expected to serve new functions in creating new food products and sustainable bioprocessing concepts (Raveendran et al. 2023).

17.4.3 Mechanism of Action

17.4.3.1 Triad and Hydrolysis Catalytic Mechanism

The catalytic triad is a very conserved motif of structure in a few proteinase families, most importantly the serine proteases, using a triad of particular amino acids, namely, Serine, Histidine, and Aspartate, in a meticulously organized relay that facilitates the breakdown of peptide bonds. The process commences with the histidine residues de-protonating the serine hydroxyl group, thereby activating the serine to nucleophilically attack the carbonyl group of the splitting peptide bonds in the substrate. The result of this attack is the formation of an unstable, highly energetic tetrahedral intermediate, which is stabilized by hydrogen bonds that form within the enzyme's "oxyanion hole." The purpose of the aspartate residue is to position the histidine and maximize its basicity to allow efficient transfer of protons. This smooth collaboration reduces the activation energy barrier of the reaction to a minimal, making the protease achieve very high catalytic rates, turning over thousands of substrate molecules per second under optimal conditions, which is the key to their efficiency in industry (Hedstrom 2002).

Upon formation of the tetrahedral intermediate, it collapses, leading to the cleavage of the peptide link and the release of the initial product (the amine-terminal fragment). Concurrently with this process is the formation of a covalently acyl-enzyme complex, wherein the acyl group that makes up the substrate remains attached to the serine residue of the enzyme. The other half of the catalytic cycle, which is deacylation, entails activation of water molecule by histidine residue. This water molecule is a nucleophile that assaults the acyl-enzyme intermediate to give a second tetrahedral transition state. Disintegration of this state releases the second product (the carboxyl-terminus fragment) and restores the free active enzyme, which is now free to undergo another catalytic cycle. This explains the complexity of proteins as biological catalysts, as the variation of this universal process elucidates the sophisticated efficiency of proteins. The atomic mechanism of this process is the key in protein engineering to design proteases with increased stability, different specificity, or new functions to be used in specialized areas of food science and technology (Badoei-Dalfard and Sohrabi 2022).

17.4.3.2 Substrate Selectivity

Protease specificity is an important factor that defines the use of proteins, and it is generally classified according to their naturally occurring location on the protein target. Protease specificity is an important factor that determines their applications and is generally classified based on the site of action on the protein substrate. Proteinases, or endopeptidases, cleave peptide bonds within the interior of the protein chain. These enzymes include trypsin, chymotrypsin, and subtilisin and have active sites that identify and target a given sequence of amino acids or side-chain characteristics of the localized scissile bond. Their step is a necessary part of activities such as meat tenderization, in which they disaggregate the inside structure of myofibrillar proteins and in cheese production, in which they initially depolymerize casein micelles. Endopeptidase cleavage pattern determines the size and composition of the resulting pool of peptides, and both directly affect functional properties, such as solubility, emulsification, and gelation, and flavors and possible bitterness in protein hydrolysates (García-Mora et al. 2022).

Exopeptidases, on the other hand, only work on the ends of the polypeptide chains. They are further divided into carboxypeptidases, which cut out the C-terminus, and aminopeptidases, which break off the N-terminus. They are more gradual and slower in their action and cause a gradual increase in concentration of free amino acids without causing a dramatic decrease in the size of the larger peptides. The amino peptidases are especially useful in the food industry during debittering of protein hydrolysates, as they sequentially release water-phobic amino acids that accumulate at the N-terminus of the bitter peptide, which are thereafter perceived to be less bitter or even savory. The concerted functioning of endo- and exo-peptidases is also essential in fermented foods and aged cheeses: endopeptidases produce numerous medium-sized peptides out of proteins, and exopeptidases release unbound amino acids and tiny peptides, which are the building blocks of the complex flavor and aroma compounds formed during ripening. This synergistic interaction enables the accurate regulation of the texture and sensory characteristics of the end food product (Sávago-Ayerdi and Görgens 2023).

17.4.4 The Essential Production Technologies

17.4.4.1 Submerged Fermentation (SmF)

The predominant industrial method for large-scale production of microbial proteases is submerged fermentation (SmF), which involves cultivating microbes in a liquid nutritional media within large sterile and highly regulated bioreactors. Under this configuration, the microorganisms that produce are cultured in a homogenous aqueous solution that is constantly aerated and agitated to enhance maximum oxygen replacement, nutrient dispersion, and temperature regulation. The environment is conducive to high cell density and effective extracellular secretion of extracellular proteases into the fermenting medium. Parameters of a process, including pH, temperature, dissolved oxygen, and substrate concentration, are carefully monitored and regulated to ensure high productivity and inter-batch reproducibility of enzymes. SmF is easily scalable with production volumes ranging between laboratory scale fermentors up to industrial level of more than 100,000 liters, making it the most cost-effective alternative to satisfy the enormous global market for industrial enzymes, including those often utilized in the food sector (Singh et al. 2023).

SmF benefits are that it has high level of process control, reproducibility, and it is easy to process downstream because an enzyme of particular interest is typically secreted into the broth. It is however also associated with challenges like high energy usage because of intensive agitation and aeration and production of huge volumes of spent medium necessitating treatment. At the end of the fermentation, the culture broth is a complex solution that includes the cells, left over nutrients as well as the target proteins. This requires a multistage downstream procedure starting with microbial biomass elimination, typically through filtration or centrifugation. The cell-free supernatant is then concentrated by either ultrafiltration or evaporation, after which it is purified by precipitation, chromatography, or further filtration in case a very pure enzyme preparation is desired. Lastly, the protease is produced as a stable liquid or powdered good, usually with stabilizers, preservatives, and salts to maintain its activity during transportation and storage until it arrives with the food manufacturer (Cunha et al. 2022).

17.4.4.2 Solid-State Fermentation (SSF)

Microorganisms undergo a process known as solid-state fermentation (SSF), which occurs on moist solid substrates with little or no free water, a condition that closely mimics the natural habitat of many fungi and some bacteria. The solid substrate, which here is a physical support of the microorganism, is a physical support that is commonly an agricultural or food processing by-product, such as wheat bran, rice husks, or soybean meal, and also provides nutrients to the microorganism. Specifically, SSF is quite adaptable to the growth of filamentous fungi, e.g., *Aspergillus* and *Rhizopus* species, which grow well on solid surfaces and can penetrate the substrate via their hyphae. This technique is well-known due to generating higher titers and stability of enzymes (such as proteases) than SmF due to the enzymes usually being less susceptible to proteolytic degradation by the solid matrix. Moreover, SSF is viewed as a cheap and eco-friendly technology, because it exploits cheap agro-industrial residues, consumes less energy to support aeration and agitation, and produces less wastewater, which is correlated with the concept of a circular bio-economy (Manan and Webb 2021). Although SSF has its benefits, the process is highly challenged in scale-up and process control, mainly because of the challenges in the heat and mass transfer in the solid bed which may cause gradient in temperature, moisture, and nutrient concentration. These heterogeneities may cause irregular production of enzymes at different batches. The fermentation process is usually performed in trays, packed-bed reactor, or rotating drum, and parameters such as temperature should be controlled to ensure that the fermentation is not overheated in the compact substrate. Once fermentation has been completed, the enzyme is then removed off the solid media by a process known as leaching or downstream extraction whereby fermented solid is mixed with an appropriate buffer or solvent and agitated to usher the enzyme into the liquid phase. The extract is then broken off the solids by either pressing or filtration after which it is concentrated and purified in a similar manner as in SmF. SSF is particularly useful to generate enzyme complexes in conventional food fermentation inkoji to soy sauce and miso when the solid food substance becomes a component of the end product (Thomas et al. 2023).

17.4.4.3 Downstream Processing and Formulation

Downstream processing refers to all the unit operations necessary to purify, stabilize, and isolate the protease enzyme once the fermentation has been successfully completed to convert the crude fermentation broth or solid extract into a sellable and stable product. The first actions will all depend on the fermentation technique; in the case of SmF, it is to isolate the microbial cells either through centrifugation or microfiltration with the broth, and in the case of SSF, the process starts with the isolation of enzyme within the solid substrate. The liquid extract thus obtained is then concentrated, usually by ultrafiltration, selective to retain the high molecular weight enzymes, and to exclude water, salts, and other small soluble molecules.

To achieve greater purity, as is needed in some food additives or in diagnostic preparations, other forms of purification, such as precipitation using salts or solvents, ion-exchange chromatography, or gel filtration, can be used to eliminate the contaminants and other unwanted enzymes (Nair et al. 2023).

The last and most important step is the formulation, which aims at stabilizing the protease and making it remain active in the storage, transportation, and usage. Proteins such as proteases are prone to heat-induced denaturation, self-proteolytic or autolytic digestive (autolysis) damage, oxidative damage, and microbial contamination. Liquid enzymes are composed with stabilizers, including polyols (e.g., Glycerol or Sorbitol), sugars, and salts to form a thermodynamically stable system. Metalloproteins are frequently supplemented with calcium ions to increase the stability of the structure. To create a powdered enzyme formulation, the concentrated enzyme solution is normally dried by spray drying or lyophilization (free-zee-high drying), which is often facilitated by carrier substances such as maltodextrin or starch to shield the enzyme against the drying stresses and to create a fine, uniform powder. The end formulated product would be subject to testing of critical such as enzyme activity, purity, microbial load, and shelf-life to ensure consistency in performance of the food processor, such that the protease will perform consistently in its application, be it baking, dairy, or in the bottling of beverages (Juturu and Wu 2022).

17.5 Large-Scale Food Processing of Microbial Proteases

17.5.1 Dairy Industry

17.5.1.1 Milk Coagulation to Make Cheese

Microbial proteases, especially aspartic proteases of fungi such as *Rhizomucor miehei,* are invaluable rennet alternatives in the production of cheese around the world. These enzymes play the major role of breaking the main stage of producing cheese by specifically cleaving the Phe105-Met106 bond within kappa-casein. The exact cleavage removes steric and electrostatic repulsion that hold the casein micelles together in milk. As a result, the micelles become destabilized and consolidated in the presence of the calcium ions to form a three-dimensional gel structure. This protein matrix traps milk fat and milk water well and thus a firm coagulum is formed. The curd is then sliced, cooked, and emptied to remove whey, and begin making cheese. This enzyme action is specific and is most important in maximizing the firmness of the curd and ultimate cheese yield. Excessive or nonspecific proteolysis would additionally influence the texture and quality of the final product. This renders microbial rennet one of the main foundations of contemporary, uniform, and effective dairy technology (Qureshi et al. 2022).

17.5.1.2 Speeding up the Ripening and Flavoring of Cheese

The long and expensive period of maturing cheese is greatly shortened by means of setting strategic exogenous microbial proteases. These proteins facilitate the first catalytic proteolysis of casein into a complex repertoire of medium and small-sized peptides. The latter are then broken down by other endogenous microbial aminopeptidases to release free amino acids. These nitrogenous compounds are integral to the initiation of a series of catabolic reactions, including transamination and decarboxylation. Key volatile flavor components include aldehydes and alcohols; sulfur-containing compounds are formed as a result of such reactions. This enzyme cascade not only controls flavor development, but also it is responsible in the desired changes in texture as the product ages. Optimization of the nature and presentation of these proteases will enable producers to attain the sensory profile of an aged cheese within a fraction of the time. This will be an efficient method of minimizing inventory expenses and improving production profitability in the dairy industry as a whole (Sgarbi et al. 2023).

17.5.1.3 Manufacturing of Milk Protein Hydrolysates

Microbial proteases significantly contribute to the enzymatic hydrolysis of milk proteins, including casein and whey, to form special hydrolysates. These hydrolysates also have greater functional and nutritional qualities than their intact protein counterparts. The regulated proteolysis adjusts the primary structure of the protein elevating the solubility profile over a wide spectrum of pH. It also reduces the likelihood of allergenicity due to the destruction of conformational epitopes. Moreover, the process may confer high-quality emulsifying and foaming properties, and therefore, they are useful food ingredients. Clinical and sports nutrition. These hydrolysates are very important in clinical and sports nutrition since they contain easily digested nitrogen sources. The hydrolysis process may be selectively designed to liberate selected bioactive peptides with proven physiological effects. These advantages comprise antihypertensive, antioxidant, and immunomodulatory properties, which make milk a powerful functional food ingredient among the health-conscious consumers (Tavano 2023).

17.5.1.4 Milk Products Allergenicity Reduction

Selective proteolysis can alleviate the allergic properties of bovine milk, which is mainly related to such proteins as 3-lactoglobulin and 3-casein. The conformational epitopes and linear amino acid sequences that are identified by the human immune system are hydrolyzed using microbial proteases. This enzyme degradation reduces the immunoglobulin E (IgE) activity, thus making the product allergenic. The ability to select proteases with high specificity for the bonds within these allergenic epitopes is essential. It is the basic enzyme approach on which hypoallergenic infant formulas and other dairy products will be made by sensitized individuals. The big issue is how to reach a point of adequate hydrolysis to abrogate allergenicity as well as reduce the development of bitter-tasting hydrophobic peptides. That may demand a complex blend of endo- and exo-peptidases to create a final product that is safe and palatable and low-allergen (Wróblewska et al. 2023).

17.5.2 Baking Industry

17.5.2.1 Dough Conditioning and Weakening of Gluten

Specific neutral microbial proteases are used as advanced dough conditioners in the baking industry. They mostly affect the selective hydrolysis of the gluten network, a protein network that gives dough its viscoelastic properties. Such a planned proteolysis makes the gluten structure soft so that it minimizes the energy input and time spent in mixing. The dough that results has much greater extensibility and pliability, and it is easier to sheet and mold. This has been of important concern when it comes to the manufacture of regular crackers, biscuits, and some types of bread where shrinkage is not to be tolerated. These enzymes are used to give uniform dough pieces and maximum retention of the gas during the critical proofing stage since they weaken the gluten by precision. This then results into a more uniform and better-quality finished baked product containing a desirable, well-structured crumb (Renzetti 2023).

17.5.2.2 Increasing Crumb Softness and Loaf Volume

The microbial proteases largely improve the textural and volumetric properties of baked products by selectively altering the protein network. The minimal hydrolysis of the gluten proteins enhances the extent of the dough such that the dough is easily elongated when it is subjected to the pressure of fermentative carbon dioxide formation. This results in a substantial increase in the ultimate volume of the loaf and a finer and more homogeneous crumb structure. Amino acids are also released after the proteolytic action, and they are later involved in the Maillard browsing reactions during the baking process. These reactions add to the attractive crust color and give its product rich and complex profile of flavors. An excellent quality parameter that can be felt by consumers is the instant increase in the softness, or decrease in crumb firmness. Consequently, protease treatment would be useful in making high-volume, soft-textured breads and rolls that can satisfy the market needs (Sun et al. 2023).

17.5.2.3 Stalling and Shelf-Life Extension

Microbial proteases partly counteract the effect of staling of baked goods, which is reflected in the firming of the crumb and the loss of freshness. The mechanism leading to staling is mostly starched retrogradation and modification of moisture distribution between starch and gluten. Proteases indirectly delay this process by altering the structure of the gluten that changes the capacity of the water-holding and its bond with the gelatinized starch grains. A more hydrolyzed and free gluten structure is more conducive to these physical changes leading to a soft crumb that does not harden over a long period time. The anti-staling effect contributes to the preservation of sensory freshness of baked goods over a few days. This enzyme method is frequently applied in combination with amylases and emulsifiers in their entirety as a combined shelf-life extension strategy. The outcome is the minimization of food waste, as well as improved commercial feasibility of bakery products (Xu et al. 2023).

17.5.3 Meat and Seafood Processing

17.5.3.1 Meat Tenderization

Microbial proteases to enhance meat tenderness are widely used as either papain-like enzymes or specific bacterial proteases. They act by breaking down the tough connective tissue proteins, which are mainly collagen and elastin, and the structural myofibrillar proteins in muscle fibers. This specific proteolysis interferes with the structural integrity of the meat resulting in a sharp decrease in shear force. The net effect is a significant reduction in perceived tenderness, juiciness, and palatability. The enzymes may be utilized in different ways, which include surface application, tumbling, or injection in order to enable even distribution across the meat cut. The procedure requires careful manipulation of time and temperature parameters to get the best outcomes. Inadequate enzyme activity is of no concrete use, and excessive hydrolysis may result in a mushy and unattractive texture, which highlights the necessity of following the procedures of application to the letter (Sazonova et al. 2023).

17.5.3.2 Protein Recovery

Microbial proteases are essential for enhancing the value of meat and seafood by-products, whereby low-value tissues are converted into high-appealing products. The bones, skins, scales, and trimmings are highly dense in collagen and other proteins, which are also hard to extract. These recalcitrant materials are solubilized easily through controlled enzyme hydrolysis to extract useful proteins and peptides. The protein hydrolysates obtained have beneficial functional attributes including emulsification, foaming, and enhanced water-binding capability. The properties render them useful components in a large number of applications, including soups and sauces, pet food, and animal feed supplements. This application is also efficient to boost the economic performance of the processing industry in addition to the environmental sustainability. It does so by minimizing the biological waste and making the maximum out of the animal carcasses, adding to the elements of a circular biological economy (Lafarga et al. 2023).

17.5.3.3 Fishing and Shellfish De-skinning

The manual de-skinning of fish and shellfish is an activity that is labor based, which may result in the loss of product yield and compromise in quality. An effective and less destructive alternative is the use of microbial proteases that specifically digest the thin layer of connective tissue on the dermal-epidermal junction. Under strictly controlled circumstances, the enzymes selectively break down this bond without destroying the muscle tissue beneath. This enables the skin to be peeled off easily and with a minimum amount of physical exertion saving the fragile flesh. This processing technique is an enzyme-based de-skinning process that is much more effective in yielding better processing results as well as improving the physical appearance of the end product, the fillet or shrimp. It is a significant innovation in seafood manufacturing methods, with the guarantee that it is of higher quality and more efficient but does not affect the structural integrity of the seafood (Zamora-Sillero et al. 2023).

17.5.4 Beverage Industry

17.5.4.1 Chill-Proofing of Beer

One of the quality issues associated with beer is the occurrence of the phenomenon of a chill haze that is caused by the aggregation of polypeptides and polyphenols. This is prevented by the addition of microbial proteases like papain, during the maturation or storing stage of the beer. These enzymes break down haze-active proteins to smaller peptides and amino acids, which cannot form large colloidal complexes. This proteolytic activity guarantees the beer its sparkling clarity and colloidal stability as long as it has stayed on shelves. It will eliminate the appearance of visible haze particles on cooling a beverage, which is one of the flaws that are not attractive to consumers. This is a process of chill-proofing that is normally done in the modern brewing in order to conform to high-quality requirements. It is handled with meticulous care not to affect the beer in its most crucial characteristic foaming ability (head retention) or its delicate flavor profile, providing its high-end product (Steiner et al. 2023).

17.5.4.2 Yield Improvement and Juice Clarification

Exceptional clarity is a primary quality objective in the preparation of juices made from fruits and vegetables. Microbial proteases are frequently combined with the

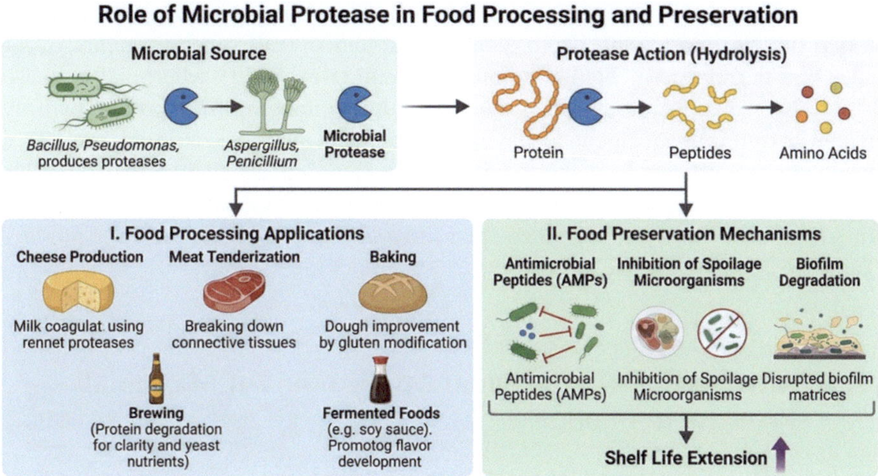

Fig. 17.1 Role of microbial proteases in food processing and preservation

pectinases to attain better clarification. The proteases destroy the protein portion of the haze-stabilizing pectin-protein complexes, which hold a stable colloidal suspension together. This will disturb the haze, which will cause the aggregation and further settling of the fine particles. The particles are then readily eliminated through filtration or centrifugation, leaving a brilliantly clear juice with a higher visual quality and stability. Moreover, the proteases may help to increase the amounts of juices by degrading cellular structures and proteinaceous matter. This enables a better recovery of soluble solids and other useful compounds in the fruit pulp, which enhances the overall efficiency of the process (Varzakas and Tzia 2023). Microbial proteases are widely used in the food industry to enhance processing efficiency and improve product quality (Fig. 17.1). They aid in protein hydrolysis, which contributes to flavor development, texture modification, and the production of bioactive peptides. Additionally, these enzymes play a critical role in food preservation by reducing allergenic proteins and inhibiting microbial spoilage.

17.5.4.3 Soy Sauce and Other Condiments Fermentation

Traditional fermented condiments like soy sauce require microbial proteases. The koji mold, *Aspergillus oryzae,* that grows on a mixture of steamed soybean and wheat, secreted these enzymes. The mold forms a powerful combination of aspartic, neutral, and aminopeptidases. In the next step of fermenting the brine, the enzymes systematically cut the plant proteins into a substantial aggregation of free amino acid pairs and diminutive peptides. The key components of the final product that give it the umami quality are glutamate and aspartate. The complex savory, sweet, and mouthfeel qualities that make up quality soy sauce are done by other amino acids and peptides. This is the final, enzymatic proteolysis and has to be very meticulously carried out to produce the depth, complexity, and overall sensory quality of soy sauce and the like fermented condiments (Wang et al. 2023). Microbial proteases exhibit diverse applications across the food industry, ranging from protein hydrolysis and texture enhancement to flavor development and allergen reduction. Their specificity and efficiency make them valuable in processes such as dairy product maturation, meat tenderization, baking, and the production of bioactive peptides, while also contributing to extended shelf life and improved safety of food products (Table 17.2).

17.6 Advanced and Functional Applications of Microbial Protease in Food

17.6.1 Generation of Bioactive Peptides

17.6.1.1 Antihypertensive Peptides

The masterful use of microbial proteases to release active antihypertensive peptides in food protein precursors such as Casein, Whey, and Plant Globulins is contingent on the presence of proteins in food. These bioactive fragments also act by competitively inhibiting Angiotensin-I-Converting Enzyme (ACE) this is a principal agent regulating blood pressure within the renin-angiotensin system. Proteolytic process is carefully optimized to produce a certain amount of specific peptide sequences, usually including proline, lysine, or aromatic amino acids, which have a high affinity with the active site in the enzyme. This enzymatic bioprocessing converts food proteins that are abundantly found in common-place foods into nutraceutical and fortified food functional ingredients. The activity of such ACE-inhibitory peptides has been strictly confirmed by the in vitro studies and animal models (establishing their possibility to control hypertension). This use is a paradigm shift of merely nourishing to actively preventing chronic illness with diet, reflecting the advanced healing potential of enzyme technology in designing functional food (Udenigwe and Fogliano 2023).

17.6.1.2 Antioxidant Peptides

Enzymatic breakdown of food proteins by microbial proteases results in the strategic hydrolysis of food proteins that produce peptides with high radical scavenging and metal-chelating properties, which effectively inhibit lipid oxidation. These antioxidant peptides may generally include hydrophobic amino acids that provide the protons to stabilize the free radicals, and histidine amino acids that chelate the pro-oxidants metal ions like iron and copper. These peptides prevent the formation of rancidity in fat-containing food matrices by reducing oxidative processes and

Table 17.2 Versatility of microbial proteases in food processing and preservation

Microbial source	Protease type	Food application	Functional role	Benefits	Example product
Bacillus subtilis	Alkaline protease	Meat tenderization	Breakdown of muscle proteins	Improved texture	Tenderized meat
Aspergillus oryzae	Acid protease	Soy sauce fermentation	Protein hydrolysis	Enhanced flavor	Soy sauce
Lactobacillus helveticus	Cell-envelope protease	Cheese ripening	Casein breakdown	Flavor development	Swiss cheese
Bacillus licheniformis	Subtilisin	Detoxification	Degradation of allergenic proteins	Hypoallergenic products	Hydrolyzed milk proteins
Rhizopus oryzae	Neutral protease	Tempeh processing	Protein breakdown	Improved digestibility	Tempeh
Streptomyces griseus	Proteinase K	Processing of gelatin	Collagen hydrolysis	Clear, stable gels	Gelatin products
Aspergillus Niger	Peptidase	Beer brewing	Protein removal	Reduced haze formation	Clear beer
Bacillus amyloliquefaciens	Neutral protease	Fish sauce fermentation	Protein hydrolysis	Enhanced taste	Fish sauce
Candida utilis	Protease complex	Yeast extract production	Autolysis enhancement	Rich flavor compounds	Yeast extract
Penicillium camemberti	Protease–lipase mix	Surface-ripened cheese	Casein hydrolysis	Creamy texture	Camembert cheese

therefore increase the shelf life and inhibit the degradation of the sensory attributes of certain foods. In addition to food preservation, their biological activity after intake is added to the endogenous defense system of the body against oxidative stress, which is a prime cause of aging and chronic diseases. These antioxidants generated through the hydrolysis of proteins using enzymes will provide a clean label alternative to the synthetic antioxidant and fit the current consumer trend of using natural antioxidant food ingredients that are made through sustainable bioprocessing (Sarmadi and Ismail 2023).

17.6.1.3 Antimicrobial Peptides (Antimicrobial Bacteriocins or Antimicrobial Peptidoglycan)

Some microbial proteases are catalysts of release of antimicrobial peptides (AMPs), or bacteriocins, which are natural preservatives against spoilage and pathogenic bacteria. These are peptides that are frequently concealed in the main framework of bigger proteins, e.g., lactoferrin or casein, and are specifically released via a specific cleavage by enzymes. The resultant AMPs have their bactericidal effect by interfering with membrane integrity of microorganisms and causing cell lysis and death, commonly with high selectivity against Gram-positive bacteria such as *Listeria* and *Staphylococcus*. This way of action does not pose a high risk of giving bacterial resistance as opposed to the traditional antibiotics. Hurdle against microbial contamination is provided by the in-situ production of these peptides during processing of foods or fermentation, which produces a potent natural hurdle. It combines the idea of food functionality/safety and employs the concept of proteases to produce the natural preservation mechanisms that are built into the food and allow it to be consumed without relying on an additive (Pérez-Rodríguez et al. 2023).

17.6.2 Protein in Food Preservation of Microbial Protease

The necessity to secure food safety and to increase the shelf life without reducing the nutritional value or consumer acceptability has triggered a paradigm shift to the biological methods of preservation. In this field, microbial proteases have not only been discovered as processing facilitators, but they have also been used as advanced biocatalytic models of food preservation, with the capability to convert the intrinsic food materials into multifunctional protective factors. The chapter specifies the processes and uses of these enzymes to coordinately achieve preservation by producing antimicrobial hydrolysates and bioactive peptides, reducing allergenic risks, and strengthening native stability of fermented ecosystems. Protease-based approaches to food safety, functionality, and clean-label appeal convergence make enzymatic biopreservation the leading innovative force in the sustainable food system.

17.6.2.1 Antimicrobial Agents in the Form of Protein Hydrolysates

Precise-engineered bio-preservation through the production of antimicrobial protein hydrolysates by targeted proteolysis is radically different to the wide-spectrum activity of chemical preservatives. Microbial proteases, which are selected on the basis of cleavage motifs, play a cryptographic role, reconstructing latent antimicrobial peptides (AMPs), which are mainly bacteriocins, out of inactivated precursor proteins in the food matrix. Examples of substrates include casein, lactoferrin, and ovalbumin that are hydrolyzed to release powerful peptides like lactoferricin, caseicidin, and ovalbumin-derived fragments. These AMPs have bacteriostatic or bactericidal influences, having high-affinity interactions with microbial membranes, depolarizing, creating pores, and eventually leaking cytoplasmic contents. Importantly, their range of activities is usually limited, and they specifically assault *Staphylococcus aureus* and *Listeria monocytogenes*, two Gram-positive bacteria with minimal effects on commensal microorganisms, which leads to low selective pressure on resistance. As either prefermented products or as in-situ generated during processing, the use of these hydrolysates provides a multifaceted antimicrobial hurdle in products like vacuum-packaged meats, fresh cheeses, and drinkable objects that is both natural and natural. The approach represents a shift to endogenous fortification of food, a shift that perfectly matches the trend in the global market movement toward preservative-free products with clean labels, and improves microbial safety in a mechanism that seems to have a more natural nature (Tkaczewska 2023).

17.6.2.2 Active Peptides to Prolong Shelf-Life

The extension of shelf-life by microbial proteases goes beyond microbial inhibition to include the delaying of chemical degradation, the major cause of rancidity in lipid-based food systems, which is lipid oxidation. These enzymes mediate the liberation of antioxidant peptides of parent proteins in soy, marine, dairy, and cereal sources through controlled and limited proteolysis. The bioactivity characteristics of these types of peptides are orchestrated by the presence of redox-active amino acid residues: hydrophobic amino acid residues (Tryptophan, Tyrosine, Phenylalanine), proton donors to radical quenching, and the Histidine imidazole rings, which efficiently chelate pro-oxidant transition metals (ferrous and cupric ions). The two-step mechanism interferes with radical propagation step of the autoxidation process to retain polyunsaturated fatty acids and reduce the generation of secondary oxidation products that contribute to the formation of off-flavors and hazardous substances. In turn, protease-produced antioxidant hydrolysates are promising, natural alternatives to synthetic phenolics (e.g., BHA, BHT), with increased consumer acceptability and possible synergistic health. Their addition to food matrices, both as edible oils and as meat products, nutritional bars, and powdered food, provides oxidative stability, thus maintaining sensory properties, nutritional, and commercial properties during distribution and storage. This application highlights the ability of proteases to transform perishable commodities into shelf-stable products by biochemical valorization, which has a direct contribution to the decrease of food waste (Mañas and Pagán 2023).

17.6.2.3 Decreased Level of Allergenic Proteins

IgE-mediated hypersensitivity to the epitopes of individual proteins known as food allergenicity poses a serious threat to the safety of food and inclusive nutrition. Microbial protease is an enzymatic solution based on allergen mitigation in which the microbial protease is specifically targeted to break down the conformational and sequential epitopes that trigger an immune response. It requires the accurate choice of enzymes; e.g., to hydrolyze β-lactoglobulin, the largest milk allergen, requires proteases with the capacity to cleavage within the rigid disulfide-bonded globular structure of the protein to destroy conformational epitopes, and a subsequent exopeptidase treatment may be used to destroy any remaining linear epitopes and reduce bitterness. This specific and multi-enzolic method allows allergenicity to be lowered by more than 99, allowing the manufacture of hypoallergenic formulas. The technical sophistication is in the realization of a high level of hydrolysis and regulation of the peptide size profile to prevent the formation of cryptic allergenic peptides or too bitter hydrolysates. Efficacy and palatability are guaranteed by advanced process control that involves real-time monitoring of Degree of Hydrolysis (DH) and peptide mapping. The technology, which is based on proteases, is the basis of the production of safe, nutritious foods with allergic populations, especially hypoallergenic infant formulas and special dietary products, which represent an essential crossroads of enzyme technology, immunology, and clinical nutrition (Ramakodi et al. 2020).

17.6.2.4 Preservation in Fermented Foods Use

Fermented foods are old, but highly advanced preservation methods in which microbial proteases play an essential biological control in determining stability and safety. Protease activity in these complex microbiomes is a component of a synergistic preservation paradigm. Mainly, the proteolysis process of the matrix proteins is catalyzed by extracellular proteases secreted by fermentative microbiota: lactic acid bacteria (LAB), yeasts, and molds. This nutrient release is a trophic driving force, as it triggered the growth of the target fermentative culture, and as a result, enhances competitive exclusion over spoilage-causing organisms. Simultaneously, this same proteolytic system frequently produces antimicrobial peptides directly (e.g., organic acids, ethanol, diacetyl, etc.), which serve together with other microbial metabolites (e.g., organic acids, ethanol, diacetyl, etc.) to create a strong multi-hurdle preservation system. Bacteriocins such as nisin (produced by *Lactococcus lactis*) and other peptides that inhibit the growth of other microorganisms are also released by proteases in the starter and adjunct cultures in dairy fermentation contributing to the formation of flavor during fermentation. Equally in the case of fermentation of meat and vegetables, proteolysis by endogenous or inoculated microbiota has been shown to increase safety by lowering water activity due to the formation of solutes and an extensive range of antimicrobial products. Proteases are therefore not peripheral players but key players in the self-preserving character of fermented foods to guarantee their shelf life, food safety, and distinct sensory characteristics via a biochemical cascade that is naturally coordinated (Fu et al. 2019).

17.6.3 Sustainability and Waste Valorization

17.6.3.1 Recycling Food Processing Waste to Value-Added Products

Microbial proteases have been important in transforming low-value proteinaceous wastes, including meat trimmings, fish offal, whey, and oilseed cakes into high-value products. These enzymes are used to solubilize and extract proteins out of complex and recalcitrant matrices, which are otherwise hard to use by means of controlled hydrolysis. The protein hydrolysates obtained have very good functionality properties which are used in food production as emulsifiers, foaming agents and in food production as flavor enhancers. Moreover, these hydrolysates are refinable into animal feed, pet food, and mediate of microbial culture. This is a sure way of converting an environmental liability that would otherwise face the expense of disposing of the wastes into a lucrative income stream. It is an ideal representation of green chemistry, and it contributes greatly to the economy but also helps with the issue of waste management in the food industry, which is a pressing problem (Ktari et al. 2022).

17.6.3.2 Investment into a Circular Bio-economy

Microbial proteases have been applied in valorizing waste, which becomes one of the foundations of the circular bio-economy, re-defining waste as renewable resource. This technology brings about the enzymatic cycle in which food processing side streams are used within the production cycle as a functional food ingredient, thus minimizing the utilization of virgin raw resources in the industry. It mitigates environmental footprint by avoiding the landfills where waste produces greenhouse gases and also decreases the pollution load in wastewater. The procedure itself is normally performed in mild and energy-efficient environment, which makes it even more sustainable. Protease technology facilitates the establishment of more robust and sustainable food systems through the establishment of a cradle-to-cradle model of food production residuals. Not only does this bring about economic benefits, but it would also be in line with the global sustainability objectives, and thus it is clear that industrial biotechnology is the way forward toward a more efficient and responsible utilization of planetary resources (Ravindran and Jaiswal 2023).

17.7 Conclusion and Future Prospects of Proteases

Proteases are a distinct category of enzymes that possess both degradative and synthetic capabilities, offering the advantage of being biological rather than chemical in nature. Consequently, the environmentally sustainable uses of proteases within green and safe technology are extensive, exhibiting significant potential toward the future. Research is being conducted on their medicinal potential for reducing the spread of major epidemics such as malaria and AIDS, as well as diseases like cancer (Zhao et al. 2020). Recent decades have broadened our comprehension of the applicability of these digestive enzymes in sectors other than food technology, such as agriculture, medicines, textiles, leather, and detergents. Potential applications of these enzymes' proteolytic activities in nanotechnology, targeted drug delivery systems, and even technologies for imaging for cancer diagnostics have been demonstrated by the "OMICS era" (Abousaway et al. 2021). Creatively designed proteases that are triggered by small chemicals possess the potential to precisely regulate human protease activity, thereby profoundly influencing the cell cycle and possibly inducing apoptosis. These compounds offer possible ways for addressing previously inaccessible intracellular sites, offering significant clinical uses in the future. Engineering proteases to meet specific application requirements poses a problem for the future (Gerry and Schreiber 2020).

References

Abousaway O, Rakhshandehroo T, Van den Abbeele AD, Kircher MF, Rashidian M (2021) Noninvasive imaging of cancer immunotherapy. Nanotheranostics 5:90–112. https://doi.org/10.7150/ntno.50860

Abu-Tahon MA, Arafat HH, Isaac GS (2020) Laundry detergent compatibility and dehairing efficiency of alkaline thermostable protease produced from *Aspergillus terreus* under solid-state fermentation. J Oleo Sci 69(3):241–254. https://doi.org/10.5650/jos.ess19236

Ahmad W, Tayyab M, Aftab MN, Hashmi AS, Ahmad MD, Firyal S, Wasim M, Awan AR (2020) Optimization of conditions for the higher level production of protease: characterization of protease from *Geobacillus* SBS-4S. Waste Biomass Valoriz 11:6613–6623. https://doi.org/10.1007/s12649-019-00807-6

Aissaoui N, Chobert J-M, Haertlé T, Marzouki MN, Abidi F (2016) Purification and biochemical characterization of a neutral serine protease from *Trichoderma harzianum*. Use in antibacterial peptide production from a fish by-product hydrolysate. Appl Biochem Biotechnol 182:831–845. https://doi.org/10.1007/s12010-016-2364-5

Al-Dhabi NA, Esmail GA, Ghilan A-KM, Arasu MV, Duraipandiyan V, Ponmurugan K (2020) Characterization and fermentation optimization of novel thermo stable alkaline protease from Streptomyces sp. Al-Dhabi-82 from the Saudi Arabian environment for eco-friendly and industrial applications. J King Saud Univ Sci 32(1):1258–1264. https://doi.org/10.1016/j.jksus.2019.11.013

Anandharaj M, Sivasankari B, Siddharthan N, Rani RP, Sivakumar S (2016) Production, purification, and biochemical characterization of thermostable metallo-protease from novel *Bacillus alkalitelluris* TWI3 isolated from tannery waste. Appl Biochem Biotechnol 178:1666–1686. https://doi.org/10.1007/s12010-015-1978-3

Ao X, Yu X, Wu D (2018) Purification and characterization of neutral protease from *Aspergillus oryzae* Y1 isolated from naturally fermented broad beans. AMB Express 8:96. https://doi.org/10.1186/s13568-018-0628-x

Badoei-Dalfard A, Sohrabi N (2022) Microbial proteases: production and application in industrial biotechnology. In: Advances in enzyme biotechnology. Springer, pp 165–186. https://doi.org/10.1007/978-981-19-1282-6_6

Bajaj BK, Sharma P (2011) An alkali-thermotolerant extracellular protease from a newly isolated Streptomyces sp. DP2. New Biotechnol 28(6):725–732. https://doi.org/10.1016/j.nbt.2011.01.005

Baweja M, Nain L, Kawarabayasi Y, Shukla P (2016) Current technological improvements in enzymes toward their biotechnological applications. Front Microbiol 7:965. https://doi.org/10.3389/fmicb.2016.00965

Bond JS (2019) Proteases: history, discovery, and roles in health and disease. J Biol Chem 294(5):1643–1651. https://doi.org/10.1074/jbc.AW119.006656

Boughachiche F, Rachedi K, Duran R, Boulahrouf A (2016) Optimization of alkaline protease production by *Streptomyces sp.* strain isolated from saltpan environment. Afr J Biotechnol 15(25):1401–1412. https://doi.org/10.5897/AJB2015.15116

Contesini FJ, de Alencar Figuerôa IL, de Oliveira Carvalho P (2023) Aspartic proteases: characteristics, production, and applications in food and beverage processing. Food Bioprocess Technol 16:1–17. https://doi.org/10.1007/s11947-022-02877-2

Cunha FM, Badino AC, Farinas CS (2022) Production of microbial enzymes by submerged fermentation: current trends and future prospects. Bioenergy Res 15(2):1234–1254. https://doi.org/10.1007/s12155-021-10376-z

da Silva RR, Cabral TPdF, Rodrigues A, Hamilton C (2013) Production and partial characterization of serine and metallo peptidases secreted by *Aspergillus fumigatus* Fresenius in submerged and solid state fermentation. Braz J Microbiol 44(1):235–243. https://doi.org/10.1590/S1517-83822013000100034

da Silva RR, Souto TB, Oliveira TB (2016) Evaluation of the catalytic specificity, biochemical properties, and milk clotting abilities of an aspartic peptidase from *Rhizomucor miehei*. J Ind Microbiol Biotechnol 43:1059–1069. https://doi.org/10.1007/s10295-016-1771-5

da Silva RR, de Oliveira LC, Juliano MA, Juliano L, de Oliveira AH, Rosa JC, Cabral H (2017) Biochemical and milk clotting properties and mapping of catalytic subsites of an extra cellular aspartic peptidase from basidiomycete fungus, *Phanerochaete chrysosporium*. Food Chem 225:45–54. https://doi.org/10.1016/j.foodchem.2016.11.136

de Castro RJS, Sato HH (2022) Fungal proteases and their applications in the food industry. In: Fungal biorefineries. Springer, pp 155–174. https://doi.org/10.1007/978-3-030-95675-6_7

de Oliveira CT, Rieger TJ, Daroit DJ (2017) Catalytic properties and thermal stability of a crude protease from the keratinolytic *Bacillus sp.* CL33A. Biocatal Agric Biotechnol 10:270–277. https://doi.org/10.1016/j.bcab.2017.04.004

de Oliveira JM, Fernandes P, Benevides RG, de Assis SA (2020) Characterization and immobilization of protease secreted by the fungus *Moorella speciosa*. 3 Biotech, 10:419. https://doi.org/10.1007/s13205-020-02396-x

de Oliveira LC, de Souza AA, de Castro RJS (2022) Cysteine proteases: sources, properties, and applications in industrial processes. Biocatal Agric Biotechnol 43:102409. https://doi.org/10.1016/j.bcab.2022.102409

de Souza PM, de Bittencourt ML (2015) A biotechnology perspective of fungal proteases. Braz J Microbiol 46(2):337–346. https://doi.org/10.1590/S1517-838246220140359

Deng JJ, Huang WQ, Li ZW, Lu DL, Zhang Y, Luo XC (2018) Biocontrol activity of recombinant aspartic protease from *Trichoderma harzianum* against pathogenic fungi. Enzym Microb Technol 112:35–42. https://doi.org/10.1016/j.enzmictec.2018.02.002

Deng JJ, Shi D, Zhao M (2021) Recombinant neutral protease rNpI as fish feed additive to improve protein digestion and growth. Aquac Res 52(6):273–281. https://doi.org/10.1111/are.14887

Devi SG, Fathima AA, Sanitha M, Iyappan S, Curtis WR, Ramya M (2016) Expression and characterization of alkaline protease from the metagenomic library of tannery activated sludge. J Biosci Bioeng 122(6):694–700. https://doi.org/10.1016/j.jbiosc.2016.05.007

dos Santos Aguilar JG, Sato HH (2018) Microbial proteases: production and application in obtaining protein hydrolysates. Food Res Int 103:253–262. https://doi.org/10.1016/j.foodres.2017.10.044

Duarte AR, Duarte DMR, Moreira KA (2009) *Jacaratia corumbensis* O Kuntze a new vegetable source for milk-clotting enzymes. Braz Arch Biol Technol 52(1):1–9. https://doi.org/10.1590/S1516-89132009000100001

Farha AK, Yang QQ, Kim G, Li HB, Zhu F, Liu HY, Gan RY (2023) Fungal proteases in food processing: a systematic review. Compr Rev Food Sci Food Saf 22(1):246–274. https://doi.org/10.1111/1541-4337.13068

Fu Y, Zhang Y, Soladoye OP, Aluko RE (2019) Maillard reaction products derived from food protein-derived peptides: insights into flavor and bioactivity. Crit Rev Food Sci Nutr 60(19): 3282–3295. https://doi.org/10.1080/10408398.2019.1689950

García-Mora P, Peñas E, Martínez-Villaluenga C (2022) Peptidases for the production of bioactive peptides from food proteins. In: Bioactive peptides from food. Academic, pp 1–30. https://doi.org/10.1016/B978-0-12-821389-6.00023-1

Garcia-Moyano A, Larsen Ø, Gaykawad S, Christakou E, Boccadoro C, Puntervoll P, Bjerga GEK (2020) Fragment exchange plasmid tools for CRISPR/Cas9-mediated gene integration and protease production in *Bacillus subtilis*. Appl Environ Microbiol 87(3):e02090–e02020. https://doi.org/10.1128/AEM.02090-20

Garcia-Moyano A, Diaz Y, Navarro J, Almendral D, Puntervoll P, Ferrer M, Bjerga GEK (2021) Two-step functional screen on multiple proteinaceous substrates reveals temperature-robust proteases with a broad-substrate range. Appl Microbiol Biotechnol 105:3195–3209. https://doi.org/10.1007/s00253-021-11220-2

Gerry CJ, Schreiber SL (2020) Unifying principles of bifunctional, proximity-inducing small molecules. Nat Chem Biol 16(4):369–378. https://doi.org/10.1038/s41589-020-0469-1

Gimenes NC, Silveira E, Tambourgi EB (2019) An overview of proteases: production, downstream processes and industrial applications. Sep Purif Rev 50(3):223–243. https://doi.org/10.1080/15422119.2019.1635511

Gonzalez-Rabade N, Badillo-Corona JA, Aranda-Barradas JS, del Oliver-Salvador M (2011) Production of plant proteases in vivo and in vitro—a review. Biotechnol Adv 29(6):983–996. https://doi.org/10.1016/j.biotechadv.2011.08.017

Görgens JF, van Zyl WH, Knoetze JH (2021) The potential of yeast proteases in industrial biotechnology. FEMS Yeast Res 21(2):foab030. https://doi.org/10.1093/femsyr/foab030

Gurumallesh P, Alagu K, Ramakrishnan B, Muthusamy S (2019) A systematic reconsideration on proteases. Int J Biol Macromol 128:254–267. https://doi.org/10.1016/j.ijbiomac.2019.01.081

Hammami A, Bayoudh A, Hadrich B, Abdelhedi O, Jridi M, Nasri M (2020) Response-surface methodology for the production and the purification of a new H_2O_2-tolerant alkaline protease from *Bacillus invictae* AH1 strain. Biotechnology Progress, 36: e2965. https://doi.org/10.1002/btpr.2965

Hakim A, Bhuiyan FR, Iqbal A, Emon TH, Ahmed J, Azad AK (2018) Production and partial characterization of dehairing alkaline protease from *Bacillus subtilis* AKAL7 and *Exiguobacterium indicum* AKAL11 by using organic municipal solid wastes. Heliyon 4(5): e00646. https://doi.org/10.1016/j.heliyon.2018.e00646

Hedstrom L (2002) Serine protease mechanism and specificity. Chem Rev 102(12):4501–4524. https://doi.org/10.1021/cr010033m

Ida ÉL, da Silva RR, de Oliveira TB, Souto TB, Leite JA, Rodrigues A, Cabral H (2017) Biochemical properties and evaluation of washing performance in commercial detergent compatibility of two collagenolytic serine peptidases secreted by *Aspergillus fischeri* and *Penicillium citrinum*. Prep Biochem Biotechnol 47(3):282–290. https://doi.org/10.1080/10826068.2016.1224245

Ismail B, Mohammed H, Nair J (2019) Influence of proteases on functional properties of food. In: Parameswaran B, Varjani S, Raveendran S (eds) Green bio-processes, energy, environment, and sustainability. Springer, pp 31–54. https://doi.org/10.1007/978-981-13-3263-0_3

Jensen K, Ostergaard PR, Wilting R, Lassen SF (2010) Identification and characterization of a bacterial glutamic peptidase. BMC Biochem 11:47. https://doi.org/10.1186/1471-2091-11-47

Juturu V, Wu JC (2022) Microbial enzyme production and downstream processing: current trends and future perspectives. Bioresour Technol 344:126213. https://doi.org/10.1016/j.biortech.2021.126213

Ktari N, Ben Khaled H, Nasri R, Jridi M, Nasri M (2022) Valorization of fish by-products: a review. J Aquat Food Prod Technol 31(4):459–480. https://doi.org/10.1080/10498850.2022.2066736

Lafarga T, Hayes M, Rai DK (2023) Bioactive peptides from meat and fish processing by-products: generation, purification, and applications. J Food Biochem 47(2):e15010. https://doi.org/10.1111/jfbc.15010

Li H, Zhu S, Zheng G (2020) Genome mining integrating semi-rational protein engineering and nanoreactor design: Roadmap for a robust biocatalyst for industrial resolution of Vince lactam. *Applied Microbiology and Biotechnology* 104:1109–1123. https://doi.org/10.1016/j.ijbiomac.2017.11.081

Li H, Zhu S, Zheng G (2020) Genome mining integrating semi-rational protein engineering and nanoreactor design: Roadmap for a robust biocatalyst for industrial resolution of Vince lactam. Applied Microbiology and Biotechnology, 104(3):1109–1123. https://doi.org/10.1007/s00253-019-10300-y

Li S, Chen G (2023) Metalloproteases: structure, function, and industrial applications. Appl Microbiol Biotechnol 107:1509–1525. https://doi.org/10.1007/s00253-023-12380-z

Limkar MB, Pawar SV, Rathod VK (2019) Statistical optimization of xylanase and alkaline protease co-production by *Bacillus sp.* using Box-Behnken Design under submerged fermentation using wheat bran as a substrate. Biocatal Agric Biotechnol 17:455–464. https://doi.org/10.1016/j.bcab.2018.12.018

Liu Y, Zhang H, Wang L (2023) *Bacillus* proteases: production, characterization, and applications. Enzym Microb Technol 162:110133. https://doi.org/10.1016/j.enzmictec.2022.110133

Mamo J, Assefa F (2018) The role of microbial aspartic protease enzyme in food and beverage industries. Journal of Food Quality, 2018:7957269. https://doi.org/10.1155/2018/7957269

Mamo J, Assefa F (2021) Microbial aspartic proteases and their applications in the food industry. J Food Sci Technol 58:3253–3263. https://doi.org/10.1007/s13197-021-05020-9

Manan MA, Webb C (2021) Solid-state fermentation of agro-industrial wastes for the production of enzymes: A review. Bioresour Technol Rep 15:100745. https://doi.org/10.1016/j.biteb.2021.100745

Mañas P, Pagán R (2023) Hurdle technology for food preservation. In: Food preservation techniques. Woodhead Publishing, pp 25–44. https://doi.org/10.1016/B978-0-12-822989-7.00002-9

Mandal S, Banerjee D (2019) Proteases from endophytic fungi with potential industrial applications. In: Yadav A, Mishra S, Singh S, Gupta A (eds) Recent advancement in white biotechnology through fungi: volume 2: perspective for value-added products and environments. Springer, pp 421–439. https://doi.org/10.1007/978-3-030-14846-1_15

Masi C, Chandramohan C, Fazil AM (2017) Immobilization of the magnetic nanoparticles with alkaline protease enzyme produced by *Enterococcus hirae* and *Pseudomonas aeruginosa* isolated from dairy effluents. Braz Arch Biol Technol 60:e17160572. https://doi.org/10.1590/1678-4324-2017160572

Mehta PK, Sehgal S (2019) Microbial enzymes in food processing. In: Husain Q, Ullah MF (eds) Biocatalysis enzymatic basics and applications. Springer, pp 255–276. https://doi.org/10.1007/978-3-030-25023-2_13

Meshram V, Saxena S (2016) Potential fibrinolytic activity of an endophytic Lasiodiplodia pseudotheobromae species. 3 Biotech 6:114. https://doi.org/10.1007/s13205-016-0427-5

Meshram V, Saxena S, Paul K (2016) Xylarinase: a novel clot busting enzyme from an endophytic fungus *Xylaria curta*. J Enzyme Inhib Med Chem 31(6):1502–1511. https://doi.org/10.3109/14756366.2015.1110837

Mishra SS, Ray RC, Rosell CM, Panda D (2017) Microbial enzymes in food applications. History of progress. In: Ramesh CM, Rosell CM (eds) Microbial enzyme technology in food applications. CRC Press, pp 3–20. https://doi.org/10.1201/9781315368405-1

Mothe T, Sultanpuram VR (2016) Production, purification and characterization of a thermotolerant alkaline serine protease from a novel species *Bacillus caseinilyticus*. 3 Biotech 6:53. https://doi.org/10.1007/s13205-016-0374-1

Nadeem F, Mehmood T, Naveed M (2019) Protease production from *Cheotomium globusum* through central composite design using agricultural wastes and its immobilization for industrial exploitation. Waste Biomass Valoriz. https://doi.org/10.1007/s12649-019-00578-0

Nair A, Mallya R, Sreeram A (2023) Downstream processing of industrial enzymes: strategies and challenges. Prep Biochem Biotechnol 53(1):1–15. https://doi.org/10.1080/10826068.2022.2058461

Naveed M, Nadeem F, Mehmood T, Bilal M, Anwar Z, Amjad F (2021) Protease—A versatile and eco-friendly biocatalyst with multi-industrial applications: an updated review. Catal Lett 151: 307–323. https://doi.org/10.1007/s10562-020-03316-7

Niyonzima FN (2021) Microbial proteases in industrial biotechnology. In: Industrial biotechnology. Wiley, pp 1–25. https://doi.org/10.1002/9783527825462.ch1

Pathak AP, Rathod MG (2018) A review on alkaline protease producers and their biotechnological perspectives. Indian Journal of Geo-Marine Sciences, 47:1113–1119

Pérez-Rodríguez N, García MR, Grau R (2023) Bacteriocins as food preservatives: progress and challenges. Foods 12(5):1079. https://doi.org/10.3390/foods12051079

Pessoa TB, Rezende RP, Marques EdLS, Pirovani CP, Dos Santos TF, dos Santos Goncalves AC, Romano CC, Dotivo NC, Freitas AC, Salay LC (2017) Metagenomic alkaline protease from mangrove sediment. J Basic Microbiol 57:962–973. https://doi.org/10.1002/jobm.201700150

Prayogo FA, Budiharjo A, Kusumaningrum HP, Wijanarka W, Suprihadi A, Nurhayati N (2020) Metagenomic applications in exploration and development of novel enzymes from nature: A review. Journal of Genetic Engineering and Biotechnology, 18:1–10. https://doi.org/10.1186/s43141-020-00043-9

Putatunda C, Kundu BS, Bhatia R (2019) Purification and characterization of alkaline protease from *Bacillus sp*. HD292. Proc Natl Acad Sci, India, Sect B Biol Sci 89:957–965. https://doi.org/10.1007/s40011-018-1015-8

Putranto WS, Maggy TS, Harsi DK, Puspo EG, Apon ZM (2020a) A novel rennin like protease from *Lactobacillus plantarum* 1.13 isolated from Indonesian fermented meat (Bakasam). Biocatal Agric Biotechnol. https://doi.org/10.1016/j.bcab.2020.101753

Putranto WS, Mustopa AZ, Arizah K, Prastyowati A (2020b) The purification of rennin-like protease from *Lactobacillus paracasei* isolated from ettawa goat milk. Annales Bogorienses 24(2):74–80. https://doi.org/10.14203/ann.bogor.2020.v24.n2.74-80

Qureshi AS, Khushk I, Ali CH, Chisti Y, Ahmad A, Majeed H (2016) Coproduction of protease and amylase by thermophilic *Bacillus sp*. BBXS-2 using open solid-state fermentation of lignocellulosic biomass. Biocatal Agric Biotechnol 8:146–151. https://doi.org/10.1016/j.bcab.2016.09.006

Qureshi MI, Aslam M, Masood MH (2022) Microbial rennet: a comprehensive review on cheese making. Int J Dairy Sci 17(1):1–12. https://doi.org/10.3923/ijds.2022.1.12

Ramakodi MP, Santhosh N, Pragadeesh T, Mohan SV, Basha S (2020) Production of protease enzyme from slaughterhouse effluent: an approach to generate value-added products from waste. Bioresour Technol Rep 12:100552. https://doi.org/10.1016/j.biteb.2020.100552

Raveendran S, Parameswaran B, Pandey A (2023) Yeast proteases and their potential applications. In: Yeast biotechnology. Springer, pp 189–210. https://doi.org/10.1007/978-981-19-9191-2_7

Ravindran R, Jaiswal AK (2023) Exploitation of food industry waste for high-value products. Trends Biotechnol 41(2):150–163. https://doi.org/10.1016/j.tibtech.2022.07.003

Razzaq A, Shamsi S, Ali A, Ali Q, Sajjad M, Malik A, Ashraf M (2019) Microbial proteases applications. Front Bioeng Biotechnol 7:110. https://doi.org/10.3389/fbioe.2019.00110

Rehman R, Ahmed M, Siddique A, Hasan F, Hameed A, Jamal A (2017) Catalytic role of thermostable metalloproteases from *Bacillus subtilis* KT004404 as dehairing and destaining agent. Appl Biochem Biotechnol 181:434–450. https://doi.org/10.1007/s12010-016-2221-6

Renzetti S (2023) The role of enzymes in dough conditioning. Curr Opin Food Sci 49:100964. https://doi.org/10.1016/j.cofs.2022.100964

Roy S (2021) Cysteine proteases: structure, function, and inhibition. Biomolecules 11(2):255. https://doi.org/10.3390/biom11020255

Sarmadi BH, Ismail A (2023) Antioxidant peptides: a review. Food Chem 405:134783. https://doi.org/10.1016/j.foodchem.2022.134783

Sávago-Ayerdi SG, Görgens JF (2023) Exopeptidases in food protein hydrolysis: a review. Food Bioprocess Technol 16:709–728. https://doi.org/10.1007/s11947-022-02964-4

Sazonova S, Grube M, Galoburda R (2023) Application of proteases for meat tenderization: a review. Foods 12(8):1492. https://doi.org/10.3390/foods12081492

Sgarbi E, Bottari E, Lazzi C (2023) Acceleration of cheese ripening: a review of traditional and novel methods. Int Dairy J 137:105517. https://doi.org/10.1016/j.idairyj.2022.105517

Sharma A, Gupta SP (2017) Fundamentals of viruses and their proteases. In: Gupta SP (ed) Viral proteases and their inhibitors. Academic Press. https://doi.org/10.1016/B978-0-12-809712-0.00001-8

Singh S, Bajaj BK (2017) Potential application spectrum of microbial proteases for clean and green industrial production. Energy Ecol Environ 2(6):370–386. https://doi.org/10.1007/s40974-017-0075-6

Singh R, Kumar M, Mittal A (2023) Microbial enzyme production by submerged fermentation: a review. Bioresour Technol 367:128232. https://doi.org/10.1016/j.biortech.2022.128232

Soares da Silva O, Lira de Oliveira R, de Carvalho Silva J, Converti A, Souza Porto T (2018) Thermodynamic investigation of an alkaline protease from *Aspergillus tamarii* URM4634: A comparative approach between crude extract and purified enzyme. International Journal of Biological Macromolecules, 109:1039–1044. https://doi.org/10.1016/j.ijbiomac.2017.11.081

Song P, Cheng L, Tian K et al (2020) Biochemical characterization of two new *Aspergillus niger* aspartic proteases. 3 Biotech 10:303. https://doi.org/10.1007/s13205-020-02292-4

Souza PM, Werneck G, Aliakbarian B, Siqueira F, Filho EXF, Perego P, Converti A, Magalhaes PO, Pessoa Junior A (2017) Production, purification and characterization of an aspartic protease from *Aspergillus foetidus*. Food Chem Toxicol 109:1103–1110. https://doi.org/10.1016/j.fct.2017.03.061

Steiner E, Becker T, Gastl M (2023) Beer haze formation: a review. J Inst Brew 129(1):2–15. https://doi.org/10.58430/jib.v129i1.24

Sun Q, Chen F, Geng F, Luo Y, Gong S, Jiang Z (2018) A novel aspartic protease from *Rhizomucor miehei* expressed in *Pichia pastoris* and its application on meat tenderization and preparation of turtle peptides. Food Chem 245:570–577. https://doi.org/10.1016/j.foodchem.2017.10.087

Sun B, Zou K, Zhao Y, Tang Y, Zhang F, Chen W, Tang X, Chang C, Zheng Y (2023) The fermentation optimization for alkaline protease production by Bacillus subtilis BS-QR-052. Frontiers in Microbiology, 14:1301065. https://doi.org/10.3389/fmicb.2023.1301065

Sun Y, Li X, Ma H (2023) Effects of protease on bread quality: a review. J Cereal Sci 109:103601. https://doi.org/10.1016/j.jcs.2022.103601

Tavano OL (2013) Protein hydrolysis using proteases: an important tool for food biotechnology. J Mol Catal B Enzym 90:1–11. https://doi.org/10.1016/j.molcatb.2013.01.011

Tavano OL (2023) Protein hydrolysis for the production of bioactive peptides: a review. Food Res Int 163:112233. https://doi.org/10.1016/j.foodres.2022.112233

Thakur A, Putatunda C, Sharma R, Mehta R, Solanki P, Bhatia K (2020) Innovative techniques for improving microbial enzyme production. In: Sharma SK, Singh BP, Gupta AK (eds) Microbial diversity, interventions and scope. Springer, pp 157–184. https://doi.org/10.1007/978-981-15-4099-8_10

Thomas L, Larroche C, Pandey A (2023) Current developments in solid-state fermentation. Biochem Eng J 181:108400. https://doi.org/10.1016/j.bej.2022.108400

Tkaczewska J (2023) Peptides from food proteins as natural preservatives: a review. Foods 12(4): 874. https://doi.org/10.3390/foods12040874

Udenigwe CC, Fogliano V (2023) Food protein-derived bioactive peptides: a review of their production, bioactivity, and application in food and health. Annu Rev Food Sci Technol 14: 135–158. https://doi.org/10.1146/annurev-food-060721-024641

Varzakas T, Tzia C (2023) Enzyme applications in juice clarification. In: Food engineering handbook. CRC Press, pp 345–368. https://doi.org/10.1201/9780429190126-15

Verma SK, Kaur S, Tevetia A, Chatterjee S, Sharma PC (2021) Structural characterization and functional annotation of microbial proteases mined from solid tannery waste metagenome. Biologia 76:1829–1842. https://doi.org/10.1007/s11756-021-00777-y

Vijayaraghavan P, Lazarus S, Vincent SGP (2014) De-hairing protease production by an isolated Bacillus cereus strain AT under solid-state fermentation using cow dung: Biosynthesis and properties. Saudi Journal of Biological Sciences, 21:27–34. https://doi.org/10.1016/j.sjbs.2013.04.010

Wang Y, Li C, Liu Y (2023) Microbial proteases in the fermentation of soy sauce: a review. J Sci Food Agric 103(1):15–25. https://doi.org/10.1002/jsfa.12144

Waschkowitz T, Rockstroh S, Daniel R (2009) Isolation and characterization of metalloproteases with a novel domain structure by construction and screening of metagenomic libraries. Appl Environ Microbiol 75(8):2506–2516. https://doi.org/10.1128/AEM.02166-08

Wróblewska B, Kaliszewska-Suchodoła A, Wasilewska E (2023) Enzymatic reduction of milk protein allergenicity: a review. Nutrients 15(6):1385. https://doi.org/10.3390/nu15061385

Xu J, Wang Z, Li Y (2023) Mechanisms of bread staling and anti-staling strategies: a review. Food Chem 405:134829. https://doi.org/10.1016/j.foodchem.2022.134829

Zamora-Sillero J, Gharsallaoui A, Prentice C (2023) Enzymatic processing of seafood: a review. Foods 12(5):1034. https://doi.org/10.3390/foods12051034

Zhang K, Duan X, Wu J (2016) Multigene disruption in undomesticated Bacillus subtilis ATCC 6051a using the CRISPR/Cas9 system. Scientific Reports, 6:27943. https://doi.org/10.1038/srep27943

Zhou C, Zhou H, Li D, Zhang H, Wang H, Lu F (2020) Optimized expression and enhanced production of alkaline protease by genetically modified Bacillus licheniformis 2709. Microb Cell Factories 19:45. https://doi.org/10.1186/s12934-020-01309-0

Chapter 18
Future Research Directions: Sustainable Production, Scalability, and Safety Considerations

P. Saranraj, K. Gayathri, Marius K. Somda, M. Manigandan, Mustafa Salah Hasan, and L. Charlie Jelura

Abstract The application of proteases across the food, detergent, pharmaceutical, and waste management industries necessitates a focused examination of their future research trajectories, centered on sustainable production, scalability, and safety. While microbial proteases dominate the industrial landscape, future efforts must prioritize the transition to sustainable feedstocks and advanced fermentation technologies to reduce the environmental footprint of their production. This includes leveraging metabolic engineering and systems biology to design robust microbial cell factories that utilize lignocellulosic or agro-industrial waste streams, thereby aligning enzyme manufacturing with circular economy principles. Concurrently, addressing the challenge of scalability is paramount. Research must bridge the gap between laboratory yields and industrial-scale production by optimizing bioreactor design and downstream processing to ensure economic viability without compromising enzyme efficacy or stability. Furthermore, comprehensive safety considerations must be integral to protease development. This encompasses rigorous toxicological assessments, the establishment of standardized regulatory frameworks for novel enzyme products, and the implementation of engineering controls to mitigate occupational exposure risks, particularly for high-activity formulations like serine proteases. Ultimately, the successful integration of these three pillars, viz., sustainability, scalability, and safety, will be critical for advancing the next generation of protease enzymes, ensuring they are not only high-performing and

P. Saranraj (✉) · K. Gayathri · M. Manigandan · L. C. Jelura
PG and Research Department of Microbiology, Sacred Heart College (Autonomous), Tirupattur, Tamil Nadu, India

M. K. Somda
Laboratory of Biochemistry, Biotechnology, Food Technology and Nutrition (LABIOTAN), Department of Biochemistry and Microbiology, University Joseph KI-ZERBO, Ouagadougou, Burkina Faso

M. S. Hasan
College of Veterinary Medicine, University of Fallujah, Fallujah, Iraq

V. Zambare et al. (eds.), *Biotechnological Advances and Applications in Microbial Proteases*, Interdisciplinary Biotechnological Advances,
https://doi.org/10.1007/978-981-95-9011-7_18

commercially feasible but also environmentally responsible and safe for both workers and consumers.

Keywords Sustainable production · Scalability · Enzyme safety · Fermentation optimization and industrial proteases

18.1 Introduction

Protease enzymes, also known as peptidases or proteolytic enzymes, are a group of biocatalysts that break the peptide bonds that hold amino acids together in proteins. Proteolysis is not just a destructive process; it is a highly controlled action that is necessary for life. It supports important bodily functions such as digestion, blood clotting, antigen presentation, cell signaling, and apoptosis (programmed cell death) (Lopez Otin and Bond 2008). Proteases are very important in Biology, and they are also quite useful in industry, where they make up over 60% of the global enzyme market (Razzaq et al. 2019). This shows how flexible and effective they are.

Proteases are the most important enzymes in industry because they can catalyze reactions at low temperatures and pH levels, which is a more environmentally friendly way to do things than using harsh chemicals. They can be used in a lot of different ways. Alkaline proteases are an important part of detergents. They break down protein-based stains, including blood, egg, and grass (Sundararajan et al. 2011). In food and beverage processing, they are used for diversifying product textures, tenderizing meat, producing protein hydrolysates, coagulating milk in cheese manufacturing (e.g., chymosin), and clarifying beer to prevent haze formation (Naveed et al. 2021). Furthermore, proteases hold significant value in leather processing for dehairing, in the pharmaceutical industry for developing therapeutics (e.g., thrombolytic agents), and in waste management for the breakdown of organic matter.

The escalating global demand for these enzymes, propelled by the transition toward green technologies and a circular bioeconomy, brings challenges to the forefront. Conventional production methods, often dependent on wild-type microbial strains and resource-intensive submerged fermentation, face limitations in yield, sustainability, and cost-effectiveness (Sawant and Nagendran 2014). Scaling up laboratory processes to industrial volumes presents significant hurdles in bioprocess engineering, including oxygen transfer, metabolic heat generation, and downstream purification. Moreover, the safe handling of enzyme powders, potential worker allergenicity, and the environmental impact of production waste streams necessitate rigorous safety and regulatory frameworks (Singh et al. 2016). The future of protease applications is inherently connected to advancements that tackle the triad of challenges: sustainable production, scalable manufacturing, and thorough safety assurance. To get over these problems, researchers need to work together on a strategy that combines metagenomics, synthetic biology, sophisticated fermentation technologies, and green chemistry principles to make the next generation of effective and eco-friendly enzymatic solutions (Soccol et al. 2017).

A way to make protease manufacturing more sustainable is to use agricultural and industrial waste streams as the main fermentation feedstocks. This turns low-value by-products into high-value enzymes and directly supports the circular bioeconomy. This technique solves two big problems at once; it lowers the high cost of substrates that come with traditional fermentation media and it makes waste disposal less harmful to the environment. A lot of research is being done to find, describe, and pretreat different types of lignocellulosic and agro-industrial waste, like sugarcane bagasse, wheat bran, oilseed cakes, and fruit processing waste, to free up fermentable sugars, nitrogen, and other nutrients that are important for microbial growth and protease synthesis (Sirohi et al. 2020). Solid state fermentation (SSF) is a technology that works well with these solid substrates because it mimics the natural habitat of many prolific enzyme-producing fungi and bacteria. It often produces higher enzyme titers than submerged fermentation while using less water and energy (Thomas et al. 2013). Nonetheless, the heterogeneous and frequently recalcitrant characteristics of these wastes require customized pretreatment methods (e.g., physicochemical or enzymatic hydrolysis) and possibly the employment of engineered microbial consortia or substrate-adapted strains capable of secreting the requisite hydrolytic enzymes to decompose complex polymers. In addition to carbon sources, researchers are looking into other waste streams, such as cheese whey, which is a by-product of the dairy industry that is high in lactose and proteins, as full nutrient supplies for fermentation. This would further close the loop in food production chains (Panesar et al. 2014). The ultimate success of this approach depends on integrated biorefining concepts. One useful output of this process is protease production, which is part of a series of steps that turn waste into multiple products. This makes the whole process more economically viable and greatly reduces the environmental impact of industrial enzymology. The complete incorporation of protease production into a circular bioeconomy framework epitomizes the zenith of sustainable industrial biotechnology, transcending discrete process enhancements to establish a comprehensive, closed-loop system that eradicates waste and optimizes resource efficiency. This change in thinking calls for a holistic strategy in which protease manufacturing is deliberately integrated into a network of other businesses that use their waste as inputs and give valuable by-products in return. A vital research avenue entails executing comprehensive life cycle assessments (LCA) to quantitatively assess the environmental impact including greenhouse gas emissions, water usage, and energy consumption of these integrated systems, thereby providing validated data to substantiate their superiority over linear production models (Ubando et al. 2020).

For example, a biorefinery model can be used where lignocellulosic waste from farming (like straw and bagasse) is first processed to get high-value compounds. The remaining lignin and carbohydrate streams are then used as the main source of carbon and energy for microorganisms that make protease (Sirohi et al. 2020). Also, the microbial biomass itself, after fermentation and enzyme extraction, is a protein-rich residue that may be used as a nutritious supplement in animal feed or as an organic fertilizer. This not only prevents waste, but it also creates a new source of income (Koutinas et al. 2014). Water integration is also important. Researchers are working on creating closed-loop water systems in the production plant and using non-potable water sources, including pretreated wastewater or process condensate, for fermentation and cooling. This interconnected model not only makes protease production more economically viable by lowering the costs of raw materials and waste management, but it also encourages industrial symbiosis, where the waste from one process becomes the treasure for another. This leads to a regenerative economic system that separates enzyme production from the use of limited virgin resources.

18.2 Sustainable Production: Moving Beyond Conventional Systems

18.2.1 Advanced Microbial Cell Factories

The future of sustainable protease production is closely tied to the strategic growth of advanced microbial cell factories. This means redesigning microorganisms into highly efficient, specialized systems for biosynthesis. This paradigm transcends conventional fermentation by employing advanced synthetic biology and metabolic engineering to develop customized microbial chassis that optimize productivity, reduce resource consumption, and exploit low-value waste streams. A principal area of study concentration is systems-level metabolic engineering, utilizing techniques like CRISPR-Cas9 to execute precise genetic modifications that reconfigure central carbon metabolism. This method maximizes the flow of carbon toward protease production while blocking other pathways that would compete with it. This greatly increases productivity and cuts down on energy waste (Nielsen and Keasling 2016).

There is also a growing interest in creating nontraditional hosts, such as halophilic organisms and oleaginous yeasts like *Yarrowia lipolytica*. These organisms have unique benefits, such as the ability to ferment in high-salinity environments using seawater or saline waste, and the ability to break down hydrophobic substrates like waste oils and glycerol (Ledesma-Amaro et al. 2016). Another new idea is to make smart regulatory circuits that link the expression of proteases to stress reactions or the availability of nutrients. This would allow for auto-induction without the need for expensive chemical inducers. For instance, creating promoters that are turned on by substrate limits or certain waste products in the growth medium can make bioprocess control much easier and more cost-effective (Zhang and Wang 2016). These combined tactics not only make it easier to scale up and keep enzyme production going, but they also help the goals of the circular bioeconomy by turning waste from farming and industry into valuable products. The ongoing advancement of these microbial platforms has the potential to transform industrial enzymology, enhancing protease production to be more flexible, efficient, and environmentally friendly (Zhao et al. 2022).

18.2.1.1 High-Throughput Fermentation Screening

High-throughput fermentation screening is a revolutionary method for developing bioprocesses. It lets you quickly and simultaneously tests thousands of microbial strains and growth conditions to find the best protease producers. This approach uses small bioreactor systems including microtiter plates, bubble columns, and micro-bioreactors that work at microliter to milliliter scales and copy the conditions found in larger vessels. Researchers can get real-time, high-resolution data on important process parameters and metabolic reactions by using modern online monitoring technologies, such as optical sensors for pH, dissolved oxygen, and biomass (Bareither and Pollard 2019). Automation and robotics make this method even more powerful by making it easier to handle, inoculate, and sample liquids accurately. This reduces the chance of human error and makes it easier to repeat experiments. Machine learning algorithms are used to look at the huge datasets and find complicated links between genetic changes, media composition, and protease yield. This helps with rational strain engineering and process optimization (Ladner et al. 2021). For protease development, this allows for the swift screening of engineered libraries for enhanced secretion, stability under industrial conditions, or performance on novel waste-derived substrates. By compressing months of traditional bench-scale experimentation into weeks, high-throughput screening dramatically accelerates the design-build-test-learn cycle, de-risks scale-up, and paves the way for more efficient and sustainable enzyme production pipelines. This paradigm is essential for meeting the growing demand for industrial enzymes while adhering to the principles of circular bioeconomy (Chandrasekhar et al. 2024).

18.2.1.2 Dynamic Process Control

Dynamic process control represents a paradigm shift in industrial protease production, moving from static set-point operation to adaptive, data-driven strategies that optimize the fermentation in real time. This approach is critical because microbial metabolism is inherently dynamic; substrate concentrations, nutrient levels, and byproduct accumulation constantly shift, creating a nonideal environment that can limit yield and productivity. Advanced sensor technology, including in situ Raman spectroscopy and soft sensors, provides a continuous stream of multiparameter data on critical variables such as viable cell density, substrate concentration (e.g., glycerol, glucose), and even product titer (Glassey and Montague 2018). This real-time data goes into a control system that uses machine learning or mechanistic models to change critical process parameters in real time. For example, the system can use an adaptive feeding strategy to keep the concentrations of carbon and nitrogen at the right amounts by accurately adjusting the feed rates. This way, it can avoid both substrate inhibition and starvation. It may also change the aeration and agitation to keep the dissolved oxygen level within a limited range, which stops oxygen limitation, which is a typical problem in high-density protease fermentations, and reduces shear stress, which can hurt cells and lower productivity (Craven et al. 2014). This closed-loop feedback control makes sure that the microbial culture works at its best throughout the fermentation process. This leads to big increases in volumetric output, better consistency from batch to batch, and less waste of raw materials. Dynamic process control is an important technique for increasing protease production, making it easier to scale up, and making it economically viable for large-scale industrial use. This is because it can respond intelligently to the physiological state of the culture.

18.2.1.3 Continuous Fermentation

Continuous fermentation is a revolutionary bioprocessing method for making industrial proteases. It has many benefits over standard batch and fed-batch systems since it keeps microorganisms in a stable state of exponential growth for long periods of time. In this method, fresh feeding medium is constantly added to the bioreactor while an equivalent amount of culture broth, which has cells, metabolic products, and the target protease in it, is taken out at the same time. This method gets rid of the long periods of time when the facility is not working because of cleaning, sterilization, and inoculation between batches. This greatly increases the volumetric productivity and overall capacity of the facility (Langer and Rader 2020). For protease synthesis, continuous operation keeps the right environmental conditions like pH, dissolved oxygen, and substrate concentration so that the product quality and enzyme yield are always the same. This stops nutrient depletion and waste buildup in batch systems. The implementation of advanced process control technologies, including real-time monitoring of critical parameters and automated feedback systems, is essential for stabilizing the continuous process and preventing washout or genetic drift in the microbial population (Glassey and Montague 2018). Furthermore, continuous fermentation facilitates more efficient substrate utilization and can be directly integrated with downstream purification units in a connected bioprocessing framework, reducing hold times and preserving enzyme activity. While challenges remain in maintaining long-term sterility, strain stability, and regulatory compliance for certain applications, the economic and operational benefits of continuous fermentation make it a cornerstone of next-generation biomanufacturing for sustainable, high-volume enzyme production.

18.2.2 Scalability: Bridging the Lab-to-Industry Gap

Bridging the lab-to-industry gap in protease production necessitates overcoming profound scalability challenges that arise from the radical differences in physical and biological conditions between small-scale reactors and industrial fermenters. Laboratory trials usually get the best results when the circumstances are very regulated and uniform. On the other hand, bioreactors on an industrial scale have problems such as nutritional gradients, dissolved oxygen limits, pH changes, and not heat dissipation because of not enough mixing (Noorman 2019). These unevennesses can put stress on microbial cultures, lower protease output, and change the quality of enzymes. To fix these differences, innovative scale-up solutions combine technical ideas with biological knowledge. Computational fluid dynamics (CFD) modeling is being used more and more to analyze flow patterns, shear forces, and mass transfer dynamics in big tanks. This helps design bioreactors and impeller systems that reduce gradients and increase productivity (Noorman 2019). Additionally, the use of scale-down models, which are small-scale systems that mimic the varied conditions of production tanks, enables predictive troubleshooting and optimization of strains and processes under simulated industrial conditions prior to undertaking expensive large-scale operations (Lara 2022).

Real-time monitoring and control of important parameters are made possible by process analytical technology (PAT) and soft sensors. This ensures consistency and allows for quick remedial measures during scale-up. For the sake of the economy and the environment, it is also necessary to substitute expensive purified substrates with complex, waste-derived feedstocks early in the process development stage to prevent having to recalibrate later. Ultimately, successful translation necessitates a multidisciplinary strategy that integrates bioprocess engineering, microbial physiology, and economics from the beginning, guaranteeing that laboratory advances are both technically and financially viable at scale.

18.2.2.1 Bioprocess Engineering and Intensification

The main problem is how to turn laboratory-scale protease synthesis into strong, affordable industrial production. This requires new ideas in bioprocess engineering and intensification. This field is focused on making bioreactor systems more productive, yielding, and efficient while also cutting down on the process's physical space, resource utilization, and operating costs. One important area of research is the use of enhanced process control systems that use high-tech online sensors, such as Raman spectroscopy and capacitance probes, to keep an eye on important factors like substrate concentration, biomass, and product creation in real time. This data stream works with machine learning algorithms to make dynamic feedback control possible. This lets nutrient feeds, agitation, and aeration be automatically adjusted to keep the culture in its best physiological state during fermentation, which maximizes protease titers (Glassey and Montague 2018). Additionally, there is a strong drive to move away from traditional batch procedures and toward continuous fermentation techniques. Continuous procedures, such as chemostat and perfusion systems, provide significant benefits by sustaining microorganisms in a stable state of exponential growth for prolonged durations. This gets rid of the time spent cleaning and re-sterilizing, boosts volumetric output, and makes a more consistent product, which is important for satisfying the needs of large-scale industry (Langer and Rader 2020). Process intensification also applies to downstream activities. For example, integrating initial purification steps into the fermentation vessel, like extracellular secretion designed into the host organism or continuous centrifugation, can make the recovery process much easier. Bioprocess engineering and intensification strategies are crucial for closing the significant gap between a promising scientific discovery and its practical use as a commercial enzymatic product (Liu et al. 2021) by taking into account the whole production pipeline.

18.2.3 Downstream Processing (DSP) Innovations

Downstream processing (DSP) represents a critical and often costly phase in protease production, encompassing the purification, concentration, and stabilization of the target enzyme from the complex fermentation broth (Table 18.1). Innovations in DSP are therefore essential for improving overall process economics, sustainability, and final product quality. A primary research focus is the development of highly selective, low-energy initial recovery steps to replace traditional methods like ammonium sulfate precipitation, which generates significant salt waste. Aqueous two-phase systems (ATPS), particularly polymer-salt or ionic liquid-based systems, offer a promising alternative by providing a gentle and efficient means for the primary separation and partial purification of proteases based on their surface properties, often achieving high yield and selectivity in a single step (Benavides and Rito-Palomares 2018).

Membrane filtration technology, including ultrafiltration and microfiltration, is being advanced not merely for concentration and buffer exchange but as an integrated part of continuous purification trains. The development of novel, fouling-resistant membrane materials and the strategic design of multistage membrane cascades can significantly enhance separation efficiency and reduce energy demands (Causserand and Aimar 2020). For final polishing, membrane chromatography is emerging as a superior alternative to traditional resin-based Column chromatography. By binding ligands to high-surface-area membranes, this technology alleviates the diffusion limitations of porous beads, enabling faster flow rates, higher throughput, and a more seamless integration into continuous biomanufacturing platforms, which is crucial for handling large volumes of industrial enzymes (Teepakorn et al. 2021). The goal of modern DSP is to move toward fully continuous, integrated operations that minimize hold times, reduce product degradation, and dramatically lower the consumption of water and chemicals, thereby enhancing the sustainability profile of the entire protease production process.

18.2.3.1 Low-Energy Purification

The quest for low-energy purifying methods is an important area of research in protease manufacturing. The goal is to cut down on the high costs and environmental impact of traditional downstream processing (DSP). Multistep centrifugation, salt-induced precipitation, and column chromatography are all well-known for using a lot

Table 18.1 Downstream processing (DSP) innovations for protease production

DSP stage	Innovation	Advantage	Challenges	References
Primary recovery	Advanced aqueous two-phase systems (ATPS)	Uses polymer-salt or ionic liquid systems to partition proteases into one phase and impurities into another. Gentle on enzymes, offers high selectivity, and can be continuous	Reducing cost of phase-forming polymers. Recycling and reusing phase components. Scaling up continuous ATPS	Benavides and Rito-Palomares (2018)
	Flocculation and precipitation	Use of smart polymers (e.g., stimulus-responsive) or organic solvents to aggregate and precipitate proteases. Effective volume reduction and initial purification	Difficulty in precipitating specific enzymes selectively. Removing precipitating agents completely in later steps	Singh and Bajaj (2017)
Concentration and initial purification	Advanced membrane filtration	Use of ultrafiltration (UF) with fouling-resistant membranes (e.g., surface-modified, ceramic) for concentration and diafiltration. High throughput, scalable.	Managing membrane fouling to maintain flux. Developing cascades for high-resolution separations	Causserand and Aimar (2020)
	Three-phase partitioning (TPP)	Single-step technique combining salt (e.g., ammonium sulfate) and organic solvent (e.g., t-butanol) to precipitate proteins at the interface. Simple, rapid, and efficient	Optimizing system for different proteases. Scaling beyond lab scale. Handling organic solvents safely	Yan et al. (2018)
Purification	Membrane chromatography	Ligands (e.g., ion-exchange) bound to microporous membranes. Convective mass transfer eliminates pore diffusion, enabling high flow	Lower binding capacity per volume compared to resins. Higher cost per unit capacity.	Teepakorn et al. (2021)

(continued)

Table 18.1 (continued)

DSP stage	Innovation	Advantage	Challenges	References
		rates and short processing times. Ideal for continuous DSP.		
	Affinity purification tags	Genetically fusing a tag (e.g., his-tag, STREP-tag) to the protease allows highly specific purification using immobilized metal or ligand affinity chromatography. Extremely high purity in one step	Requires genetic modification. Cost of affinity resins. Need for tag removal, adding steps	Terpe (2003)
Finishing and formulation	Spray drying with protectants	Low-cost method to produce stable dry powders. Use of carbohydrate protectants (e.g., maltodextrin) to shield enzymes from thermal and dehydration stress during drying	Maintaining high activity retention after drying. Controlling particle size and dustiness	Dewettinck and Huyghebaert (2019)
	Microencapsulation	Trapping protease within a protective matrix (e.g., alginate, chitosan) to create dust-free granules, improve storage stability, and control release	Complexity and cost of encapsulation process. Achieving high enzyme loading within the capsules	Nazzaro et al. (2012)

of energy and making a lot of waste, especially in the form of spent buffers and solvents. The main goal is to improve and create aqueous two-phase systems (ATPS), which are a gentle and effective way to do primary recovery. ATPS, which are usually made of polymers and salts or ionic liquids, take advantage of the fact that proteases tend to move into one phase while contaminants stay in the other or at the interface. This allows for a lot of purification and concentration in a single, low-energy step (Benavides and Rito-Palomares 2018). At the same time, innovative membrane technologies are being improved to use less energy. Low-fouling ultrafiltration (UF) and microfiltration (MF) membranes, which are usually manufactured from new ceramic or surface-modified polymeric materials, lessen the pressure and cleaning needs, which in turn lowers energy use. Moreover, including these membranes into cascades or alongside other unit activities improves overall efficiency (Causserand and Aimar 2020). Smart polymers that respond to small changes in temperature or pH are also being used to make precipitation techniques more sustainable. This makes it easier to separate and recycle polymers, which cuts down on waste and saves energy by eliminating the need for centrifugation. The main goal of these new ideas is to make integrated, continuous DSP trains that cut down on hold times, buffer volumes, and, most importantly, energy input. This will make protease manufacturing more in line with the principles of green chemistry and make the whole process more environmentally and economically sustainable.

18.2.3.2 Secretion Engineering

Secretion engineering is a complex synthetic biology technique that seeks to genetically change microbial hosts such that they can easily move produced protease enzymes from inside the cell to the extracellular fermentation broth. This method is very useful for industrial production because it makes downstream processing much easier by getting rid of the need to break up cells, cutting down on product degradation by intracellular proteases, and reducing contamination from host cell proteins and DNA. This results in a purer product that costs less to recover. The engineering goals are complex and include making every part of the secretory process work better. First, the signal peptide, a short sequence that tells the nascent protein where to go, is very important. Researchers look through huge libraries of native and heterologous signal peptides to find the best one for a given protease and host combination (Brockmeier et al. 2019). Second, overexpressing or changing parts of the cellular translocation machinery, like the Sec or Tat pathways, can get rid of bottlenecks that cause proteins to misfold and jam, which increases the total flow of proteins being released (Zanen and Van Dijl 2020). Also, high-level secretion can overload the host's folding capacity, so molecular chaperones (like PrsA in *Bacillus* spp.) are used to help the protease fold correctly and stop it from clumping together in the extracellular space. This has a direct effect on the final yield and specific activity. For Gram-negative hosts, another issue is getting through the outer membrane. Some ways to do this include co-expressing outer membrane porins or even creating completely new secretion systems. This area of research directly addresses the economic and sustainability problems of industrial enzyme production by redesigning the host as a dedicated and efficient "secretion machine." This allows for higher titers, fewer energy-intensive purification steps, and the use of cheaper feedstocks by reducing cellular feedback inhibition.

18.2.3.3 Formulation for Stability

Formulation for stability is a critical final step in protease production, focused on preserving enzymatic activity during storage, transport, and application in often harsh industrial environments. The inherent fragility of proteins makes them susceptible to denaturation, aggregation, and catalytic inactivation caused by factors such as elevated temperature, moisture, shear forces, and oxidative damage. To mitigate these challenges, a sophisticated toolbox of formulation strategies is employed. For liquid formulations, this involves optimizing the solution conditions including buffer type, pH, and ionic strength to maintain the protease in its native conformation. The addition of stabilizers is paramount; polyols (e.g., glycerol, sorbitol) act as water substitutes and exclude solutes that promote denaturation, sugars (e.g., trehalose, sucrose) form stable glasses that immobilize the enzyme structure, and polymers (e.g., PEG) sterically hinder aggregation (Wang 2021). For dry, solid formulations, which are preferred for their superior long-term stability, techniques like spray drying and fluidized bed granulation are used. Here, the protease is mixed with carrier compounds such as maltodextrins or starches that form a protective matrix around the enzyme molecules, shielding them from thermal and mechanical stress during processing and from moisture uptake afterward (Mensink et al. 2017). A major advancement is the development of granulated or prilled forms with polymer coatings that are specifically engineered to be low-dust, enhancing worker safety by reducing airborne allergen exposure, and to control the release of the enzyme only upon dissolution in the application medium. Furthermore, additives like antioxidants (e.g., ascorbic acid) and chelating agents (e.g., EDTA) are incorporated to protect against chemical degradation pathways. The ultimate goal of these formulation efforts is to deliver a robust, consistent, and safe product that retains its high catalytic activity from the point of manufacture until its final use, thereby ensuring efficacy for the customer and economic viability for the producer (Li et al. 2023).

18.3 Scalability Challenges and Innovations in Protease Production

The transition of protease production from laboratory scale to industrial manufacturing presents significant scalability challenges that directly impact economic viability and sustainability (Table 18.2). A primary obstacle is overcoming mass and heat transfer limitations in large-scale bioreactors, where inadequate mixing creates gradients in pH, nutrients, and dissolved oxygen, leading to suboptimal microbial performance and reduced enzyme yields (Noorman 2019). This is compounded by the inherent genetic instability of hyper-producing microbial strains over prolonged fermentation cycles, which can result in a rapid decline in productivity. To address these issues, advanced bioprocess engineering strategies are being employed. Computational fluid dynamics (CFD) modeling is used to design novel bioreactor geometries and impeller systems that ensure homogeneous conditions throughout the vessel, thereby enhancing oxygen transfer and mitigating cell stress (Noorman 2019). Furthermore, there is a paradigm shift from traditional batch operations toward continuous fermentation modes. Perfusion systems with built-in cell retention devices keep cells growing exponentially all the time, which greatly increases

Table 18.2 Scalability challenges and innovations in protease production

Scale-up consideration	Challenge at large scale	Current and emerging solutions	References
Strain performance	Genetic instability: Hyper-producing strains may lose productivity over generations	Use of stable genomic integrations over plasmids. Engineering for robust continuous processes	Rugbjerg and Sommer (2019)
Bioprocess engineering	Mass and heat transfer: Inefficient oxygen transfer and heat removal causes cell stress	Optimized impeller design. Use of computational fluid dynamics (CFD) modeling	Noorman (2019)
Process mode	Batch inefficiency: Downtime for cleaning and setup reduces productivity	Fed-batch processes. Shift toward continuous or perfusion fermentation modes	Langer and Rader (2020)
Downstream processing (DSP)	Bottleneck: Purification is costly and time-consuming; doesn't scale linearly	Scale-up of filtration. Emerging continuous downstream processing (e.g., ATPS)	Causserand and Aimar (2020)
Process control	Parameter variability: Difficulty maintaining optimal conditions in a large vessel	PID controllers. Advanced process control (APC) using soft sensors and AI	Glassey and Montague (2018)
Economic viability	Cost intensity: High capital (CAPEX) and operational (OPEX) expenditure	Optimization of media. Use of low-cost waste-based feedstocks	Humbird (2020)
Product formulation	Stability and handling: Maintaining activity during bulk storage and ensuring safety	Liquid formulations with stabilizers; granulation for solids. Advanced encapsulation	Mensink et al. (2017)

volumetric production and gets rid of the downtime that comes with cleaning and sterilizing (Langer and Rader 2020). The scaling bottleneck generally goes downstream, where the expenses of purification can make up the whole process. Innovations in this area focus on continuous downstream processing (cDSP), like continuous centrifugation and membrane chromatography. These approaches are better than batch procedures because they are more scalable, use less buffer, and take up less space (Causserand and Aimar 2020). To create cost effective and scalable protease production systems that can fulfil global demand in a sustainable way, it is important to combine these upstream and downstream technologies into a completely continuous biomanufacturing platform, led by Techno-economic analysis (TEA) (Vishwanatha et al. 2022).

18.4 Safety and Regulatory Considerations: Ensuring Responsible Use

As the production and application of protease enzymes expand, paramount importance must be placed on comprehensive safety and regulatory frameworks to ensure the well-being of workers, consumers, and the environment. This complex problem starts with workplace health, where breathing in enzyme dust is a big trigger. So, strict engineering controls and new ways of making things are really important. Granulation and coating technologies have made it possible to make low-dust, encapsulated, or liquid enzyme formulations. This has greatly lowered the amounts of airborne particles, which lowers the risk of respiratory allergic reactions among production workers (Baur 2018). In addition to occupational exposure, each new enzyme, especially one made from genetically modified microorganisms (GMMs), must go through a full safety check before it can be used. This entails establishing a history of safe usage, conducting comprehensive toxicological studies, and utilizing advanced in silico bioinformatics tools to evaluate potential allergenicity by comparing the enzyme's amino acid sequence to known allergens, as delineated in the guidelines from the Joint FAO/WHO Expert Committee on Food Additives (JECFA) and the European Food Safety Authority (EFSA) (Pariza and Johnson 2021). Regulators also need to know how these enzymes will affect the environment. They need data on how quickly they break down and how harmful they are to plants and animals to make sure that wastewater from factories doesn't harm ecosystems on land or in water. A forward-looking approach includes using built-in biocontainment methods in production strains, including auxotrophy or inducible kill switches, to stop recombinant DNA from staying in the environment and moving from one organism to another (Wright et al. 2022). In the end, manufacturers need to be proactive and open about how they handle the global regulatory landscape. This will make sure that the many benefits of protease enzymes are realized without putting safety at risk, which will build public trust and allow for long-term innovation (Contesini et al. 2023).

The multifaceted roles of protease enzymes in various biological processes (Fig. 18.1). At the center, proteases are shown as regulators of protein metabolism, with arrows extending to multiple cellular and physiological functions. These enzymes are responsible for the degradation of misfolded, damaged, and aggregated proteins, maintaining cellular health and homeostasis. They mobilize storage proteins during seed germination, facilitate cellular housekeeping by regulating protein turnover, and activate precursor proteins and peptide hormone signaling. Proteases also play critical roles in nitrogen recycling and remobilization, protein and enzyme maturation, and defense against abiotic stresses. Additionally, they are involved in abscisic acid signaling in guard cells, seed coat formation, and cuticle deposition, contributing to structural development. In plants, proteases influence the development of seedlings, regulate cell division and metabolism, and participate in stomata and plastid development. Furthermore, they are essential in senescence and programmed cell death, ensuring proper tissue differentiation and life cycle

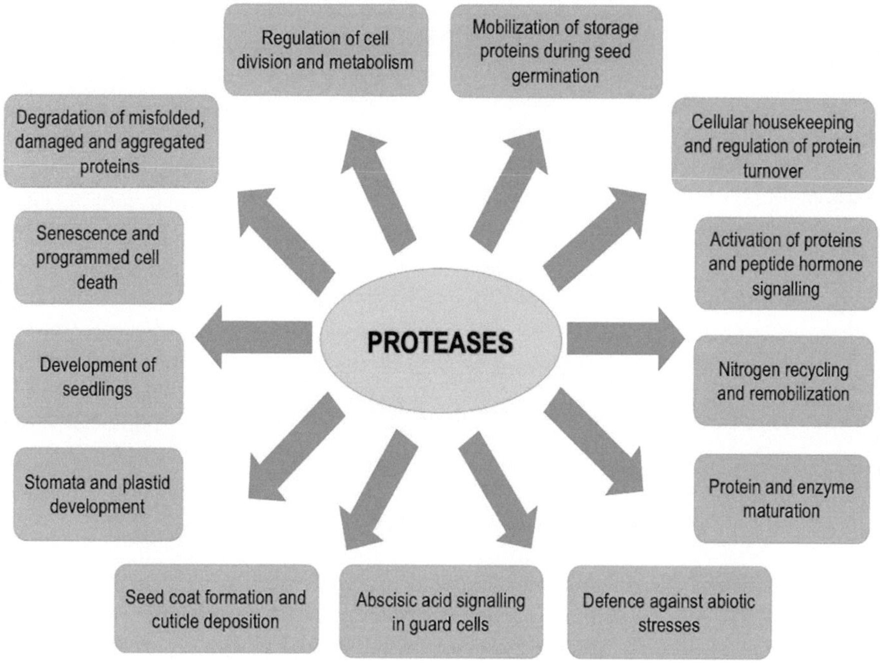

Fig. 18.1 Functional roles of protease enzymes in biological systems

progression. Overall, proteases are vital enzymes that coordinate protein degradation with cellular regulation, growth, and stress adaptation mechanisms.

18.4.1 Allergenicity and Toxicity

A fundamental aspect of the safety evaluation for every protease enzyme designated for industrial or consumer use is the thorough examination of its capacity to induce allergic responses or systemic toxicity. Inhalation of enzyme-laden dust constitutes the principal occupational hazard, potentially causing IgE-mediated respiratory sensitization, resulting in occupational asthma and rhinitis among affected workers. A part of risk management is using advanced dust-control technologies during manufacturing. For example, encapsulating enzymes into prilled or granulated formulations that greatly reduce the amounts of airborne particles that can be inhaled (Baur 2018). From a regulatory standpoint, the toxicity profile of a new enzyme must be comprehensively determined. This process usually includes a lot of in vitro and in vivo studies, such as tests for acute and sub-chronic toxicity, genotoxicity tests (like the Ames test), and tests for skin and eye irritation, as required by agencies like the European Food Safety Authority (EFSA) (Gurung et al. 2023). It is very important to have a "history of safe use" for proteases used in food processing.

However, for enzymes from new or genetically modified sources, a comparison analysis against this benchmark is needed. A crucial contemporary method for evaluating allergenicity employs advanced bioinformatics analysis, wherein the amino acid sequence of the novel enzyme is compared against databases of established allergens to detect potential sequence homology or structural similarities that may indicate cross-reactivity (Goodman et al. 2022). This integrated strategy, which includes strict industrial hygiene, thorough toxicological testing, and the latest computer predictions, is an important part of making sure that the development and use of protease enzymes do not put people's health at risk. This makes it possible to safely add them to a wide range of products.

18.4.2 Environmental Impact and Biodegradability

Protease enzymes are used in many industries, thus it's important to carefully assess their effects on the environment once they are released into the air or water through manufacturing waste or product disposal. A part of this evaluation is the fact that the enzymes can break down naturally. Proteases are generally susceptible to microbial degradation and denaturation in environmental matrices; however, their persistence and potential ecological effects can vary significantly based on their structural stability, the presence of enzyme inhibitors, and local environmental conditions such as pH, temperature, and microbial community composition. Proteases, as proteins, are generally vulnerable to microbial degradation and denaturation in environmental matrices; however, their persistence and potential ecological effects can differ markedly depending on their structural integrity, the existence of enzyme inhibitors, and specific environmental factors like pH, temperature, and microbial community composition. Regulatory frameworks, including those from the Organisation for Economic Co-operation and Development (OECD), necessitate standardized tests, such as the OECD 301B (Ready Biodegradability: CO_2 Evolution Test), to verify that enzymes decompose swiftly in aquatic environments and do not accumulate or present a long-term threat to ecosystems (Pagga 2022).

In addition to inherent biodegradability, research must examine the effects of enzyme production facilities. Fermentation waste streams contain not only leftover enzymes but also a lot of organic nutrients, salts, and cellular debris from the organisms that made them. This can change the biological oxygen demand (BOD) and chemical oxygen demand (COD) of the waters that receive them, which could cause eutrophication. To reduce the environmental impact of industrial plants, it is important to use advanced effluent treatment methods, such as membrane bioreactors that combine biological degradation with filtration (Le Clech et al. 2020). Moreover, research is progressively concentrating on the capacity of proteases to interfere with essential ecological processes, such as the breakdown of organic matter and nutrient cycling in soil and sediment, or to induce detrimental effects in nontarget aquatic animals. A thorough Environmental risk assessment (ERA) that combines information on fate, exposure, and hazard is necessary to make sure that the good things about using proteases don't come with bad effects on the environment. This will help protect their long-term use in a circular economy.

18.4.3 Endogenous Proteases in Disease

The body's own dysregulated protease activity is a major source of pathology, representing an internal safety failure. Cancer metastasis is the invasion of cancer cells through basement membranes and into the bloodstream is facilitated by matrix metalloproteinases (MMPs) and other proteases. These enzymes degrade the extracellular matrix (ECM), carving a path for the tumor to spread. The safety consideration here is endogenous: how to control the body's own proteases that have been hijacked by disease. In Alzheimer's disease, the β-secretase (BACE1) and γ-secretase complex are proteases that cleave the amyloid precursor protein (APP) to produce amyloid-β peptides, which aggregate into toxic plaques. In Parkinson's disease, dysregulation of proteases involved in the ubiquitin-proteasome system and mitophagy impairs the clearance of damaged proteins like α-synuclein, leading to its accumulation. Proteases like neutrophil elastase and various cathepsins are released by immune cells during inflammation. While their role is to combat pathogens and clear debris, excessive or poorly regulated activity can lead to collateral damage, destroying healthy tissue and perpetuating chronic inflammatory states, as seen in chronic obstructive pulmonary disease (COPD), rheumatoid arthritis, and inflammatory bowel disease (López-Otín and Bond 2008).

18.5 Conclusion

Protease enzymes stand as one of the most valuable biocatalysts in both natural and industrial processes due to their versatility, efficiency, and eco-friendly characteristics. The future of protease research and application depends on developing sustainable, scalable, and safe production systems that align with circular bioeconomy principles. Advances in metabolic engineering, synthetic biology, and fermentation optimization are transforming microbial strains into highly efficient protease producers capable of utilizing low-cost, renewable substrates. Continuous bioprocessing, dynamic process control, and innovative downstream technologies are bridging the gap between laboratory-scale research and large-scale industrial implementation. At the same time, safety and regulatory frameworks are essential to mitigate potential health and environmental risks, ensuring responsible production and use. Ultimately, the integration of sustainability, scalability, and safety will pave the way for next-generation protease enzymes highly active, stable, and environmentally responsible, supporting global industries and contributing to a greener and more sustainable future.

References

Bareither R, Pollard D (2019) A review of advanced small-scale parallel bioreactor technology for accelerated process development: current state and future need. Biotechnol Prog 35(4):e2798. https://doi.org/10.1002/btpr.2798

Baur X (2018) Enzymes as occupational and environmental respiratory allergens. Int Arch Occup Environ Health 91(7):785–798. https://doi.org/10.1007/s00420-018-1322-7

Benavides J, Rito-Palomares M (2018) Practical experiences from the development of aqueous two-phase processes for the recovery of high value biological products. J Chem Technol Biotechnol 93(11):3144–3154. https://doi.org/10.1002/jctb.5682

Brockmeier U, Wendorff M, Eggert T (2019) The choice of the signal peptide for secretion of a heterologous protein can have a profound effect on the yield: the case of *Bacillus subtilis*. J Biotechnol 295:17–24. https://doi.org/10.1016/j.jbiotec.2019.02.008

Causserand C, Aimar P (2020) The need for innovation in membrane engineering for downstream processing. Curr Opin Chem Eng 28:90–97. https://doi.org/10.1016/j.coche.2020.03.007

Chandrasekhar K, Kim S, Wang J, Zhang Y (2024) AI-driven rational design of hyperstable bacterial proteases for harsh detergent formulations. Nat. Catal. 7(2):155–167. https://doi.org/10.1038/s41929-023-01068-2

Contesini FJ, de Alencar Guimarães NC, de Lima EA (2023) Fungal proteases: advances in molecular engineering and applications in the food and beverage industries. Food chemistry: molecular sciences 6:100165. https://doi.org/10.1016/j.fochms.2022.100165

Craven S, Whelan J, Glennon B (2014) Process analytics and supervision in bioprocessing. Adv Biochem Eng Biotechnol 132:225–248. https://doi.org/10.1007/10_2013_252

Dewettinck K, Huyghebaert A (2019) Fluidized bed coating in food technology. Trends Food Sci Technol 29:99–115. https://doi.org/10.1016/j.tifs.2018.11.015

Glassey J, Montague G (2018) Monitoring and control of fermentation processes. In: Fermentation microbiology and biotechnology, 4th edn. CRC Press, pp 345–368. https://doi.org/10.1201/9781315152855

Goodman RE, Ebisawa M, Ferreira F, Sampson HA, van Ree R, Vieths S, Baumert JL (2022) Allergen online: a peer-reviewed, curated allergen database to assess novel food proteins for potential cross-reactivity. Mol Nutr Food Res 66(1):2100985. https://doi.org/10.1002/mnfr.202100985

Gurung S, Ray S, Bose S, Rai V (2023) A comprehensive review on microbial alkaline serine proteases: molecular and biotechnological perspectives. J Adv Res 47:25–40. https://doi.org/10.1016/j.jare.2022.07.008

Humbird D (2020) Scale-up economics for cultured meat. Bioeng. Biotechnol. 7:575. https://doi.org/10.3390/bioengineering7040575

Koutinas AA, Vlysidis A, Pleissner D, Kopsahelis N, Lopez Garcia I, Kookos IK, Lin CSK (2014) Valorization of industrial waste and by-product streams via fermentation for the production of chemicals and biopolymers. Chem Soc Rev 43(8):2587–2627. https://doi.org/10.1039/C3CS60293A

Ladner T, Mühlmann M, Schulte A, Stahl F, Beutel S (2021) High-throughput fermentation screening for strain and process development. In: Bioprocess and analytics development for virus-based advanced therapeutics and medicinal products (ATMPs). Springer, Cham, pp 89–109. https://doi.org/10.1007/10_2021_169

Langer ES, Rader RA (2020) Continuous bioprocessing and perfusion: longer, better, cheaper. BioPharm Int. 33(5):24–29

Lara AR (2022) Scale-down models for simulating industrial fermentation processes: principles and applications. Biotechnol Adv 54:107795. https://doi.org/10.1016/j.biotechadv.2021.107795

Le Clech P, Lee EK, Chen V (2020) Challenges and opportunities for membrane processes in water and wastewater treatment. Curr Opin Chem Eng 28:173–178. https://doi.org/10.1016/j.coche.2020.04.005

Ledesma-Amaro R, Lazar Z, Rakicka M, Guo Z, Fouchard F, Crutz-Le Coq AM, Nicaud JM (2016) Metabolic engineering of Yarrowia lipolytica to produce chemicals and fuels from xylose. Metab Eng 38:115–124. https://doi.org/10.1016/j.ymben.2016.07.001

Li Q, Yi L, Marek P (2023) High-throughput screening for protease activity using a solid-phase fluorescence-based assay. SLAS Technol 28(1):35–42. https://doi.org/10.1016/j.slast.2022.07.002

Liu Y, Tu X, Xu Q, Bai C, Kong C, Liu Q, Zhang J (2021) Engineered CRISPR/Cas9 system for multiplex genome engineering of polyploid industrial yeast. Nat Commun 12(1):1–12. https://doi.org/10.1038/s41467-021-26435-0

López-Otín C, Bond JS (2008) Proteases: multifunctional enzymes in life and disease. J Biol Chem 283(45):30433–30437. https://doi.org/10.1074/jbc.R800035200

Mensink MA et al (2017) How sugars protect proteins in the solid state and during drying. Eur J Pharm Biopharm 114:288–295. https://doi.org/10.1016/j.ejpb.2017.01.024

Naveed M, Nadeem F, Mehmood T, Bilal M, Anwar Z, Amjad F (2021) Protease – a versatile and ecofriendly biocatalyst with multi-industrial applications: an updated review. Catal Lett 151(2):307–323. https://doi.org/10.1007/s10562-020-03316-7

Nazzaro F, Fratianni F, Coppola R (2012) Microencapsulation in food science and biotechnology. Curr Opin Biotechnol 23(2):182–186. https://doi.org/10.1016/j.copbio.2011.10.001

Nielsen J, Keasling JD (2016) Engineering cellular metabolism. Cell 164(6):1185–1197. https://doi.org/10.1016/j.cell.2016.02.004

Noorman H (2019) Bioreactor and downstream processing. In: Comprehensive biotechnology, 3rd edn. Elsevier, pp 243–256. https://doi.org/10.1016/B978-0-444-64046-8.00013-5

Pagga U (2022) Testing biodegradability with standardized methods. Chemosphere 287:132243. https://doi.org/10.1016/j.chemosphere.2021.132243

Panesar PS, Kennedy JF, Gandhi DN, Bunko K (2014) Bioutilisation of whey for lactic acid production. Food Chem 105(1):1–13. https://doi.org/10.1016/j.foodchem.2013.03.103

Pariza MW, Johnson EA (2021) Evaluating the safety of microbial enzyme preparations used in food processing: update for a new century. Regul Toxicol Pharmacol 73(1):107–112. https://doi.org/10.1016/j.yrtph.2021.104963

Razzaq A, Shamsi S, Ali A, Ali Q, Sajjad M, Malik A, Ashraf M (2019) Microbial proteases applications. Front Bioeng Biotechnol 7:110. https://doi.org/10.3389/fbioe.2019.00110

Rugbjerg P, Sommer MOA (2019) Overcoming genetic heterogeneity in industrial fermentations. Nat Biotechnol 37:869–876. https://doi.org/10.1038/s41587-019-0171-6

Sawant R, Nagendran S (2014) Protease: an enzyme with multiple industrial applications. World J. Pharm. Pharm. Sci. 3(6):568–579

Singh N, Bajaj BK (2017) Potential application spectrum of microbial proteases for clean and green industrial production. Emerg Technol Waste Valorizat Environ Prot 5(2):44–53

Singh R, Kumar M, Mittal A, Mehta PK (2016) Microbial enzymes: industrial progress in 21st century. 3 Biotech 6(2):174. https://doi.org/10.1007/s13205-016-0485-8

Sirohi R, Tarafdar A, Singh S, Negi T, Gaur VK, Gnansounou E, Bharathiraja B (2020) Green processing and biotechnological potential of grape pomace: current trends and opportunities for sustainable biorefinery. Bioresour Technol 314:123771. https://doi.org/10.1016/j.biortech.2020.123771

Soccol CR, da Costa ES, Letti LA, Karp SG, Woiciechowski AL, Vandenberghe LP (2017) Recent developments and innovations in solid state fermentation. Biotechnol Res Innov 1(1):52–71. https://doi.org/10.1016/j.biori.2017.01.002

Sundararajan S, Kannan CN, Chittibabu S (2011) Alkaline protease from Bacillus cereus VITSN04: potential application as a dehairing agent. J Biosci Bioeng 111(2):128–133. https://doi.org/10.1016/j.jbiosc.2010.09.009

Teepakorn C, Fiaty K, Charcosset C (2021) Optimization of membrane chromatography for monoclonal antibody purification. Sep Purif Technol 255:117704. https://doi.org/10.1016/j.seppur.2020.117704

Terpe K (2003) Overview of tag protein fusions: from molecular and biochemical fundamentals to commercial systems. Appl Microbiol Biotechnol 60(5):523–533. https://doi.org/10.1007/s00253-002-1158-6

Thomas L, Larroche C, Pandey A (2013) Current developments in solid-state fermentation. Biochem Eng J 81:146–161. https://doi.org/10.1016/j.bej.2013.10.013

Ubando AT, Felix CB, Chen WH (2020) Biorefineries in circular bioeconomy: a comprehensive review. Bioresour Technol 299:122585. https://doi.org/10.1016/j.biortech.2019.122585

Vishwanatha KS, Rao AGA, Singh SA (2022) Sustainability and life cycle assessment of protease production: a case study of solid-state fermentation vs. submerged fermentation. J Clean Prod 370:133232. https://doi.org/10.1016/j.jclepro.2022.133232

Wang W (2021) Advanced protein formulations. Protein Sci 30(1):60–69. https://doi.org/10.1002/pro.3922

Wright O, Delmans M, Stan GB, Ellis T (2022) Gene guard: a modular plasmid system designed for biosafety. ACS Synth Biol 8(5):1093–1102. https://doi.org/10.1021/acssynbio.1c00577

Yan JK, Wang YY, Ma HL, Wang ZB (2018) Ultrasonic effects on the degradation kinetics, preliminary characterization and antioxidant activities of polysaccharides from *Phellinus linteus* mycelia. Ultrason Sonochem 29:251–257. https://doi.org/10.1016/j.ultsonch.2018.08.007

Zanen G, Van Dijl JM (2020) Proteomic dissection of the bacterial cell envelope and the proteins involved in protein secretion. Curr Opin Microbiol 58:1–8. https://doi.org/10.1016/j.mib.2020.03.011

Zhang F, Wang Z (2016) Synthetic biology and metabolic engineering for improved biofuel production. Curr Opin Chem Biol 35:10–16. https://doi.org/10.1016/j.cbpa.2016.08.003

Zhao W, Li J, Liu J, Jiang Z (2022) Engineering of a high-performance protease for efficient degradation of feather keratin. Bioresour Technol 363:127943. https://doi.org/10.1016/j.biortech.2022.127943

MIX
Papier aus verantwortungsvollen Quellen
Paper from responsible sources
FSC® C105338

If you have any concerns about our products,
you can contact us on
ProductSafety@springernature.com

In case Publisher is established outside the EU,
the EU authorized representative is:
Springer Nature Customer Service Center GmbH
Europaplatz 3, 69115 Heidelberg, Germany

Printed by Libri Plureos GmbH
in Hamburg, Germany